Principles of Mass Transfer

Kal Renganathan Sharma, PE
Professor
School of Chemical and Biotechnology
SASTRA Deemed University
Thanjavur

PHI Learning Private Limited
Delhi-110092
2025

In fond memory of ***Shri Asoke K. Ghosh*** *(October 1942 – February 2024), Founder Chairman and Managing Director of PHI Learning, whose vision endlessly inspires.*

The Legacy Continues....

Published by Pushpita Ghosh, PHI Learning Private Limited, Rimjhim House, 111, Patparganj Industrial Estate, Delhi-110092 and Printed by Syndicate Binders, A-20, Hosiery Complex, Noida, Phase-II Extension, Noida-201305 (N.C.R. Delhi).

₹895.00

PRINCIPLES OF MASS TRANSFER
Kal Renganathan Sharma, PE

ISBN-978-81-203-3142-6 (Print Book)
ISBM-978-93-5443-072-5 (e-Book)

To
my son
Chi. R. Hari Subrahmanyan Sharma (alias Ramkishan)

Contents

Preface

This textbook is designed for the two core subjects, Mass Transfer 1 and Mass Transfer 2, in the undergraduate curriculum of chemical engineering. It is also suitable for a course in Downstream Processing for biotechnology students and for a course in Drug Delivery for bioengineering discipline. Besides, the book can be used for the laboratory courses in the separation processes and unit operations.

Since the introduction of the currently used textbooks several years ago, which are mostly those used in US universities and recommended in the syllabus of major Indian universities, many new mass transfer operations have been developed. The degree of emphasis has also changed considerably. In designing such different mass transfer operations such as vapour distillation, gas absorption, fluid adsorption, liquid extraction, solid drying, gas humidification, solids leaching, ion-exchange, reverse osmosis, membrane separations, ultrafiltration, molecular sieves, chromatography, foam separation, solution stripping, solution crystallization, carburizing steel, bleaching, polymer devolatilization, gel electrophoresis, and other novel separation operations, a four-step design procedure is used. The first step is to construct an operating line from a mass balance. The second step is to obtain an equilibrium line from equilibrium thermodynamics for a given application either from theory or more commonly from experimental data. The third step is to include the rate effects into a lumped parameter called efficiency. The fourth step is to calculate the number of stages required for a desired level of separation.

The erection of desalination plants using reverse osmosis technology at Kalpakkam and at Koodankulum highlights the need to provide the engineers in this peninsular region with tools and methods to evaluate the feasibility and economic viability of available methods.

The Fick's laws of diffusion have been generalized and some examples of the manifestation of the acceleration term in transient applications illustrated. The applications of mass transfer are discussed in a user friendly manner. Novel separation methods are discussed with applications in the chemical, biotechnological, pharmaceutical, nanotechnological, bioengineering, bioinformatics, metallurgical, mechanical, electronic, environmental and computer engineering. Equipment selection is discussed for different operations. A table of industrial applications for each and every mass transfer unit operation is provided.

End of chapter problems emphasize grasping of the fundamentals and application of the theories to a wide variety of scenarios. The use of calculus is made when necessary. There are 164 figures used as illustrations. The problems and worked examples are from case studies from the chemical process and biotechnology industries.

Mass transfer used to be one of my favorite subjects as an undergraduate student at Indian Institute of Technology Madras and during graduate school at West Virginia University, Morgantown, WV, USA.

I have instructed over 1250 students in deemed universities and universities in USA such as George Mason University and West Virginia University, Morgantown, WV. The feedback received from the students and the classroom tested lectures in mass transfer I, mass transfer II and modern separations operations have been used in preparing this textbook.

Acknowledgements are extended to Prof. E.L. Cussler, whose department seminar I had the good fortune of attending. He kept the audience laughing and to this day I remember his reference to artificial gill. Special mention goes to Prof. John W. Zondlo, Joeseph A. Shaewitz, Eugene V. Cilento, R. Yang, R. Turton of West Virginia University, Morgantown, WV, USA, who motivated me to develop a concentration in transport phenomena and mass transfer.

My kind thanks to M.S. Ananth, Director, IIT Madras, Prof. Ramachandra Rao for teaching me Staged Cascades and Phase Equilibria, Prof. Krishnaih, Prof. Durgaprasad Rao and Prof. Neelakantan of IIT Madras. Special mention is made of R. Sethuraman, Vice Chancellor at SASTRA deemed university, S. Swaminathan, S. Vaidya Subramanian and K.N. Somasekharan, Deans at SASTRA deemed university, for many a valuable interaction.

Kudos to president Victoria Franchetti Haynes, President at Research Triangle Institute, Research Triangle Park, NC, USA who I consider my mentor in industrial research. Many thanks to my parents Mr. S. Kalyanaraman and Mrs. Shyamala Kalyanaraman.

Thought goes towards the maintenance of the bioinformatics laboratory where the manuscript was prepared on a Wipro Pentium IV, 3.0 GHz PC and all those who have stood by my side, but cannot be mentioned by name.

Special thanks to Hari, my son, and the rest of the family for letting me stay late in the evenings and the weekends with the endeavour.

I do not want to forget Prof. Y.B.G. Varma, formerly Head of Chemical Department at IIT Madras, for his guidance during my B.Tech. project, and for the support provided to me by P.G. Murthy, former Principal of Vellore Engineering College, Vellore.

Special thanks to editorial and production staff at PHI for their timeliness in bringing the manuscript to print.

The blessings of his Holiness Sri. Jayendra Saraswati swamigal and His Holiness Pujyasri Sankara Vijayendra Saraswati swamigal of the Kancheepuram Sankara Mutt through the years are deeply valued.

Kal Renganathan Sharma, PE

CHAPTER 1

Fick's Laws of Diffusion

Nomenclature

a_o	spacing between atoms (m)
A	area across which diffusion occurs (m)
A_N	Avagadro number (6.023 E23) molecules/mol
C	concentration (mol/m^3)
d	diameter of molecule (m)
d_{hair}	diameter of the hair (m)
D	diffusion coefficient (m^2/s)
D_{AB}	binary diffusion coefficient of A in B (m^2/s)
D_{BA}	binary diffusion coefficient of B in A (m^2/s)
f	friction coefficient of the solute
k_B	Boltzmann's constant (J/molecule/K)
J	molar flux (mol/s)
J''	molar flux per unit area (mol/m^2/s)
J_0	constant flux (mol/s)
k'''	first order reaction rate constant (l/s)
L	length of the cylinder (m)
m	mass of molecule (kg)
M	molecular weight (kg/mol)
n	number of molecules per unit volume (/m^3)
N_m	number of molecules
N	total flux per unit area (mol/m^2/s)
N_f	fraction of the sites vacant in the crystal
R_o	solute radius (m)
p	partial pressure (N/m^2)
P	permeability coefficient (barrer)
Pe_m	Peclect number (mass) (Ud_{hair}/D)
P_{tot}	total system pressure (N/m^2)
S	solubility coefficient (barrer.s/m^2)
T	temperature (K)
v_1	velocity of molecule (m/s)

$<v_x^2>$	average of square of the velocity of the molecules (m^2/s^2)
V	molar volume (m^3/mol)
x	mole fraction
X	dimensionless distance (x/d_{hair})
Z	dimensionless distance (z/d_{hair})
z	space coordinate (m)

Greek

μ	viscosity (kg/m/s)
μ_1	chemical potential
ρ	density (kg/m^3)
ρ_m	molar density (mol/m^3)
τ_r	relaxation time (s)
ϕ	Thiele modulus, $R\,(k'''/D_{AB})^{1/2}$
δ	thickness of the oxide layer (m)
θ	cone half-angle
σ_{12}	collision diameter (m)
Ω	molecular interaction parameter
ω	jump frequency (/s)

Subscripts

A	species A
B	species B
1	location 1
2	location 2
tot	total

1.1 DIFFUSION PHENOMENA

Diffusion is a phenomenon of migration of a species from a region of a higher concentration to a region of a lower concentration under the driving force of a concentration gradient. There can be other driving forces such as temperature difference, large concentration gradient of a second species, osmotic potential, steam sweep, centripetal forces, pressure drop, electromotive forces, surface tension gradient, surface forces, etc. that can cause the transfer of species from one point to another, often times in a secondary manner.

The term *molecular diffusion* refers to the Brownian motion of molecules (as observed by Einstein) and the movement from a region of higher concentration to a region of lower concentration. This is in obeyance of the second law of thermodynamics, the Clausius inequality. The movement of species from a region of lower concentration to a region of higher concentration in a spontaneous manner is infeasible. This is because not all heat can be converted to work without some heat being rejected to the atmosphere. Heat always flows from a hotter temperature to a colder temperature by the phenomenon of molecular conduction. By analogy, the molar species also moves from a region of higher chemical potential to a region of lower chemical potential. The direction of transfer is to equalise the concentration.

There can be no negative concentration. As the third law of thermodynamics states that the lowest attainable temperature is 0 K, by the same analogy the lowest concentration attainable is zero mol/m^3. This will be used in later discussions to obtain a plane of zero concentration or a penetration length etc.

Diffusion is central to separation operations widely used in the chemical and biotechnology industries. It is used to better understand the transport of solutes in the living cells and to design artificial organs. It plays a pivotal role in the sequence distribution analysis in genome projects. The efficiency of distillation and dispersal of pollutants can be derived from the principles of diffusion. As cities around the world face a drought crisis, the desalination of sea water for potable water needs is going to be increasingly relied upon. This is the method of choice in the deserts of the Middle East where the energy is in abundance and cheap but the drinking water scarce. The Bhabha Atomic Research Centre, BARC, at Trombay in Mumbai has set up the world's largest desalination plant at the atomic power plant at Kalpakkam about 50 km from Chennai. This plant has two sections. One section produces 18 lakh litres of potable water a day from sea water using the reverse osmosis method and the other section, 45 lakh litre a day using the thermal method.

Another desalination plant with two units at 50,000 litres per hour was inaugurated at Koodankukulum in Tirunelveli, Tamil Nadu using reverse osmosis technology. In order to desalinate sea water, several transfer operations such as reverse osmosis, electrodialysis, ion exchange, extraction, flash vaporisation, molecular sieve filtration, pervaporation can be used. The principles of diffusion and mass transfer can help evaluate the technical feasibility of each operation at large scale at the lowest possible cost without much harm to the environment and in a safe manner. The chemical reactions performed on a large scale during commercial manufacture of products are often conducted in the presence of a catalyst. During the reaction, the critical reactant has to diffuse through the catalyst and approach the active site for reaction and encounter the other reactant prior to reacting and forming the product. Diffusion in the catalyst needs to be understood for better design. The useful product has to be separated from the unreacted reactants and other by-products using mass transfer separation operations where diffusion is a critical governing phenomenon.

Albert Einstein observed that a cube of sugar placed in the bottom of a hot tea cup diffuses and a uniform concentration of sugar in the entire cup results at the final state. The term *thermophoresis* refers to those processes where the primary driving force of diffusion is the temperature difference. *Diffusophoresis* is when a large drop in concentration of a second species drives the transfer of the first species. *Osmosis* is where osmotic potential drives the flow of solvent from a region of low solute concentration to a region of higher solute concentration. In *reverse osmosis,* the solvent is pumped from a region of higher solute concentration to a region of lower solute concentration. The wilting of lettuce when salted is a good example of osmosis phenomenon as the water oozes out of the leafy vegetable and the turgor pressure gives in to a shrunken mass. In *sweep diffusion*, steam sweeps away the solute with it. The centripetal forces during *centrifugation* results in different forces upon different masses resulting in separation. *Pressure diffusion* is characterised by a pressure drop ΔP in the direction of transfer. *Electrolysis* refers to the movement of charged particles subject to an electromotive force. *Surface diffusion* is the movement of species of interest on the surface of the solid. Surface tension gradient can be utilised in separation by *foaming*.

In the mid-1800s, Fick introduced two differential equations that provide a mathematical framework to describe the otherwise random phenomenon of molecular diffusion. The flow of mass by diffusion across a plane was proportional to the concentration gradient of the diffusant across the plane. The components in a mixture are transported by a driving force during diffusion. The molecular motion is Brownian. The ability of the diffusant to pass through a body is dependent on both the diffusion coefficient, D, and the solubility coefficient S. The permeability coefficient P is given by:

$$P = DS \tag{1.1}$$

Fick's laws of diffusion were proposed in the year 1855. Adolf E. Fick, the youngest of five children, was born on 3 September, 1829 to a civil engineer. During his secondary schooling, Fick was interested in mathematics and was enamored by the work of Poisson. His brother, a professor of anatomy at the University of Marlburg, persuaded him to switch to medicine despite his desire for a career in mathematics. Carl Ludwig was Fick's tutor at Marlburg. Ludwig strongly believed that medicine and life itself have a basis in mathematics, physics and chemistry. His thesis dealt with the visual errors caused by astigmatism. Ironically, most of Fick's accomplishments do not depend on diffusion studies at all, but on his more general investigations of physiology. He did outstanding work in mechanics, hydrodynamics and hemorheology and in the visual and thermal functioning of the human body.

In his first diffusion paper (1855 a), Fick interpreted the experiments from Graham with interesting theories, analogies and quantitative experiments. He showed that diffusion can be described on the same mathematical basis as Fourier's law of heat conduction (1822) and Ohm's law of electricity. The Fick's first law of diffusion can be written as:

$$J = -AD\frac{\partial C}{\partial x} \tag{1.2}$$

where J is defined as the one-dimensional molar flux. The diffusivity is the proportionality constant that depends on the material under consideration.

Fick's second law of diffusion can be derived by considering a thin shell of thickness Δx with constant cross-sectional area A across which the diffusion is considered to occur. A mass balance in the incremental volume considered, $A\Delta x$ for an incremental time, Δt neglecting any reaction or accumulation of the species, can be written as:

$$\text{(mass in)} - \text{(mass out)} \pm \text{(mass reacted/generated)} = \text{mass accumulated} \tag{1.3}$$

$$\Delta t(J_x - J_{x+\Delta x}) = A\Delta x\Delta C \tag{1.4}$$

Dividing Eq. (1.4) throughout by $A\Delta x\Delta t$ and obtaining the limits as Δx and Δt goes to zero;

$$-\frac{\partial J}{\partial x} = \frac{A\partial C}{\partial t} \tag{1.5}$$

Combining Eq. (1.2) and Eq. (1.5) the governing equation for the diffusing species becomes when the area across which the diffusion occurs is a constant;

$$D\frac{\partial^2 C}{\partial x^2} = \frac{\partial C}{\partial t} \tag{1.6}$$

Eq. (1.6) is sometimes referred to as Fick's second law of diffusion. This is a fundamental equation that describes the transient, one-dimensional diffusion of diffusing species. Fick attempted to integrate Eq. (1.6) and was discouraged by the numerical effort needed. He found it difficult to measure the second derivative experimentally and discovered that the second difference increases exceptionally the effect of experimental errors. Finally he demonstrated in a cylindrical cell the steady state linear concentration gradient of sodium chloride. He used a glass cylinder containing crystalline sodium chloride in the bottom and a large volume of water in the top. By periodically changing the water in the top volume, he was able to establish a steady state concentration gradient in the cylindrical cell. He confirmed his equation with this steady state gradient.

In three dimensions, the Fick's first law of diffusion can be written as:

$$J'' = -D\nabla C \tag{1.7}$$

where the differential operator ∇ is given by:

$$\nabla = i\frac{\partial}{\partial x} + j\frac{\partial}{\partial y} + k\frac{\partial}{\partial z} \tag{1.8}$$

such that,

$$\nabla C = i\frac{\partial C}{\partial x} + j\frac{\partial C}{\partial y} + k\frac{\partial C}{\partial z} \tag{1.9}$$

In the special case of Eq. (1.7) the one-dimensional case of Eq. (1.2) results. Diffusion may be viewed as a process by which molecules intermingle as a result of their kinetic energy of random motion.

In liquid systems on earth density differences within a liquid system often result in convective mixing of the fluid. This gravity-induced convection, coupled with gravity-independent diffusion, contributes to the overall mass transfer within the system. In space, the convective contribution is greatly reduced and a closer examination of the diffusion contribution can be observed.

The skylab science demonstration was the first in a series of investigations designed by Facimire et al (1973) to study low gravity diffusive mass transfer. The specific objective of the demonstration was to photographically document the diffusion of tea in water in spacecraft. In preparation for the experiment, skylab pilot Jack Lousma filled 3/4th of a 1/2" diameter, 6" long transparent tube, with water. A highly concentrated tea solution was then delivered to the water surface (via a 5 cc syringe) through a synthetic fibre wad. The tube was then capped. The fibre pad was employed to try to bring the tea and water in contact without entrapped air. Three attempts to produce the wad were unsuccessful. During the fourth attempt, a "a good bubble free interface" was realised. The next day, Lousma reported that no diffusion of the tea in the liquid had occurred. Thus the experiment was initiated again. During this new experimental run, the wad was removed and the tea was delivered on top of the water. After an air bubble between the tea and water was removed via the syringe, a "smooth continuous interface" was achieved. The tea was allowed to diffuse during the next 3 days. Post flight, 16 mm photographs of the diffusion were analysed. In 51.15 hours, the visible diffusion front advanced 1.96 cm. It was noted that the diffusion front had become increasingly parabolic during the demonstration. It was noted that very little diffusion occurred near the container wall. A similar ground-based experiment was performed for comparison to the space investigation. After 45.5 hours, three different zones were visible: (a) dark area, (b) an area of medium darkness, and (c) very light area. The medium coloured area had advanced 1.6 cm in 45.5 hours.

If a few crystals of K_2CrO_4, potassium chromate, are placed at the bottom of a tall bottle filled with triple distilled water, the yellow colour will slowly spread through the bottle. At first the colour will be concentrated in the bottom of the bottle. After a day it will penetrate a few centimetres upwards. After several years the solution will appear homogeneous. The process responsible for the movement of the coloured material is diffusion. Diffusion was studied by Albert Einstein. He noted that as sugar dissolves in water, the viscosity of the solution increases. Stokes Einstein equation is used for estimating diffusion coefficients for molecules in liquid phase. Diffusion is a molecular phenomenon. At a microscopic level, molecules do undergo Brownian random motion. However, the driving force sets some direction to the transfer of the species under consideration. In gases, diffusion progresses at a rate of about 10 cm/min and in liquids its rate is about 0.05 cm/min, in solids, while its rate is about 100 nm/min.

Diffusion varies less with temperature although for polymers, Arrhenius relationships have been reported for change of diffusion coefficient with temperature. The slow rate of diffusion makes it a rate limiting step in cases where it occurs sequentially with other phenomena. The rates of distillation are limited by diffusion and that of industrial reactions on porous catalysts. It limits the speed of absorption of nutrients in the human intestine and the control of micro-organisms in the production of penicillin. The rate of corrosion of steel, splat cooling of metallic glasses, dopant diffusion in silicon chip manufacture, and the release of flavour from food, delivery of drugs to tumor cells are limited by diffusion. The equalizing effect of diffusion needs to be distinguished from other methods of producing a uniform mixture such as bulk convective mixing. Agitation also is used in homogenisation. The energy for movement from diffusion comes from the thermal energy of the molecules. The rate of evaporation of water at 25°C into complete vacuum was calculated as 3.3 kg/m^2/s. Placing a layer of stagnant air at 1 STP and 100 microns thickness above the water surface reduces the rate of evaporation by a factor of about 600.

Worked Example 1.1: *Effect of Area in Diffusion Path*
Consider a conical funnel of radius R_0 filled with water. The cone half-angle with the vertical is θ. When a cube of sugar is placed in the throat of the funnel and at steady state, the concentration profile of the sugar in the funnel is obtained. Assume that there is a constant supply of sugar at the throat of the funnel and at the top of the funnel zero concentration of sugar is achieved by a constant removal of sugar at a flux of J_0 mol/s.

The area across which the diffusion occurs is not constant but varies in this problem. Hence Eq. (1.2) and Eq. (1.5) are combined in this manner for the z coordinate with the positive direction facing the top of the funnel:

$$-\frac{\partial J}{\partial z} = A\frac{\partial C}{\partial t} \tag{1.10}$$

At steady state,

$$AD\frac{\partial^2 C}{\partial z^2} + D\left(\frac{\partial A}{\partial z}\right)\left(\frac{\partial C}{\partial z}\right) = 0 \tag{1.11}$$

$$A = \pi z^2 \tan^2\theta; \quad \frac{\partial A}{\partial z} = 2\pi z \tan^2\theta \tag{1.12}$$

$$\frac{\partial^2 C}{\partial z^2} + \frac{2}{z}\left(\frac{\partial C}{\partial z}\right) = 0 \tag{1.13}$$

Let

$$\frac{\partial C}{\partial z} = p; \text{ then, } \frac{\partial p}{p} = -2\frac{\partial z}{z} \tag{1.14}$$

Integrating both sides,

$$\ln p = -2\ln z + c' \tag{1.15}$$

or

$$\frac{\partial C}{\partial z} = \frac{c}{z^2} \tag{1.16}$$

Integrating,

$$C = c' - \frac{c}{z} \tag{1.17}$$

From the boundary condition at the top of the funnel for zero concentration,

$$C' = c\frac{\tan\theta}{R_0} \tag{1.18}$$

$$C = c\left(\frac{\tan\theta}{R_0} - \frac{1}{z}\right) \tag{1.19}$$

In order to keep the concentration zero at the top of the funnel, a finite flux is needed. Hence from the flux condition and Fick's first law of diffusion;

$$c' = -\frac{J_0}{\pi D \tan^2\theta} \tag{1.20}$$

Hence for $z > 0$,

$$C = \frac{J_0}{\pi D \tan^2\theta}\left(\frac{1}{z} - \frac{\tan\theta}{R_0}\right) \tag{1.21}$$

Thus the steady state concentration profile varies as inverse of the height z from the throat of the funnel.

1.2 BULK MOTION, MOLECULAR MOTION AND TOTAL MOLAR FLUX

Consider two containers of CO_2 gas and He gas separated by a partition as shown in Figure 1.1. The molecules of both the gases are in constant motion and make numerous collisions with the partition. If the partition is removed, the gases will mix due to the random velocities of their molecules. In sufficient time, a uniform mixture of CO_2 and He molecules will result in the container.

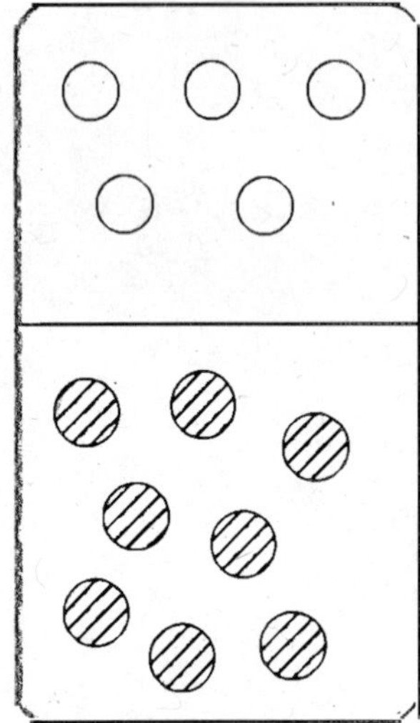

FIGURE 1.1 Container with a partition separating helium and carbon dioxide gases.

The helium molecules will move to the bottom and the carbon dioxide molecules will move to the top. If the two species are denoted by A and B, and the total fluxes as N_A and N_B, then the net flux N can be written as:

$$N = N_A + N_B \tag{1.22}$$

The movement of A comprises two components, one due to the bulk motion N and the fraction x_A of N which is A and the second component resulting from the diffusion of A, J''_A.

$$N_A = Nx_A + J''_A \tag{1.23}$$

$$N_B = Nx_B + J''_B \tag{1.24}$$

Adding Eqs. (1.23, 1.24)

$$N = N + J''_A + J''_B \tag{1.25}$$

or

$$-D_{AB}\frac{\partial C_A}{\partial z} = D_{BA}\frac{\partial C_B}{\partial z} \tag{1.26}$$

If $C_A + C_B$ is constant then, $D_{AB} = D_{BA}$.

Worked Example 1.2: *Unimolar Diffusion*
Consider the diffusion of a liquid A evaporating into a gas B in a partially filled tall tube. Assume that the liquid level is maintained at $z = z_1$. At the top of the tube at $z = z_2$ a stream of gas mixture of A–B flows steadily past, thereby maintaining the mole fraction of A at X_{A2}. At the liquid gas interface, the gas phase concentration of A expressed as mole fraction is X_{A1}. This is taken to be the gas-phase concentration of A corresponding to the equilibrium with the liquid at the interface, i.e. X_{A1} is the vapour pressure of A divided by the total pressure, p_A^{vap}/P_{tot} provided that A and B form an ideal gas mixture. It is further assumed that the solubility of B in liquid A is negligible. The entire system is presumed to be held at constant temperature and pressure. Gases A and B are assumed to be ideal. When this evaporating surface attains steady state, there is a net motion of A away from the evaporating surface and vapour B is stationary. Obtain the concentration profile at steady state for A in the head space.

$$N_B = 0$$

$$N_A = N_A x_A - D_{AB}\frac{\partial C}{\partial z} \tag{1.27}$$

In terms of mole fractions,

$$N_A(1 - x_A) = -C_{tot}D_{AB}\frac{\partial x_A}{\partial z} \tag{1.28}$$

A mass balance over an incremental volume of height Δz at steady state across a constant cross-sectional area A

$$-\frac{\partial N_A}{\partial z} = 0 \tag{1.29}$$

Combining Eqs. (1.28, 1.29) and integrating the resulting second order differential equation with respect to z gives,

$$\frac{1}{(1 - x_A)} \frac{\partial x_A}{\partial z} = c_1 \tag{1.30}$$

A second integration then gives;

$$-\ln(1 - x_A) = c_1 z + c_2 \tag{1.31}$$

The two integration constants can be solved by the use of the information given as boundary conditions at locations 1 and 2.

$$\ln\left(\frac{1 - X_{A1}}{1 - x_A}\right) = \left(\frac{z - z_1}{z_2 - z_1}\right) \ln\left(\frac{1 - X_{A1}}{1 - X_{A2}}\right) \tag{1.32}$$

$$\left(\frac{1 - X_{A1}}{1 - x_A}\right) = \left(\frac{1 - X_{A1}}{1 - X_{A2}}\right)\left(\frac{z - z_1}{z_2 - z_1}\right) \tag{1.33}$$

$$N_{Az} = \left(\frac{cD_{AB}}{z_2 - z_1}\right) \ln\left(\frac{X_{B2}}{X_{B1}}\right)$$

The above expressions are used during the experimental measurement of gas diffusivities.

1.3 DIFFUSIVITY IN GASES

The diffusion coefficients of binary hydrocarbon-hydrocarbon gas systems at low pressures below about 3.5 MPa can be predicted using the method of Gilland (1934):

$$D_{12} = \frac{0.1014 T^{1.5}\left(\dfrac{1}{M_1} + \dfrac{1}{M_2}\right)^{0.5}}{P_{tot}\left(V_1^{1/3} + V_2^{1/3}\right)^2}$$

where 1 is the solute and 2 is the solvent. The units of T, P and V are ^0R, psia, and cm^3/gmole respectively and diffusivity is given by ft^2/hr.

There is no one universal theory that predicts a priori all the diffusion coefficients. Experimental measurements have to be relied upon. Sometimes the experimental measurements are difficult to make and the quality of the results is not adequate. The estimates of diffusion coefficients of gases at room temperature are about 0.1 cm^2/s, that of liquids 10^{-5} cm^2/s and that of solids, 10^{-10} cm^2/s. Diffusion coefficients in polymers lie between that of solids and liquids. Binary diffusion coefficients in gases for pairs of gases are given in Table 1.1.

The most widely cited method for theoretical estimation of gaseous diffusion is the one developed independently by Chapman and Enskog. This theory accurates to an average of about 8% yields:

TABLE 1.1 Measured values of Diffusion Coefficients in Gases at 1 atm Pressure and data available at nearest temperature to 298 K (cm^2/s)

	CO_2	H_2	He	Ar	O_2	H_2O	N_2	Air	CH_4
CO_2		0.646	0.597	0.133	0.156	0.202	0.165	0.400	0.00215
H_2			1.706	0.902	0.891	0.915	0.779	0.71	0.726
He				0.742	0.822	0.908	0.794	0.658	0.494
Ar							0.216		0.675
O_2						0.282	0.181	0.176	1.1
H_2O							0.293	0.260	0.212
N_2									0.148
Air									0.196
CH_4									

$$D = \frac{\left(1.86\text{E}-3\,T^{3/2}\left(\frac{1}{M_1}+\frac{1}{M_2}\right)\right)^{1/2}}{p\sigma_{12}^2\,\Omega} \tag{1.34}$$

where D is the diffusion coefficient in cm^2/s, T is the absolute temperature in degrees K, p is the pressure in atmospheres and M the molecular weight. The quantities σ_{12} and Ω are molecular properties. σ_{12} is the collision diameter and the arithmetic average of the diameters of the atoms of the species present. Ω is the collision integral. Some other correlations for diffusivities for gases are available in the literature. Reid, Sherwood and Prausnitz (1977) compared predictions from different correlations with 68 experimental values of D_{AB}. The amount of error, ease of calculations, range of applicability, empirical parameters uses, underlying mechanistic theory from which the correlation were derived are some of the considerations prior to using one or the other available correlations.

For non-polar binary mixtures at low pressure, Wilkee-Lee (1955)

$$D_{AB} = \frac{(0.0027 - 0.0005\,M_{AB}^{1/2})T^{3/2}M_{AB}^{1/2}}{P\sigma_{AB}^2\,\Omega_D} \tag{1.35}$$

For polar binary mixtures and low pressure, Brokaw (1969)

$$D_{AB} = \frac{(0.001858\,T^{3/2}\,M_{AB}^{1/2})}{P\sigma_{AB}^2\,\Omega_D} \tag{1.36}$$

For self diffusivity, high pressure, $\rho_r \le 1.5$, Mathur-Thodos (1965)

$$D_{AA} = \left(\frac{10.7\text{E}-5\,T_r}{\beta\rho_r}\right) \tag{1.37}$$

For supercritical mixtures, Catchpole and King (1994)

$$D_{AB} = \frac{5.152 D_c T_r (\rho^{-.667} - 0.4510)\left(1 + \dfrac{M_A}{M_B}\right) R}{\left(\left(1 + \dfrac{V_{cB}}{V_{cA}}\right)^{0.333}\right)^2} \tag{1.38}$$

Observations show that for many polyatomic gases and mixtures, upto 1000 K and 700 atm $D_{AB}P$ is constant. The characteristic length σ_{AB} is in *A*.

1.4 DIFFUSION COEFFICIENTS IN LIQUIDS

Diffusion coefficients in liquids such as common organic solvents, mercury and molten metal fall in the order of magnitude of 10^{-5} cm^2/s. The diffusion can even be slower, sometimes even 100 times slower for high molecular weight solutes like polystyrene and polybutadiene. Diffusion in liquids are slower compared with gases and can become the rate limiting step in simultaneous reaction and diffusion.

A demonstration experiment was performed to illustrate diffusion phenomenon in liquids. 100 ml of 3 solutions were poured into a large filter funnel with adequate care taken not to mix the layers. The top layer was made up of hydrochloric acid HCl dissolved in toluene. The middle layer was universal indicator in water. The bottom layer was ammonia NH_3 dissolved in chloroform $CHCl_3$. The top HCl from the top layer diffused into the middle layer and colour in the second layer changed indicating an acid. The bottom ammonia layer was also diffusing into the middle layer and the colour changed indicating a base. Eventually when the ammonia from the bottom layer and HCl from the top layer that has diffused into the second layer has mixed, the indicator turned into the colour for a neutral solution.

1.4.1 Stokes-Einstein Equation for Dilute Solutions

The Stokes-Einstein equation can be used to calculate diffusion coefficients in liquids:

$$D = \frac{k_B T}{f} = \frac{k_B T}{6\pi\mu R_o} \tag{1.39}$$

Eq. (1.39) can be derived as follows. A rigid solute sphere is assumed for the molecule diffusing in a common solvent. The frictional drag force acting on the molecule opposing its motion is proportional to the velocity of the sphere;

$$\text{Drag force} = f v_1 \tag{1.40}$$

where f is the friction coefficient. From Stokes law (1850) for a sphere moving in a fluid, $f = 6\ \pi\mu R_o$. The driving force was taken by Einstein (1905) to be the negative of the chemical potential gradient defined per molecule.

$$-\nabla\mu_1 = (6\,\pi\mu R_o)v_1 \tag{1.41}$$

Eq. (1.41) is valid when the molecule reaches a steady state velocity. This is when the net force acting on the molecule is zero.

Solution is assumed to be ideal and dilute.

$$\mu_1 = \mu_1^0 + k_B T \ln(x_1) = \mu_1^0 + k_B T \ln c_1 - k_B T \ln c_2 \tag{1.42}$$

For dilute solutions, c_2 far exceeds the solute concentration and can be taken as constant. The gradient at constant temperature is:

$$\nabla\mu_1 = k_B T\,\frac{\nabla c_1}{c_2} = -\,(6\,\pi\mu R_o)v_1 \tag{1.43}$$

or

$$-\frac{k_B T}{(6\pi\mu R_o)}\,\nabla c_1 = c_1 v_1 = J'' \tag{1.44}$$

Comparing Eq. (1.44) with the Fick's law of diffusion given in Eq. (1.7), the Stokes-Einstein relationship of Eq. (1.39) results.

As it will be discussed in the next chapter, it can be seen that Eq. (1.41) is valid only at steady state. Often times, in transient applications a sudden step-change in concentration, i.e. the driving force, is imposed on the system. The molecule will experience an accelerating regime prior to reaching steady state. During the accelerating regime,

$$-\nabla\mu_1 - (6\pi\mu R_o)v_1 = \frac{m\,dv_1}{dt} \tag{1.45}$$

or

$$k_B T\nabla c_1 + \frac{mc_1 dv_1}{dt} = -\,(6\,\pi\mu R_o)\,c_1 v_1 \tag{1.46}$$

or

$$-\left(\frac{k_B T}{6\,\pi\mu R_o}\right)\nabla c_1 = \frac{m}{6\,\pi\mu R_o}\,\frac{\partial J''}{\partial t} + J'' \tag{1.47}$$

Eq. (1.47) is the generalised Fick's law of diffusion that accounts for the acceleration regime of the molecule as well as the steady state regime and will be discussed further in Chapter 2. The relaxation time is then given by:

$$\tau_r = \frac{m}{6\,\pi\mu R_o} = \frac{mD}{k_B T} \tag{1.48}$$

In terms of P_{tot} the system pressure for ideal gas, the relaxation time can be written as:

$$\tau_r = \frac{MD\rho_m}{P} \tag{1.49}$$

The velocity of mass diffusion is given by:

$$v_m = \left(\frac{D}{\tau_r}\right)^{1/2} = \left(\frac{k_B T}{m}\right)^{1/2} \tag{1.50}$$

Eq. (1.49) can be rewritten in terms of the molar gas constant and molecular weight as:

$$v_m = \left(\frac{D}{\tau_r}\right)^{1/2} = \left(\frac{RT}{M}\right)^{1/2} \tag{1.51}$$

The kinetic representation of pressure can be written after observing that a molecule moving in a cube of dimensions l with a velocity of v_x undergoes a momentum change of $2mv_x$ upon one collision with the wall. The number of collisions on the wall can be estimated by first calculating the time taken by the molecule to move the roundtrip from the wall after a collision to the opposite wall and return as $2l/v_x$. The number of collisions undergone by a molecule is $v_x/2l$. The rate of transfer of momentum to the surface from the molecular collisions is then, mv_x^2/l. The total force exerted by all the molecules colliding can be obtained by summing the contribution from each molecule and the pressure is obtained by dividing the sum by the area of the wall and is given by:

$$P_{tot} = \frac{m}{l^3}(v_{x1}^2 + v_{x2}^2 + v_{x3}^2 + \ldots) \tag{1.52}$$

Let N_m be the number of molecules in the system and n the number of molecules per unit volume. Then Eq. (1.51) can be rewritten after multiplying the numerator and denominator by N_m,

$$P_{tot} = mn <v_x^2> = \rho <v_x^2> = \frac{1}{3\rho} <v^2> \tag{1.53}$$

As the molecules treated as particles move in random motion, there is no preferred direction in the box. Hence $v^2 = v_x^2 + v_y^2 + v_z^2$. The square root of v^2 is called the *root mean squared speed* of the molecule and is an widely accepted average molecular speed. From the ideal gas law $P_{tot} = \rho RT/M$. Combining this with Eq. (1.53);

$$\frac{1}{3} <v^2> = \frac{RT}{M} = \frac{A_N k_B T}{M} \tag{1.54}$$

Comparing Eqs. (1.54) and (1.51) it can be seen the velocity of mass is 1/3rd of the root mean square velocity. This could be due to the one-dimensional diffusion.

The Stokes-Einstein formula for diffusion coefficients is limited to cases in which solute is larger than the solvent. As a result, other correlations have been derived for cases when the solute and solvent size are similar. Predictions in liquid are not as accurate as that in gases. The Wilke and Chang (1955) correlation for diffusion in liquids was an empirical correlation and is given by:

$$D = 7.4\text{E}{-8}\frac{(\phi M_2)^{1/2}T}{\mu V_1^{0.6}} \tag{1.55}$$

Worked Example 1.3: *Effect of Temperature on Relaxation Time*
Write an expression for the relaxation time during diffusion of the considered species in liquids. Combine this expression with that of the effect of temperature on diffusion coefficient and obtain the dependence of relaxation time on temperature.

Squaring Eq. (1.50) and re-arranging the terms,

$$\tau_r = \frac{DM}{RT} \tag{1.56}$$

Combining Eq. (1.56) with Eq. (1.55);

$$\tau_r = 7.4\text{E}{-8}\frac{(\phi M_2)^{3/2}}{R\mu V_1^{0.6}} \tag{1.57}$$

It can be seen that the relaxation time becomes independent of temperature and depends only on the viscosity of the fluid, and molecular size parameters.

Hildebrand adapted a theory of viscosity to self-diffusivity,

$$D_{AA} = \frac{B(V - V_{ms})}{V_{ms}} \tag{1.58}$$

where V is the molar volume and V_{ms} is the molar volume when fluidity is zero. The Siddiqi and Lucas correlation (1986) for aqueous solutions can be written as:

$$D^0_{Aw} = 2.98\text{E}{-7}\; V_A^{-0.5473}\; \mu_w^{-1.026}\; T \tag{1.59}$$

For hydrocarbon mixtures, the Haydeek-Minhas [1982] correlation can be used;

$$D^0_{AB} = 13.3\text{E}{-8}\; T^{1.47} \mu_B^{(10.2/V_A\, -\, 0.791)}\; V_A^{-.71} \tag{1.60}$$

When electrolytes are added to a solvent they dissociate to a certain degree. It would appear that the solution contains at least three components: solvent, anions and cations. If the solution is to remain neutral in charge at each point assuming the absence of any applied electric potential field, the anions and cations diffuse effectively as a single component as for molecular diffusion. The diffusion of the anionic and cationic species in the solvent can thus be treated as a binary mixture. The theory of dilute diffusion of salts is well developed and has been experimentally verified. For dilute solutions of a single salt, the Nernst-Haskell equation is applicable;

$$D^0_{AB} = \frac{RT}{F^2}\left(\frac{|1/n^+| + |1/n^-|}{1/\lambda^0_+ + 1/\lambda^0_-}\right) \tag{1.61}$$

where D^0_{AB} is diffusivity based on molarity rather than normality of dilute salt A in solvent B in cm²/s.

1.4.2 Diffusion in Concentrated Solutions

The correlations discussed above pertain to the diffusion in dilute solutions. With increased concentration, some things are different and the considerations will be different. Diffusion coefficients vary with the volume fraction of the solute, often times in a complex manner with extremas. Diffusion coefficients are no longer a proportionality constant but do vary with the concentration and become concentration dependent. In one approach, the hydrodynamic interaction of the spheres was taken into account, and the friction factor f corrected for as: (Batchelor, 1972):

$$f = 6\pi\mu R_o(1 + 1.5\phi_1 + ...) \tag{1.62}$$

in which ϕ_1 is the volume fraction of the solute. Substituting Eq. (1.58) in Eq. (1.45);

$$-\nabla\mu_1 - (6\pi\mu R_o)(1 + 1.5\,\phi_1 + ...)\,v_1 = \frac{mdv_1}{dt} \tag{1.63}$$

$$k_B T\nabla c_1 + mc_1\frac{dv_1}{dt} = -(6\,\pi\mu R_o)(1 + 1.5\phi_1 + ...)\,c_1 v_1 \tag{1.64}$$

or

$$-\frac{\left(\dfrac{k_B T}{6\pi\mu R_o}\right)}{(1 + 1.5\phi_1 + ...)}\nabla c_1 = \frac{m}{[6\pi\mu R_o](1 + 1.5\phi_1 + ...)}\left[\frac{\partial J''}{\partial t} + J''\right] \tag{1.65}$$

and $D = \dfrac{k_B T}{6\pi\mu R_o}(1 + 1.5\phi_1 + ...)$

$$\tau_r = \left(\frac{m}{6\pi\mu R_o}\right)(1 + 1.5\,\phi_1 + ...) \tag{1.66}$$

For non-ideal solution, the chemical potential can be written as:

$$\mu_1 = \mu_1^0 + k_B T\ln(c_1\gamma_1) \tag{1.67}$$

where γ_1 is the activity coefficient,

$$\nabla\mu_1 = \frac{k_B T}{c_1\gamma_1}(\gamma_1\nabla c_1 + c_2\nabla\gamma_1)$$

Substituting Eq. (1.49) in Eq. (1.45)

$$-D\nabla c_1\left(1+\frac{\partial \ln \gamma_1}{\partial \ln c_1}\right)=J''+\frac{m}{6\pi\mu R_o}\frac{\partial J''}{\partial t} \tag{1.68}$$

The correction for diffusion coefficient given in Eq. (1.65) may be attributed to cluster of molecules in the solution.

1.5 DIFFUSION IN SOLIDS

1.5.1 Mechanisms of Diffusion

Atomic diffusion in solids is of increased interest because of the phenomenal growth in VLSI (very large scale integration) of transistors on the silicon chip. Interstitial or substitutional mechanism of diffusion is said to occur when atoms occupy specific sites in a lattice. In interstitial mechanism of diffusion, impurity jumps from one interstitial site to the next. In substitutional mechanism of diffusion, impurity jumps from one lattice site to the neighbouring vacant lattice site. Since the concentration of vacancies are low, substitutional diffusion is much slower than interstitial diffusion. In concentrated diffusion, the atom replaces the lattice atom and moves through the interstices.

The mechanism of diffusion vary greatly depending upon the crystalline structure and the nature of the solute. For crystals with lattices of cubic symmetry the diffusivity is isotropic but not so for non-cubic crystals. Interstitial mechanism of diffusion refers to small diffusing solute atoms passing through from one interstitial site to the next. The matrix atoms of the crystal lattice move apart temporarily to provide the necessary space. When there are vacancies when lattice sites are unoccupied, an atom from an adjacent site may jump into such a vacancy. This mechanism is called *vacancy mechanism*.

NEC corporation (1998) has developed an interstitial concentration simulation method. Here a mesh is set in a simulation region of a semi-conductor device. Under a condition when an area outside of the simulation region is infinite, a provisional interstitial diffusion flux at the boundary of the simulation region are calculated. Then, an interstitial diffusion rate at the boundary of the simulation is calculated by a ratio of the provisional interstitial diffusion flux to the provisional interstitial concentration. Finally an interstitial diffusion equation is solved for each element of the mesh using the interstitial diffusion rate at the boundary.

Crowd ion mechanism refers to the displacement of an extra atom that can displace several atom positions thus producing a diffusion flux. The diffusivity in a single crystal is always substantially smaller than that of a multi-crystalline sample because the latter has diffusion along the grain boundaries.

For diffusion in metals, Stark (1976) and Franklin (1975) gave the following expression:

$$D = a_0^2 N_f \omega \tag{1.69}$$

where a_0 is the spacing between the atoms, N_f the fraction of sites vacant in the crystal and ω is the jump frequency, i.e, the number of jumps per unit time from one position to the next.

Worked Example 1.4 *Steady Diffusion in a Hollow Cylinder*
Develop the concentration profile in a hollow cylinder when a species is diffusing without any chemical reaction. Consider the concentration of the species to be held constant at the inner and outer surface of the cylinder at C_{Ai} and C_{Ao} respectively.

A mass balance on a thin shell of thickness Δr at radius r in the cylinder would yield:

$$J''_A 2\pi r L - J''_A 2\pi(r + \Delta r)L = 0 \tag{1.70}$$

In the limit when $\Delta r \to 0$,

$$-\frac{\partial(rJ''_A)}{\partial r} = 0 \tag{1.71}$$

Upon integration,

$$J''_A = -\frac{c_1}{r} \tag{1.72}$$

or,

$$-\frac{D_{AB}\partial C_A}{\partial r} = -\frac{c_1}{r} \tag{1.73}$$

Upon integration,

$$C_A = \frac{c_1 \ln(r)}{D_{AB}} + c_2 \tag{1.74}$$

From the boundary conditions c_1 and c_2 can be solved for:

$$r = R_o,\ C_A = C_{Ao} \tag{1.75}$$

$$r = R_i,\ C_A = C_{Ai} \tag{1.76}$$

$$c_1 = D_{AB}\frac{(C_{Ao} - C_{Ai})}{\ln(R_o/R_i)} \tag{1.77}$$

$$c_2 = C_{Ao} - \ln(R_o)\frac{(C_{Ao} - C_{Ai})}{\ln(R_o/R_i)} \tag{1.78}$$

$$\left(\frac{C_A - C_{Ao}}{C_{Ao} - C_{Ai}}\right) = \frac{\ln(r)}{\ln(R_o/R_i)} - \frac{\ln(R_o)}{\ln(R_o/R_i)} \tag{1.79}$$

Defining a log mean radius, $\langle R_l \rangle = \dfrac{(R_o - R_i)}{\ln(R_o/R_i)}$ (1.80)

$$\left(\frac{C_{Ao} - C_A}{C_{Ao} - C_{Ai}}\right) = \langle R_1 \rangle \left(\frac{\ln(R_o/r)}{R_o - R_i}\right) \tag{1.81}$$

1.5.2 Diffusion in Porous Solids

Solute movement by diffusion can be by virtue of concentration difference or by means of pressure difference. Micropores, mesopores and macropores can be distinguished by means of the pore sizes. There are a lot of publications where pore diffusion is discussed along with gas-solid reactions and catalysis. A Knudsen diffusion may be identified when the mean free path of the molecule is comparable to the pore size. When the pore size to the mean free path of the molecule ratio is about 20, molecular diffusion prevails. When $d/\lambda < 0.2$, rate of diffusion is a function of the collision of the gas molecules and wall, Knudsen diffusion is said to occur,

$$N_A = \frac{du_A \Delta p}{3RTl} \tag{1.82}$$

where u_A is the molecular velocity of A. The Knudsen diffusion coefficient;

$$D_{KA} = \frac{d}{3}\left(\frac{8RT}{\pi M_A}\right)^{1/2} \tag{1.83}$$

and mean free path λ expressed as:

$$\lambda = 3.2\,\mu\left(\frac{RT}{2\pi M}\right) \tag{1.84}$$

In the range of d/λ from roughly 0.2–20, both molecular diffusion and Knudsen diffusion are important:

$$N_A = \left(\frac{D_{AB,\,eff}\,p_t}{RTz}\right) \ln\left(\frac{N_A/N((1 + D_{AB,\,eff}/D_{KA,\,eff}) - y_{A2})}{N_A/N((1 + D_{B,\,eff}/D_{KA,\,eff}) - y_{A1})}\right) \tag{1.85}$$

Hydrodynamic flow of gases will occur when there is a difference in absolute pressure across a porous solid. Consider a solid consisting of uniform straight capillary tubes of diameter d_c and length l reaching from the high pressure to the low pressure side. Assuming laminar flow, the Hagen-Poiseuille's law for a compressible fluid that obeys the ideal gas law can be written as;

$$N_A = d^2 p_{t,\,av} \frac{(p_{t,1} - p_{t,2})}{32\,\mu lRT} \tag{1.86}$$

The entire pressure difference is assumed as the result of friction in the pores and ignores entrance and exit losses and kinetic energy effects.

1.6 DIFFUSION IN POLYMERS

The diffusion coefficients for polymers lie between that of solids and liquids. Different systems where diffusion of high molecular weight substances become of importance is when the polymer forms a solute of a dilute solution or one component of a polymer-polymer blend. A polymer blend can be miscible, immiscible, compatible or incompatible. When two polymers are mixed to yield a product with improved property then it is said to be *compatible* blend. When two polymers mix at a molecular level then they are said to be *miscible* blend. Concentrated systems, where the volume fraction of the polymer solute is large, is another category where diffusion has to be treated in a different manner compared with other systems.

A polymer molecule dissolved in a solvent can be envisioned as a necklace comprising spherical beads connected by string (Vrentas and Duda, 1980). The polymer molecules are separated and only interact through the solvent. The Stokes-Einstein equation for the diffusion coefficient of the polymer can be used for a Flory teta solvent.The root-mean-square radius of gyration used as a measure of the size of the polymer can be used as the radius of the solute in the Stokes-Einstein formula given by Eq. (1.39). These values can be measured by light scattering. For concentrated solutions, the diffusivity is given by:

$$D = D_0\left(1 + \frac{\partial \ln \gamma_1}{\partial \ln \phi_1}\right) \tag{1.87}$$

The D_0 includes the solute's activation energy. This must be sufficient to overcome any attractive forces whose constrain is near neighbouring polymer segments. This coefficient can be expected to vary with free volume of the polymer chains. Only the fraction of the free volume, will be accessed by the solute.

$$D_0 = D_0' \exp\left(-\frac{E}{RT}\right) \exp\left(\frac{-\omega_1 V_{10} + \omega_2 V_{20}}{\omega_1 k_1 + \omega_2 k_2}\right) \tag{1.88}$$

where, E is the solute-polymer attractive energy; ω_i the mass fractions; V_{i0} the specific critical free volumes, and K_i the additional free volume parameters. These parameters are strong functions of the actual temperature minus the glass transition temperatures.

For polymer blends, the Rouse model is suggested;

$$D = \frac{k_B T}{N_p \zeta} \tag{1.89}$$

where, N_p is the degree of polymerisation and ζ the friction coefficient characteristic of the interaction of a bead with its surroundings.

SUMMARY

Diffusion is a phenomenon of migration of a species from a region of a higher concentration to a region of a lower concentration under the driving force of a concentration gradient. The skylab science demonstration by Facimire that photographically documents the diffusion of tea in water in spacecraft was discussed.

By law it is stated that the lowest achievable concentration is zero mol/m^3.

The Fick's first law of diffusion is stated as;

$$J = -\frac{AD\partial C}{\partial x}$$

The mass flux is given by the product of the concentration gradient times the diffusion coefficient and the area across the path of diffusion. Fick's second law of diffusion deals with the transient concentration profile and is obtained by combining the mass balance on a thin shell of the medium and the Fick's first law of diffusion and can be stated as:

$$\frac{D\partial C^2}{\partial x^2} = \frac{\partial C}{\partial t}$$

The movement of species A comprises of two components, one due to the bulk motion N and the fraction x_A of N which is A and the second component resulting from the molecular diffusion of A, J''_A. Thus,

$$N_A = Nx_A + J''_A$$

$$N_B = Nx_B + J''_B$$

The N and J fluxes are distinguished from each other. The diffusivity variation with temperature in gases was discussed. The correlations of Chapman and Enskog, Wilkee-Lee, Mathur and Thodos and Catchpole and King were presented. Values of diffusion coefficient for binary pairs of gases were given in Table 1.1. In liquids the Stokes Einstein relation for diffusion coefficients was derived. During the derivation, when accounting for the acceleration regime of the solute molecule, the generalized Fick's laws of diffusion were derived.

$$J'' + \left(\frac{\mathrm{mD}}{\mathrm{k_B T}}\right)\frac{\partial J''}{\partial t} = -\left(\frac{\mathrm{k_B T}}{6\pi\mu R_o}\right)\frac{\partial C}{\partial x}$$

Correlations of Nernst-Haskell for electrolytes were mentioned. The effect of concentration, i.e., dilute solutions and concentrated solutions were separately discussed. Correlations of Wilkee-Chang, Sidiqui-Lucas, Haydeek-Minhas were included. The diffusion in solids was discussed. The different mechanisms of diffusion such as vacancy mechanism, interstitial mechanism, substitutional mechanism, crowd ion mechanism were outlined. The Knudsen diffusion when the mean free path of the molecule is greater than the diffusion path such as in pore diffusion was discussed. The diffusion in polymers and the Arrhenius dependence of diffusion coefficient with temperature was discussed.

REVIEW QUESTIONS

1. During Brownian motion, the molecules follow a random zig-zag path and sometimes move in the opposite direction compared with the imposed concentration difference driving the diffusion. Is this a violation of second law of thermodynamics?
2. What is self-diffusivity? Is a concentration difference needed for movement of the species that defines self-diffusivity?
3. What is the difference between binary and ternary diffusion coefficients?
4. In Eq. (1.20) why is there a singularity at $z = 0$. Discuss.
5. What are the differences between multi-component diffusion and binary diffusion?
6. What happens to the formula for total flux during equimolar counter diffusion compared with that for molecular diffusion?
7. Correlations for diffusion in gases, liquids and solids were discussed. What would be appropriate for liquid diffusing in a solid or gases diffusing in a liquid?
8. Discuss the units of each term in Eq. (1.1) $P = DS$.
9. Explain the effect of temperature on the mass propagation velocity. What happens to the diffusion coefficient and relaxation time at high pressure?
10. Why are insects in the tropics larger in size when compared with the insects in the artic region?
11. Is the forces of gravity taken into account in the derivation of Stokes-Einstein relationship for diffusivity coefficients?
12. Can you expect a plane of zero concentration or null transfer during drug delivery in the tissue region? If so, how?
13. Diffusion coefficient is a proportionality constant in the Fick's first law of diffusion independent of concentration. For concentrated solutions, it is said to vary with concentration. How can this be interpreted?

PROBLEMS

1. *Estimate of Diffusion Coefficient of Argon in Hydrogen*
Calculate the diffusion coefficient of Argon in Hydrogen at 1 atm and 195°C. Compare this with the experimental values reported in the literature.
(**Ans.** 1.775 E-4 m^2/s)

2. *Parabolic Law of Oxidation*
During the corrosion of metals a oxide layer is formed on the metals. Assume that the oxygen diffuses through the oxide layer and show that the thickness of the oxide layer δ can be given by $(C_{\text{bulk}}\ D_{AB}t/\rho_m)^{1/2}$ using Fick's law of diffusion. A gentle breeze is blowing at a constant velocity of U over the corroded

layer. Is this going to increase the rate of corrosion due to the convection contribution?

3. *Krogh Tissue Cylinder*
Capillaries through which blood flows in the human anatomy is surrounded by tissue space. The oxygen and other drugs that are dissolved in the blood stream need to diffuse through the capillary walls into the tissue region. Write the governing equations for the concentration of the solute in the capillary and in the tissue as;

$$-V\frac{dC}{dz} = \frac{2}{r_c}K_o(C - C_{T|r_c+t_m})$$

Considering the effects of diffusion in x direction only in the tissue and assuming a zeroth order reaction rate

$$\frac{D_{AB}\partial^2 C_T}{\partial x^2} = R_o$$

Integrating, and substituting for the boundary conditions;

$$x = x_c + t_m,\ C_T = C_T|_{x_c+t_m}$$

$$x = x_T,\ \frac{dC_T}{dx} = 0$$

Show that the concentration profile is;

$$C_T - C_{T|x_c+t_m} = \left(\frac{R_o}{2D_{AB}}\right)(x^2 - (x_c + t_m)^2) - \frac{R_o x_T}{D_{AB}}(x - (x_c + t_m)).$$

Show that the variation of concentration as a function of z can be calculated as;

$$C = C_0 - \frac{R_o z A_T}{VA}$$

Combine the two equations and show that;

$$C_T - C_0 = \frac{R_o z A_T}{VA} + K_o x_c\left(\frac{R_o A_T}{2A}\right) + \left(\frac{R_o}{2D_{AB}}\right)(x^2 - (x_c + t_m)^2) - \frac{R_o x_T}{D_{AB}}(x - (x_c + t_m)$$

Show that at a critical distance from the capillary wall, the concentration in the solute will become zero. This can be solved from the above equations. At and beyond the critical distance,

$$\frac{dC_T}{dx} = 0 = C_T$$

Replacing x_T with $x_{critical}$,

$$0 = C_0 + \frac{R_o z A_T}{VA} + K_o x_c \frac{R_o A_T}{2A} + \left(\frac{R_o}{2D_{AB}}\right)(x^2 - (x_c + t_m)^2) - \frac{R_o x_{critical}}{D_{AB}(x - (x_c + t_m))}$$

$$x_{critical^2}\left(-\frac{R_o}{2D_{AB}}\right) = C_0 + \frac{R_o z A_T}{VA} + K_o x_c \frac{R_o A_T}{2A} - \left(\frac{R_o}{2D_{AB}}\right)(x_c + t_m)^2 - \frac{R_o x_{critical}}{D_{AB}(x - (x_c + t_m))}$$

The quadratic equation in $x_{critical}$ is then,

$$A\, x^2_{critical} + B\, x_{critical} + C = 0;$$

where

$$A = -\left(\frac{R_o}{2D_{AB}}\right)$$

$$B = +(x_c + t_m)\left(\frac{R_o}{D_{AB}}\right)$$

$$C = C_0 + \frac{R_o z A_T}{VA} + K_0 x_c \frac{R_o A_T}{2A} - ((R_o/2D_{AB})x_c + t_m)^2 + R_o \frac{(x_c + t_m)}{D_{AB}}$$

When the solution of the quadratic expression for the critical distance in the tissue are real, and found to be less than the thickness of the tissue then the onset of zero concentration will occur prior to the periphery of the tissue. This zone can be seen as the anorexic or oxygen depleted regions in the tissue.

4. *Sacred Pond at Vaitheeswaran Koil*
Evaporation from ponds is retarded by the introduction of lotus leaves in the sacred ponds in temples such as in Vaitheeswarn Koil. Assume that in a pond of area 9 m × 9 m, 4130 leaves each with a diameter of 3" were placed. Calculate the reduction in diffusion rate on account of the reduction in area in the path of evaporation.

(**Ans.** 23.25% reduction)

5. *Diffusion Coefficient of Tobacco Mosaic Virus*
Estimate the diffusion coefficient of tobacco mosaic virus that is shaped as a cylinder if 0.35 μm length and 9 nm in diameter in water at 20°C. Its molecular weight is 42 million and partial specific volume 0.53 cm^3/g.

(**Ans.** 1.05 E-11 m^2/s)

6. *Diffusion of Oxygen through Spiracles*
Many insects breathe through spiracles. Spiracles are open tubes that extend into an insect's body. Oxygen diffuses from the surrounding air and gas exchange

takes place through the walls. For every mole of oxygen diffusing in, there is one mole of CO_2 diffusing out. To prevent water loss the walls of the spiracle are coated with cuticle of 10 μm thickness. The oxygen concentration outside the cuticle is constant and is 5% of the equilibrium concentration. What is the local oxygen flux in the spiracle to the tissue ? Derive an oxygen concentration profile within the tissue. In the spiracle an efficient method of respiration?

spiracle radius = 100 μm
oxygen solubility in tissue C_t = 0.2 m mol/l
spiracle length = 9 mm
$D_{o,cuticle}$ = 3 E-5 cm^2/s
$D_{o,air}$ = 0.15 cm^2/s

(**Ans.** 0.194 $mol/cm^2/s$)

7. *Scrubbing of SO_2*
During coal combustion, the emission of sulphur dioxide from power plants can be reduced by using CaO scrubbers. In the scrubber,

$$2CaO + 2SO_2 + O_2 \rightarrow 2CaSO_4$$

Consider the diffusion of SO_2 into a spherical particle of CaO. Show that governing equation can be derived from the shell balance as:

$$D_{AB}\left[\frac{1}{r^2}\frac{\partial}{\partial r}\left(r^2\frac{\partial C_A}{\partial r}\right)\right] = k'''C_A$$

Show that the concentration profile of SO_2 in the spherical lime particle can be written as;

$$\frac{C_A}{C_{AS}} = X^{-1/2}\frac{I_{1/2}\left(r\dfrac{k'''}{D_{AB}}\right)^{1/2}}{I_{1/2}\left(R\dfrac{k'''}{D_{AB}}\right)^{1/2}}$$

The Thiele modulus is $\phi = R\left(\dfrac{k'''}{D_{AB}}\right)^{1/2}$

8. *Co-Extrusion*
In the manufacture of the casings of the solid rocket motor SRM, the material requirements are bifunctional. They have to have high hoop strength on one side and high ablation resistance on the other. In order to prepare such materials the technology of coextrusion is utilized. In a twin-screw extruder both the materials are co-extruded together. During the residence time of the polymers in the extruder, the interdiffusion of either material in the other occurs. Calculate the interlayer thickness as a function of the extruder residence time, diffusivities of the two materials.

(**Ans:** $\delta_{eff} = (D_{12}t_{res})^{1/2} + (D_{21}t_{res})^{1/2}$

9. *Diffusion Coefficient of Milk in the Refrigerator*

Estimate the diffusion coefficient of lactic acid in the refrigerator. Compare this with the value at room temperature and that of the milk through the plastiç container.

(**Ans:** 0.21 cm^2/s at 25°C. 0.1943 cm^2/s at 10°C; 0.183 cm^2/s in plastic container)

10. *Wilting of the Lettuce*

Lettuce leaves in a salad wilt. The process of wilting is accelerated if the lettuce is salted. The water droplets on the surface of the leaves come from the interior of the lettuce plant cells. Consequently the turgor pressure and internal rigidity of the leaves is lowered and they wilt. The process of water transport out of the cells caused by increase in external salt concentration is an example of osmosis.Dutrochet made systematic observations of osmotic pressure in the 1800s. He observed that small animal bladders filled with dense solution and completely closed and plunged in water became turgid and swollen excessively. Water flowed into the bladder so as to dilute the solution inside. Van't Hoff noted that the osmotic pressure was proportional to the product of the solute concentration and the absolute temperature with a constant of proportionality that equalled the molar gas constant R. The Darcy's law provided the solvent flux as a function of the pressure gradient and the constant of proportionality called hydraulic permeability:

$$J_{solv} = -\kappa \frac{\partial(p - \pi)}{\partial x}$$

where, J_{solv} is the solvent flux, κ is the hydraulic permeability and π is the osmotic pressure that can be written as RTC_{sol}, where C_{sol} is the solute concentration. For the wilting of the lettuce show that the governing equations can be written assuming the salt can permeate through the lettuce and neglecting the hydraulic pressure gradient at steady state as,

$$0 = \kappa RT \frac{\partial^2 C_{solv}}{\partial x^2}$$

with the following space conditions,

$$x = \delta,\ C_{solv} = 0$$

$$x = 0,\ C_{solv} = C_{solv0}$$

Show that if the lettuce is semi-permeable membrane, at steady state, the solvent transport can be given by;

$$p - p_0 - (\pi - \pi_0) = -\frac{J_{solv}}{\kappa(x - x_0)}$$

where x_0 is the reference location at which the hydraulic and osmotic pressures are known.

11. *Restriction Mapping*

Endonucleases or restriction enzymes cut the unmethylated DNA at several sites and restrict their activity. About 300 restriction enzymes are known and they act upon 100 distinct restriction sites that are palindromes. Some cut leaves blunt ends and others leave them sticky. The restriction fragment lengths can be measured by using the technique of gel electrophoresis. The solid matrix is the gel usually agarose or polyacrylamide which is permeated with liquid buffer. As DNA is a negatively charged molecule when placed in a electric field the DNA migrates toward the positive pole. DNA migration is a function of its size. Calibration is used to relate the migration distance as a function of size. Migration distance of DNA under a field for a set time is measured. The DNA molecule is made to fluorescence and made visible under ultraviolet light by stainingthe gel with ethidium bromide. A second method is to tag the DNA with a radioactive label and then to expose the X-ray film to the gel. Show that the migration under gel electrophoresis can be given by:

$$J_{\text{frag}} = -(z_A u_A F)^* \frac{\partial E}{\partial x} - D_{\text{frag}} \frac{\partial C_A}{\partial x}$$

Show that the governing equation can be written in 1 dimension as:

$$0 = D_{\text{frag}} \frac{\partial^2 C_A}{\partial x^2} + \left(-(z_A u_A F)^* \frac{\partial^2 E}{\partial x^2} \right)$$

12. *Pheromone and Insect Control*

During insect control, controlled release of pheromones are used. Pheromones are sex attractants released by insects. When mixed with an insecticide and used, it annihilates all of one sex of a particular insect pest. The pheromone sublimation rate in the impermeable holder can is given as:

$$S_0 = 9\text{E}{-}16(1 - 10^6 C_1)$$

where C_1 is the concentration in the vapour. The diffusivity through the polymer is 1.2E–11 cm^2/s. It can be assumed that the pheromone outside the chamber is 0. If the polymeric diffusion barrier is 600 microns thick and has an area of 1.6 cm^2, what is the concentration of pheromone in the vapour? How fast is the pheromone released by the device?

(**Ans:** 0.722 mmol/lit; 2.502E–16 gm/s)

13. *Oxygen Transport in the Eye*

The cornea is a unique, living tissue and is a transparent window through which light enters the eye to be focused on the retina thus forming the images of our surroundings and enabling sight. When the eye is open, it receives all of its oxygen requirements from the surrounding air. Other nutrients are likely to be delivered via the tear duct fluid that bathes the outer surface of the cornea or the aqueous humor which fills the chamber behind the cornea and in front of the lens. Some oxygen may enter the aqueous humor from vasculature in the muscle at around the periphery of the lens. When the eye is closed, it is cut off from the O_2 source in the air. There is a rich microvascular bed (well perfused) with high vascular density on the inner surface of the eyelid which supplies the cornea with oxygen (and possibly other nutrients). What is the pO_2 at the surface of the cornea when the eye is closed?

Table of Model Parameters

Layer	*Thickness* (μm)	*Diffusion Coefficient* (cm^2/s)	VO_2 mL O_2.ml tissue^{-1}s^{-1}
epithelium	40	3.8E–10	2.0E–4
stroma	450	3.8E–10	1.0E–5
endothelium	10	3.8E–4	2.0E–4

(**Ans:** 12.6 mm Hg for a cornea of 5 cm)

14. *Loss from Beverage Containers*

Coca-Cola bottles are made out of plastic. The contents diffuse at a slow rate through the walls of the container and out into the air and result in some losses. It has been suggested to coat the inner wall of the container to reduce the losses. With a coating thickness of 25 μm and a diffusion coefficient in the coating of 1E–9 m^2/s, what would be the benefit to the manufacturer? Assume a thickness of 1.5 mm for the plastic container and a diffusion coefficient of the contents in the plastic container as 1E–6 m^2/s.

(new flux is 5.68% of the old flux)

15. *Dasangam*

Dasangam is offered to the God during special pooja. Idealise a *dasangam* into a cone. Consider the reaction between oxygen and *dasangam* upon ignition to be of first order. As the reaction proceeds consider an added ash layer through which the oxygen will have to diffuse. Obtain the concentration profile of oxygen in the ash layer. Make suitable assumptions and estimate the time taken for consumption two *dasangams* of a cone height of 3 cm and a diameter of 1.5 cm.

Consider the diffusion coefficient of oxygen in the ash layer to be 1E–12 cm^2/s.

(**Ans:** $C/C_s = \Sigma_1^\infty c_n \sin(n^2\pi^2 (z/z_{ash}) \exp(-Dt\lambda_n^2/H^2)$; $c_n = 2(1-(-1)^n)/n\pi\lambda_n = (H^2n^2\pi^2/z_{ash}^2 + H^2k'''/D)^{0.5}$; time taken $= 2\pi H^3\rho_m/\tan^2\theta/3J_{z=0}$)

16. *Reaction and Diffusion in a Nuclear Fuel Rod*

In autocatalytic reactions such as during nuclear fission the neutrons can be studied by a first order reaction. The mass balance in a long cylindrical rod with first order auto catalytic reaction can be written at steady state as;

$$\frac{1}{r}\frac{\partial(rJ_r)}{\partial r} + k'''C = 0$$

The long cylindrical rod is at zero initial concentration of autocatalytic reactant *A*. The surface of the rod is maintained at a constant concentration C_s for times greater than zero. The boundary conditions are,

$$r = 0,\ \partial C/\partial r = 0$$

$$r = R,\ C = C_s$$

Show that the steady state solution can be obtained as follows after re-defining $u^s = C/C_s$.

$$\frac{\partial^2 u^s}{\partial X^2} + \frac{1}{X}\frac{\partial u^s}{\partial X} + k^* u^s = 0$$

$$X^2 \frac{\partial^2 u^s}{\partial X^2} + X\frac{\partial u^s}{\partial X} + X^2 k^* u^s = 0$$

The above equation can be recognized as the Bessel equation. The solution,

$$u^s = c_1 J_0 (X\sqrt{k^*}) + c_2 Y_0 (X\sqrt{k^*})$$

It can be seen that $c_2 = 0$ as the concentration is finite at $X = 0$. The boundary condition for surface concentration is used to obtain c_1. Thus

$$c_1 = J_0 \left(\frac{R\sqrt{k^*}}{D\tau_r}\right)$$

Thus,

$$u^s = \left(\frac{J_0(X\sqrt{k^*})}{J_0(R\sqrt{k^*/D\tau_r}}\right)$$

17. *Grooming Hair with Oil*

In order to keep the hair on the human skull from becoming dehydrated, it is oiled or hair cream is applied every day. During the course of the day estimate the loss of the oil from the human hair by diffusion. Show that there are two contributions. One is from the molecular diffusion from the head to the atmosphere in the vertical direction and the other is by convection from a wind blowing in the horizontal direction. Show that the governing equation can be given by;

$$\frac{\partial^2 u}{\partial z^2} = \left(\frac{Ud_{hair}}{D}\right)\frac{\partial u}{\partial x}$$

Show that the solution for the concentration profile of the oil in the surrounding region of the human skull at steady state can be given by;

$$u = 1 - erf\ Z\left(\frac{Pe_m}{4X}\right)^{1/2}$$

Assuming that the diameter of the hair is 2 microns, the velocity of air is 1 m/s and the diffusivity is 1E–5 m^2/s, estimate the time taken for the layer of cream of 1 micron to be replaced. Make suitable assumptions such as the cranial area of 2500 cm^2 and the length of the hair is 5 cm.

(**Ans:** 9.60 min/million hair)

18. *Dyeing of the Wool*
A dye bath at a concentration C_0 and a volume V is used to dye wool that is bathed in it. The dye diffuses into the wool. Measuring the concentration of the dye in the wool as a function of time, can you (a) estimate the diffusion coefficient of the dye, If so how? (b) estimate the relaxation time?

19. *Dopant Profile by Ion Implantation*
Ion implantation is used to introduce dopant atoms into the semiconductor material to alter its electrical conductivity. During ion implantation a beam of ions containing the dopant is directed at the semiconductor surface. For example, boron atoms are implanted into silicon wafers by Lucent Technologies, Murray Hill, NJ, USA. Assume that the transfer of boron into the silicon surface is on account of both the convection and diffusion contributions at steady state. Show that the governing equation for the transfer of boron at the gas-solid interface is given by:

$$-\frac{\partial C_A}{\partial z} = D_{AB}\frac{\partial^2 C_A}{\partial z^2}$$

Given a characteristic length l, show that the equation can be reduced to

$$-Pe_m\frac{\partial u}{\partial Z} = \frac{\partial^2 u}{\partial Z^2}$$

and the solution is,

$$u = 1 - \frac{J^*_{ss}}{Pe_m}\exp(-Pe_m Z)$$

20. *Soot from the Steam Engine*
The steam engine that powers the train that takes you from Chennai to New Delhi in 31 hours discharges coal dust at steady rate of 68 kg-mol/hr. The train moves at a velocity of 90 km/hr. Estimate the thickness of soot that will deposit on a passenger sitting near the window of S6 during the entire journey. S6 is about 200 feet from the engine. Assume that the diffusion coefficient of the soot in air is 1E–6 m^2/s. Repeat the analysis for a wind speed of 10 km/hr.

(*Hint:* Bulk concentration of soot in surrounding air can be calculated by considering a basis of time as that taken for the passenger to move 600 ft to the discharge point in fixed space, and in that time the discharge amount calculated from the discharge rate and the dispersed region from the penetration length in all three directions).

(**Ans:** 6.3 cm; 7.0 cm)

21. *Steady Diffusion in a Hollow Sphere*
Develop the concentration profile in a hollow sphere when a species is diffusing without any chemical reaction. Consider the concentration of the species to be held constant at the inner and outer surface of the cylinder at C_{Ai} and C_{Ao} respectively. Show that:

$$\left(\frac{C_A - C_{Ai}}{C_{Ai} - C_{Ao}}\right) = \left(\frac{1 - R_i/r}{1 - R_i/R_o}\right)$$

22. *Determination of Diffusivity*
Unimolar diffusion can be used to estimate the binary diffusivity of a binary gas pair. Consider the evaporation of CCl_4 Carbon tetrachloride into a tube containing oxygen. The distance between the CCl_4 level and the top of the tube is 16.5 cm. The total pressure in the system is 760 mm Hg and the temperature –5°C. The vapour pressure of CCl_4 at that temperature is 29.5 mm Hg. The area of the diffusion path in the diffusion tube may be taken as 0.80 cm^2. Determine the binary diffusivity of O_2–CCl_4 when in an 11-hour period after steady state 0.026 cm^3 of CCl_4 is evaporated.

(**Ans:** 8.61E–6 cm^2/s)

23. *Helium Separation from Natural Gas*
McAfee (1958) proposed a method to separate helium from natural gas. He noted that pyrex glass is almost impermeable to all gases but helium. The diffusion coefficient of helium is 25 times the diffusion coefficient of hydrogen. Consider a pyrex tubing of length L and inner and outer radii, R_i and R_o. Show that the rate at which helium will diffuse through the pyrex can be given by:

$$J_{He} = 2\pi L\, D_{He\text{-}pyrex} \left(\frac{C_{He,1} - C_{He,2}}{\ln (R_o/R_i)}\right)$$

24. *Solid Dissolution into a Falling Film*
A liquid is flowing in laminar motion down a vertical wall. The wall consists of a species that is slightly soluble in the liquid. Show that the governing equation for species diffusing into the liquid from the wall can be written as:

$$\frac{\partial^2 u}{\partial z^2} = \left(\frac{UL}{D}\right)\frac{\partial u}{\partial x}$$

Show that an error function solution results for the above PDE.

25. *Carburizing Steel*
Low carbon steel can be hardened in order to improve the wear resistance by carburizing. Steel is carburized by exposing it to gas, liquid or solid that provides a high carbon concentration at the surface. Given the % carbon vs. depth graphs for various times at 930°C, how can the diffusion coefficient be estimated from the graphs.

REFERENCES

Batchelor, G.K., 1972, *J. of Fluid Mechanics*, 52, 245, 71, 1.

Bird, R.B., Stewart, W.E. and Lightfoot, E.N., 2002, *Transport Phenomena,* John Wiley and Sons, New York.

Brokaw, 1969, *Ind. Eng. Chem. Process. Des. & Dev*, **8**, 2, 240.

Catchpole and King, 1994, *Ind. Eng. Chem. Process. Des. & Dev.*, **33**, 1828.

Chapman, S. and Cowling, T.G., 1970, *The Mathematical Theory of Non-Uniform Gases*, Cambridge University Press, UK.

Crank, J., 1975, *The Mathematics of Diffusion*, Oxford University Press, UK.

Cussler, E.L., 1997, *Diffusion Mass Transfer in Fluid Systems,* Cambridge University Press, UK.

Einstein, A., 1905, *Annalen der Physik,* **7**, 549.

Fascimire, B., 1973, NASA Marshall Flight Center, AL.

Fick, A.E., 1855a, *Poggendorff's Annelen der Physik*, **94**, 59.

Fick, A.E., 1855b, *Philosophical Magazine*, **10**, 30.

Franklin, W.M, 1975, *in Diffusion in Solids: Recent Developments*, Academic Press, New York.

Fourier, J.B., 1822, *Theorie Analytique de la Chaleur*, Paris.

Gilland, E.R., 1934, *Ind. Eng. Chem.*, **26,** 681.

Haydeek and Minhas, 1982, *Can. J. of Chem. Eng.*, 60, 195.

McAfee, K.B., 1958, *Scientific American,* 199, 1, 52.

Mathur and Thodos, 1965, *AIChE,* **11,** 613.

NEC Corporation, 1998, US Patent 5,784, 300.

Perry, R.H. and Green, D.W., 1997, *Perry's Chemical Engineers Handbook,* McGraw Hill, New York.

Reid, R.C., Sherwood, T.K. and Prausnitz, J.M., 1977, *Properties of Gases and Liquids,* McGraw Hill, New York.

Resnick, R., and Halliday, D., 1991, *Physics*, Part I, 38th Wiley Eastern Reprint, New Delhi.

Sharma, K.R., 2005, *Damped Wave Transport and Relaxation,* Elsevier, Amsterdam, Netherlands.

________, October 1998, *Process Design for Hydrogen Production using Semi-Permeable Membrane Technology, 50th ACS Southeast Regional Meeting*, Research Triangle Park, NC, USA.

________, October 1998, *Kinetics of Spinodal Decomposition and Nucleation of Bituminous Coal Extract in Polycarbonate Matrix, 50th ACS Southeast Regional Meeting*, Research Triangle Park, NC, USA.

Sharma, K.R., November 1998, On *the Use of Thermocapillary Drag Coefficient, 90th AIChE Annual Meeting,* Miami, FL.

________, November 1998, *Mesoscopic Mixing Efficiency in Gas-Liquid-Solid Fluidized Bed Process for Ozone Bleaching of Pulp, 90th AIChE Annual Meeting*, Miami, FL.

________, May 1999, *Fluidized Bed Degasification for Removing Unreacted Monomers in Continuous Mass Polymerization, Annual Technical Conference for Society of Plastics Engineers,* ANTEC 1999, New York.

________, R. Chawla, June 1999, *Fluidized Bed Degasification Technology, 31st ACS Central Regional Meeting*, Columbus, OH.

________, *Mesoscopic Characterization of Mass Transfer under Turbulence in Circulating Fluidized Beds*, October/November 1999, *91st AIChE Annual Meeting*, Dallas, TX, USA.

________, *Gas-Solid Mass Transfer in Fluidized Bed Degasifier*, November 1999, *International Mechanical Engineering Congress and Exposition, IMECE 99*, Nashville, TN, USA.

Siddiqi and Lucas, 1986, *Can. J. of Chem. Eng.,* 64, 839.

Stark, J.P., 1976, *Solid State Diffusion*, Wiley, New York.

Stokes, R.H., 1950, *J. of Amer. Chem. Soc.,* **72**, 763, 2243.

Treybal, R.E., *Mass-Transfer Operations*, McGraw Hill, New York.

Vrentas, J.S., and Duda, J.L., 1980, *J. of Appl. Polym. Sci.*, **25**, 1297.

Wilke and Lee, 1955, *Ind. Eng. Chem.,* **47,** 1253.

Wilke, C.R. and Chang, P.C, 1955, *AIChE J*, 1264.

CHAPTER 2

Generalized Fick's Laws of Diffusion

Nomenclature

a_w	half-width of the slab (m)
D_{AB}	binary diffusivity (m^2/s)
C	concentration (mol/m^3)
erf	error function $\int \exp(-s^2)ds$
f	function of time
g	function of the spatio-temporal variable, η
$\mathbf{g}$	acceleration due to gravity (m/s^2)
I_0	modified Bessel function of the first kind and zeroth order
J_0	Bessel function of the first kind and zeroth order
K_0	modified Bessel function of the second kind and zeroth order
J''	mass flux (mol/m^2/s)
J^*	dimensionless mass flux $(J''/(D_{AB}/\tau_r)^{1/2}/(C_s - C_0)$
m	mass of the molecule (kg)
n	number of molecules per unit volume (/m^3)
u	dimensionless concentration $(C - C_s)/(C_0 - C_s)$
v_m	velocity of mass (m/s)
x	distance (m)
X	dimensionless distance $(x/(D\tau_r)^{1/2})$
X_a	dimensionless half-width $(a/(D\tau_r)^{1/2})$
V	function of time only
Y_0	Bessel function of the second kind and zeroth order

Greek

δ	kronecker delta function
η	transformation variable, $\eta = X^2 - \tau^2$
λ_n	lamda function
τ	dimensionless time (t/τ_r)
τ_r	relaxation time (s)

2.1 LIMITATIONS TO FICK'S LAWS OF MOLECULAR DIFFUSION

Onsager (1931) argued that the Fourier's law of heat conduction (1822) and by analogy the Fick's law of diffusion contradicts the principle of microscopic reversibility. Fick's law is only an approximation of the description of transient diffusion and neglects the time needed for the acceleration of mass flow. The time needed for acceleration may be small and could be of the same order of magnitude as that of the collision time between molecules. In order to remove the paradox in the Fick diffusion model, assuming an infinite speed of propagation, the damped wave diffusion and relaxation equation can be written by analogy with that suggested by Maxwell (1867), Morse and Feshbach(1953) and postulated independently by Cattaneo (1958) and Vernotte (1958) for heat conduction. Reviews on heat waves have been presented by Joseph and Preziosi (1989, 1990) and Osizik and Tzou (1992). Very little work has been reported on the connection between macroscopic laws and microscale phenomena and the manifestation of the microscale phenomena in the macroscale. A comprehensive insight into the analytical solutions using the manifestations of the generalized transport equation is given in Sharma (2005).

The Cattaneo and Vernotte equation was found to be admissible within the framework of the second law of thermodynamics (Tzou 1997). Bai and Lavine (1995) discussed the damped wave equation in the context of second law of thermodynamics. The relaxation time was suggested to be viewed as the time-lag between the heat flux vector and the temperature gradient when the response time is short. Some investigators have related the relaxation time to the electron-phonon collisions and the volumetric heat capacities of the electrons and the metal lattice (Ozisik and Tzou 1994).

Nernst (1917) suggested that at low temperatures in good thermal conductors, heat may have sufficient "inertia" to give rise to oscillatory discharge. Landau (1941) found two speeds, one for ordinary sound, and one for a second sound, which describe propagation waves of temperature. In Landau's theory there is no damping or dissipation and both sound speeds are associated with wave equations rather than telegraph equations. Propagation of Landau waves in Helium II is specifically a quantum phenomenon. Other forms of non-Fourier heat conduction is the ballistic transport equations due to Chen (2001), dual phase lag model by Tzou (1997), Jeffrey (1924) and the EPRT, equation of phonon transport proposed by Majumdar (1991) and the microscopic two-step model suggested by Qiu and Tien (1992). An infinite order PDE was presented by Sharma (2003) using Taylor series expansion to fully describe the transient events. Experimentally measured values of relaxation times were reported in the order of a few microseconds by Tzou (1997) for steel at 400°C. Mitra, et al. (1995) and Kaminski (1990) have measured relaxation time for materials with a non-homogeneous inner structure of the order of 15–20 seconds.

The motivation for seeking a generalized law for mass diffusion where Fick's law becomes the particular case is six-fold. The first being the theory of Onsager. When transient concentration events are described using the Fick's law, a "blow-up" occurs during short contact times in the expression for surface flux. This is so for the cases of (Sharma, 2005); (a) surface flux expression in a semi-infinite body subject to a step change in one of the boundary concentration, (b) surface flux for a finite slab subject to constant wall concentration on either of its edges, (c) concentration term in the constant wall flux problem in cylindrical coordinates in a semi-infinite medium solved for by the Boltzmann transformation leading to a solution in exponential integral, (d) in the short time limit, the solution from the parabolic diffusion equations for a semi-infinite sphere using the similarity transformation. The second is some of the singularities found in the description of transient mass diffusion using the parabolic diffusion equations. The third reason for seeking alternate forms to Fick's law of mass diffusion is because light is the speediest of velocities. On examining the solution for the transient concentration, Landau and Lifshitz (1987) noted that for times greater than zero the concentration is finite at all points in the infinite medium except at infinite location. It can be inferred that the mass pulse has travelled at infinite speed. This is in conflict with the light speed barrier stated by the theory of relativity of Einstein.

The fact that any speed of a moving object, including the thermal wave, must be less than the speed of light was examined by Kelly (1968) for diffusion. The fourth reason is the realisation of the empirical nature of the development of the Fick's law of mass diffusion from observations at steady state. Use of it in the transient state is an extrapolation. The fifth is the over prediction of theory to experiment in important industrial process chromatography, adsorption, gel electrophoresis, restriction mapping, during manufacture of semi-conductor devices and drug delivery systems. The sixth reason is that the Flick's law breaks down also at small scales (Bejan, 1988; Casimir, 1938). In this limit the flux is described by an expression similar to the one used in radiation heat transfer (Swartz and Pohl, 1989). The mass transport, for example, in dielectric crystalline materials is believed to be primarily by atomic or crystal vibrations. These vibrations travel as waves and the energy of the waves quantitated is phonon, (Kittel, 1986).

An attempt is made to improve the solutions presented for the semi-infinite medium by Baumeister and Hamill (1971) by developing a transformation, $\eta = \tau^2 - X^2$. This is seen to transform the hyperbolic PDE in two variables of space and time after removal of the damping term to a Bessel differential equation. Readily usable solutions can be obtained by realising that the integration constant from the Bessel solution is with respect to the transformation variable which is a function of two variables. As in the case of ordinary differential equations where the integration constants need to be solved for, in partial differential equations of a higher order, functions of variables less then the order of PDE, need to be solved for. This way the discontinuity seen in the Baumeister and Hamill solution is removed and the general solution is improved upon. The finite slab problem is revisited and bounded solutions obtained by two different methods.

In the first method, the separation of variables and the final condition in time for the wave concentration are used to obtain bounded Fourier series solution. In this approach the hyperbolic PDE is first divided by an $e^{-\tau/2}$ and then the resulting equation in wave concentration is solved for. In the second method, the damped wave diffusion and relaxation equation is directly solved for without dividing the equation with the $e^{-n\tau}$ where $n = 1/2$, by a hybridised method of relativistic transformation of coordinates and the method of separation of variables. A hybrid method of substitution and separation of variables is used to obtain an exact solution to the hyperbolic equation for a finite slab. The solution to the equation is shown to be within the bounds of the second law of thermodynamics. The solution from both the methods is well bound and does not overshoot as indicated by Taitel (1972). The periodic boundary condition is also studied using the method of complex concentration.

2.2 DERIVATION OF DAMPED WAVE DIFFUSION AND RELAXATION EQUATION FROM FREE ELECTRON THEORY

The damped wave conduction and relaxation equation is derived from the free electron theory. The derivation of Ohm's law of electric conduction is revisited to obtain the damped wave momentum transfer and relaxation equation by analogy. The electrical resistivity of materials differs by 30 orders of magnitude. So a single theory to explain the behaviour of all materials may be difficult to develop. In the free electron model, the outermost electrons of the atoms can take part in conduction. They are not bound to the atom but are free to move through the whole solid. These electrons have been variously called the free electron cloud, the free electron gas or the Fermi gas. The assumption is that the potential field due to the ion cores is uniform throughout the solid. The free electrons have the same potential energy everywhere in the solid. Due to the electrostatic attraction between a free electron and the ion core, this potential energy will be a finite negative value. Only energy differences are important and the constant potential can be taken as zero. Then the only energy that has to be considered is the kinetic energy. The kinetic energy is substantially lower than that of the bound electrons in an isolated atom as the field of motion for the free electron is considerably enlarged in the solid as compared to the field around an isolated atom. The free electron theory can be used to better understand electrical conduction. By Lorenz analogy, the heat conduction can also be predicted in a similar manner. The independent electron assumption was developed by Drude in 1905. Some of the assumptions in the free electron theory are that electrons are responsible for all of the conduction. The electrons behave like an ideal gas, occupy negligible volume, undergo collisions, and are perfectly elastic. Electrons are free to move in a constrained flat bottom well. Electron distribution of energy is a continuum.

The general equation of motion for the drift velocity of the free electron on account of an applied temperature gradient driving force can be given by the following expression from the Drude theory,

$$m\frac{dv_e}{dt} + m\frac{v_e}{\tau} = -\frac{3}{2}\frac{k_B dT}{dx} \tag{2.1}$$

where m is the mass of the electron, v_e the drift velocity, τ is the collision time of the electron with an obstacle, k_B the Boltzmann constant and dT/dx the applied temperature gradient. The drift velocity of the electron is different from the random velocities associated with it. It is superimposed on the random motion. It is in the net direction of the superimposed field. This leads to a net flow of charge and the passage of electric current. The electron encounters obstacles during drift and the directional motion is lost and reduced to the random motion. The memory gained is lost and the clock is set back to zero. Collisions occur in the time interval, τ. The rate of destruction of momentum by virtue of the collision is given by mv_e/τ. This slows down or drags down the electron. The drag force will balance the applied force due to the temperature gradient at steady state to yield the Fourier's law of heat conduction. This can be seen in the following steps:

$$v_e = -\frac{3\tau}{2m}\frac{k_B dT}{dx} \tag{2.2}$$

The heat flux can be defined as:

$$q = n\left(\frac{3}{2}k_B T\right) v_e \tag{2.3}$$

where n is the number of electrons per unit volume, $(3/2\ k_B T)$ is the average energy of the electron from the equipartition energy theorem. Using the Boltzmann relation, the heat flux can also be written as:

$$q = n\frac{1}{2}mv_e^3 \tag{2.4}$$

Multiplying Eq. (2.2) throughout by $n\ (3/2\ k_B T)$ and using Eq. (2.3), Eq. (2.2) becomes

$$q = -\left(\frac{9nT\tau}{4m}\right)k_B^2\left(\frac{dT}{dx}\right) = -k\frac{\partial T}{\partial x} \tag{2.5}$$

where the thermal conductivity can be written as:

$$k = \left(\frac{9nT\tau}{4m}k_B^2\right) \tag{2.6}$$

During transient heat conduction the acceleration term may become important. Rewriting Eq. (2.1) as:

$$\frac{\tau dv_e}{dt} + v_e = -\frac{3\tau}{2m}k_B\frac{dT}{dx} \tag{2.7}$$

Multiplying Eq. (2.7) throughout by $n(3/2\ k_B T)$ and combining with Eqs. (2.3, 2.6),

$$n\left(\frac{3}{2}k_B T\right)\tau\frac{dv_e}{dt} + q = -k\frac{\partial T}{\partial x} \tag{2.8}$$

Using the Boltzmann relation $(1/2\ mv_e^2 = 3/2\ k_B T)$, Eq. (2.8) becomes,

$$\frac{1}{2}n\tau mv_e^2\frac{dv}{dt} + q = -k\frac{\partial T}{\partial x} \tag{2.9}$$

Differentiating Eq.(2.4) wrt to t,

$$\frac{\partial q}{\partial t} = \frac{3}{2} n m v_e^2 \frac{dv_e}{dt} \tag{2.10}$$

Combining Eq. (2.10) with Eq. (2.9):

$$\frac{\tau}{3} \frac{\partial q}{\partial t} + q = -k \frac{\partial T}{\partial x} \tag{2.11}$$

Eq. (2.11) is equivalent to the Cattaneo and Vernotte equation given by Eq. (2.12) when $\tau/3 = \tau_r$.

$$\tau_r \frac{\partial q}{\partial t} + q = -k \frac{\partial T}{\partial x} \tag{2.12}$$

By analogy for mass diffusion, the generalized Fick's law of diffusion and relaxation can be written as:

$$\tau_r \frac{\partial J''}{\partial t} + J'' = -D_{AB} \frac{\partial C}{\partial x} \tag{2.13}$$

The governing equation for concentration in Cartesian, cylindrical and spherical coordinates taking into account the generalized Fick's law of mass diffusion and relaxation is given in Table 2.1.

2.3 TRANSIENT CONCENTRATION PROFILE IN A FINITE SLAB SUBJECT TO CONSTANT WALL CONCENTRATION

2.3.1 Taitel Paradox and the Final Time Condition

Previous reports, Taitel (1972), Barletta and Zanchini (1997) have raised some concerns about the second law of thermodynamics and the Cattaneo and Vernotte equation. Taitel considered heat conduction in an infinitely wide parallel slab with thickness $2L$ such that the thermal conductivity k, the thermal diffusivity α, the specific heat at constant volume and the thermal relaxation time τ_r of the slab can be considered constant. It was noted that at times zero, $\partial T/\partial t = 0$ and use it as one of the time conditions and $T = T_0$ at times zero as the second time condition. For times greater than zero, the temperature distribution on the two sides of the slab is kept uniform with a value $T_w \neq T_0$. By symmetry at the centre of the slab, $\partial T/\partial x = 0$, is the fourth space condition. A second order hyperbolic partial differential equation, PDE can be completely described by two space and two time conditions. Upon obtaining the transient temperature, Taitel points out that the absolute value of the temperature change $(T - T_0)$ may exceed $|T_w - T_0|$.

Barletta and Zanchini developed a solution for the finite slab problem by the method of separation of variables. They showed by a plot of $1 - u$ vs. X for Vernotte

TABLE 2.1 Damped Wave Conduction and Relaxation Equation for Concentration including the Convective Effects

Cartesian Coordinates

$$\tau_{mr}\left[\frac{\partial^2 C_A}{\partial t^2} + \frac{v_x \partial^2 C_A}{\partial t \partial x} + \frac{v_y \partial^2 C_A}{\partial t \partial y} + \frac{v_z \partial^2 C_A}{\partial t \partial z}\right] + \frac{\partial C_A}{\partial t} \left(\tau_{mr}\frac{\partial v_x}{\partial t} + v_x\right)\frac{\partial C_A}{\partial x}$$

$$+ \left(\tau_{mr}\frac{\partial v_y}{\partial t} + v_y\right)\frac{\partial C_A}{\partial y} + \left(\tau_{mr}\frac{\partial v_z}{\partial t} + v_z\right)\frac{\partial C_A}{\partial z}$$

$$= D_{AB}\left[\frac{\partial^2 C_A}{\partial x^2} + \frac{\partial^2 C_A}{\partial y^2} + \frac{\partial^2 C_A}{\partial z^2}\right] + R_A \quad \text{(T2.1.1)}$$

Polar Coordinates

$$\tau_{mr}\frac{\partial^2 C_A}{\partial t^2} + \frac{v_r \partial^2 C_A}{\partial t \partial r} + \left(\frac{v_\theta}{r}\right)\left(\frac{\partial^2 C_A}{\partial t \partial \theta}\right) + \frac{v_z \partial^2 C_A}{\partial t \partial z} + \frac{\partial C_A}{\partial r}\left(\tau_{mr}\frac{\partial V_r}{\partial t} + V_r\right)$$

$$+ \frac{1}{r}\frac{\partial C_A}{\partial \theta}\left(\tau_{mr}\frac{\partial v_\theta}{\partial t} + v_\theta\right) + \left(\tau_{mr}\frac{\partial v_z}{\partial t} + v_z\right)\left(\frac{\partial C_A}{\partial z}\right) + \left(\frac{\partial C_A}{\partial t}\right)$$

$$= D_{AB}\left(\frac{1}{r}\frac{\partial}{\partial r}\right)\left(\frac{r\partial C_A}{\partial r}\right) + \frac{1}{r^2}\frac{\partial^2 C_A}{\partial \theta^2} + \frac{\partial^2 C_A}{\partial z^2} + R_A \quad \text{(T2.1.2)}$$

Spherical Coordinates

$$\tau_{mr}\frac{\partial^2 C_A}{\partial t^2} + \frac{v_r \partial^2 C_A}{\partial t \partial r} + \left(\frac{v_\theta}{r}\right)\left(\frac{\partial^2 C_A}{\partial t \partial \theta}\right) + v_\phi \frac{1}{r\sin\theta}\frac{\partial^2 C_A}{\partial t \partial \phi} + \frac{\partial C_A}{\partial r}\left(\tau_{mr}\frac{\partial v_r}{\partial t} + v_r\right)$$

$$+ \frac{1}{r}\frac{\partial C_A}{\partial \theta}\left(\tau_{mr}\frac{\partial v_\theta}{\partial t} + v_\theta\right) + \frac{1}{r\sin\theta}\left(\tau_{mr}\frac{\partial v_\phi}{\partial t} + v_\phi\right)\left(\frac{\partial C_A}{\partial \phi}\right) + \left(\frac{\partial C_A}{\partial t}\right)$$

$$= D_{AB}\left(\frac{1}{r^2}\frac{\partial}{\partial r}\right)\left(\frac{r^2\partial C_A}{\partial r}\right) + \frac{1}{r^2\sin\theta}\frac{\partial}{\partial \theta}\left(\sin\theta\frac{\partial C_A}{\partial \theta}\right) + \left(\frac{1}{r^2\sin^2\theta}\right)\left(\frac{\partial^2 C_A}{\partial \phi^2}\right) + R + R_A \quad \text{(T.2.1.3)}$$

number 1 ($\alpha\tau_r/4L^2$) and Fourier number of 0.7 ($\alpha t/4L^2$), that $|T - T_0|$ may exceed $|T_w - T_0|$ as pointed out by Taitel. In another plot of $1 - u$ vs. X for Vernotte number 1 and Fourier number of 0.25, the equilibrium value for the temperature was attained by an oscillatory process. The parabolic conduction predicts a continuous increase in temperature from zero to one at any internal position. The solution obtained by Taitel, for the centreline temperature of the finite slab is given below. They considered a constant wall temperature and the initial time conditions included a $\partial T/\partial t = 0$ term in addition to the initial temperature condition. The exact solution presented by Taitel is as follows:

$$u = \Sigma_0^\infty b_n \exp\left(\frac{\tau}{2}\right) \exp\left(-\left(\frac{\tau}{2}\right)\left(\sqrt{1 - \frac{4(2n+1)^2\pi^2\alpha\tau_r}{a^2}}\right)\right)$$

$$+ \Sigma_0^\infty c_n \exp\left(\frac{\tau}{2}\right) \exp\left(+\left(\frac{\tau}{2}\right)\left(\sqrt{1 - \frac{4(2n+1)^2\pi^2\alpha\tau_r}{a^2}}\right)\right) \tag{2.14a}$$

Multiplying both sides of the equation by $\exp(\tau/2)$,

$$u \exp\left(\frac{\tau}{2}\right) = W = \Sigma_0^\infty b_n \exp\left(-\left(\frac{\tau}{2}\right)\left(\sqrt{1 - \frac{4(2n+1)^2\pi^2\alpha\tau_r}{a^2}}\right)\right)$$

$$+ \Sigma_0^\infty c_n \exp\left(\exp\left(+\left(\frac{\tau}{2}\right)\left(\sqrt{1 - \frac{4(2n+1)^2\pi^2\alpha\tau_r}{a^2}}\right)\right)\right) \tag{2.14b}$$

At infinite times, the LHS of Eq. (2.14b) is 0 times ∞ and is zero. The RHS does not vanish. Thus the expression given by Taitel and later discussed as a temperature overshoot, may be as a result of the growing exponential term in the above expression.

2.3.2 Method of Separation of Variables and Exact Solution

Sharma (2005) considered a finite slab of width $2a$ with an initial concentration at C_0. The sides of the slab are maintained at constant concentration of C_s. The governing equation in the dimensionless form is then,

$$\frac{\partial u}{\partial \tau} + \frac{\partial^2 u}{\partial \tau^2} = \frac{\partial^2 u}{\partial X^2} \tag{2.15}$$

where $\left(\frac{C - C_s}{C_0 - C_s}\right)$; $\tau = \frac{t}{\tau_r}$; $X = \frac{x}{\sqrt{\alpha\tau_r}}$ (2.16)

The initial condition is given as follows:

$$t = 0, \ Vx, \ C = C_0; \ u = 1 \tag{2.17}$$

Boundary Conditions in space,

$$t > 0, \quad x = 0, \quad \partial C/\partial x = 0; \quad \partial u/\partial x = 0 \tag{2.18}$$

$$t > 0,\ x = \pm a,\ C = C_s;\quad u = 0 \tag{2.19}$$

The fourth and final condition in time,

$$t = \infty,\ \forall x,\ C = C_s;\quad u = 0 \tag{2.20}$$

The governing equation was obtained by a one-dimensional mass balance (in – out + reaction = accumulation). This is achieved by eliminating J'' between the damped wave diffusion and relaxation equation and the equation from mass balance $(-\partial J''/\partial x = \partial C/\partial t)$. This is achieved by differentiating the constitutive equation with respect to x and the mass balance equation with respect to t and eliminating the second cross derivative of J'' with respect to x and time. This equation is then non-dimensionalized. The solution is obtained by the method of separation of variables.

Let,

$$u = V(\tau)\,\phi(X) \tag{2.21}$$

Eq. (2.15) becomes,

$$\frac{\phi''(X)}{\phi(X)} = \frac{V'(\tau) + V''(\tau)}{V(\tau)} = -\lambda_n^2 \tag{2.22}$$

$$\phi(X) = c_1 \sin(\lambda_n X) + c_2 \cos(\lambda_n X) \tag{2.23}$$

From the boundary conditions,

At $X = 0,\ \dfrac{\partial \phi}{\partial X} = 0,$

So,

$$c_1 = 0 \tag{2.24}$$

$$\phi(X) = c_1 \cos(\lambda_n X) \tag{2.25}$$

$$0 = c_1 \cos(\lambda_n X_a) \tag{2.26}$$

$$\frac{(2n - 1)\pi}{2} = \lambda_n X_a \tag{2.27}$$

$$\lambda_n = (2n - 1)\pi \sqrt{\frac{D\tau_{mr}}{2a}},\quad n = 1, 2, 3, \ldots \tag{2.28}$$

The time domain solution would be,

$$V = \exp\left(-\frac{\tau}{2}\right) c_3 \exp\left(\sqrt{\left(\frac{1}{4} - \lambda_n^2\right)}\ \tau\right) + c_4 \exp\left(-\sqrt{\left(\frac{1}{4} - \lambda_n^2\right)}\ \tau\right) \tag{2.29}$$

or

$$V \exp\left(\frac{\tau}{2}\right) = c_3 \exp\left(\sqrt{\left(\frac{1}{4} - \lambda_n^2\right)}\ \tau\right) + c_4 \exp\left(-\sqrt{\left(\frac{1}{4} - \lambda_n^2\right)}\ \tau\right) \tag{2.30}$$

From the final condition $u = 0$ at infinite time, so is $V\phi \exp(\tau/2) = W$, the wave concentration at infinite time. The wave concentration is that portion of the solution that remains after dividing the damping component either from the solution or the governing equation. For any non-zero ϕ, it can be seen that at infinite time the LHS of Eq. (2.30) is a product of zero and infinity and a function of x and is zero. Hence the RHS of Eq. (2.30) is also zero and hence in Eq. (2.30) c_3 need be set to zero. Hence,

$$u = \Sigma_1^{\infty} c_n \exp\left(-\frac{\tau}{2}\right) \exp\left(-\sqrt{\left(\frac{1}{4} - \lambda_n^2\right)}\,\tau\right) \cos(\lambda_n X) \tag{2.31}$$

where λ_n is described by Eq. (2.28). C_n can be shown using the orthogonality property to be $4(-1)^{n+1}/(2n-1)\pi$. It can be seen that Eq. (2.31) is bifurcated. As the value of the thickness of the slab changes, the characteristic nature of the solution changes from monotonic exponential decay to subcritical damped oscillatory. For $a < \pi\sqrt{(D_{AB}\tau_r)}$, even for $n = 1$, $\lambda_n > 1/2$. This is when the argument within the square root sign in the exponentiated time domain expression becomes negative and the result becomes imaginary. Using Demovrie's theorem and taking real part for small width of the slab,

$$u = \Sigma_1^{\infty} c_n \exp\left(-\frac{\tau}{2}\right) \cos\left(\sqrt{\left(\lambda_n^2 - \frac{1}{4}\right)}\,\tau\right) \cos(\lambda_n X) \tag{2.32}$$

Eqs (2.31, 2.32) can be seen to be well bounded. Eq. (2.32) becomes zero after some time. This would be time taken to reach steady state. Thus, for a $\geq \pi\sqrt{(D_{AB}\tau_r)}$

$$u = \Sigma_1^{\infty} c_n \exp\left(-\frac{\tau}{2}\right) \exp\left(-\sqrt{\left(\frac{1}{4} - \lambda_n^2\right)}\,\tau\right) \cos(\lambda_n X) \tag{2.33}$$

where $c_n = \dfrac{4(-1)^{n+1}}{(2n-1)\pi}$ and $\lambda_n = (2n-1)\pi\dfrac{\sqrt{(D_{AB}\tau_r)}}{2a}$

The centreline concentration for a particular example is shown in Figure 1.1. Eight terms in the infinite series given in Eq. (2.31) were taken and the values calculated on a 1.9 GHz Pentium IV desktop personal computer. The number of terms was decided on the incremental change or improvement obtained by doubling the number of terms. The number of terms was arrived at a 4% change in the dimensionless temperature. The results for the case of a small slab are shown in Figure 2.1. The subcritical damped oscillations can be seen. The time taken to steady state can be read from the x intercept. In Figure 3.1 a parametric study of the relaxation time is shown. A small slab of thickness of 1 cm and binary diffusivity of 10^{-5} m^2/s is considered. Twelve terms were taken in the infinite series solution and four different relaxation times were calculated. The accuracy of the data was less than 4%. For the case when the relaxation time was small, i.e. when Eq. (2.33) was applicable for the solution, the centreline concentration decayed monotonically with the x axis as its asymptote. When the relaxation time considered are large, such that Eq. (2.32) is applicable, the subcritical damped oscillations can be seen. The time taken to steady state can be read from the x intercept in such cases. This happens when:

$$\tau_r > \left(\frac{a^2}{\pi^2 D_{AB}}\right) \tag{2.34}$$

At infinite relaxation time, the governing equation will revert to the wave equation (Sharma, 2005) and the D'Alambert solution will result. For a wide range of mass relaxation times this approach can be seen to be viable.

The Taitel paradox is obviated by examining the final steady state condition and expressing the state in mathematical terms. The W term, which is the dimensionless concentration upon removal of the damping term, needs to go to zero at infinite time. This resulted in a well bounded solution. The use of final condition is may be what is needed for this problem to be used extensively in engineering analysis without being branded as violating second law of thermodynamics. The conditions are the touted violations of second law and are not physically realistic. A bifurcated solution results. For small width of the slab, $a < \pi\sqrt{(D_{AB}\tau_r)}$, the transient concentration is sub critical damped oscillatory.

An exact well bounded solution that is bifurcated depending on the width of the slab is provided. The transient solution to the damped wave non-Fick hyperbolic wave propagative and relaxation equation is obtained by the method of separation of variables. A well bounded infinite series expression is provided. The temperature overshoot identified by Taitel (1972) is obviated by examining the final steady state condition and expressing the state in mathematical terms. A bifurcated solution results. For small width of the slab, $a < \pi\sqrt{(D\tau_r)}$, the transient concentration is sub critical damped oscillatory. In both (Taitel, 1972 and Barletta and Zanchini, 1997); four conditions were used for initial and boundary constraints. The two in space domain and the initial concentration at time zero are retained. However, the slope with the time domain of the concentration at time zero is replaced with the final condition for the time domain, i.e. at steady state the transient concentration will decay out to a constant value or to zero in the dimensionless form. This consideration is shown to change the nature of the solution considerably to a well bounded expression that is bifurcated. For small values of the slab, the transient concentration is sub critical damped oscillatory. For other values, the Fourier series representation is augmented by a modification to the exponential time domain portion of the solution. In this section, the use of the final condition at steady state as the fourth condition to give a bounded solution in obeyance of Clausius inequality was achieved.

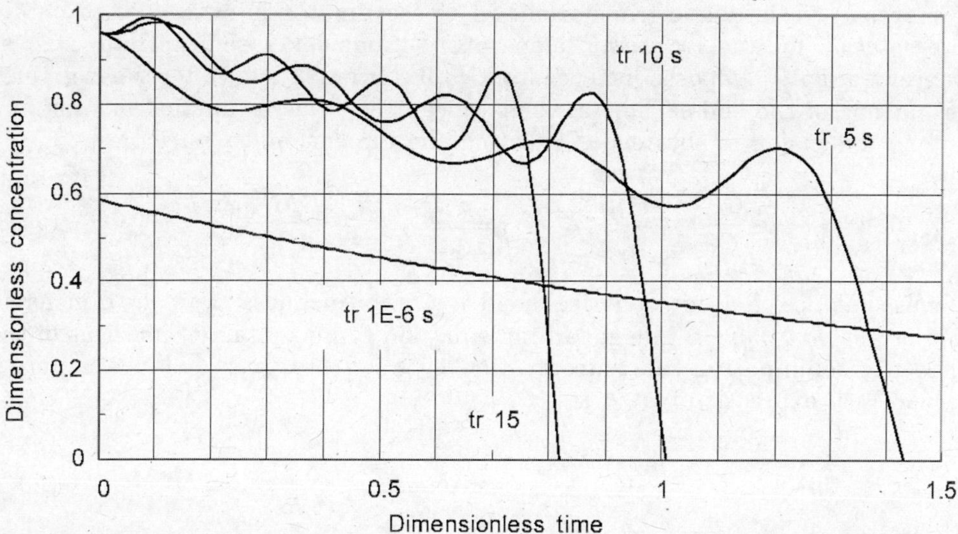

FIGURE 2.1 Dimensionless concentration for a finite slab at different relaxation times.

2.4 SEMI-INFINITE MEDIUM SUBJECT TO CONSTANT WALL CONCENTRATION

The semi-infinite medium is considered to study the spatio-temporal patterns that the solution of the non-Fick damped wave diffusion and relaxation equation exhibits. This kind of consideration has been used in the study of Fick mass diffusion. The boundary conditions can be different kinds such as the constant wall concentration, the constant wall flux, CWF, pulse injection, convective, impervious and exponential decay. The similarity or Boltzmann transformation worked out well in the case of parabolic PDE where an error function solution can be obtained in the transformed variable. The conditions at infinite width and zero time are the same. The conditions at zero distance from the surface and at infinite time are the same.

2.4.1 Relativistic Transformation of Coordinates

Baumeister and Hamill (1971) solved the hyperbolic heat conduction equation in a semi-infinite medium subjected to a step change in temperature at one of its ends using the method of Laplace transform. The space integrated expression for the temperature in the Laplace domain had the inversion readily available within the tables. This expression was differentiated using the Leibniz's rule and the resulting temperature distribution was given for $\tau > X$ as:

$$u = \left(\frac{C - C_0}{C_s - C_0}\right) = \exp\left(-\frac{X}{2}\right) + X\int_X^{\tau} \exp\left(-\frac{p}{2}\right)\frac{I_1\sqrt{p^2 - X^2}}{\sqrt{p^2 - X^2}}\,dp \qquad (2.35)$$

The method of relativistic transformation of coordinates is evaluated to obtain the exact solution for the transient temperature. Consider a semi-infinite slab at initial concentration C_0, imposed by a constant wall concentration, C_s for times greater than zero at one of the ends. The transient concentration as a function of time and space in one dimension is obtained. Obtaining the dimensionless variables:

$$u = \left(\frac{C - C_0}{C_s - C_0}\right);\ \tau = \frac{t}{\tau_r};\ X = \frac{x}{\sqrt{(D_{AB}\tau_r)}};\ J^* = \frac{J''/(D_{AB}/\tau_r)^{1/2}}{(C_s - C_0)} \tag{2.36}$$

The mass balance on a thin spherical shell at x with thickness Δx in one dimension is written as $-\partial J^*/\partial X = \partial u/\partial t$. The governing equation can be obtained in terms of the mass flux after eliminating the concentration between the mass balance equation and the non-Fick expression.

$$\frac{\partial J^*}{\partial \tau} + \frac{\partial^2 J^*}{\partial \tau^2} = \partial^2 J^*/\partial X^2 \tag{2.37}$$

It can be seen that the governing equation for the dimensionless mass flux is identical in form with that of the dimensionless concentration. The initial condition is:

$$\tau = 0,\quad J^* = 0 \tag{2.38}$$

The boundary conditions are;

$$X = \infty,\quad J^* = 0 \tag{2.39}$$

$$X = 0,\quad C = C_s;\quad u = 1 \tag{2.40}$$

Let us suppose that the solution for J^* is of the form $W \exp(-n\tau)$ for $\tau > 0$ where W is the transient wave flux. Then,

$$\exp(-n\tau)\,W(-n + n^2) + \exp(-n\tau)\frac{\partial W}{\partial \tau}(1 - 2n) + \exp(-n\tau)\frac{\partial^2 W}{\partial \tau^2} = \exp(-n\tau)\left(\frac{\partial^2 W}{\partial X^2}\right) \tag{2.41}$$

For $n = 1/2$ Eq. (2.37) becomes,

$$\frac{\partial^2 W}{\partial \tau^2} - \frac{W}{4} = \frac{\partial^2 W}{\partial X^2} \tag{2.42}$$

The solution to Eq. (2.42) can be obtained by the following relativistic transformation of coordinates, for $\tau > X$. Let $\eta = (\tau^2 - X^2)$. Then Eq. (2.42) becomes,

$$\frac{\partial^2 W}{\partial \tau^2} = \frac{4\tau^2 \partial^2 W}{\partial \eta^2} + \frac{2\partial W}{\partial \eta} \tag{2.43}$$

$$\frac{\partial^2 W}{\partial X^2} = \frac{4X^2 \partial^2 W}{\partial \eta^2} - \frac{2\partial W}{\partial \eta} \tag{2.44}$$

Combining Eqs. (2.43 and 2.44) into Eq. (2.42),

$$4(\tau^2 - X^2)\frac{\partial^2 W}{\partial \eta^2} + \frac{4\partial W}{\partial \eta} - \frac{W}{4} = 0 \tag{2.45}$$

or

$$\frac{\eta^2 \partial^2 W}{\partial \eta^2} + \frac{\eta \partial W}{\partial \eta} - \frac{\eta W}{16} = 0 \tag{2.46}$$

Eq. (2.46) can be seen to be a special differential equation in one independent variable. The number of variables in the hyperbolic PDE has thus been reduced from two to one. Comparing Eq. (2.46) with the generalized form of Bessel's equation (Sharma, 2005) it can be seen that $a = 1$, $b = 0$, $c = 0$, $s = 1/2$, $d = -1/16$. The order of the solution is calculated as 0 and the general solution is given by:

$$W = c_1 I_0\left(\frac{1}{2}\sqrt{\eta}\right) + c_2 K_0\left(\frac{\sqrt{\eta}}{2}\right) \tag{2.47}$$

The wave flux W is finite when $\eta = 0$ and hence it can be observed that c_2 can be seen to be zero. The c_1 can be solved from the boundary condition given in Eq. (2.40). The expression for the dimensionless mass t flux for times τ, greater than X is thus,

$$J^* = c_1 \exp\left(-\frac{\tau}{2}\right) I_0\left(\frac{\sqrt{\tau^2 - X^2}}{2}\right) \tag{2.48}$$

For large times, the modified Bessel's function can be given as an exponential and reciprocal in square root of time by asymptotic expansion. Consider the surface flux, i.e. when in Eq. (2.48) X is set as zero.

$$J^* = c_1 \frac{\exp\left(-\frac{\tau}{2}\right)\exp\left(\frac{\tau}{2}\right)}{\sqrt{(2\pi\tau)}} = \frac{c_1}{\sqrt{(2\pi\tau)}} \tag{2.49}$$

For times, when $\exp(\tau)$ is much greater than the mass flux, it can be seen that the second derivative in time of the dimensionless flux in Eq. (2.37) can be neglected compared with the first derivative. The resulting expression is the familiar expression for surface flux from the Fourier parabolic governing equation for constant wall concentration in a semi-infinite medium and is given by:

$$J^* = \frac{1}{\sqrt{2\pi\tau}} \tag{2.50}$$

Comparing Eq. (2.50) and Eq. (2.49) it can be seen that c_1 is 1. Thus the dimensionless heat flux is given by:

$$J^* = \exp\left(-\frac{\tau}{2}\right) I_0\left(\frac{\sqrt{\tau^2 - X^2}}{2}\right) \tag{2.51}$$

The solution for J^* needs to be converted to the dimensionless concentration u and then the boundary conditions applied. From the mass balance,

$$-\frac{\partial J^*}{\partial X} = \frac{\partial u}{\partial \tau} \tag{2.52}$$

Thus differentiating Eq.(2.51) wrt to X and substituting in Eq. (2.52) and integrating both sides wrt τ. For $\tau > X$,

$$u = \int X \exp\left(-\frac{\tau}{2}\right) I_1\left(\frac{1/2\,(\tau^2 - X^2)^{1/2}}{(\tau^2 - X^2)^{1/2}\, d\tau}\right) + C(X) \tag{2.53}$$

It can be left as an indefinite integral and the integration constant can be expected to be a function of space. The $C(X)$ can be solved for by examining what happens at the wave front. At the wave front, $\eta = 0$ and time elapsed equals the time taken for a mass disturbance to reach the location x given the wave speed $\sqrt{(D_{AB}/\tau_r)}$. The governing equations for the dimensionless mass flux and dimensionless concentration are identical in form. At the wave front, Eq. (2.46) reduces to:

$$\frac{\partial W}{\partial \eta} = \frac{W}{16} \tag{2.54}$$

or

$$W = c' \exp\left(\frac{\eta}{16}\right) = c' \tag{2.55}$$

$$u = c' \exp\left(-\frac{\tau}{2}\right) = c' \exp\left(-\frac{X}{2}\right) \tag{2.56}$$

Thus $C(X) = c' \exp(-X/2)$.

Thus,

$$u = \int X \exp\left(-\frac{\tau}{2}\right) \frac{I_1\left(\frac{\sqrt{\tau^2 - X^2}}{2}\right) d\tau}{\sqrt{\tau^2 - X^2}} + c' \exp\left(-\frac{X}{2}\right) \tag{2.57}$$

From the boundary condition in Eq. (2.40) it can be seen that $c' = 1$. Thus, for $\tau > X$,

$$u = \int X \exp\left(-\frac{\tau}{2}\right) \frac{I_1\left(\frac{\sqrt{\tau^2 - X^2}}{2}\right) d\tau}{\sqrt{\tau^2 - X^2}} + \exp\left(-\frac{X}{2}\right) \tag{2.58}$$

It can be seen that the boundary conditions are satisfied by the Eq. (2.58) and describes the transient concentration as a function of space and time that is governed by the hyperbolic wave diffusion and relaxation equation. The flux expression is given by Eq. (2.51).

2.4.2 Regimes of Mass Flux at an Interior Point Inside the Medium

It can be seen that expressions for dimensionless mass flux and dimensionless concentration given by Eq. (2.51) and Eq. (2.58) are valid only in the open interval for $\tau > X$. When $\tau = X$, the wavefront condition results and the dimensionless mass flux and concentration are identical and is:

$$J^* = u = \exp\left(-\frac{X}{2}\right) = \exp\left(-\frac{\tau}{2}\right) \tag{2.59}$$

When $X > \tau$, the transformation variable can be redefined as:
$\eta = X^2 - \tau^2$. Eq. (2.42) becomes:

$$\frac{\eta^2 \partial^2 W}{\partial \eta^2} + \frac{\eta \partial W}{\partial \eta} + \frac{\eta W}{16} = 0 \tag{2.60}$$

The general solution for this Bessel equation is given by:

$$W = c_1 J_0\left(\frac{\sqrt{\eta}}{2}\right) + c_2 Y_0\left(\frac{\sqrt{\eta}}{2}\right) \tag{2.61}$$

The wave temperature W is finite when $\eta = 0$ and hence it can be seen that c_2 can be seen to be zero. The c_1 can be solved from the boundary condition given in Eq. (2.50). The expression in the open interval or the dimensionless heat flux for times τ, smaller than X is thus

$$J^* = c_1 \exp\left(-\frac{\tau}{2}\right) J_0\left(\frac{\sqrt{X^2 - \tau^2}}{2}\right) \tag{2.62}$$

On examining the Bessel function in Eq. (2.62), it can be seen that the first zero of the Bessel occurs when the argument becomes 2.4048. Beyond that point, the Bessel function will take on negative values indicating a reversal of heat flux. There is no good reason for the mass flux to reverse in direction at short times. Hence the Eq. (2.62) is valid from the wave front down to where the first zero of the Bessel function occurs. Thus the plane of zero transfer explains the initial condition verification from the solution.

By using the expression at the wave front for the dimensionless mass flux, c_1 can be solved for and found to be 1. The Eq. (2.62) can also be obtained directly from Eq. (2.51) by using $I_0(\eta) = J_0(i\eta)$. The expression for temperature in a similar vein for the open interval $X > \tau$ is thus;

$$u = \int \frac{X \exp\left(-\frac{\tau}{2}\right) J_1\left(\frac{\sqrt{X^2 - \tau^2}}{2}\right)}{\sqrt{X^2 - \tau^2}} d\tau + \exp\left(-\frac{X}{2}\right) \tag{2.63}$$

Consider a point X_p in the semi-infinite medium. Three regimes can be identified in the mass flux at this point from the surface as a function of time. The series expansion of the modified Bessel composite function of the first kind and zeroth order was used using a Microsoft Excel spreadsheet on a Pentium IV desktop microcomputer. The three regimes and the mass flux at the wave front are summarized as follows:

1. The first regime is a thermal inertia regime when there is no transfer.
2. The second regime is given by expression as in Eq. (2.62) for the mass flux, and

$$J^* = \exp\left(-\frac{\tau}{2}\right) J_0\left(\frac{\sqrt{X^2 - \tau^2}}{2}\right) \tag{2.64}$$

The first zero of the zeroth order Bessel function of the first kind occurs at 2.4048. This is when,

$$2.4048 = \frac{\sqrt{X^2 - \tau^2}}{2} \quad \text{or} \quad \tau_{lag} = \sqrt{X^2 - 23.132} \tag{2.65}$$

Thus τ_{lag} is the inertial lag that will ensure before the mass flux is realised at an interior point in the semi-infinite medium at a dimensionless distance X from the surface. As demonstration, one value of X, i.e., 5 is used. Thus for points closer to the surface the time lag may be zero. Only for dimensionless distances greater than 4.8096, the time lag is finite. For distances *closer than 4.8096 sqrt($\alpha\tau_r$)*, the thermal lag experienced *will be zero*. For distances,

$$x > 4.8096\sqrt{\alpha\tau_r} \tag{2.66}$$

The time lag experienced is given by Eq. (2.65) and is $\sqrt{(X^2 - 4\beta_1^2)}$ where, β_1 is the first zero of the Bessel function of the first kind and zeroth order and is 2.4048. In a similar fashion, the penetration distance of the disturbance for a considered instant in time, beyond which the change in initial temperature is zero can be calculated as:

$$X_{pen} = (23.132 + \tau_i^2)^{1/2} \tag{2.67}$$

3. The third regime starts at the wavefront and is described by Eq. (2.59).

$$J^* = c_1 \exp\left(-\frac{\tau}{2}\right) I_0\left(\frac{\sqrt{\tau^2 - X^2}}{2}\right) \tag{2.68}$$

4. At the wave front, $J^* = u = \exp(-X/2) = \exp(-\tau/2)$

2.4.3 Approximate Solution

The expressions for transient concentration derived in the above section need integration prior to use. More easily usable expressions can be developed by making suitable approximations. Realising that for PDE a set of functions instead of constants as in the case of ODE needs to be solved from the boundary conditions, the c in Eq. (2.48) is allowed to vary with time. This results in an expression for transient concentration that is more readily available for direct use of the practitioner. Extensions to three dimensions in space are also straight forward in this method.

In this section, the exact solution for the constant wall concentration problem in semi-infinite medium in one-dimension is revisited since the discussion by the method of Laplace transforms by Baumeister & Hamill. An expression that does not need further integration is attempted to be derived in this section. Consider a semi-infinite slab at initial concentration, C_0, subjected to sudden change in concentration at one of the ends to C_s. The mass propagative velocity $V_m = \sqrt{(D_{AB}/\tau_r)}$. The initial condition:

$$t = 0,\ Vx,\ C = C_0 \tag{2.69}$$

$$t > 0,\ x = 0,\ C = C_s \tag{2.70}$$

$$t > 0,\ x = \infty,\ C = C_0 \tag{2.71}$$

Obtaining the dimensionless variables:

$$u = \left(\frac{C - C_0}{C_s - C_0}\right);\ \tau = \frac{t}{\tau_r};\ X = \frac{x}{\sqrt{(D_{AB}\tau_r)}} \tag{2.72}$$

The mass balance on a thin spherical shell at x with thickness Δx is written. The governing equation can be obtained after eliminating J'' between the mass balance equation and the derivative with respect to x of the flux equation and introducing the dimensionless variables. Suppose $u = \exp(-n\tau)\ w(X, \tau)$. By choosing $n = 1/2$, the damping component of the equation is removed. Thus for $n = 1/2$, the governing equation becomes,

$$-\frac{w}{4} + \frac{\partial^2 w}{\partial \tau^2} = \frac{\partial^2 w}{\partial X^2} \tag{2.73}$$

By inspecting the flux solution obtained by method of Laplace transforms, (Sharma[10]), consider the transformation variable η as:

$$\eta = \tau^2 - X^2, \qquad \text{for } \tau > X \tag{2.74}$$

$$\frac{\partial w}{\partial \tau} = \left(\frac{\partial w}{\partial \eta}\right) 2\tau \tag{2.75}$$

$$\frac{\partial^2 w}{\partial \tau^2} = \left(\frac{\partial^2 w}{\partial \eta^2}\right) 4\tau^2 + 2\left(\frac{\partial w}{\partial \eta}\right) \tag{2.76}$$

In a similar fashion,

$$\frac{\partial^2 w}{\partial X^2} = \left(\frac{\partial^2 w}{\partial \eta^2}\right) 4X^2 + 2\left(\frac{\partial w}{\partial \eta}\right) \tag{2.77}$$

Substituting Eqs. (2.76–2.77) into Eq. (2.73)

$$\left(\frac{\partial^2 w}{\partial \eta^2}\right) 4(\tau^2 - X^2) + 4\left(\frac{\partial w}{\partial \eta}\right) - \frac{w}{4} = 0 \tag{2.78}$$

$$\eta^2 \partial^2 \frac{w}{\partial \eta^2} + \eta \partial \frac{w}{\partial \eta} - \frac{\eta w}{16} = 0 \tag{2.79}$$

Eq. (2.79) can be recognized as the modified Bessel equation.

$$w = c_1 I_0 \left(\sqrt{\frac{\eta}{2}}\right) + c_2 K_0 \left(\sqrt{\frac{\eta}{2}}\right) \tag{2.80}$$

For $X = 0$, u is 1 or finite and hence it can be seen that w is also finite and hence $C_2 = 0$. Writing the expression for u,

So,

$$u = c_1 \exp\left(-\frac{\tau}{2}\right)\left(I_0\sqrt{\frac{\eta}{2}}\right) \tag{2.81}$$

From BC, Eq. (2.81) becomes,

$$1 = c_1 \exp\left(-\frac{\tau}{2}\right) I_0\left(\frac{\tau}{2}\right) \tag{2.82}$$

c_1 can be eliminated by dividing Eq. (2.81) by Eq. (2.82) to yield:

$$u = \frac{I_0\left(\left(\sqrt{\tau^2 - X^2}\right)/2\right)}{(I_0(\tau/2))} \tag{2.83}$$

for $\tau > X$. for $\tau < X$ or $\tau < x/v_h$

$$u = \frac{J_0\left(\left(\sqrt{X^2 - \tau^2}\right)/2\right)}{(I_0(\tau/2))} \tag{2.84}$$

It can be inferred that an expression in time is used for c_1. A domain restricted solution for short and long times may be in order. For long times, $I_0(\tau/2)$, approximates as $\exp(\tau/2)/\sqrt{2\pi\tau}$.

$$1 = c_1 \exp\left(-\frac{\tau}{2}\right)\left(I_0\left(\frac{\tau}{2}\right)\right) = c_1 \exp\left(-\frac{\tau}{2}\right)\frac{\exp(\tau/2)}{\sqrt{2\pi\tau}} \tag{2.85}$$

$$c_1 = \sqrt{(2\pi\tau)} \tag{2.86}$$

or,
$$u = \sqrt{(2\pi\tau)}\exp\left(-\frac{\tau}{2}\right)\left(\frac{I_0\sqrt{\tau^2 - X^2}}{2}\right) \tag{2.87}$$

Thus the concentration solution in a semi-infinite medium when subject to a step change in concentration at one of the ends is obtained as an open interval solution. The dimensionless flux as derived earlier is:

$$J^* = \frac{J''}{\dfrac{\sqrt{\dfrac{D_{AB}}{\tau_r}}}{C_s - C_0}} = \exp\left(-\frac{\tau}{2}\right) I_0\left(\frac{\sqrt{\tau^2 - X^2}}{2}\right) \tag{2.88}$$

Further the instantaneous surface flux is when $X = 0$;

$$J^* = \exp\left(-\frac{\tau}{2}\right) I_0\left(\frac{\tau}{2}\right) \tag{2.89}$$

2.4.4 Extension to 3-Dimension

The governing equation in 3 dimensions can be written as:

$$\frac{\partial u}{\partial \tau} + \frac{\partial^2 u}{\partial \tau^2} = \frac{\partial^2 u}{\partial X^2} + \frac{\partial^2 u}{\partial Y^2} + \frac{\partial^2 u}{\partial Z^2} \tag{2.90}$$

After dividing by $\exp(-\tau/2)$, the equation becomes,

$$-\frac{w}{4} + \frac{\partial^2 w}{\partial \tau^2} = \frac{\partial^2 w}{\partial X^2} + \frac{\partial^2 w}{\partial Y^2} + \frac{\partial^2 w}{\partial Z^2} \tag{2.91}$$

Let
$$\eta = \tau^2 - X^2 - Y^2 - Z^2 \tag{2.92}$$

$$\left(\frac{\partial^2 w}{\partial \eta^2}\right) 4(\tau^2 - X^2 - Y^2 - Z^2) + 8\left(\frac{\partial w}{\partial \eta}\right) - \frac{\omega}{4} = 0 \tag{2.93}$$

$$\eta^2 \frac{\partial^2 w}{\partial \eta^2} + 2\eta\frac{\partial w}{\partial \eta} - \frac{\eta w}{16} = 0 \tag{2.94}$$

Comparing Eq. (2.94) with the generalised Bessel equation:

$$b = 0;\ a = 2;\ c = 0;\ s = 1/2;\ d = -1/16;\ p = 2\sqrt{(1/4)} = 1 \text{ (order)}$$

$$w = c_1/\eta^{1/2}\, I_1\sqrt{\eta/2} + c_2/\eta^{1/2}\, K_1\sqrt{\eta/2} \tag{2.95}$$

It can be deduced that c_2 is zero as w is finite when $\eta = 0$. The boundary condition as a point temperature at the origin gives rise to $\tau > \sqrt{X^2 + Y^2 + Z^2}$

$$u = \left[\frac{\tau}{\sqrt{\tau^2 - X^2 - Y^2 - Z^2}}\right] \frac{I_1\left(\dfrac{\sqrt{\tau^2 - X^2 - Y^2 - Z^2}}{2}\right)}{I_1(\tau/2)} \tag{2.96}$$

For $\sqrt{X^2 + Y^2 + Z^2} > \tau,\ \tau > 0$

$$u = \left[\frac{\tau}{\sqrt{X^2 + Y^2 + Z^2 - \tau^2}}\right] \frac{J_1\left(\sqrt{\dfrac{X^2 + Y^2 + Z^2 + \tau^2}{2}}\right)}{I_2(\tau/2)} \tag{2.97}$$

2.5 PERIODIC BOUNDARY CONDITION

Consider a semi-infinite slab at initial concentration C_0, imposed by a periodic concentration at one of the ends by $C_0 + C_1 \cos(\omega\tau)$. The transient concentration as a function of time and space in one dimension is obtained. Obtaining the dimensionless variables;

$$u = \frac{C - C_0}{C_1};\ \tau = \frac{t}{\tau_r};\ X = \frac{x}{\sqrt{D\tau_r}} \tag{2.98}$$

The mass balance on a thin shell at x with thickness Δx is written. The governing equation is obtained after eliminating J between the mass balance equation and the derivative with respect to x of the flux equation and introducing the dimensionless variables. The initial condition is:

$$t = 0,\ C = C_0;\ u = 0 \tag{2.99}$$

The Boundary Conditions are:

$$X = \infty,\ C = C_0;\ u = 0 \tag{2.100}$$

$$X = 0,\ C = C_0 + C_1 \cos(\omega\tau);\ u = \cos(\omega^*\tau) \tag{2.101}$$

Let us suppose that the solution for u is of the form $f(x) \exp(-i\omega^*\tau)$ for $\tau > 0$, where ω is the frequency of the concentration wave imposed on the surface and C_1 is the amplitude of the wave. Then,

$$(-i\omega^*)\, f \exp(-i\omega^*\tau) + (i^2\omega^{*2})\, f \exp(-i\omega\tau) = f'' \exp(-i\omega^*\tau) \tag{2.102}$$

$$i^2 f(\omega^{*2} + i\omega^*) = f''$$

$$f(X) = c \exp(-iX\omega^* \sqrt{(\omega^* + i)}) \tag{2.103}$$

d can be seen to be zero as at $X = \infty$, $u = 0$.

$$u = c \exp(-iX\omega^* \sqrt{(\omega^* + i)}) \exp(-i\omega^*\tau) \tag{2.104}$$

From the boundary condition at $X = 0$,

$$\text{Cos}(\omega^*\tau) = \text{Real part } (c \exp(-i\omega^*\tau)) \text{ or } c = 1 \tag{2.105}$$

$$u = \exp(-X\omega^*(A + iB) \exp(-i\omega^*\tau) = \exp(-A\omega^*X) \exp(-i(BX\omega^* + \omega^*\tau) \tag{2.106}$$

where,

$$A + iB = i\sqrt{(\omega^* + i)} \tag{2.107}$$

Squaring both sides,

$$A^2 - B^2 + 2ABi = i^2 (\omega^* + i) = -\omega^* - i \tag{2.108}$$

$$A^2 - B^2 = -\omega^*;\ 2AB = -1 \text{ or } B = -1/2A$$

or

$$A^2 - 1/4A^2 = -\omega^* \tag{2.109}$$

$$A^2 = (-\omega^* \pm \frac{\sqrt{(\omega^{*2} + i)}}{2};\ B = -\frac{1}{2} A \tag{2.110}$$

Obtaining the real part,

$$\frac{C - C_0}{C_1} = u = \exp(-A\omega^*X) \cos(\omega^*(BX + \tau)) \tag{2.111}$$

The time lag in the propagation of the periodic disturbance at the surface is captured by the above relation. Thus the boundary conditions can be seen to be satisfied by Eq. (2.111). In a similar vein to the supposition of $f(x) \exp(-i\omega^*\tau)$ the mass flux J'', can be supposed to be of the form, $J^* = g(x) \exp(-i\omega^*\tau)$. Thus,

$$g = \frac{f'}{1 - i\omega^*} \tag{2.112}$$

Combining f from Eq. (2.103) into Eq. (2.112)

$$J^* = -\omega^*(A + iB) \exp(-X\omega^*(A + iB) \exp(-i\omega^*\tau) \tag{2.113}$$

$$= -\omega^*(A + iB) \exp(-A\omega^*X) \exp(-i(BX\omega^* + \omega^*\tau)$$

$$= -\omega^*(A + iB) \exp(-A\omega^*X) (\text{Cos}(BX\omega^* + \omega^*\tau) + i \sin(BX\omega^* + \omega^*\tau))$$

Obtaining the real part:

$$J'' = \sqrt{D/\tau_r}\ \omega^* \exp(-A\omega^*X) (B \sin(\omega^*(BX + \tau))) - A \cos(\omega^*(BX + \tau)) \tag{2.114}$$

where $J^* = J''/\sqrt{D/\tau_r}$. Thus the sustained part of the solution is periodic with a time lag from the periodic boundary condition imposed on the surface. The mass flux is determined by the negative temperature gradient and the accumulation term in the Cattaneo and Vernotte Equation. For certain values it can be seen that the mass flux can reverse in direction.

SUMMARY

The damped wave diffusion and relaxation equation was derived from the free electron theory by taking into account the acceleration regime of the electron prior to it reaching the steady drift velocity. The relaxation time was found to be a third of the collision time between the electron and obstacle in a given material.

Six different reasons were given to seek a generalized Fick's law of mass diffusion; (a) It contradicts the principle of microscopic reversibility of Onsager. (b) "blow-up" occurs in expressions that depicts transient events at some instances. (c) light is the speediest of velocity, speedier than the speed of mass. (d) empirical summary of observations and extrapolation from steady state events. (e) overprediction of theory to experiment in a variety of industrially important applications. (f) breaks down at small geometric scales.

The hyperbolic governing equation was solved for by four different methods for three different boundary conditions. The reports in the literature of a temperature overshoot were revisited. For a small slab, $a < \pi(D\tau_r)^{1/2}$ the concentration was shown to exhibit subcritical damped oscillations. The exact solution was found as an infinite series as;

$$u = \Sigma_1^{\infty}\, c_n \exp\left(-\frac{\tau}{2}\right) \cos\left(\sqrt{\left(\lambda_n^2 - \frac{1}{4}\right)}\,\tau\right) \cos(\lambda_n X)$$

In the case of the semi-infinite medium reports in the literature about a wave discontinuity was revisited. A substitution variable that is symmetric in space and time, i.e., $\eta = \tau^2 - X^2$ was proposed to transform the governing equation into a Bessel differential equation. Three regimes were recognized in the solution—an inertial lagging zero transfer regimes, a rising and a third falling regime. Expressions for the penetration length and inertial lag time were derived. The solution for the transient cocentration was found as:

$$u = \int X \exp\left(-\frac{\tau}{2}\right) \frac{I_1\left(\dfrac{\sqrt{\tau^2 - X^2}}{2}\right) d\tau}{\sqrt{\tau^2 - X^2}} + \exp\left(-\frac{X}{2}\right)$$

The manifestation of the relaxation time for the case of the periodic boundary condition was studied using the method of complex concentration. In some cases the mass flux was found to reverse in direction. The storage of concentration is an important consideration. The solution is an over damped system. The finite slab problem subject to a step change in concentration on either sides was solved by the method of separation of variables. The second order hyperbolic PDE in two variables was fully described by two space and two time conditions. The final condition in time for the wave concentration was used to obtain a well bounded solution for the transient concentration compared with an overshoot from previous reports. The final condition is more physically realistic of the transient events. The problem in the semi-infinite medium was solved for by the method of retavistic transformation of coordinates both after dividing the hyperbolic PDE by exp(-ô/2) and before dividing the hyperbolic PDE. The solution from one dimension was extended to three dimensions.

REVIEW QUESTIONS

1. State the Onsager reciprocal relations. Show that $D_{12} = D_{21}$.
2. What was Landau's observation of infinite speed of propagation?
3. What is an overshoot?
4. What is the drag force experienced by the electron compared with the acceleration term?
5. What is penetration length?
6. What is inertial lag time?
7. What is the first zero of the Bessel function of the first order? How is this used in the derivation of the penetration length and inertial lag time in a three-dimensional medium?
8. What is the physical significance of the maxima in Figure 2.1?
9. What is the physical significance of the x intercept in Figure 2.1? Can an expression for the time taken to steady state be derived from these x-intercept values?
10. Examine $I_0(\tau/2)\exp(-\tau/2)$ in terms of extremamas, asymptotic limits and under what conditions can $I_0(\tau/2)$ be reduced to a simpler expression?
11. What is the meaning of a negative mass flux? What happens to the ratio of the accumulation and diffusion terms?
12. It was shown that for large relaxation times, the transient concentration in a finite slab exhibit subcritical damped oscillations. What is the critical size of the slab below which the oscillations can be seen? What is the value of diffusion coefficients when the oscillations can be seen?
13. Contrast subcritical damped oscillations from critical and under damped oscillations. What is resonance mean for this problem?
14. Scale the governing equation and show that when the temporal derivative of the dimensionless concentration exceeds the $\exp(\tau)$ the hyperbolic PDE reduces to the wave equation. Further when $\exp(\tau)$ is greater than the temporal derivative the hyperbolic PDE reverts to the parabolic PDE identical to that of Fick's second law of diffusion.

    Why is there a maxima in the dimensionless flux as a function of time?

PROBLEMS

1. *Pulse in Infinite Medium*
A pulse of solute is injected in the centre of an infinite medium. Show that using the Fick's second law of diffusion, the concentration profile as a function of space time can be shown to be a Gaussian curve.

$$C_A = \frac{M/A}{(4\pi Dt)^{1/2}} \exp\left(-\frac{z^2}{4Dt}\right)$$

Use the damped wave diffusion and relaxation equation and obtain the spatio-temporal concentration profile of the solute.

$$u = J^* \exp\left(-\frac{\tau}{2}\right) I_0\left(\frac{1}{2}(\tau^2 - X^2)\right)$$

2. *Finite Speed Diffusion and Simultaneous Fast Reaction in a Semi-infinite Catalyst in Cartesian Coordinates*
Consider a semi-infinite medium at an initial concentration of zero. For times greater than zero, a step change in concentration is effected at one of the surfaces. The species reacts by a first order reaction as it comes in contact with the solid medium. Show that the governing equation for mass transfer can be written as:

$$\frac{\partial^2 u}{\partial X^2} = \frac{\partial^2 u}{\partial \tau^2} + (1 + k^*)\frac{\partial u}{\partial \tau} + k^* u$$

where, $k^* = (k'''\tau_{mr})$; $u = \dfrac{C}{C_s}$; $\tau = \dfrac{t}{\tau_{mr}}$; $X = \dfrac{x}{\sqrt{D\tau_{mr}}}$

The boundary conditions and initial time conditions are:

$$X = 0, \quad u = 1$$

$$X = \infty, \quad u = 0$$

$$t = 0, \quad u = 0$$

Show that the hyperbolic PDE of the second order can be solved by the method of relativistic transformation of coordinates. The damping term is first removed by a $u = w \exp(-n\tau)$ substitution. Choosing $n = (1 + k^*)/2$ the governing equation becomes

$$\frac{\partial^2 w}{\partial X^2} = -\frac{\omega(1 - k^*)^2}{4} + w_{\tau\tau}$$

Let the transformation be: $\eta = \tau^2 - X^2$ for $\tau > X$. Show that the solution can be written as:

$$u = \frac{I_0\left(\dfrac{|1-k^*|\sqrt{(\tau^2 - X^2)}}{2}\right)}{I_0\left(\dfrac{|1-k^*|\tau}{2}\right)}$$

This is valid for $\tau > X$, $k^* \neq 1$. For $X > \tau$,

$$u = \frac{J_0\left(\dfrac{|1-k^*|\sqrt{(X^2 - \tau^2)}}{2}\right)}{I_0\left(\dfrac{|1-k^*|\tau}{2}\right)}$$

At the wave front, $\tau = X$, $u = \exp(-\tau(1 + k^*)/2) = \exp(-X(1 + k^*)/2)$.

3. *Penetration Length, Inertial Lag Time* Show that the inertial lag time for problem 2 can be written as:

$$\tau_{\text{inertia}} = \sqrt{\left(\frac{X_p^2 - 23.1323}{|1-k^*|^2}\right)}$$

4. Show that for k^* is 1 the governing equation in Problem 2 can be written as:

$$\frac{\partial^2 u}{\partial X^2} = \frac{\partial^2 u}{\partial \tau^2} + 2\frac{\partial u}{\partial \tau} + u$$

Show that the solution to the above equation is:

$$u = \exp(-X)\, S_X(\tau) = 0 \qquad \text{when } 0 < \tau < X$$
$$u = \exp(-X)\, S_X(\tau) = \exp(-X) \qquad \text{when, } \tau > X$$

5. *Zeroth Order Reaction*
Repeat Problem 2 for a zeroth order reaction. Show that the governing equation can be written as:

$$\frac{\partial^2 u}{\partial X^2} = \frac{\partial^2 u}{\partial \tau^2} + \frac{\partial u}{\partial \tau} + k^*$$

Show that the solution can be obtained by the method of Laplace transforms as:

$$u = X\int_X \exp\left(-\frac{p}{2}\right)\frac{[I_1\, 1/2(p^2 - X^2)^{1/2}]}{(p^2 - X^2)^{1/2}}\,dX + \exp\left(-\frac{X}{2}\right) - k^*\int_X^{\tau}(\tau - p)\exp(-p)\,dp$$

$$+\int_X^{\tau}\exp\left(-\frac{p}{2}\right) I_0\frac{1}{2}(p^2 - X^2)^{1/2}\,(\tau - p)\exp\left(-\frac{\tau - p}{2}\right) I_0\frac{1}{2}(\tau - p) + I_1\left(\frac{1}{2}(\tau - p)\right)dp$$

This is valid for $\tau > X$.
For $0 < \tau < X$, $u = 0$.

6. *Finite Speed Diffusion with Fast Chemical Reaction in Infinite Catalyst Medium in Cylindrical Coordinates*
Consider a cylindrical catalyst medium at an initial reactant concentration of zero. The surface of the solid non-catalyst cylinder is maintained at a constant concentration of C_s for times greater than zero. The mass propagative velocity is given as the square root of the ratio of the binary diffusivity and mass relaxation time, $V_m = \sqrt{D/\tau_{mr}}$. The two boundary conditions and initial condition are:

$$t = 0, \quad r > R, \quad C = 0$$

$$t > 0, \quad r = R, \quad C = C_s$$

$$r = \infty, \quad t > 0, \quad C = 0$$

Show that the governing equation in concentration is obtained by eliminating the second cross-derivative of mass flux wrt to r and t between the damped wave mass diffusion and relaxation equation and the mass balance equation in cylindrical coordinates. Let

$$u = \left(\frac{C}{C_s}\right); \tau = \left(\frac{t}{\tau_{mr}}\right); X = \frac{r}{\sqrt{D_{AB}\tau_{mr}}} \quad k^* = k'''\tau_{mr};$$

$$k^*u = \left(\frac{\partial u}{\partial \tau}\right)(1 + k^*) + \frac{\partial^2 u}{\partial \tau^2} = \frac{\partial^2 u}{\partial X^2} + \frac{1}{X}\frac{\partial u}{\partial X}$$

Let $u = w \exp(-n\tau)$. Show that by choosing $n = (1 + k^*)/2$, the governing equation becomes:

$$\frac{1}{X}\frac{\partial w}{\partial X} + \frac{\partial^2 w}{\partial X^2} = -\frac{w(1 - k^*)^2}{4} + w_{\tau\tau}$$

Show that the governing equation can be solved by using the method of relativistic transformation of coordinates. Consider the transformation variable η for $\tau > X$ as; $\eta = \tau^2 - X^2$. Show that the solution can be written as:

$$u = \left[\frac{(\tau^2 - X_R^2)^{1/4}}{(\tau^2 - X^2)^{1/4}}\right]\left[\frac{I_{1/2}(|1 - k^*|)/2\sqrt{(\tau^2 - X^2)}}{I_{1/2}(|1 - k^*|)/2\sqrt{(\tau^2 - X_R^2)}}\right]$$

This is for $\tau > X$, $k^* \neq 1$. For $X > \tau$, $k^* \neq 1$.

$$u = \left[\frac{(\tau^2 - X_R^2)^{1/4}}{(X^2 - \tau^2)^{1/4}}\right]\left[\frac{J_{1/2}(1/2(|1 - k^*|)\sqrt{(X^2 - \tau^2)})}{I_{1/2}(1/2\sqrt{(\tau^2 - X^2)})}\right]$$

Show that the inertial lag time and penetration length are related by:

$$\tau_{lag} = \sqrt{\frac{(X_p^2 - 4\pi^2)}{(|1 - k^*|^2)}}$$

Show that the steady state solution for Eq. (3.67) can be written as:

$$u^{ss} = \frac{K_0(k^*)^{1/2}\, X}{K_0(k^*)^{1/2}\, X_R}$$

7. *Finite Speed Diffusion with Fast Chemical Reaction in Infinite Catalyst in Spherical Coordinates*
Consider a spherical catalyst medium at an initial reactant concentration of zero. The surface of the solid non-catalyst sphere is maintained at a constant concentration of C_s for times greater than zero. The mass propagative velocity is given as the square root of the ratio of the binary diffusivity and mass relaxation time, $V_m = \sqrt{(D/\tau_{mr})}$. The two boundary conditions and the initial condition are:

$$t = 0,\ r > R,\ C = 0$$
$$t > 0,\ r = R,\ C = C_s$$
$$r = \infty,\ t > 0,\ C = 0$$

Show that the governing equation in dimensionless form can be given by

$$u = \left(\frac{C}{C_s}\right);\ \tau = \left(\frac{t}{\tau_{mr}}\right);\ X = \frac{r}{\sqrt{D_{AB}\tau_{mr}}}\quad k^* = k'''\tau_{mr};$$

$$k^*u + \left(\frac{\partial u}{\partial \tau}\right)(1 + k^*) + \frac{\partial^2 u}{\partial \tau^2} = \frac{\partial^2 u}{\partial X^2} + \frac{2}{X}\frac{\partial u}{\partial X}$$

The damping term is removed from the governing equation. This is done realizing that the transient concentration decays with time in an exponential fashion. Let $u = w\exp(-n\tau)$. Choosing $n = (1 + k^*)/2$, the above equation becomes:

$$\frac{2}{X}\frac{\partial w}{\partial X} + \frac{\partial^2 w}{\partial X^2} = -\frac{w(1 - k^*)^2}{4} + w_{\tau\tau}$$

Show that this can be solved by using the method of relativistic transformation of coordinates. Consider the transformation variable η for $\tau > X$, $\eta = \tau^2 - X^2$. Show that the solution can be written as:

$$u = \left[\frac{(\tau^2 - X_R^2)^{1/2}}{(\tau^2 - X^2)^{1/2}}\right]\left[\frac{I_1(|1 - k^*|)/2\sqrt{(\tau^2 - X^2)}}{I_1(|1 - k^*|)/2\sqrt{(\tau^2 - X_R^2)}}\right]$$

This is for $\tau > X$, $k^* \neq 1$. For $X > \tau$, $k^* \neq 1$.

$$u = \left[\frac{(\tau^2 - X_R^2)^{1/2}}{(X^2 - \tau^2)^{1/2}}\right]\left[\frac{J_1(1/2(|1 - k^*|)\sqrt{(X^2 - \tau^2)})}{I_1(|1 - k^*|)2\sqrt{(\tau^2 - X^2)}}\right]$$

Show that the inertial lag time and penetration length can be given by:

$$\tau_{lag} = \sqrt{\frac{(X_p^2 - 7.6634^2)}{(|1 - k^*|^2)}}$$

Show that the steady state solution can be written as:

$$u^{ss} = \left(\frac{X_R}{X}\right)^{1/2} \frac{I_{1/2}[(k^*)^{1/2} X]}{I_{1/2}[(k^*)^{1/2} X_R]}$$

8. *Simultaneous Reaction and Finite Speed Diffusion in a Finite Slab*
Consider a finite slab of width $2a$ with an initial concentration at C_0. The sides of the slab are maintained at constant concentration C_s. Show that the governing equation during simultaneous reaction and finite speed diffusion can be written as:

Let $\tau = \dfrac{t}{\tau_r}$; $X = \dfrac{x}{\sqrt{(D\tau_r)}}$; $k^* - (k'''\tau_r)$

$$k^*C + \frac{\partial C}{\partial \tau}(1 + k^*) + \frac{\partial^2 C}{\partial \tau^2} = \frac{\partial^2 C}{\partial X^2}$$

Show that the solution can be assumed to consist of a steady state part and a transient part, i.e. $C = C^t + C^{ss}$. The steady state part and boundary conditions can be selected in such a fashion that the transient portion becomes homogeneous.

$$\frac{\partial^2 C^{ss}}{\partial X^2} = k^* C^{ss}$$

The Boundary Conditions are:

$$X = 0, \quad \frac{\partial C^{ss}}{\partial X} = 0$$

$$X = \pm \frac{a}{\sqrt{(D\tau_r)}}, \quad C^{ss} = C_s$$

Show that the steady state solution can be written as:

$$\frac{C^{ss}}{C_s} = \frac{\cosh(\sqrt{k^*}X)}{\cosh(\sqrt{k^*} - X_a)}$$

Show that the equation and time and space conditions for the transient portion of the solution can be written as:

$$\frac{\partial^2 C^t}{\partial \tau^2} + (1 + k^*)\frac{\partial C^t}{\partial \tau} + k'''C^t = D\frac{\partial^2 C^t}{\partial X^2}$$

Initial Condition: $\tau = 0,\ C^t = C_0$
Final Condition: $\tau = \infty,\ C^t = 0$

The Boundary Conditions are now homogeneous after the expression of the result as a sum of steady state and transient parts and are:

$$\frac{\partial C^t}{\partial X} = 0,\ X = 0$$

$$X = X_a,\ C^t = 0$$

Show that the solution can be obtained by the method of separation of variables. Initially the damping term is eliminated using a substitution such as $C^t = W \exp(-n\tau)$. The governing equation, at $n = 1/2\ (1 + k^*)$, then becomes

$$W_{xx} = -(|1 - k^*|)^2\ W/4 + W_{\tau\tau}$$

Show that the solution by the method of separation of variables, after applying the final condition, can be written as:

$$C^t = \Sigma_0^\infty \frac{C_n 2(-1)^n}{(n + 1/2)\pi} \exp\left(-\left(\frac{1 + k^*}{2}\right)\tau\right) \exp\left(-\tau\frac{\sqrt{|1 - k^*|^2}}{4} - \lambda_n^2 \cos(\lambda_n X)\right)$$

where, $\lambda_n = (n + 1/2)\pi\ \dfrac{\sqrt{(D\tau_r)}}{a}$, $n = 0, 1, 2, 3, \ldots$

C_n can be derived using the orthogonality property and can be shown to be C_0 $2(-1)^n/(n + 1/2)\pi$.

9. In Problem 8 examine the solution. Show that model solutions given are bifurcated, i.e. the characteristics of the function change considerably when a parameter such as the width of the slab is varied. Here a decaying exponential becomes exponentially damped cosinous. This is referred to as subcritical damped oscillatory behaviour. For $a < \pi\sqrt{(D\tau_r)}/(|1 - k^*|)$ even for $n = 1$, show that all the terms in the infinite series will pulsate. This is when the argument within the square root sign in the exponentiated time domain expression becomes negative and the result becomes imaginary. Using Demovrie's theorem and taking real part for small width of the slab, show that for small slabs,

$$C^t = \Sigma_0^\infty C_n \exp\left(\frac{-(1 + k^*)}{2}\tau\right) \cos\left(\tau\sqrt{\frac{\lambda_n^2}{4} - (|1 - k^*|)^2}\right) \cos(\lambda_n X)$$

10. *Simultaneous Reaction and Finite Speed Diffusion in a Finite Cylinder*
Consider the simultaneous reaction and finite speed diffusion in a finite cylinder of radius R with an initial concentration of C_0. The sides of the cylinder are maintained at a finite concentration of C_s. Show that the governing equation can be written as:

$$\frac{\partial^2 C}{\partial \tau^2} + (1 + k^*)\frac{\partial C}{\partial \tau} + k^* C = \frac{\partial^2 C}{\partial X^2} + \frac{1}{X}\frac{\partial C}{\partial X}$$

where $\tau = t/\tau_r$; $X = r/\sqrt{(D\tau_r)}$; $k^* = k''' \tau_r$

The initial condition and boundary conditions are given by:

$$\tau = 0,\ \forall r,\ C = C_0$$

$$\tau > 0,\ X = 0,\ \partial C/\partial X = 0$$

$$\tau > 0,\ X = R/\sqrt{(D/\tau_r)},\ C = C_s$$

Show that the solution can be expressed as a sum of a steady state part and a transient part.

$$C = C^{ss} + C^t$$

Show that the steady state part can be written as:

$$\frac{C^{ss}}{C_s} = \frac{I_0(k^{*1/2}X)}{I_0(k^{*1/2}X_s)}$$

Show that the transient portion of the concentration can be written as:

$$C^t = \Sigma_1^\infty A_n \exp\left(\frac{-(1+k^*)\tau}{2}\right) \exp\left(-\tau\left(\frac{1}{2} + \frac{\sqrt{(1-k^*)^2}}{4} - \lambda_n^2\right) J_0(\lambda_n X)\right)$$

where A_n can be obtained from the principle of orthogonality and the initial condition:

$$A_n = \frac{\int_0^{X_R} J_0(\lambda_n X)\, dX}{\int_0^{X_R} J_0^2(\lambda_n X)\, dX}$$

Confirm that the nature of the solution is bifurcated. For values of $R <$

$$\frac{\sqrt{(D\tau_r)}(4.8096)}{|1 - k^*|}$$

$$C^t = C_0 \Sigma_1^\infty A_n \exp\left(-\frac{(1+k^*)}{2}\tau\right) \cos\left(\tau\sqrt{\lambda_n^2 - \frac{|1-k^*|^2}{4}}\right) J_0(\lambda_n X)$$

where A_n, is given by Eq. (3.146) and λ_n given by $\lambda_n = n\pi\sqrt{(D\tau_r)}/R$ where $n = 0, 1, 2, 3, \ldots$

11. *Simultaneous Reaction and Finite Speed Diffusion in a Finite Sphere* Consider the simultaneous reaction and finite speed diffusion in a sphere of finite sphere of radius R with an initial concentration of C_0. The sides of the sphere are maintained at a finite concentration of C_s. Show that the governing mass balance equation from mass balance and the constitutive relation for wave diffusion and relaxation which can be written after non-dimensionalization as:

$$\frac{X^2\partial^2 C}{\partial X^2} + \frac{2X\partial C}{\partial X} - k^* C = \frac{\partial^2 C}{\partial \tau^2} + \frac{(1+k^*)\partial C}{\partial \tau}$$

where, $\tau = t/\tau_r$; $X = r/\sqrt{(D\tau_r)}$; $k^* = (k'''\tau_r)$

The boundary conditions and initial time condition are:

$$t = 0,\ Vr,\ C = C_0$$

$$t > 0,\ X = 0,\ \partial C/\partial X = 0$$

$$t > 0,\ X = R/\sqrt{(D\tau_r)}\ C = C_s$$

Show that the solution to the concentration can be expressed as a sum of a transient portion and a steady state portion in order to ensure that the transient boundary values are homogeneous.

Let, $$C = C^t + C^{ss}$$

Show that the steady state part and boundary conditions can be selected in such a fashion that the transient portion becomes homogeneous. The steady state portion of the concentration and the boundary conditions are:

$$\frac{X^2\partial^2 C^{ss}}{\partial X^2} + \frac{2X\partial C^{ss}}{\partial X} - k^* C^{ss} = 0$$

$$X = 0,\ \frac{\partial C^{ss}}{\partial X} = 0$$

$$X = \frac{R}{\sqrt{(D\tau_r)}},\ C^{ss} = C_s$$

Show that the solution at steady state can be written as:

$$\frac{C^{ss}}{C_s} = \frac{X_R^{1/2} I_{1/2}(Xk^{*1/2})}{I_{1/2}(X_R k^{*1/2})\, X^{1/2}}$$

Show that the solution of the transient part is the rest of the problem:

$$\frac{(1+k^*)\partial C^t}{\partial \tau} + \frac{\partial^2 C^t}{\partial \tau^2} + k^* C^t = \frac{\partial^2 C^t}{\partial X^2} + \frac{2}{X}\frac{\partial C^t}{\partial X}$$

The initial, final time conditions and space boundary conditions are now:

Initial Condition : $C^t = C_0$
Final Condition : $C^t = 0$
$X = 0$, C^t = finite
$X = X_R$, $C^t = 0$

Show that the solution can be obtained by the method of separation of variables.

$$C^t = \Sigma_0^\infty A_n \exp\left(\frac{-(1+k^*)}{2}\right) - \left(\sqrt{\frac{(|1-k^*|)^2}{4}} - \lambda_n^2\right) \frac{\tau J_{1/2}(\lambda_n X)}{\sqrt{X}}$$

where A_n can be obtained from the principle of orthogonality at the initial condition.

$$C_0 = \Sigma_0^\infty A_n \frac{J_{1/2}(\lambda_n X)}{\sqrt{X}}$$

Multiplying both sides of the equation by $J_{1/2}\,(\lambda_m X)$ and integrating between the limits of 0 and $R/\sqrt{(D\tau_r)}$ and using the principle of orthogonality:

$$A_n = C_0 \frac{\int_0^{X_R} \frac{J_{1/2}(\lambda_n X)}{\sqrt{X}}\, dX}{\int_0^{X_R} \frac{J_{1/2}^2(\lambda_n X)}{\sqrt{X}}\, dX} \tag{3.115}$$

For terms in the infinite series where the ($\lambda_n > |1 - k^*|/2$) the exponentiated time domain terms become negative within the square root and using the De Movrie's theorem the real part gives the $\cos((\lambda_n^2 - (|1 - k^*|)^2/4)^{1/2}\tau)$.

12. *Parabolic Diffusion in Semi-Infinite Slab*
Consider transient diffusion in a semi-infinite slab in cartesian coordinates at an initial concentration of C_0 and suddenly at times greater than zero subject to a steady wall concentration at one of the surfaces at C_s. Assume that the Fick's second law of diffusion is valid and show that the governing equation can be written as:

$$\frac{\partial u}{\partial t} = \frac{D\partial^2 u}{\partial x^2}$$

where,

$$u = \left(\frac{C - C_0}{C_s - C_0}\right)$$

subject to the following initial and boundary conditions;

$$t = 0,\ u = 0, \quad 0 < x \leq \infty$$

$$x = 0,\ u = 1, \quad t > 0$$

$$x = \infty,\ u = 0, \quad t > 0$$

Consider the Boltzmann transformation, $\eta = x/(4Dt)^{1/2}$
Show that the governing equation becomes;

$$-\frac{\eta \partial u}{\partial \eta} = \frac{1}{4}\frac{\partial^2 u}{\partial \eta^2}$$

Upon integration show that,

$$u = 1 - erf\left(\frac{x}{(4\,Dt)^{1/2}}\right)$$

13. *Perfume in the King of Prussia Shopping Mall*
A container containing perfume was opened in the shopping mall in King of Prussia, at Pennsylvannia in USA. Write the governing equation for transient damped wave mass diffusion and relaxation in three dimensions using spherical coordinates. Consider an infinite sphere. Discuss the spatiotemporal concentration in 3 dimensions. What approximations are needed ?
Show that the governing equation can be written as:

Let $u = \dfrac{C}{C_s}$; $X = \dfrac{r}{\sqrt{(D\tau_r)}}$; $\tau = \dfrac{t}{\tau_r}$;

Then, $$\frac{\partial^2 u}{\partial \tau^2} + \frac{\partial u}{\partial \tau} = \frac{2/X\partial u}{\partial X} + \frac{\partial^2 u}{\partial X^2} + \frac{1/X^2\partial^2 u}{\partial \theta^2} + \frac{1/X^2\sin^2\theta\,\partial^2 u}{\partial \phi^2} + \frac{\cot\theta/X^2\partial u}{\partial \theta}$$

Show that the solution can be obtained by the method of relativistic transformation of coordinates and can be written as:

$$u = \left[\frac{\tau^2}{\tau^2 - X^2 - \xi^2 - \psi^2}\right]\frac{I_2\left(\dfrac{\sqrt{\tau^2 - X^2 - \xi^2 - \psi^2}}{2}\right)}{I_2(\tau/2)}$$

This is for $\tau > \sqrt{X^2 - \xi^2 - \psi^2}$

For $X^2 + \xi^2 + \psi^2 > \tau^2$, in a similar fashion,

$$u = \left[\frac{\tau^2}{X^2 + \xi^2 + \psi^2 - \tau^2}\right] \frac{J_2\left(\dfrac{\sqrt{X^2 + \xi^2 + \psi^2 - \tau^2}}{2}\right)}{I_2(\tau/2)}$$

Further show that the inertial lag time and penetration length can be obtained from the first zero of Bessel function of the second order and first kind.

$$\tau_{\text{lag}} = \sqrt{X_{\text{p}}^2\,(1 + \theta_{\text{p}}^2 + \phi_{\text{p}}^2\,\sin^2\theta p - 105.498)}$$

14. *Sequential Separation of Whey Proteins by Radial Chromatography* The sequential separation of whey proteins is often effected by radial chromatography. Two major categories of proteins are casein and whey proteins. Casein is a colloid and whey is soluble. In the cheese industry, two types of precipitation techniques are most commonly used to separate the total milk proteins separated into caseins and whey proteins, i.e. rennet precipitation and acid precipitation. The cheese industry produces large amounts of whey, much of which is used to make whey protein concentrate Commercial-scale fractionation of different whey proteins has been hampered by the lack of an economical fractionation technology. It would be desirable, therefore, to provide a method for the continuous and sequential separation of various proteins from whey in a simple one or two step separation process. Sepragen Corp. San Jose in USA, provides a process for the sequential separation of at least five different proteins from whey and incorporating these separated whey proteins into pharmaceutical and food formulations. The process of the invention is directed to the continuous, sequential separation of whey proteins by chromatography, comprising adsorbing the proteins in liquid whey on a suitable separation medium packed in a chromatographic column and sequentially eluting immunoglobulin (e.g, IgG), .beta.-lactoglobulin (.beta.-Lg), .alpha.-lactalbumin (.alpha.-La), bovine serum albumin (BSA), and lactoferrin (L-Fe) fractions with buffers at suitable pH and ionic strength. Even though both axial and *radial flow chromatography* may be utilized, a horizontal flow column is particularly suitable. The liquid whey is selected from the group consisting of pasteurised sweet whey, pasteurised acid whey, non-pasteurised acid whey, and whey protein concentrate. Discuss the spatio-temporal concentration in the radial flow chromatographic column.

Show that the governing equation can be written as:

$$\left(\frac{D_{\text{AB}}}{r}\right)\frac{\partial}{\partial r}\left(r\frac{\partial C}{\partial r}\right) - \frac{V_r}{r}\frac{\partial(rC)}{\partial r} - k'''C = (1 + k'''\tau_{\text{mr}})\frac{\partial C}{\partial t} + \tau_{\text{mr}}\left(\frac{V_r}{r}\right)\frac{\partial^2(rC)}{\partial r\partial t} + \frac{\partial^2 C}{\partial t^2}$$

Obtain the dimensionless form by the following substitutions:

$$X = \frac{r}{\sqrt{(D\tau_r)}};\ \tau = \frac{t}{\tau_r};\ u = \left(\frac{C}{C_s}\right);\ k^* = k'''\tau_{mr};\ Pe_m = \frac{V_r}{\sqrt{(D/\tau_{mr})}}$$

$$\left(\frac{1}{X}\right)\left(\frac{\partial}{\partial X}\left(X\frac{\partial u}{\partial X}\right)\right) - \left(\frac{Pe_m}{X}\right)\left(\frac{\partial(Xu)}{\partial X}\right) - k^*u = (1 + k^*)\frac{\partial u}{\partial \tau} + \left(\frac{Pe_m}{X}\right)\left(\frac{\partial^2(uX)}{\partial X\partial \tau}\right) + \frac{\partial^2 u}{\partial \tau^2}$$

Show that the solution can be written as:

$$u = c\exp\left(-(1 + k^*)\frac{\tau}{2}\right)\exp\left(-\tau\sqrt{(1/4\ |1 - k^*|^2 - \lambda_n^2)}\right)\exp\left(-X(1 + k^*)\frac{Pe_m}{4}\right)$$
$$[c_1 J_1(2X|Pe_m|^{1/2})]$$

Show that at steady state the solution can be written as:

$$u = \exp\left(-\frac{Pe_m X}{2}\right) I_0\,(Xk^{*1/2})$$

Show that the critical distance is when,

$$R_{crit} > 1.916\sqrt{\left(\frac{D\tau_r}{Pe_m}\right)}$$

The design of the radial flow chromatography column greater than the R_{crit} will not result in any further separation. This is the critical length of diffusion.

15. *Wave Diffusion in Moving Film*

A thin liquid film flows slowly and without ripples down a flat surface. One side of the film wets the surface and the other side is in contact with a gas which is sparingly soluble in the liquid. How much gas dissolves in the liquid?

$$-\frac{\partial J}{\partial z} = \frac{V_x \partial C}{\partial z} + \frac{\partial C}{\partial t}$$

Using the Fick's second law for the diffusion flux the governing equation for the concentration of the solute gas in the liquid is given by

$$\frac{D_{AB}\partial^2 C}{\partial z^2} = \frac{V_x \partial C}{\partial z} + \frac{\partial C}{\partial t}$$

Using the damped wave diffusion and relaxation equation, the governing equation for the solute gas in the liquid is given by

$$\frac{D\partial^2 C}{\partial z^2} - \frac{V_x \partial C}{\partial z} = +\,\tau_r\left(\frac{V_x \partial^2 C}{\partial z \partial t} + \frac{\partial^2 C}{\partial t^2}\right) + \frac{\partial C}{\partial t}$$

Obtain the spatio-temporal concentration profile using (a) the governing equation derived from Fick's second law and (b) the governing equation derived from damped wave diffusion and relaxation equation.

16. *Derivation of Maxwell Speed Distribution of Molecules*
As the altitude in the atmosphere is increased the decrease in n with z can be used to deduce the speed distribution law. Show that:

$$-dp = nmgdz$$

From the ideal gas law, for constant temperature, $p = nk_BT$, $= dp = dnk_BT$. Show that,

$$\frac{dn}{n} = -\frac{mgdz}{k_BT} = \frac{dp}{p}$$

Upon integration,

$$n = c' \exp\left(-\frac{mg}{k_BT}\right) z$$

Show that,

$$v_z n(v_z)\, dv_z = c' \exp\left(-\frac{mgz}{k_BT}\right) dz$$

From conservation of energy the special molecules have the property that;

$$\frac{1}{2} mv_z^2 = mgz$$

$$mv_z dv_z = mgdz$$

Substituting this equation into the one for velocity of molecule,

$$v_z n dv_z = \frac{c'}{g} \exp\left(-\frac{mv_z^2}{2k_BT}\right) dv_z$$

Writing similar expressions for the other two components, show that the

$$N(v) = 4\pi\left(\frac{m}{2\pi k_BT}\right)^{3/2} v^2 \exp\left(-\frac{mv^2}{2k_BT}\right)$$

17. *Penetration Length, Inertial Lag Time*
Fascimire's experiments aboard the space shuttle in diffusive mass transfer on the tea wad was discussed in section 1.1. It showed that the penetration length of diffusion was 1.96 cm in 51.15 hr. Using Eq. 2.65, find the relaxation time including finite speed diffusion effects. Assume a binary diffusivity of 9.2E–8 m^2/s.

Ans: 104 secs

REFERENCES

Bai, C. and Lavine, A.S., 1995, "On Hyperbolic Heat Conduction and the Second Law of Thermodynamics", *ASME J. of Heat Transfer*, **117**, 256–263.

Barletta, A. and Zanchini, E., 1997, "Hyperbolic Heat Conduction and Local Equilibrium: A Second Law Analysis", *Int. J. Heat and Mass Transfer*, **40**, 5, 1007–1016.

Baumeister, K.J. and Hamill, T.D., 1971, "Hyperbolic Heat Conduction Equation—A Solution for the Semi-Infinite Body Problem", *ASME J. of Heat Transfer*, **93,** 126–128.

Bejan, A., 1988, *Advanced Engineering Thermodynamics,* John Wiley, New York.

Cattaneo, C., 1958, "A Form of Heat Conduction which Eliminates the Paradox of Instantaneous Propagation", *Comptes Rendus*, **247**, 431–433.

Casimir, H.B.G., 1938, "Note on the Conduction of Heat in Crystals", *Physica,* **5,** 495–500.

Chen, G., "Ballistic Diffusive Conduction Equations", 2001, *Phys. Rev. Lett.,* **86,** 11, 2297–2300.

Fourier, J.B., 1955, *Theorie analytique de la chaleur*, English Translation by A. Freeman, 1955, Dover Publications, New York.

Jeffreys, H., 1924, *The Earth*, Cambridge University Press, UK.

Joseph, D.D. and Preziosi, L., 1989, " Heat Waves", *Reviews of Modern Physics*, **61**, 41–73.

__________, 1990, "Addendum to Heat Waves" *Reviews of Modern Physics*, **62**, 375–391.

Kaminski, W., 1990, "Hyperbolic Heat Conduction Equation for Materials with a Non-homogeneous Inner Structure", *ASME J. of Heat Transfer*, **112**, 555–560.

Kelly, D.C., 1968, "Diffusion "A Retavistic Appraisal", *American J. Physics*, **36**, 585–591.

Kittel, C., 1986, *Introduction to Solid State Physics*, John Wiley, New York.

Landau, L., 1941, "The Theory of Superfluidity of Helium II", *J. Phys.*, **5**, 71.

Landau, L. and Liftshitz, E.M., 1987, *Fluid Mechanics*, Pergamon, UK.

Maxwell, J.C., 1867, "On the Dynamical Theory of Gases", *Phil. Trans. Roy. Soc.*, **157**, 49.

Majumdar, A., 1991, "Microscale Heat Conduction in Dielectric Thin Films", *ASME Conference Paper, Heat Transfer Division,* **184**, 33–42.

Mitra, K, Kumar, S., Vedavarz, A. and Moallemi, M.K., 1995, "Experimental Evidence of Hyperbolic Heat Conduction in Processed Meat", *ASME J. of Heat Transfer,* **117**, 568–573.

Morse, P.M. and Feshbach., H., 1953, *Methods of Theoretical Physics*, McGraw Hill, New York.

Nernst, W., 1917, *Die Theoretischen Grundalgen des n Warmestazes,* Knapp Halle.

Onsager, L, 1931, "Reciprocal Relations in Irreversible Processes", *Phys. Review*, **37**, 405-426.

Ozisik, M.N. and Tzou, D.Y., 1994, "On the Wave Theory of Heat Conduction", *ASME J. of Heat Transfer*, **116**, 526–535.

________, 1992, *On the Wave Theory in Heat Conduction,* ASME Winter Annual Meeting, Anaheim, CA.

Qiu, T.Q. and Tien, C.L., 1992, "Short-Pulse Laser Heating on Metals", *Int. J. Heat and Mass Transfer,* **35**, 719–726.

Sharma, K.R., 2005, *Damped Wave Transport and Relaxation*, Elsevier, Amsterdam.

________, March 2003, Surface *Renewal Theory for Mass Transfer Coefficient using* Infinite Order PDE Inorder to Capture non-Fickian Diffusion, 225th ACS National Meeting, New Orleans, LA.

________, "Manifestation of Acceleration during Transient Heat Conduction", *J. of Thermophysics and Heat Transfer* (in press).

________, February 2006, *A Fourth Mode of Heat Transfer called Damped Wave Conduction*, 42nd Annual Convention of Chemists Meeting, Santiniketan.

________, March 2006, *On the Second Law Violation in Fourier Conduction*, 231st ACS National Meeting, Atlanta, GA.

________, March 2006, *Bounded Expression for Surface Flux by Damped Wave Conduction and Relaxation,* 231st ACS National Meeting, Atlanta.

________, March 2006, *On the Second Law Analysis of non-Fourier Heat Conduction*, 231st ACS National Meeting, Atlanta.

________, March 2006, *On the Pulse Decay during Simultaneous Reaction and Damped Wave Diffusion and Relaxation,* 231st ACS National Meeting, Atlanta.

________, March 2006, *Damped Wave Conduction and Relaxation Equation– Solution to the Finite Slab Problem by the Hybridized Method of Retavistic Transformation and Separation of Variables,* 231st ACS National Meeting, Atlanta.

________, March 2006, *On the Measured Values of Relaxation Time*, 231st ACS National Meeting, Atlanta.

________, March 2006, *On the Time Taken to Steady State during Heating a Finite Slab*, 231st ACS National Meeting, Atlanta.

________, March 2006, *On the Cooling Time of an Orange in the Refrigerator*, 231st ACS National Meeting, Atlanta.

________, March 2006, *Critical Thickness of Coating and Onset of Subcritical Oscillations*, 231st ACS National Meeting, Atlanta.

________, March 2006, *On the Second Law Violation in Fourier Conduction*, 231st ACS National Meeting, Atlanta.

________, March 2006, *On the Use of Final Condition in Time to Obtain Solution of Cattaneo & Vernotte Damped Wave Transport and Relaxation Equation*, 231st ACS National Meeting, Atlanta.

Swartz, E.T. and Pohl, R.O., 1989, "Thermal Boundary Resistance" *Reviews of Modern Physics,* **61**, 605–668.

Taitel, Y., 1972, "On the Parabolic, Hyperbolic and Discrete Formulation of the Heat Conduction Equation", *Int. J. Heat and Mass Transfer*, **15**, 369–371.

Tzou, D.Y., 1993, "An Engineering Assessment to the Relaxation Time in Thermal Wave Propagation", *Int. J. Heat Mass Transfer*, **36**, 1845–1850.

_________, 1997, *Macro-to Micro Scale Heat Transfer: The Lagging Behaviour*, Taylor and Francis, Washington, DC.

Vernotte, P., 1958, "Les Paradoxes de la Theorie Continue de l'equation de la Chaleur", *C. R. Hebd. Seanc. Acad. Sci. Paris*, 246, **22**, 3154–3155.

CHAPTER 3

Mass Transfer Coefficients

Nomenclature

a	interfacial area for mass transfer (m^2/gm)
a_a	reference value of interfacial area (m^2/gm)
a_w	half-width of the slab (m)
A	area across the mass transfer path (m^2)
C	concentration (mol/m^3)
d_t	diameter of tube (m)
D	binary diffusivity (m^2/s)
g_m	mass flow rate (kg/s)
H	Henry's law constant (cm^3.atm/mol)
H_g	height equivalent to one transfer unit (m)
H_{og}	overall height equivalent to one transfer unit (m)
j_D	Chilton-Colburn analogy 'j factor' mass
j_H	Chilton-Colburn analogy 'j factor' heat
k	mass transfer coefficient based on concentration (cm/s)
k_x	mass transfer coefficient based on mole fraction (mol/cm^2/s)
k_p	mass transfer coefficient based on pressure (mol/cm^2/s/atm)
k'''	first order reaction rate (/s)
K_p	overall mass transfer coefficient base on pressure (mol/cm^2/s/atm)
N	mass flux (mol/s)
N''	mass flux per unit area (mol/m^2/s)
r	distance from the sphere (m)
r_s	radius of the sphere (m)
t_c	contact time of the eddy at the interface (s)
u	dimensionless concentration $(C - C_0)/(C_s - C_0)$
v_0	superficial velocity of fluid (m/s)
Q	discharge rate (m^3/s)
w	film thickness (m)

Greek

α	thermal diffusivity (m^2/s)
ρ_m	molar density (mol/m^3)

δ_{film} film thickness (m)
δ_h hydrodynamic boundary layer thickness (m)
δ_c concentration boundary layer thickness (m)
v kinematic viscosity (m^2/s)
μ viscosity (kg/m/s)

Dimensionless Groups

Grashoff number	Gr ($d^3 g \Delta\rho/\rho/v^2$) buoyancy force/viscous force
Lewis number	Le (α/D) thermal diffusivity/mass diffusivity
Mawell number (mass)	$Max_m(D\tau_r/a^2)$
Peclect number (mass)	$Pe_m(dv/D)$ convection/diffusion
Prandtl number	Pr (v/α) momentum diffusivity/thermal diffusivity
Reynolds number	Re ($\rho vd/\mu$) inertial forces/viscous forces
Schmidt number	Sc (v/D) momentum diffusivity/mass diffusivity
Sherwood number	Sh (kl/D) convective mass transfer/diffusion
Stanton number	St (k/v) mass transfer velocity/flow velocity
Sharma number	Sha ($k\tau_r/a$) mass bulk velocity/relaxational transfer
Thiele modulus	ϕ ($d(k'''/D)^{1/2}$) reaction rate/diffusion rate

Subscripts

b bulk
cl chlorine
f feed
i interface
p product
1 component 1

3.1 MASS TRANSFER OPERATIONS

A second method of modelling or representing the movement of species from a region of higher concentration to a region of lower concentration is by the use of mass transfer coefficients. Mass transfer coefficients can be used to design equipment for mass transfer operations at the lowest cost, and set to operate in an environmentally sound and safe manner. They can be developed from careful experimentation over a wide range of conditions. A large number of unit operations in chemical, biotechnological, nuclear, pharmaceutical and metallurgical industries are concerned with separating a component from a mixture or formation of an alloy or a mixture during which a critical step is the mass transfer of one or more species from one phase to another across an interface. It can be guessed as much and verified that in every chemical produced in the industry, the raw materials have to be purified and the final product has to be separated from the unreacted raw materials and by-products and inert diluent or medium of the reactions, such as enzyme, catalyst, solvent, etc.

For example, in the petrochemical refinery, naphtha needs to be cracked and separated into various useful fractions by a mass transfer method called *distillation* and in the plastic industry, the unreacted monomers are separated from the polymer mass in a multi-stage devolatiliser of either the falling strand or wiped film types. In the sugar industry, the cane juice is removed in a triple effect evaporator to produce the crystalline sugar.

Often times during mass transfer the composition of the considered system changes. The cost incurred during separations is the bulk of the cost of production of most chemicals. Often times the species has to transfer through an interface. Depending on the nature of the interface, the mass transfer operations can be classified accordingly. Thus interface mass transfer across six types of interfaces are possible for binary systems, as solid-liquid and liquid-solid are the same. Three phase systems form yet another type of interface. These are:

A. Gas-Liquid
B. Gas-Solid
C. Liquid-Solid
D. Gas-Gas
E. Liquid-Liquid
F. Solid-Solid

Some important mass transfer operations in each of the six types are listed below:

Gas-Liquid
Absorption
Humidification
Drying (Constant Rate)
Distillation
Foam separation
Pervaporation

Gas-Solid
Fluidised bed coating
Falling Rate Drying
Adsorption

Solid-Liquid
Leaching
Adsorption
Ion-Exchange
Crystallisation
Floatation
Drug Delivery

Gas-Gas
Thermal Diffusion
Pressure Swing Adsorption
Sweep Diffusion
Atomolysis
Centrifugation

Solid-Solid
Carburising Steel
Polymer Colouring

Liquid-Liquid
Reverse Osmosis
Dialysis
Electrodialysis
Extraction

Besides these six types of mass transfer operations, there is another set of mass transfer operations where three phases are involved.

Gas-Liquid-Solid
Gas-Liquid-Solid Fluidised Bed

3.2 DIFFERENT DEFINITIONS OF MASS TRANSFER COEFFICIENT

Like the Newton's law of cooling and the definition of a heat transfer coefficient, a mass transfer coefficient can be similarly defined usually during convective mass transfer as:

$$N = k_l a\ (C_{1f} - C_{1i}) = k_l a\ (C_{1i} - C_{1p}) \tag{3.1}$$

$$N'' = k_l\ (C_{1f} - C_{1i}) = k_l\ (C_{1i} - C_{1p}) \tag{3.2}$$

In a typical mass transfer operation, two types of fluxes can be defined. One is in terms of mol/s (N) and the other is in terms of mol/m^2/s (N''). Depending on the problem at hand, sometimes the interfacial area or the effective area allowed by the contacting equipment becomes a salient consideration. In these cases Eq. (3.1) is used. In other cases only the initial and feed and the final product concentrations are important and the use of Eq. (3.2) can simplify matters and apply focus on the relevant concentrations. For example, the porous membranes and the raschig rings in a distillation tower may influence the operating efficiency and can be accounted for by the effective interfacial area a. When writing units for a, careful attention must be paid. Some typical mass transfer coefficients used are as follows:

$$N'' = k(C_{1b} - C_{1i}) \tag{3.3}$$

k has the units of cm/s and is defined based upon a concentration difference.

$$N'' = k_p[p_b - p_i] \tag{3.4}$$

k_p has units of mol/cm^2/s/atm and is used in absorption and biological problems and is defined in terms of a pressure differential. For gases, partial pressures are a useful representation of the concentration.

$$N'' = k_x\ (x_b - x_i) \tag{3.5}$$

k_x has units of mol/cm^2/s and is defined in terms of mole fraction differences instead of a concentration difference. It can be expected to come in handy in liquid-liquid operations or liquid-solid operations.

$$N'' = k(C_{1b} - C_{1i}) + C_1 v^0 \tag{3.6}$$

Here the effects of convection and molecular diffusion are added together to provide a total flux. It is useful in gases.

Worked Example 3.1 *Mass Transfer from Sea Water through a Packed Bed of Nanoporous Carbon*

Consider sea water pumped across a packed bed of nanoporous carbon with molecular sieve pore diameter of 4 nm. Assume a voidage of 0.65 and material density of carbon of 1.8 gm/cm^3 and show that the interfacial area is 5.416802 million cm^2/gm. Sea water inlet concentration can be taken to be 3.6%. Pure water is filtered from the column at 5 cm/s. The packed bed is saturated after it has passed through 100 cm of the bed. What is the mass transfer coefficient, should the saturation concentration of sodium chloride is 30 ppm?

$$N'' = k_x(x_A - x_i) \tag{3.7}$$

The mass balance of salt in the liquid phase pumped across the packed column at steady state can be written by:

– Amount filtered + flow in – flow out = accumulation

$$k_x(x_A - 0) + v\rho_m \frac{\partial x_A}{\partial z} = 0 \tag{3.8}$$

$$z = 0,\ x_A = 0.0114 \tag{3.9}$$

$$z = \infty,\ x_A = 0$$

Integrating Eq. (3.8),

$$\frac{\partial x_A}{x_A} = -\frac{k_x}{v\rho_m}\partial z$$

Upon integrating both sides,

$$x_a = c' \exp\left(-\frac{zk_x}{v\rho_m}\right) \tag{3.10}$$

It can be seen from the boundary condition that $c' = 0.0114$.

Thus, $$x_a = 0.0114 \exp(-z\ 3.5\ k_x) \tag{3.11}$$

At saturation, $z = 100$ cm, $x_i = 0.003$

Thus, $$k_x = 0.381 \text{ mol/m}^3\text{/s} \tag{3.12}$$

$$k_l = k_x a$$

Thus, $$k_l = 206 \text{ mol/m/s/g.}$$

Worked Example 3.2 *Etching of Silicon Wafers*

During the manufacture of transistors, a coating called photoresist is applied to the silicon wafers of thickness 540 nm. The density of the polymer is 0.94 g/cm^3. After etching the photoresist is removed by extraction using an organic solvent. Take the solubility of the photoresist in the organic solvent is as 2.4 E-3 g/cm^3. If the photoresist dissolves in a little more than 7 minutes, what is the mass transfer coefficient?

$$N'' = k(C_i - C_b) = 540\text{ E-7*0.94/420/1E5} = k(0.0024 - 0)$$

$$k = 5.036\text{ E-5 cm/s} \tag{3.13}$$

Note the flux is given in terms of gm/s.

3.2.1 Local and Overall Transfer Coefficient

Consider the evaporation of chlorine from a swimming pool into air. At the gas-liquid interface

$$N'' = k(C_{\mathrm{cl,b}} - C_{\mathrm{cl,i}}) \tag{3.14}$$

where, k is the liquid phase mass transfer coefficient defined in terms of concentration of chlorine and represents and has units of cm/s. Using the Henry's law, the interface equilibrium partial pressure of chlorine can be calculated as:

$$p_{\mathrm{cl,i}} = HC_{\mathrm{cl,i}} \tag{3.15}$$

The gas phase flux can be estimated as:

$$N'' = k_{\mathrm{p}}(p_{\mathrm{cl,i}} - p_{\mathrm{b}}) \tag{3.16}$$

where k_{p} is the mass transfer coefficient.

At steady state, Eqs. (3.14) and (3.16) are equal;

$$k(C_{\mathrm{cl,b}} - C_{\mathrm{cl,i}}) = k_{\mathrm{p}}(p_{\mathrm{cl,i}} - p_{\mathrm{b}}) = k_{\mathrm{p}}(HC_{\mathrm{cl,i}} - p_{\mathrm{b}})$$

The interfacial chlorine concentration can thus be calculated by combining the information in Eq. (3.16) in the above to give,

$$C_{\mathrm{cl,i}} = \frac{p_{\mathrm{cl,i}}}{H} = \frac{(kC_{\mathrm{cl,b}} + k_{\mathrm{p}}p_{\mathrm{b}})}{(k_{\mathrm{p}}H + k)} \tag{3.17}$$

and the flux,

$$N'' = \frac{(C_{\mathrm{cl,b}}H - p_{\mathrm{b}})}{\left(\dfrac{H}{k} + \dfrac{1}{k_{\mathrm{p}}}\right)} \tag{3.18}$$

It can be seen that Eq. (2.18) is analogous to an electrical circuit with two resistances in series. The flux corresponds to the electric current and the concentration difference that serves as the driving force for mass transfer corresponds to voltage. The effective resistance is then

$$\frac{1}{K_{\mathrm{p}}} = \frac{H}{k} + \frac{1}{k_{\mathrm{p}}} \tag{3.19}$$

where K_{p} is the overall gas phase mass transfer coefficient. Thus,

$$N'' = k_{\mathrm{p}}(p_{\mathrm{i}}^{*} - p_{\mathrm{b}}) \tag{3.20}$$

where p_{i}^{*} is the hypothetical pressure that would be in equilibrium with the bulk liquid. Thus the overall coefficient gives a lumped parameter with different components forming the gas and liquid phases. Examples of local coefficients are k, k_{x}, k_{p} etc. The local coefficients can even be defined at a particular point.

3.2.2 HTU and NTU

HTU, is the height equivalent to one transfer unit. In terms of the length dimension in one parameter, the HTU represents the height of the equipment required to accomplish a separation of desired purity level. The mass transfer coefficients are often times a function of the flow rates. The ratio of the mass transfer coefficient and the flow rate is nearly a constant and is the HTU.

$$H_g = \frac{g_m}{k_g a} \tag{3.21}$$

$$H_{og} = \frac{g_m}{k_g a} \tag{3.22}$$

The NTU, number of transfer units, gives the number of theoretical stages or plates required to carry out the same separation in a stage-wise or plate-type apparatus.

3.3 MASS TRANSFER CORRELATIONS

Several mass transfer correlations have been developed from experimental data empirically and some from analytical modelling from the governing equations. Most correlations have the general form of a Sherwood number depending on the Reynolds number and other dimensionless groups. The Sherwood number is the ratio of the mass transfer coefficient and the diffusion coefficient in the same phase and the Reynolds number represents the ratio of the inertial forces to the viscous forces. The correlations are in terms of dimensionless groups. Correlations can be developed in one of these three methods (Sharma, 2003):

(a) Buckingham-Pi method
(b) Rayleigh method
(c) Non-dimensionalisation of governing equations

In the Rayleigh method, the dependent variable is written as proportional to the independent variables that are found to be of relevance to the problem. The units of each of the variable are then written in terms of the fundamental dimensions of mass, length, time, temperature, candela, magnetic units, coulomb, etc. The power exponents are equated on the fundamental dimensions and then the independent variables are grouped into dimensionless groups and correlation is obtained.

In the Buckingham-Pi method, the number of dimensionless groups are arrived at by subtracting the number of independent variables and the fundamental dimensions and using the theorem of Buckingham and Pi. The repeating variables are identified and each dimensionless group is arrived at by equating the power exponents on each fundamental dimension.

For a given problem, the governing equations from mass balance are written. Equation of momentum, continuity and conservation of mass and energy are sometimes used to completely describe the critical aspects to the problem. The equations are non-dimensionalised. Along the way, the dimensionless groups that can be used to develop a correlation for the given situation fall out of the analysis.

Some of the frequently used correlations are as follows:

For laminar flow past a flat plate, the mass transfer coefficient is given by the following correlation:

$$Sh = 0.323\ Re^{1/2}Sc^{1/3} \tag{3.23}$$

$$j_D = j_H = \frac{f}{2} = 0.664\ (Re_l)^{-1/2} \tag{3.24}$$

where, Sh is the Sherwood number, Re is the Reynolds number, Sc is the Schmidt number, j_D is the Chilton and Colburn j factor for mass transfer and f is the friction factor. The j_D = St. $Sc^{2/3}$ where, St is the Stanton number and Sc is the Schmidt number. The 0.323 coefficient is the Blausius approximate solution. It is applicable at low mass transfer rates. The natural convection effects can be included in the correlations as follows:

$$Sh = 0.508\ Sc^{1/2}(0.952 + Sc)^{-1/4}Gr^{1/4} \tag{3.25}$$

where Gr is the Grashoff number. For laminar flow past a stationary disc, the mass transfer coefficient can be given by:

$$Sh = \frac{8}{\pi} \tag{3.26}$$

For a spinning disc,

$$Sh = 0.879Re^{1/2}\ Sc^{1/3} \tag{3.27}$$

For turbulent flow past the plate,

$$Sh = 0.0292\ Re^{-.8} \tag{3.28}$$

$$j_D = j_H = \frac{f}{2} = \frac{0.037}{Re_L^{0.2}} \tag{3.29}$$

For intermediate flow,

For 3×10^5 to 3×10^6

$$Sh = 0.037\ Sc^{1/3}(Re^{0.8} - 15{,}500) \tag{3.30}$$

The natural convection effects during turbulent flow can be accounted for turbulent flow part a vertical plate as:

$$Sh = 0.0299\ Gr^{2/5}Sc^{7/15}(1 + 0.494\ Sc^{2/3})^{-2/5} \tag{3.31}$$

For turbulent flow past a spinning disc,

$$Sh = 5.6\ Re^{1.1}Sc^{1/3} \tag{3.32}$$

Mass transfer to a flat plate membrane in a stirred vessel,

$$\text{Sh} = a \text{ Re}^{\text{b}} \text{ Sc}^{\text{c}} \tag{3.33}$$

$$a = 0.0443; \; b \text{ is often } 0.65 - 0.70; \; c = 0.33.$$

Mass transfer correlation in gas-liquid systems are important in several applications. For wetted wall columns, the falling film is considered in the correlations. For laminar flow in a vertical wetted wall column,

$$\text{Sh}_{\text{avg}} = 3.41 \frac{x}{\delta_{\text{film}}} \tag{3.34}$$

where the film thickness is given by;

$$\delta_{\text{film}} = \left(\frac{3\mu Q}{w\rho g}\right)$$

For turbulent flow in a vertical wetted wall column,

$$\text{Sh}_{\text{avg}} = 0.02 \text{ Re}^{0.83}\text{Sc}^{0.44} \tag{3.35}$$

Mass transfer correlations for flow in pipes and ducts are as follows:

For laminar flow in the regime of developing concentration profile,

$$\text{Sh} = 3.66 + \frac{(0.0668\, d_{\text{t}}/x) \text{ ReSc}}{(1 + 0.04(d_{\text{t}}/x\text{ReSc})^{2/3}} \tag{3.36}$$

In tubes, for fully developed concentration profile;

$$\text{Sh} = 3.66 \tag{3.37}$$

For laminar flow when the concentration and velocity profiles are fully developed,

$$\text{Sh} = 4.3636 \tag{3.38}$$

For flow in vertical tubes in laminar flow,

$$\text{Sh}_{\text{avg}} = 1.62 \left(\frac{\text{Pe}_{\text{m}} d}{L}\right)^{1/3} \left(\frac{1 \pm 0.0742(\text{Gr})^{3/4}}{\text{Re}(\text{Sc}d/L)^{1/4}}\right)^{1/3} \tag{3.39}$$

For laminar flow in tubes in Reverse Osmosis systems,

$$\text{Sh}_{\text{avg}} = 1.632 \left(\frac{u d_{\text{t}}^2}{\text{DL}}\right)^{1/3} \tag{3.40}$$

For fully developed pipe flow in pipes and ducts from wall to the fluid, the correlations for mass transfer can be given by:

$$\text{Sh} = 7.6 \tag{3.41}$$

For turbulent flow in pipes,

$$\mathrm{Sh_{avg}} = 0.023\ \mathrm{Re}^{0.83}\mathrm{Sc}^{1/3} \tag{3.42}$$

Mass transfer correlations for flow past submerged objects are as follows:

For a single sphere,

$$\mathrm{Sh} = \frac{2r}{(r - r_s)} \tag{3.43}$$

For single sphere during creeping flow, with forced convection,

$$\mathrm{Sh}^2 = 4.0 + 1.21\ \mathrm{Pe}_m^{2/3} \tag{3.44}$$

For single sphere at low flow rates for molecular diffusion and forced convection;

$$\mathrm{Sh} = 2.0 + a\ \mathrm{Re}^{1/2}\mathrm{Sc}^{1/3} \tag{3.45}$$

a varies with the application, depending on whether the fluid is a liquid or solid. A lies between 0.5 to 0.62 for Re < 200. The Ranz and Marshall correlation gives an a value of 0.60. For liquids a is taken as 0.95.

At high flow rates,

$$\mathrm{Sh} = 0.347\mathrm{Re}^{0.6}\mathrm{Sc}^{1/3} \tag{3.46}$$

For non-spherical objects such as cubes, cylinders, prisms,

$$j_D = \frac{0.692}{\mathrm{Re}^{0.456}} \tag{3.47}$$

Correlations for mass transfer from bubbles and drops have been developed. The continuous phase mass transfer coefficient for stagnant drops that are spherical,

$$\mathrm{Sh} = 0.74 \left(\frac{\rho_c}{M_c}\right)_{av} \mathrm{Re}^{1/2}\mathrm{Sc}^{1/3} \tag{3.48}$$

Mass transfer correlation for fluidised beds to immersed surfaces,

$$\mathrm{Sh} = \left(\frac{fd_p^2}{D}\right)^{1/2} \text{ for slugging beds} \tag{3.49}$$

For bubbling fluidised beds,

$$\mathrm{Sh} = \left(\frac{D\tau_r}{d_p^2}\right)^{-1/2} I_o\left(\frac{1}{2f\tau_r}\right) \exp\left(-\frac{1}{2f\tau_r}\right) \tag{3.50}$$

where, $(D\tau_r/d_p^2)$ is the Maxwell number (mass) and represents the ratio of the relaxation time to the diffusion time.

3.4 MASS TRANSFER THEORIES

Several theories have been developed to explain the observations of the way the mass transfer coefficients correlates with the Reynolds number and Schmidt number. The *film theory*, one of the first models to be developed, provides a picture of the physical significance of mass transfer coefficient. When a fluid flows past a surface, a certain concentration profile will be established. Upon observation of the concentration change with distance, it can be inferred that most of the drop in concentration can be attributed to a short region that can be denoted as an 'effective film thickness'. The mass transfer coefficient can be calculated as:

$$N'' = k(C_{1i} - C_{1b}) = \frac{D(C_{1i} - C_1 b)}{\delta} \qquad k = \frac{D}{\delta} \tag{3.51}$$

The *penetration theory* was developed by Higbie (1935), 70 years ago. Higbie pointed out in most mass transfer applications the contact time is short. During transfer the process is transient. Consider a semi-infinite medium at a uniform initial concentration of C_0. At times greater than zero one of the surfaces is maintained at a constant surface concentration C_s. Using the Fick's second law of mass diffusion the governing equation can be written as:

$$\frac{\partial u}{\partial t} = \frac{D\partial^2 u}{\partial z^2} \tag{3.52}$$

where,

$$u = \left(\frac{C - C_0}{C_s - C_0}\right)$$

The governing equation given in Eq. (3.52) can be solved using the Boltzmann transformation such as $\eta = z/(4Dt)^{1/2}$. Eq. (3.51) becomes

$$-\frac{\eta \partial u}{\partial \eta} = \frac{\partial^2 u}{\partial \eta^2}$$

$$u = 1 - \text{erf}\left(\frac{z}{(4Dt)^{1/2}}\right) \tag{3.53}$$

The mass transfer coefficient can be calculated from Eq. (2.62). The surface flux is written by:

$$\frac{D}{(4\pi Dt)^{1/2}}(C_s - C_0) = k(C_s - C_0) \tag{3.54}$$

Thus,

$$k = \frac{1}{2}\left(\frac{D}{\pi t_c}\right)^{1/2} \tag{3.55}$$

Higbie assumed a constant time of exposure for all the eddies of fluid at the interface. Instead of the Fick's second law of diffusion when the generalised Fick's law of diffusion and relaxation is used, it was shown by Sharma (2005) that for the semi-infinite medium subject to constant wall concentration the surface flux by the method of Laplace transforms can be given by:

$$J^* = \exp\left(-\frac{\tau}{2}\right) I_0\left(\frac{\tau}{2}\right) \tag{3.56}$$

The mass transfer coefficient then can be written as:

$$k = \left(\frac{D}{\tau_r}\right)^{1/2} \exp\left(-\frac{\tau}{2}\right) I_0\left(\frac{\tau}{2}\right) \tag{3.57}$$

where,

$$\tau = \frac{t_c}{\tau_r} \tag{3.58}$$

Dankwerts (1955) developed the *surface renewal* theories. Here he derived the mass transfer coefficient for the general case where the eddies stay at the surface for varying lengths of time and Higbie's penetration theory is a particular case where the contact times of the eddies is constant.

$$k = (\mathrm{Ds})^{0.5} \tag{3.59}$$

where s is the surface renewal rate. Dobbins (1964) noted that whereas the film theory assumes that the concentration profile has reached steady state in the time of mass transfer and the Higbie and Dankwerts theories account for the transient nature of diffusion during the eddy contact time, they use a semi-infinite boundary condition for zero transfer. Dobbins used a finite length boundary condition and modified Eq. (2.68) as:

$$k = (\mathrm{Ds})^{0.5} \coth\left(\frac{\mathrm{s}z_b^2}{D}\right)^{1/2} \tag{3.60}$$

A revisit of Dobbins set of boundary conditions to the governing equation derived by using the generalised Fick's law of diffusion for finite width can result in the solution given earlier in chapter 2. The dimensionless concentration when the relaxation times are large can be found from Eq. (2.32) as:

$$u = \Sigma_1^\infty c_n \exp\left(-\frac{\tau}{2}\right) \cos\left[\left(\lambda_n^2 - \frac{1}{4}\right)\tau\right] \cos(\lambda_n X) \tag{3.61}$$

$$\frac{\partial u}{\partial \tau} = -\frac{\partial J^*}{\partial X} \quad \text{(from mass balance)} \tag{3.62}$$

$$\frac{\partial u}{\partial \tau} = -\Sigma_1^\infty c_n \exp\left(-\frac{\tau}{2}\right)\left[\sqrt{\lambda_n^2 - \frac{1}{4}} \sin\left(\sqrt{\lambda_n^2 - \frac{1}{4}}\tau\right) + \cos\frac{\left(\sqrt{\lambda_n^2 - \frac{1}{4}}\tau\right)}{2}\right]\cos(\lambda_n X)$$

$$= -\frac{\partial J^*}{\partial X} \tag{3.63}$$

$$J^* = c' + \Sigma_1^\infty\left(\sqrt{\lambda_n^2 - \frac{1}{4}}\right)\sin\left(\sqrt{\lambda_n^2 - \frac{1}{4}}\right)\tau + \frac{1}{2}\cos\left(\sqrt{\lambda_n^2 - \frac{1}{4}}\right)\tau \sin(\lambda_n X)\exp\left(-\frac{\tau}{2}\right)\frac{c_n}{\lambda_n}$$

where $c_n = 4(-1)^{n+1}/(2n-1)\pi$ and $\lambda_n = (2n-1)\pi\sqrt{\dfrac{(D\tau_r)}{2a_w}}$

From the final condition, $c' = 0$ (3.64)

$$\sin(\lambda_n X_a)\frac{c_n}{\lambda_n} = \frac{8a_w}{\pi^2}\sqrt{\frac{(D\tau_r)}{(2n-1)^2}}$$

Thus,

$$\frac{k\tau_r}{a_w} = \exp\left(-\frac{\tau}{2}\right)\frac{8}{\pi^2}\frac{\sqrt{\lambda_n^2 - \frac{1}{4}}}{(2n-1)^2}\left[\sin\sqrt{\lambda_n^2 - \frac{1}{4}}\tau + \frac{\cos\sqrt{\frac{\lambda_n^2 - 1}{4}}\tau}{2(2n-1)^2}\right] \tag{3.65}$$

Taking only the first term in the infinite series,

$$\frac{k\tau_r}{a_w} = \frac{8}{\pi^2}\exp\left(-\frac{\tau}{2}\right)\left(\frac{D\tau_r\pi^2 - a_w^2\pi^2}{4a_w^2}\right)^{1/2}\left[\sin\left(\frac{D\tau_r\pi^2 - a_w^2}{4a_w^2}\right)^{1/2}\tau + \frac{1}{2}\cos\left(\frac{D\tau_r\pi^2 - a_w^2}{4a_w^2}\right)^{1/2}\tau\right] \tag{3.66}$$

In the zero time limit $(k\tau_r/a_w)$ becomes $4/\pi^2 = 0.41$
Rewriting,

$$\frac{ka_w}{D} = \text{Sh} = 0.41\left(\frac{a_w^2}{D\tau_r}\right) = \frac{0.41}{\text{Max}_m} \tag{3.67}$$

$(k\tau_r/a_w)$ is dimensionless group that is de no vo. It has been named as *Sharma number*. It gives the ratio of the bulk mass transfer rate and the relaxational transfer rate. Max_m is the *Maxwell number* (mass) and gives the ratio of the diffusion rate and relaxational rate. The *Sherwood number* gives the ratio of the mass transfer rate in bulk compared with the diffusion rate. At large relaxation times when the above correlation is expected to hold good, it can be seen that the dimensionless mass transfer rate becomes independent of the diffusivity.

The surface-stretch theory was proposed by Lightfoot (in Treybol 1980). Lightfoot noted that when a buoyant rise of a drop is accompanied by some dissolution from it the interfacial area changes in a periodic fashion. For instance, when methyl acetate rises in water, the spherical drop becomes oval in shape and back to sphere and so on and so forth. With the spherical shape there is more mass transfer making it change shape to oval when the mass transfer rate is lower. After sufficient time, the spherical shape is regained again and thus oscillatory shape can be expected. The energy for the shape change possibly comes from the interfacial tension and the temperature of the fluid. The mass transfer coefficient can then be written as:

$$k = \left(\frac{a}{a_r}\right)\left(\frac{D}{\pi t_c}\right)^{1/2}\left(\int\left(\frac{a}{a_r}\right)^2 dt\right)^{1/2} \tag{3.68}$$

where a is the time dependent transfer surface area and a_r is the reference value.

The *boundary layer* theory combines the results from fluid mechanics and the concentration profile to yield a concentration boundary layer for flow past a plate or other surfaces (Schlichting, 1979). Consider the laminar flow past a soluble flat plate. The results from the governing equation obtained from the equations of continuity, momentum and mass balance can be used to estimate how the Sherwood number varies with the Reynolds and Schmidt numbers. A smoothly flowing fluid is disrupted by the drag caused by the flat plate. The region of disruption called the boundary layer becomes larger as the flow proceeds down the plate. The *boundary layer* is usually defined as the locus of distances over which 99% of the disruptive effect occurs. As the flow profile develops, the dissolution from the soluble plate produces a concentration profile. The boundary layer for concentration is not identical with that of the boundary layer for fluid flow. The mass transfer coefficient can be calculated using the information from the concentration profile and velocity profile obtained. It can be assumed that the flow varies as a power series in the boundary layer thickness δ_h. Similarly, the concentration boundary layer thickness δ_c can be derived in terms of the hydrodynamic boundary layer thickness δ_h. Thus it can be shown,

$$\frac{v_x}{v_0} = \frac{3}{2}\left(\frac{y}{\delta}\right) - \frac{1}{2}\left(\frac{y}{\delta}\right)^3 \tag{3.69}$$

From the momentum balance and some steps it can be shown that (Cussler, 1997),

$$\frac{\delta_h}{x} = \left(\frac{280}{13}\right)^{1/2}\left(\frac{\rho v_0 x}{\mu}\right)^{-1/2} \tag{3.70}$$

$$f = \frac{0.646}{\text{Re}_x^{-1/2}} \tag{3.71}$$

The concentration is expressed as a power series,

$$C = c_1 + c_2 y + c_3 y^2 + \ldots$$

The boundary conditions can be written as:

$$y = 0,\ C = C_i;\ \partial^2 C/\partial y^2 = 0 \tag{3.72}$$

$$y = \infty,\ C = 0;\ \partial C/\partial y = 0 \tag{3.73}$$

$$y = \delta_c,\ C = 0;\ \partial C/\partial y = 0 \tag{3.74}$$

Then it can be shown,

$$\frac{C}{C_i} = 1 - \frac{3}{2}\left(\frac{y}{\delta_c}\right) + \frac{1}{2}\left(\frac{y}{\delta_c}\right)^3 \tag{3.75}$$

From the mass balance and some steps it can be shown (Cussler, 1997) that,

$$\left(\frac{\delta_c}{\delta}\right)^3 = \left(\frac{D\rho}{\mu}\right) + 0\left(x - \frac{3}{4}\right) \tag{3.76}$$

$$\left(\frac{\delta_c}{x}\right) = 4.64\left(\frac{\mu}{x v_0 \rho}\right)^{1/2}\left(\frac{D\rho}{\mu}\right)^{1/3} \tag{3.77}$$

Thus the Schmidt number begins to appear in the estimation. The mass transfer coefficient can be derived as:

$$\frac{kl}{D} = 0.646\left(\frac{\mu}{\rho D}\right)^{1/3}\left(\frac{v_0 l \rho}{\mu}\right)^{1/3} 2 \tag{3.78}$$

Thus the Sherwood number can be equated with the Reynolds number and the Schmidt number.

3.5 EQUILIBRIUM LINE, OPERATING LINE AND EFFICIENCY

The design of different mass transfer operations such as distillation, adsorption, absorption, ion-exchange, leaching, drying, humidification, evaporation, reverse osmosis, extraction, drug delivery, membrane separations, molecular sieves, electrophoresis, different kinds of chromatography, crystallisation, foaming, zone refining, carburising steel, dialysis, electrodialysis, pervaporation, pressure swing adsorption, thermal diffusion, sweep diffusion, iso-electric focusing etc., consists of four important steps. These are shown in Figure 3.1.

1. Construction of equilibrium line from thermodynamics
2. Construction of operating lines from mass balance
3. Rate and efficiency
4. Number of stages

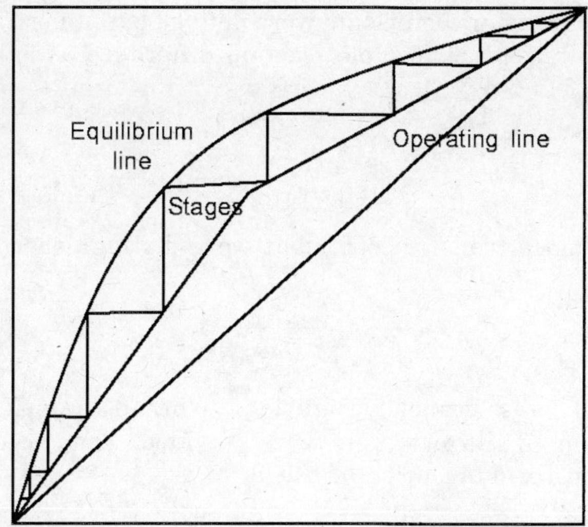

FIGURE 3.1 Equilibrium line, operating line and stages.

The interface through which mass transfer occurs is assumed to be in equilibrium and the equilibrium characteristics have been studied and archived in physical property databases. Theories are available to predict the phase behaviour or construct an equilibrium line. As will be shown later, consider the example of distillation of a system such as ethylbenzene and styrene. The phase equilibrium line at operating pressure can be constructed as shown in Figure 3.1. It denotes the vapour phase mole fraction of the more volatile component and the liquid phase mole fraction in equilibrium with each other at the operating pressure. It is a phase diagram. The operating line can be derived from mass balance between the feed, product and residue. There are two lines; one is the rectifying section and the other is the stripping section. Both the overall balances and the component balances can be written. In order to achieve a certain desired degree of separation, often times several stages are required to complete the operation. At each stage, at best, the equilibrium will be reached. The number of stages can be stepped from the graph similar to the one shown in Figure 3.1. For instance industrially 34 trays would be sufficient to separate styrene and ethylbenzene. Most operations in the industry will never reach equilibrium. So a real line can be calculated. The mass transfer rate effects can be lumped into a parameter called efficiency. The number of stages will be larger for the real line.

SUMMARY

Twenty-six different mass transfer operations were classified according to the nature of the interface through which the mass transfer occurs. Thus six different interfaces—gas-liquid, solid-liquid, gas-solid, liquid-liquid, gas-gas and solid-solid systems were identified in the classification. Different forms of the mass transfer coefficient were defined. Separate mass transfer coefficients were defined based upon the concentration difference, pressure difference, and mole fraction difference as follows:

$$N'' = k(C_{1b} - C_{1i}) \qquad \text{(concentration difference)}$$

$$N'' = k_p(p_b - p_i) \qquad \text{(pressure difference)}$$

$$N'' = k_x(x_b - x_i) \qquad \text{(mole fraction difference)}$$

Local and overall mass transfer coefficient were distinguished from each other.

$$\frac{1}{K_p} = \frac{H}{k} + \frac{1}{k_p}$$

where K_p is the overall mass transfer coefficient. Worked examples in desalination of sea water and etching of silicon wafers were provided. The general form for mass transfer correlations were found to be of the form

$$\text{Sh} = c_1 \text{Re}^n \text{Sc}^m$$

where the Sherwood number depends on the Reynolds and Schmidt numbers in a power exponential fashion. The use of interfacial area was discussed as a critical step in the description of mass transfer coefficient. HTU and NTU the height equivalent of the transfer unit and the number of transfer units were defined.

$$H_{og} = \frac{g_m}{k_g a}$$

Three methods of obtaining correlations between dimensionless groups for mass transfer in different systems are the Rayleigh method, Buckingham–Pi theorem and from the equations of continuity, momentum, and mass balances.

Different mass transfer correlations used were given for flow past a plate in laminar and turbulent flow, flow past a spinning disk, flow past a sphere, flow in fluidised beds, bubbles and drops, etc. Film theory, Higbie's penetration theory, Dankwerts surface renewal theory, Dobbins model, surface stretch theory, and boundary layer theory were outlined. He generalized Fick's law of mass diffusion that was used to modify the results from the penetration and surface renewal theories. The dimensionless group Sharma number was introduced.

$$\text{Sha} = \left(\frac{k\tau_r}{a_w}\right)$$

Sharma number can be seen to be Sha = $(ka_w/D)(D\tau_r/a_w^2)$ which is a product of the Sherwood number and the Maxwell number (mass). It gives the ratio of the mass transfer in bulk to the relaxational transfer during acceleration of molecules. A general 4-step procedure for the design of the different mass transfer operations was described. These were the equilibrium line, operating line, efficiency and stage calculations in sequence.

REVIEW QUESTIONS

1. Give some examples of applications in mass transfer with a solid-solid interface.
2. Can interfacial area have units of m^2/m^3, m^2/mole?
3. Why is adsorption listed under both the gas-solid interface and liquid-solid interface?
4. How many interfaces are present in a gas-solid-liquid fluidised bed?
5. In worked example 3.1, what is the salt concentration of the packed bed. Is any additional information needed?
6. What happens to the mass transfer coefficient when the temperature is changed?
7. Is the mass transfer coefficient independent of the concentration for even the liquid concentrated systems?
8. Write an expression for the overall mass transfer coefficient during gas separations using membranes. Write a feed side mass transfer coefficient, a membrane diffusion profile and a product side concentration profile.
9. What happens to Eq. (3.43) when the distance of approach reaches the order of magnitude or less than the radius of the sphere ? Should another correlation for Sherwood number be used?
10. Rewrite Eq. (3.40) in terms of Reynolds number and Schmidt number.
11. Show that the film thickness in the film theory for transient diffusion would be proportional to $(Dt)^{0.5}$. Substitute this in Eq. (3.51). How does this compare with Eq. (3.55), Eq. (3.59)?
12. Can you repeat the analysis in Q. 11 including the generalised Fick's law of mass diffusion?
13. Is the premise of the surface stretch theory in violation of the second law of thermodynamics? Why?
14. Can you repeat the analysis used to derive Eq. (3.71) and Eq. (3.78) using the generalised Fick's law of diffusion? Can you predict the form of correlations expected to evolve?
15. In the four-step design procedure for stagewise contact separation mass transfer operations, can the efficiency be calculated from the transient concentration profile derived from either the Fick's second law of diffusion or by the generalised Fick's law of mass diffusion?

PROBLEMS

1. *Hydrometallurgical extraction of Gold*
   Gold ores are treated with a cyanide solution and adsorbed onto an anion exchange resin. The gold cyanide will leach the gold from its ore and adsorb onto the resin. Later the gold cyanide in the saturated adsorbent bed is eluted

and the eluted solution is subjected to electrowining prior to obtaining the gold. For a concentration of 1% of gold cyanide at a flow rate of 4.5 cm/s in the feed solution and a saturation concentration of 100 ppm at a bed depth of 170 cm, what is the mass transfer coefficient?

(**Ans:** 220.4 mol/m^3/s)

2. *Purification of Gold by Smelting*
In a smelt furnace, 100 gm of gold is manufactured per hour and the purity of gold is increased from 413 ppm to 2 ppb. What is the mass transfer coefficient if the length of the smelt furnace is 1.5 m?

(**Ans:** 133.3 gm/hr/m)

3. *Iron Ore Beneficiation*
The Hanna mining at Dickinson, Michigan, USA concentrates magnetite and hematite from 34%. Fe to 64.4% Fe using water at 40,000 gpm using a flotation technique that generates bubbles at a frequency of 2 hertz. What is the mass transfer coefficient? Can the interfacial area of mass transfer be estimated? Assume a bubble size of 2 cm.

(**Ans:** 184.7 kmol/m^2/s)

4. *Fractional Freezing Process for Purification of Hydrazine*
Martin Marrietta has developed a fractional freezing process. Anhydrous hydrazine is frozen slowly at a temperature of 262 K to allow pure crystals to settle and contaminants to accumulate in supernatant fluid which is decanted. Pure hydrazine crystals are allowed to melt and are re-frozen. After 3–4 freeze thaw cycles, the aniline content is reduced and the carbon content is lowered from approximately 7000 ppm to less than 20 ppm. Repetitive crystallisation-separation thaw cycles were able to produce hydrazine of purity level of up to 99.99%. Should the mass transfer coefficient be 1.5 mol/m^2/s, what is the throughput of the system.

(**Ans:** 0.01047 mol/m^2/s)

5. *Separation of Heavy Water from Isotopic Chromatography*
Heavy water is used as a moderator for nuclear reactors in power plants. Heavy water is present in the concentration of 108 ppm in sea water.
An isotopic chromatographic column is setup with an interfacial area of 1 million cm^2/g. With a mass transfer coefficient of 413 mol/m^3/s expected saturation of 413 ppb at a bed depth of 80 cm, what is the velocity of sea water the column is capable of handling.

(**Ans:** 1.1 mm/s)

6. *Preparation of Ultra-Clean Coal using Supersolvent*
NMP, N-methyl-2-pyrrolidone, has a boiling point of 202°C at atmospheric pressure. Along with TMU tetramethyl urea, HMPA, hexamethyl phosphoramaide, DMSO di-methyl sulphoxide these form a class of super solvents that can extract the carbonaceous matter in coal. It is dipolar and aprotic and can mix with water making the recovery of the solvent under mild conditions. When a West Virginia Bakerstown coal is contacted with boiling NMP for 20 minutes, about 70% of the coal on an ash free was found to dissolve in the

solvent. What is the mass transfer coefficient assuming the diffusion coefficient in the solid of carbonaceous matter of 1 E-5 m^2/s and a surface renewal time of 4 sec. What throughput can a commercial extractor handle? An interfacial area of 1000 cm^2/g may be assumed.

(**Ans:** 2.5 E-6 m/s; 200 g/s)

7. *Intermediate Yield in Reactions in Series*
Reactions such as chlorination of methane shown below take the form of simple reactions in series A $\rightarrow$ R $\rightarrow$ S $\rightarrow$ T $\rightarrow$ U scheme.

$$CH_4 + Cl_2 \rightarrow CH_3Cl \text{ (methyl chloride)} + HCl$$

$$CH_3Cl + Cl_2 \rightarrow CH_2Cl_2 \text{ (dimethylene dichloride)} + HCl$$

$$CH_2Cl_2 + Cl_2 \rightarrow CHCl_3 \text{ (Chloroform)} + HCl$$

$$CHCl_3 + Cl_2 \rightarrow CCl_4 \text{ (Carbon tetrachloride)} + HCl$$

In order to improve the yield of any of the intermediate products such as methyl chloride, dimethylene dichloride, chloroform the liquid phase reactants and product are sometimes contacted with a gas phase in a tubular reactor. Mass transfer from the liquid phase to the vapour phase of the intermediate product is thus effected. The further reactions in the gas phase are not favoured. This results in a better yield for the intermediate product. Consider chloroform as the intermediate product and show that for $A \rightarrow I \rightarrow P$:

$$\frac{D_I \partial^2 C_{Ib}}{\partial z^2} - k_{1a}(C_{Ib} - C_{Ii}) - \frac{AdC_1}{dt} = 0$$

$$\frac{dC_I}{dt} = k_1 C_A - k_2 C_I$$

$$C_A + C_I + C_P = C_{A0}$$

$$\frac{dC_A}{dt} = -k_1 C_A$$

$$\frac{dC_p}{dt} = k_2 C_I$$

where k_1 and k_2 are the reaction rates of the simple consecutive reactions, A $\rightarrow$ I $\rightarrow$ P. Neglecting the diffusion effects as it is expected to be dominated by bulk flow and assuming a constant interface concentration that depends on the temperature and pressure of the reactor, show that the desired concentration can be solved for by a set of simultaneous differential equations and simultaneous unknowns. (Renganathan and Turton, 1990).

8. *Popcorn Polymer*
Monsanto two CSTR, continuous stirred tank reactor in series process is used to manufacture the copolymer of alphamethylstyrene and acrylonitrile. During the liquid phase polymerisation reaction, some of the monomer in the vapour

phase polymerise to form a popcorn polymer. What is the overall mass transfer coefficient of acrylonitrile from the liquid to the vapour phase including the polyrate effects? 70 kg/hr of reactor syrup at 65% conversion is collected at end of the second reactor. The residence time in the reactor is 2.5 hr and the relative volatility of AMS can be taken as 4. The feed is 57.14% acrylonitrile. The reactor is half filled and is a cylindrical tank of 1.5 m height.

(**Ans:** 1.96 m/s)

9. *Resistances in Parallel*
Crisco corn oil and water flow down an inclined plane at an angle of 30° with the horizontal in layered laminar flow. The layer thicknesses are 16 and 21 μm respectively. Show that the dissolved SO_2 in water diffuses both to the surrounding air and to the oil and that the mass transfer resistances are parallel to each other. What is the overall mass transfer coefficient when the discharge rate of both water and oil is 413 cc/sec. The solubility of SO_2 in water and oil are 100 g/lit and 12 g/lit respectively. The equilibrium concentration of SO_2 in air is 0.05 mol/litre.

(**Ans:** 1.01 E-6 m/s)

10. *Injection of Initiator as Bubbles in Helical Ribbon Agitated CSTR during Continuous Mass Polymerisation of ABS*
Polybutadiene was dissolved in a feed stream of styrene, acrylonitrile, and methyl ethyl ketone to form a feed mixture. The mixture was polymerised in a continuous process under agitation using the Monsanto two continuous stirred tank reactor in series process. The polymerisation occurred in a two-stage reactor system over an increasing temperature profile. During the polymerisation process, some of the forming copolymer grafts to the rubber particles while some of it does not graft, but instead form matrix copolymer. The resulting polymerisation product was then devolatilised, extruded, and pelletised. The peroxy initiators are used to initiate the free radical chain polymerisation reactions. These have been found to yield better results in terms of the degree of grafting and desired molecular weight etc. The first reactor is agitated using a close clearance helical ribbon agitator. For the same power, draw the HRA provides a better degree of agitation and flow throughput with a circulation pattern with flow of fluid down the wall and up at the central axis. The mixing time in this reactor is given as a constant time the agitator RPM. For a CSTR at 35 RPM, the mixing time can be taken as 4 minutes. The half-life of the initiators can be less than the mixing time can cause non-uniform grafting. So initiators with higher half-lives are selected. In any case the initiator can be introduced in the reactor as a bubble train at a frequency of 2.5 Hz. The bubble diameter is 2 cm. For aninitiator half-life of 7 minutes, what is the interfacial area of the bubbles at the half-way point. At a feed rate of 45.5 kg/hr with a monomer concentration of 20 mol/lit and a conversion level of 15%, what is the mass transfer rate of the initiator to the graft phase?

(**Ans:** 0.21 mol/m^2/s)

REFERENCES

Bird, R.B., Stewart, W. and Lightfoot, E., 2002, *Transport Phenomena,* Wiley, New York.

Blatt, Dravid, Michales and Nelson, 1070, *Membrane Science and Technology*, **47**, 70.

Cornet and Kaloo, 1966, *Proc. 3rd Int. Congr. Metallic. Corr.*, **3**, 83.

Cussler, E.L., 1997, *Diffusion: Mass Transfer in Fluid Systems*, Cambridge University Press, UK.

Danckwerts, P.V., 1955, *AIChE J*, **1**, 456.

Dobbins, W.E., 1964, *Int. Conf. Water Pollution Res.*, London, Pergamon, New York.

Faust, Wenzel, Clump, Maus and Andersen, 1980, *Principles of Unit Operations,* Wiley, New York.

Gilland and Sherwood, 1934, *Ind. Engr. Chem.*, **26**, 516.

Graetz, L., 1880, *Zeitschrift fur Mathematik and Physik*, **56**, 1.

Higbie, R, 1935, *Trans. AIChE*, **31**, 365.

Hines and Maddos, 1985, *Mass Transfer: Fundamentals and Applications*, Prentice Hall, NJ.

Johnstone and Pigford, 1942, *Trans. AIChE*, **38**, 25.

Kafesjina, Plank and Gerhard, 1961, *AIChE J*, **7**, 463.

King, 1980, *Separations Processes,* McGraw Hill, New York.

Kirwan, 1987, *Mass Transfer Principles in* Rousseau, *Handbook of Separation Process Technology*, Wiley, New York.

Levenspiel, O., 1989, *Chemical Reactor Omnibook,* Oregon State University Press, USA.

Levich, 1962, *Physicochemical Hydrodynamics,* Prentice Hall, NJ.

Nusselt, W., 1909, *Zeitschrift des Vernines der Deutcher Ingenieure,* **53**, 1750.

Perry, R.H., and Green, D.W., 1997, *Perry's Chemical Engineer's Handbook*, McGraw Hill, New York.

Renganathan, K. and Turton, R., 1990, *Ind. Eng. Chem. Res.*, 709–711.

Renganathan, K. Zondlo J.W., 1993, *Fuel Science and Technology*, **11**, 5, 6, 677–697.

Ruckenstein, E., and Rajagopalan, R., 1980, *Chem. Eng. Commn.*, **4**, 15.

Schlichting, H., 1979, *Boundary Layer Theory*, McGraw Hill, New York.

Sharma, 2003, *Fluid Mechanics and Machinery*, Anuradha Publishers, Kumbakonam, India.

Sharma, K.R., December 2001, *Heavy Water Beneficiation by Isotopic Chromatographic Separation, CHEMCON 2001, Chemical Engineering Congress*, Chennai.

Sharma, K.R., 2005, *Damped Wave Transport and Relaxation*, Elsevier, Amsterdam.

_________, 2006, *One the Pulse Decay During Simultaneous Reaction and Damped Wave Diffusion and Relaxation*, 231st ACS National Meeting Meeting, Atlanta, GA, March.

Sherwood, Pigford and Wilke, 1975, *Mass Transfer*, McGraw Hill, New York.

Sissom and Pitts, 1972, *Elements of Transport Phenomena*, McGraw Hill, New York.

Skelland, 1974, *Diffusional Mass Transfer*, Wiley, New York.

Stewart, W., Angelo, J.B., and Lightfoot, E. N., 1970, *AIChE J.*, **16**, 771.

Taylor and Krishna, 1993, *Multicomponent Mass Transfer*, Wiley, New York.

Toor, R.L. and Marchello, J.M., 1958, *AIChE J.*, **4**, 97.

Treybal, R.E., 1980, *Mass Transfer Operations*, McGraw Hill, New York.

Wankat, 1990, *Rate-Controlled Separations*, Chapman-Hall, New York.

CHAPTER 4

Distillation

Nomenclature

A	cross-sectional area (m^2)
a	interfacial area (m^2/gm)
A_f	(L/mG) absorption factor
B	reboil ratio (L′/W)
Dis	distillate rate (mol/hr)
f	ratio of residue to distillate
F	feed rate (mol/hr)
F_f	number of degrees of freedom
G	vapour phase flow rate, rectifying section (mol/hr)
G''	vapour phase flow rate, rectifying section per unit area (mol/m^2/hr)
G'	vapour phase flow rate, stripping section (mol/hr)
G_{Np+1}	steam flow rate (mol/hr)
H_F	enthalpy of feed (J/mol)
H_D	enthalpy of distillate (J/mol)
H_W	enthalpy of residue (J/mol)
H_{Ln}	enthalpy of liquid at plate n (J/mol)
H_{Gn+1}	enthalpy of vapour at plate n (J/mol)
K_y	overall mass transfer coefficient (mol/m^2/s)
l	depth of liquid-vapour froth (m)
L	liquid phase flow rate, rectifying section (mol/hr)
L'	liquid phase flow rate, striping section (mol/hr)
L_f	liquid phase flow rate, feed (mol/hr)
N	total number of stages in the column
N_p	number of plates in the column
N_c	number of chemical species in the system
m	Henry's law of constant
m_1	distribution coefficient of component 1
m_2	distribution coefficient of component 2
m_i	distribution coefficient of component i
p_A	partial pressure of component A (N/m^2)

p_A^*	pure component vapour pressure of A at that temperature (N/m^2)
P_{split}	product split (x_DDis/x_wW)
P	pressure (N/m^2)
P_c	critical pressure (N/m^2)
q	feed ratio of liquid flow rate to total flow rate
Q	heat input (watts)
Q_c	condenser heat load (watt)
Q_B	reboiler heat load (watt)
Q_{Ln}	heat loss from plate n (watt)
Q'	heat removed in the condenser, (kJ/mol) $Q' = (Q_c + \text{Dis}\ H_D)/\text{Dis}$
Q''	heat supplied at the reboiler, (kJ/mol) $Q'' = (WH_W - Q_B)/W$
R	(L/Dis) reflux ratio
R_m	minimum reflux ratio
S_W	reboiler stripping factor ($G_{Np+1}m_w/x_w$)
T	temperature (K)
T_c	critical temperature (K)
v_v	molar volume of vapour (m^3/mol)
v_L	molar volume of liquid (m^3/mol)
x_0	top plate liquid phase composition
x_F	feed composition of the species of the interest
x_D	distillate composition of the species of interest
x_D^*	equilibrium distillate composition of the species of interest
$\langle x_D \rangle$	composited distillate average
x_w	bottoms composition of the species of interest
x_n	liquid phase composition in the nth stage
x_N	liquid phase composition exiting the bottom plate
x_{int}	intersection point, abscissa
$x_{intercept}$	intercept of the straight line with the x axis
$x_{j,D}$	distillate composition of jth species of multi-component mixture
y_{n+1}	vapour phase composition entering the nth tray of the species of interest
y_n	vapour phase composition in the nth stage
y_0	vapour phase composition that exits the top of the column
y_1	top plate vapour phase composition
y_{N+1}	vapour composition entering the bottom of the column
y_{int}	intersection point, ordinate
W	residue (mol/hr)
z_f	composition of feed
$z_{j,F}$	composition of the jth component of the feed in multicomponent distillation

Greek

α	$(y^*(1-x))/((1-y^*)x)$ relative volatility
α_j	relative volatility of the jth species

$<\alpha_{lk}>$	average relative volatilities
λ_{vap}	latent heat of vaporisation (J/mol)
ϕ	number of phases
ρ	density (kg/m^3)
η	Murphree stage efficiency
Δ_D	difference point (z_D, Q')
Δ_W	difference point (x_W, Q'')
δ	Riccati transformation variable

4.1 INTRODUCTION

Early references to distillation in the history of mankind dates back to the 4th millennium BC in Babylonia. Back then, specially shaped clay pots were used to extract small amounts of distilled alcohol through natural cooling for use in perfumes. Continuous distillation is the most important form of separation technology used in the chemical process industry and biotechnology enterprises. The most widely used industrial applications of continuous, steady-state distillation are in petroleum refineries, petrochemical plants and natural gas processing plants. *Distillation* is a process where a more volatile component is separated from a less volatile component by supply of energy to heat the liquid. The vapour is collected and condensed. The residue is reheated and recycled back to the distillation apparatus. There can be several different kinds of distillation such as batch, differential, vacuum, steam stripping, flash, multi-stage continous contact operation, multi-component distillation, azeotropic distillation, extractive distillation, reactive distillation, freeze distillation, etc.

Distillation is the oldest and most common refinery process (Figure 4.1). It is based on the principle that liquids with different boiling points can be separated by vaporisation and condensation. Crude oil distillation involves feeding crude oil at its boiling point of 400°C into a large fractionation column. The oil is split up into a series of product fractions depending on desired boiling point ranges. Vaporised oil on its way up through the column encounters condensed liquid that is on its way down. They are made to come in contact and exchange components at a number of sieve trays in the column. This contact causes light fluid molecules to vaporise and heavy gas molecules to condense. The further up the column, lighter is the fluid found at the sieve tray. Part of the crude oil that cannot be vaporised is extracted as heavy oil at the bottom of the column. Exxon-Mobil is the world's largest company that explores oil and in its refineries is fuel manufactured. A schematic of the overall process used by Exxon-Mobil is shown in Figure 4.2.

In India the major oil companies are Indian Oil, Bharat Petroleum, Hindustan Petroleum, Oil and Natural Gas Corporation, Chennai Petroleum, Kochi Refineries, Bangagoan Refinery, Mangalore Refinery and IBP. The largest refinery in India owned by Reliance has a capacity of 31 million tonnes. Indian oil companies have been increasing their refinery capacity. For example, recently the 3 million tonne refining capacity expansion and modernisation project of Chennai Petroleum Corporation Ltd. CPCL, in Manali near Chennai, in order to improve the overhead distillates yield and

FIGURE 4.1 Typical distillation towers in oil refineries.

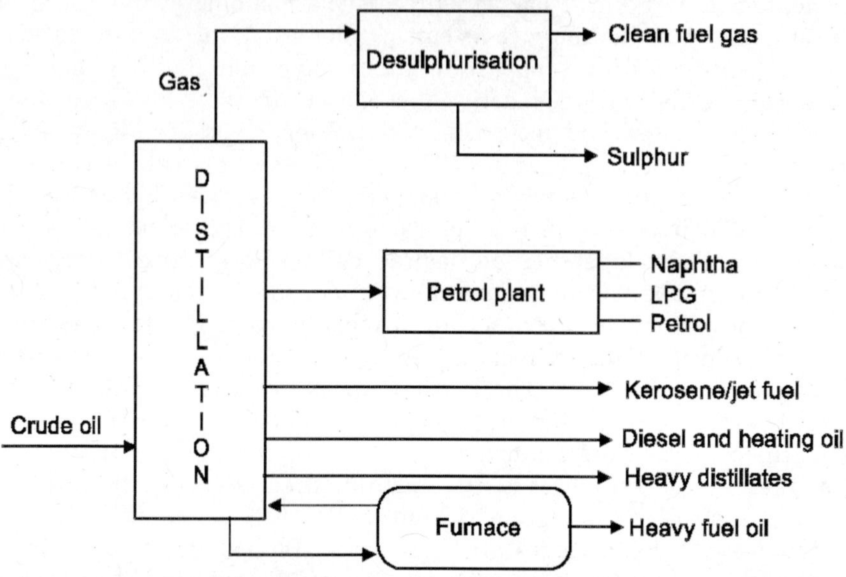

FIGURE 4.2 Exxon-Mobil corporation's refining of crude oil into fractions.

produce cleaner fuel, came on stream. It is a 9.5 million ton refinery and the expansion has been on 120 acres at 2360 crore rupees. It is a first of its kind facility to process 100% hydrocracker bottom and as a result the liquefied petroleum gas production has increased from 1.8 lakh tonnes to about 4 lakh tonnes. The petrol production has improved from 3.9 lakh tonnes to 7 lakh tonnes. The hydrocracker unit is a crucial component in the production of automobile fuels with low sulphur content.

Industrial distillation is typically performed in large, vertical cylindrical columns known as "distillation towers" or "distillation columns" with diameter ranging from about 65 cm to 6 m and height ranging from about 6 m to 60 m. The distillation towers have liquid outlets at intervals up the column which allow the withdrawal of different *fractions* or products having different boiling ranges. The 'lightest' products (those with the lowest boiling point) exit from the top of the columns and the 'heaviest' products (those with the highest boiling point) exit from the bottom of the column. Large-scale industrial towers also use reflux to achieve more complete separation of products. Fractional distillation is also used in air separation, producing liquid oxygen, liquid nitrogen, and high purity argon. Distillation of cholorosilanes also enable the production of high-purity silicon for use as a semicinductor. In industries, sometimes a packing material is used in the column instead of trays, especially when low pressure drops across the column are required, as when operating under vacuum. This packing material can either be random dumped packing (1–3" wide) or structured sheet metal. Typical manufacturers include companies like Koch and Sulzer. Liquids tend to wet the surface of the packing and the vapours pass across this wetted surface, where mass transfer takes place. Unlike conventional tray distillation in which every tray represents a separate point of vapour liquid equilibrium, the vapour liquid equilibrium curve in a packed column is continuous. However, when modelling packed columns it is useful to compute a number of 'theoretical stages' to denote the separation efficiency of the packed column with respect to more traditional trays. Differently shaped packings have different surface areas and void space between packings. Both these factors affect packing performance. Typical number of plates required for a desired level of separation for different aromatic, organic and aqueous sytems are given in Table 4.1.

4.2 METHOD OF MCCABE AND THIELE FOR MULTI-STAGE BINARY DISTILLATION

The widely accepted method popular with the practioniers in the chemical process industry is that of McCabe and Thiele to calculate the number of theoretical stages needed to perform a desired level of separation of a binary system. A schematic of a multi-stage binary distillation column showing the key components is shown in Figure 4.3. The components are:

1. Central Feed
2. Vertical Cascade of Stages
3. Enriching Section
4. Overhead Condenser
5. Reflux
6. Distillate
7. Stripping Section
8. Reboiler

TABLE 4.1 Typical number of trays used in the real world of chemical plants

S.No.	*System*	*Typical Number of Trays*
1.	Ethylene/Ethane	73
2.	Propylene/Propane	138
3.	Propyne/1,3–Butadiene	40
4.	1,3 Butadiene/Vinyl Acetylene	130
5.	Benzene/Toluene	34, 53
6.	Benzene/Ethyl Benzene	20
7.	Toluene/Xylene	45
8.	Ethyl Benzene/Styrene	34
9.	o-Xylene/m-Xylene	130
10.	Methanol/Formaldehyde	23
11.	Dichloroethane/Trichlorethane	30
12.	Acetic Acid/Acetic Anhydride	50
13.	Vinyl Acetate/Ethyl Acetate	90
14.	Ethylene Glycol/Diethylene Glycol	16
15.	Acetic Anhydride/Ethylene Diacetate	32
16.	Cumene/Phenol	38
17.	Phenol/Acetophenone	39, 54
18.	Water/HCN	15
19.	Water/Acetic Acid	40
20.	Water/Methanol	60
21.	Water/Ethanol	60
22.	Water/Isopropanol	12
23.	Water/Vinyl Acetate	35
24.	Water/Ethylene Oxide	50
25.	Water/Ethylene Glycol	16

In the enriching section, the vapour rises above the feed and is washed with the liquid to strip the less volatile component. The washing liquid is prepared in the overhead condenser and part of the distillate is returned to the distillation column as reflux. A reflux ratio is calculated and comes in handy during further design estimates. The distillate is the product that is removed from the condenser and contains the more volatile component in high concentration. The stripping section is that section below the feed where the vapour that boils sweeps the more volatile component with it and moves upward through the column. The vapour is prepared in the reboiler by partial vaporisation. The liquid that is removed from the reboiler is the residue and has predominantly the less volatile component. The reboiler can be viewed as an additional stage. The liquid and vapour at each of the n trays are at their dew and bubble points. The highest temperature is at the bottom and the lowest temperature is at the top. The purity levels achieved for the distillate and residue depend on the liquid/gas ratios used and the number of theoretical stages

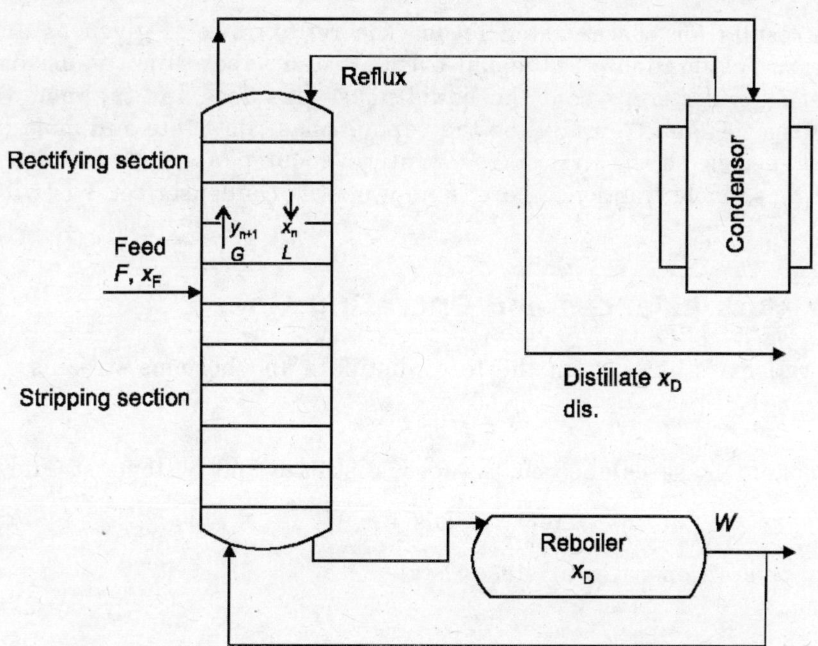

FIGURE 4.3 Schematic of typical multi-stage distillation column.

provided in the two sections of the tower and the operating efficiency resulting from mass transfer diffusion resistances. The cross-sectional area of the distillation column varies with the quality and price of the material of interest. The interrelationship between the purity levels of the distillate, residue, feed and the number of stages required ideally was outlined by McCabe and Thiele.

The most common problem in distillation is designing a distillation column. Given the flow rate and concentration of entering stream, and the flow rate and concentration of the product stream, the size of column needs to be calculated. The industrial columns consist of a series of plates of the sieve tray or bubble cap variety and are used to separate the stages. The gas and liquid phases are made to come in contact with each other until it can reach equilibrium.

4.2.1 Main Steps of Design

The main steps of the design involve;

1. Mass balance—Derive operating line
2. Energy balance—Obtain equilibrium curve
3. Transient mass transfer rate equations—Derive murphree efficiency
4. Step off the ideal and real stages required (staircase graph)

The method is less rigorous compared with other methods such as the Ponchon Savarit method that uses the enthalpy information as well. This method was found to give good results for concentrated feeds. The reflux ratio is given as an input parameter in the estimations. Equimolal overflow and vaporisation is assumed. At each stage let L_n and x_n represent the liquid phase flow rate and the mole fraction of the desired species and G_n and y_n be the vapour phase flow rate and mole fraction of the desired species. In the first stage y_1 is the vapour phase mole fraction and x_0 is the liquid phase mole fraction. There is no y_0 as the condenser returns will be all liquid.

4.2.2 The Mass Balances and Operating Lines

From an overall mass balance on the feed, distillate and bottoms streams,

$$F = \text{Dis} + W \tag{4.1}$$

From a component mass balance on the feed, distillate and bottoms stream,

$$x_F F = x_D \text{Dis} + x_w W \tag{4.2}$$

When the column operates at steady state,

$$L_{n-1} + G_{n+1} = L_n + G_n \tag{4.3}$$

The component balance would then be,

$$Lx_0 + Gy_{n+1} = Lx_n + Gy_1 \tag{4.4}$$

The vapour flux G and the liquid flux L are assumed constant between the top of the tower and the nth plate. Rewriting Eq. (4.4)

$$y_{n+1} = \left(y_1 - \left(\frac{L}{G}\right)x_0\right) + \left(\frac{L}{G}\right)x_n \tag{4.5}$$

It can be seen that y_{n+1} varies linearly with x_n. The slope can be seen to be (L/G) and the intercept $(y_1 - L/Gx_0)$. In the special case of the total condenser, $(\text{Dis} = G + L)$,

$$x_D = x_0 = y_1 \tag{4.6}$$

Noting that the vapour phase entering the nth plate is at the composition y_{n+1} and the liquid exiting the nth stage is at composition x_n. From a mass balance on the nth plate and the condenser exit streams;

$$x_D\text{Dis} = -x_n L + y_{n+1} G \tag{4.7}$$

or

$$y_{n+1} = x_n(L/G) + x_D(\text{Dis}/G) \tag{4.8}$$

Defining a reflux ratio, $R = L/\text{Dis}$

Then,

$$\frac{\text{Dis}}{G} = \frac{\text{Dis}}{(L + \text{Dis})} = \frac{1}{(R + 1)} \tag{4.9}$$

$$\frac{L}{G} = \frac{L}{(L + \text{Dis})} = \frac{R}{(R + 1)} \tag{4.10}$$

Combining Eqs. (4.9 and 4.10) with Eq. (4.4) the operating line of the rectifying section is then,

$$y_{n+1} = x_n\left(\frac{R}{R + 1}\right) + x_D\left(\frac{1}{R + 1}\right) \tag{4.11}$$

The operating line of the rectifying section is essentially a mass balance relationship between the components of the different inlet and outlet streams. The reflux ratio R can be used as a lever to achieve the desired objectives during the operation of the distillation column. In a similar fashion, analysis of the reboiler section and a plate in the column leads to the derivation of the operating line of the stripping section,

$$y_{n+1} = x_n\left(\frac{W}{G' - 1}\right) - x_w\left(\frac{W}{G'}\right)$$

$$y_{n+1} = x_n\left(\frac{B}{B - 1}\right) - \left(\frac{x_w}{B - 1}\right) \tag{4.12}$$

where B is the reboil ratio. The *reboil ratio* is that ratio of the vapour produced by the reboiler to the residue withdrawn (L'/W).

The nature of feed can be several kinds ranging from all vapours, all liquid or a partial mixture of both liquid and vapour. The feed can be found at the intersection of the two operating lines from the rectifying and stripping sections. For any feed,

$$F + L + G' = G + L' \tag{4.13}$$

where G' and L' are the vapour and liquid phase flow rates in the stripping section. q can be defined as:

$$q = \frac{(L' - L)}{F} = \frac{L_F}{F} \tag{4.14}$$

When the feed is all liquid below the bubble point, the q value is > 1.0. For a saturated liquid feed $q = 1$, and for a mixture of liquid and vapour in the feed, the q value lies between 0 and 1.0. For the saturated vapour q is 0 and for superheated vapour the q value is less than 0 and is negative. Later an expression for q in terms of the enthalpies will be shown.

From the component mass balance,

$$Fz_F = \text{Dis } x_D + Wx_w \tag{4.15}$$

The intersection of the two operating lines can be found as,

$$y_{n+1} = x_n\left(\frac{L}{G}\right) + x_D\left(\frac{\text{Dis}}{G}\right) \tag{4.16}$$

$$y_{n+1} = x_n\left(\frac{L'}{G'}\right) - x_W\left(\frac{W}{G'}\right) \tag{4.17}$$

Subtracting Eq. (4.16) from Eq. (4.17)

$$y_{n+1}\,(G' - G) = x_n(L' - L) - (x_W W + \text{Dis}\;x_D) \tag{4.18}$$

Rewriting Eq. (4.17) by using Eq. (4.15) and Eq. (4.13)

$$y_{n+1} = \frac{qx_n}{q-1} - \frac{z_D}{q-1} \tag{4.19}$$

Eq. (4.19) is the feed line. The slope of the line gives the energy required to convert the feed to the vapour phase. The feed line for different feed conditions is shown in Figure 4.4.

4.2.3 Feed Location, Pinch Point, Minimum Reflux and Minimum Stages

The information from the VLE, vapour-liquid equilibrium data and the operating lines derived from the mass balances as shown above are combined to calculate the number of theoretical stages needed to achieve a desired level of separation. A step by step illustration of this method is shown in *worked example 4.1*. The Pentium 4, 3 GHz PC has been used. The graph of the VLE data called the equilibrium line relates the vapour composition on a particular stage to the liquid composition in the same stage. For a given reflux ratio R, the two operating lines from the rectifying and stripping section can be shown to intersect with the feed line for any feed (Figure 4.3). Each stage can be seen as a staircase graph and stepped off from the graph. This is shown in the *worked example 4.1*. Thus by a combined graph of the equilibrium line and operating lines, the number of stages can be stepped off with ease and is an important procedure. The feed can be located in a stage at the intersection of the operating lines. This is the *optimal location of the feed*. Other locations are suboptimal.

The *minimum number of stages* required can be calculated when the reflux ratio becomes infinite. At infinite reflux ratio ($R = \infty$), the operating lines can be seen to revert to $y_{n+1} = x_n$. Both the slopes of the rectifying and stripping sections become equal. At infinite reflux, zero distillate is produced. Thus for a lot of energy supplied, in a few stages, that can be stepped off from the graph, only a few drops of pure product is recovered (Figure 4.4). The *minimum reflux ratio* can be calculated by letting the operating line from the rectifying section to intersect with the equilibrium line and the q feed line and the number of stages stepped off from the graph (Figure 4.5). The reflux ratio used is often 1.2 to 1.6 times the minimum reflux ratio that is needed. This ratio depends on the relative capital costs and operating costs of energy. If capital is cheap, a low reflux ratio with more number of stages can be used and where the energy is cheap a large reflux with fewer stages is preferred. The reboiler load will be low for the former and large for the latter. The condition of minimum

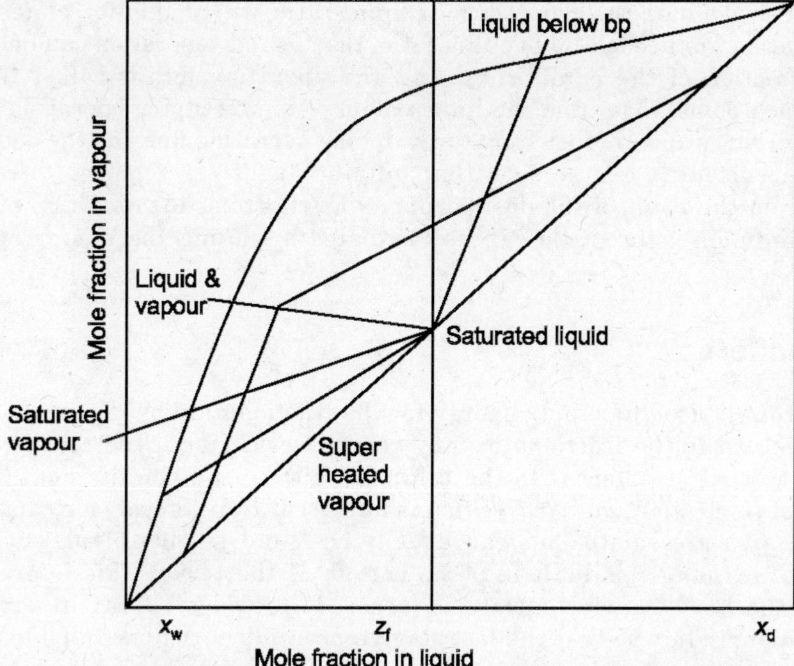

FIGURE 4.4 *q* Line for different feed condition.

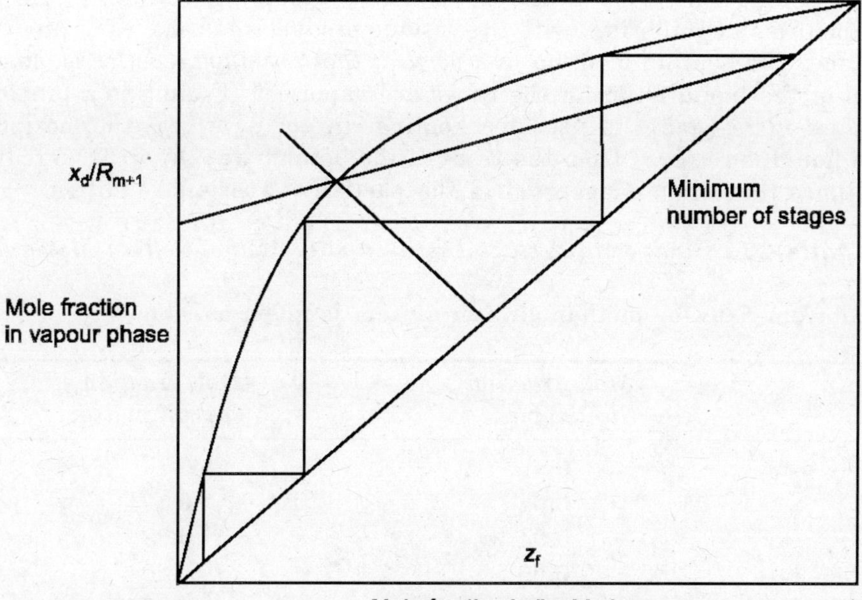

FIGURE 4.5 Minimum reflux ratio and minimum number of stages.

reflux ratio also denotes the point of maximum ratio where infinite plates for the desired separation are needed. Above this ratio, the desired separation can be effected.

The intersection of the equilibrium line and operating line is called the *pinch point.* Two pinch points, one from the intersection of the rectifying operating line and the equilibrium curve and another from the stripping operating line and the equilibrium line, can be expected. It can be seen that infinite stages are required to reach the pinch point from the point of the desired purity level. From this condition of infinite plates, the minimum reflux ratio can be calculated by noting the y intercept of the graph. $(x_D/(R_m+1))$.

4.2.4 Reboilers

The reboiler that is used to supply energy for the fractionation by vaporisation of the residue and return to the fractionator may be of several kinds. The reboiler may be built either internal or external to the column. This depends on the heat load. For low capacity of separation, *jacketed kettle* may be used. Here the vapour capacity and heat transfer area are small. This can usually be found in pilot plants. Sometimes a *tubular heat exchanger* is built into the bottom of the tower. This is an internal reboiler and the heat transfer surface is larger. It provides vapour in equilibrium with the residue product so that the last stage represents enrichment due to reboiler. The problem is when cleaning the heat exchanger requires the shut down of the entire distillation operation. In order to avoid this maintenance routine, external reboilers can be used. They can be arranged with spares for cleaning. This can be used for large installations. The *kettle reboiler* has the heating medium inside the tubes. The vapour is in equilibrium with the residue product and behaves as an extra theoretical stage. The heating medium in a *vertical thermosiphon reboiler* is outside the tubes. All of the liquid entering the tubes are vaporised. Fouling is a problem. In *horizontal reboilers,* steam is used for heating. In some vertical thermosiphon reboilers, the liquid is received from the traps of the bottom tray. In order to reduce fouling sometimes the boiler is operated in the partially vaporised condition.

Worked Example 4.1 *Binary Multi-stage Distillation of Methanol from Methanol / Water System*

The equilibrium date for methanol/water system is given as follows.

Temp. (°C)	*Mole fraction* (*liquid phase*)	*Mole fraction* (*vapour phase*)
100	0	0
96.4	0.02	0.134
93.5	0.04	0.23
91.2	0.06	0.304
89.3	0.08	0.365
87.7	0.1	0.418
84.4	0.15	0.517

(*Contd.*)

Temp. (°C)	*Mole fraction* (*liquid phase*)	*Mole fraction* (*vapour phase*)
81.7	0.2	0.579
78.0	0.3	0.665
75.3	0.4	0.729
73.1	0.5	0.779
71.2	0.6	0.825
69.3	0.7	0.870
67.5	0.8	0.915
66.0	0.9	0.958
65.0	0.958	0.979
64.5	1.0	1.0

Use a reflux ratio of 2.5 and determine the number of stages required to achieve a distillate composition of 96.5%. The residue composition of methanol as 15% and a W/Dis ratio of 0.8 can be assumed.

The operating lines can be derived from Eqs. (4.11 and 4.12). The equilibrium line has been constructed using the above data in MS excel spreadsheet on a Pentium IV, 3.0 GHz PC. The staircase graph was constructed by dragging the chart into MS paint. The VLE, vapour liquid equilibrium data has been drawn as a chart using a smoothing routine available. The risers of the staircase reflect VLE and the treads represent the mole balances. Thus the McCabe-Thiele method has been used with the assumption of equi-molar overflow and the number of ideal stages calculated as 5. The stripping line has a positive x intercept. The slope of the operating line in the rectifying section is positive. The feed can be seen to have about 60% methanol from the graph in Figure 4.6.

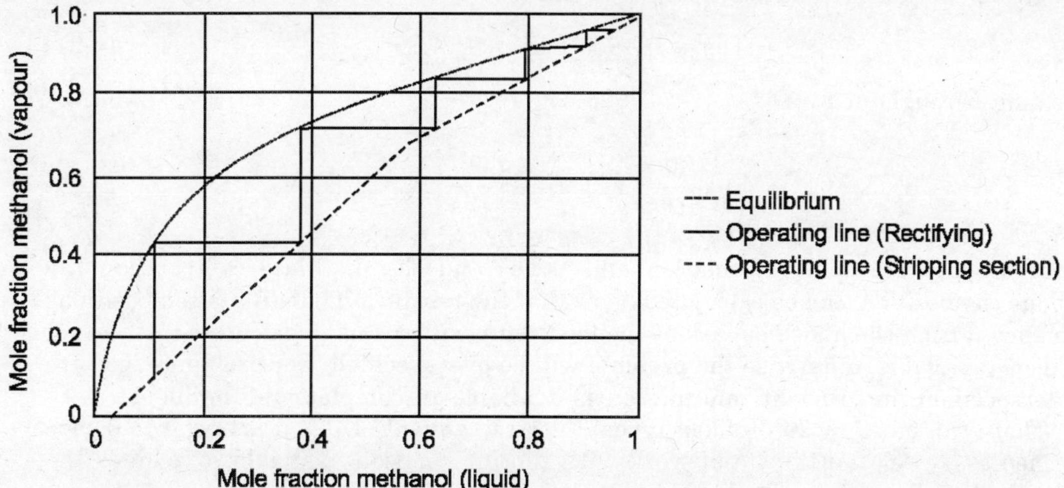

FIGURE 4.6 Multi-stage equilibrium separation of methanol-water binary system.

4.2.5 Specifications Limitations

The parameters of the design of a multi-stage binary distillation column at the beginning are usually available:

1. Temperature, pressure, composition and flow rate of feed
2. Pressure of distillation
3. Optimal feed location
4. Heat losses

It can be shown that only 3 more design specifications besides the above is needed to completely specify the problem. (Treybal, 1980). The rest can be calculated from the other parameters. In the above sections it can be seen that a lot of the parameters are related to each other by the mass balances and equilibrium line from VLE data. The other 6 parameters are:

1. Total number of trays N
2. Reflux ratio R
3. Reboil ratio B
4. Composition of product
5. Product split, P_{split} ($x_D\text{Dis}/x_w W$)
6. Ratio of total distillate to total residue (Dis/W)

There are some pitfalls in specifying the problem completely. Sometimes the problem may be overspecified or underspecified. For example, it can be seen that three of the above six parameters are given, i.e. the product split P_{split}, W/Dis and distillate composition. Then, from a component balance of the three main streams of feed, distillate and product,

$$Fz_F = \text{Dis}\, x_D + Wx_W \tag{4.20}$$

Rewriting the above equation using $F = \text{Dis} + W$

$$(\text{Dis} + W)z_F = \text{Dis}\, x_D + Wx_W \tag{4.21}$$

Dividing throughout by Dis,

$$\left(1 + \frac{W}{\text{Dis}}\right) z_F = x_D + \left(\frac{W}{\text{Dis}}\right) x_w \tag{4.22}$$

From the product split information and W/Dis and the distillate composition the residue composition, can be calculated. Now that the residue and distillate compositions are known from the above equation, the feed composition can be calculated. It cannot be made available, otherwise the problem will be overspecified. For problems that are underspecified, insufficient information is available to complete the problem.

There can be ${}^6C_3 = 20$ problem types. With the advent of PC, it is easier to derive the necessary relationships and work with the information available to achieve the desired objectives. These problem types and the solution approach are as follows:

I. *Given Reflux Ratio R, Reboil Ratio B and* Dis / W, *find the number of trays, N_p, product split and the bottoms composition.*
The intersection point (x_{int}, y_{int}) can be seen to obey Eqs. (4.11, 4.12, 4.19)

$$y_{int} = x_{int}\left(\frac{R}{R+1}\right) + x_D\left(\frac{1}{R+1}\right) \tag{4.23}$$

$$y_{int} = x_{int}\left(\frac{B}{B-1}\right) - \left(\frac{x_w}{B-1}\right) \tag{4.24}$$

$$y_{int} = \left(\frac{qx_{int}}{q-1}\right) - \left(\frac{1}{q-1}\right) z_F \tag{4.25}$$

Further from Eq. (4.15);

$$\left(\frac{\text{Dis}}{W} + 1\right) z_F = \left(\frac{\text{Dis}}{W}\right) x_D + x_w \tag{4.26}$$

There are four equations and 4 unknowns, x_w, x_D, x_{int}, y_{int} which can be solved for simultaneously. Once the intersection point is found the same procedure as shown in worked example 4.1 to step of the number of ideal stages required for a given desired level of separation can be stepped from the staircase graph.

II. *Given W/*Dis, *product split R, calculate N_p, x_w, B*
Eq. (4.26) can be written in terms of x_w only given the information in the product split and W/Dis. Once the x_w, x_D can be calculated from the product split information, eqs. (4.25) and (4.23) can be solved simultaneously as they form a set of two equations and two unknowns. From this and Eq. (4.24) the reboil ratio B can be calculated.

III. *Given Product Split, P_{split} (x_DDis/x_w W), Reboil Ratio B, Residue composition x_w, calculate N_p, R, W/*Dis.
Straightforward solution. Given the residue composition and reboil ratio, the operating line for the stripping section can be constructed in the graph. The intersection with the feed line can be identified. Rewriting Eq. (4.22),

$$(P_{split}\, x_w/x_D + 1)z_F = P_{split}\, x_w + x_w \tag{4.27}$$

Thus the distillate composition can be calculated given the product split and residue mole fraction. Now from Eq. (4.23), the reflux ratio can be calculated. This completes the specification of the problem.

IV. *Given Product Split, P_{split} (x_DDis/x_wW), Reboil Ratio, B and Reflux Ratio R, calculate Residue composition, x_w, N_p, W/Dis*
The four equations in Eqs. (4.23–4.26) can be solved for by simultaneous equations, simultaneous unknowns.

V. *Given Product Split P_{split} (x_DDis/x_wW), Reflux Ratio R, Residue Composition x_w, calculate N_p, W/Dis and Reboil Ratio*

Straightforward solution. Given product split and residue composition, the distillate composition can be calculated from Eq. (4.27). Now the W/Dis can be calculated from the product split information. The intersection points of the two operating lines of rectification and stripping can be calculated by solving for Eqs. (4.23 and 4.25). The reboil ratio can then be calculated from Eq. (4.24). This completes the description of the specifications. The slope of the rectifying line can be calculated and graphically it is straight forward to identify the intersection point.

VI. *Given Product Split, Residue Composition, and W/Dis, calculate the N_p, Reflux Ratio R, and Reboil Ratio, B*

Given the product split, residue composition and W/Dis, the distillate composition can be calculated. The feed composition can be calculated and cannot be specified. If stated it would be over specification. Between the feed line, operating lines of rectification and stripping as given in Eqs. (4.23–4.25) and the feed line, the abscissa and ordinate of the intersection points and the R and B can be calculated by solving a system of 4 equations and 4 unknowns.

VII. *Given the Product Split, x_DDis/x_wW, W/Dis and Reboil Ratio, B, calculate N_p, R, x_w.*

Given the product split and W/Dis the distillate composition can be written in terms of the residue composition. Using this information in Eq. (4.20) there remains three simultaneous equations, Eqs. (4.23, 4.24 and 4.25), and three unknowns to solve for the intersection point ordinates and the residue composition. This completes the description of the problem.

VIII. *Given W/Dis, Residue composition x_w, Reflux ratio R, calculate N_p, Reboil ratio, B and Product Split, x_DDis/x_wW*

The distillate composition can be calculated from Eq. (4.26) given the W/Dis and residue composition as the feed is completely specified. Now the product split can be calculated. Graphically the operating line for the rectifying section can be constructed. The intersection point with the feed line is identified in the graph. Given the residue composition, the operating line of the stripping section is constructed. From the slope of the line, the reboil ratio can then be calculated. The number of ideal stages can be stepped off the staircase graph.

IX. *Given W/Dis, Reboil Ratio, B, Residue composition x_w, calculate N_p, Reflux Ratio, R, Product Split, P_{split} x_DDis/x_wW*

The distillate composition can be calculated from Eq. (4.26) given the W/Dis and residue composition as the feed is completely specified. Now the product split can be calculated. Graphically the operating line for the stripping section can be constructed. The intersection point with the feed line is identified in the graph. As the distillate composition is now available, the operating line of the rectifying section is constructed. From the slope of the line the reflux ratio can then be calculated. The number of ideal stages can be stepped off the staircase graph.

X. *Given the Reflux Ratio R, Reboil Ratio, B and Residue Composition x_w, Calculate N_p, Product Split, P_{split} x_DDis/x_wW, W/Dis.*
Given the residue composition and reboil ratio the operating line of the stripping section can be constructed. The intersection point with the feed line can be identified in the combination graph of both the equilibrium and operating lines. Given the reflux ratio, the slope of the operating line of the rectifying section is calculated and constructed. The intersection of the operating line of the rectifying section gives the distillate composition. From the information gained, the product split is now calculable. N_p, the number of ideal stages needed for a desired level of separation could be stepped off from the staircase graph.

XI. *Given N_p, Reboil Ratio B, Reflux Ratio R calculate* W/Dis *Residue composition x_w, Product Split P_{split}* (x_DDis/Wx_w)
This problem requires an iterative solution. Here is a suggested procedure:

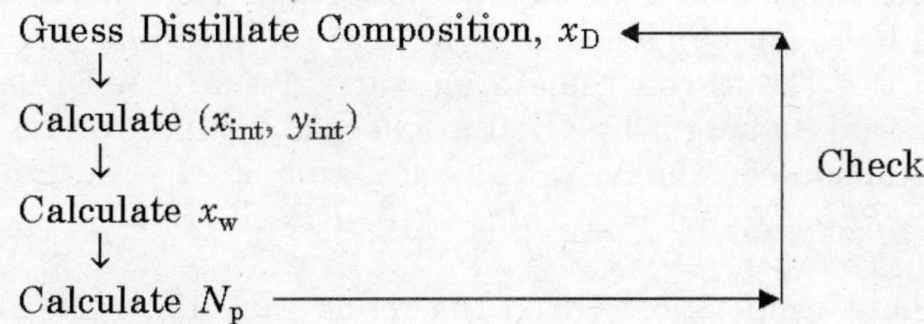

XII. *Given N_p, Residue composition x_w, P_{split}* (x_DDis/Wx_w) *calculate Reflux Ratio, R,* W/Dis, *Reboil Ratio, B*
Guess B, and given residue composition the operating line of the stripping section can be constructed. The intersection point with the feed line can be identified in the graph. Given the product split and residue composition, the distillate composition can be calculated from Eq. (4.27). From the distillate composition and the intersection point, the operating line of the rectifying section can be constructed. The N_p can be stepped off from the graph. This is compared with the given N_p. Then the reboil ratio is changed and the routine repeated until the calculated and given N_p are identical within numerical error.

XIII. *Given N_p, Reboil Ratio, B P_{split} (x_DDis / Wx_w) calculate Reflux Ratio R Calculate* W/Dis, *Residue composition, x_w*
Guess the residue composition x_w and given the reboil ratio the operating line of the stripping section can be constructed. The intersection point with the feed line can be identified in the graph. Given the product split and residue composition, the distillate composition can be calculated from Eq. (4.27). From the distillate composition and the intersection point, the operating line of the rectifying section can be constructed. The N_p can be stepped off from the graph. This is compared with the given N_p. Then the residue composition is changed and the routine repeated until the calculated and given N_p are identical within numerical error.

XIV. *Given* N_p, *Reboil Ratio, B,* W/Dis *calculate Residue composition* x_w, *calculate Reflux Ratio, R,* P_{split} (x_DDis/Wx_w)

Guess the residue composition, x_w and given the reboil ratio the operating line of the stripping section can be constructed. The intersection point with the feed line can be identified in the graph. Given the W/Dis and residue composition the distillate composition can be calculated from Eq. (4.26). From the distillate composition and the intersection point the operating line of the rectifying section can be constructed. The N_p can be stepped off from the graph. This is compared with the given N_p. Then the residue composition is changed and the routine repeated until the calculated and given N_p are identical within numerical error.

XV. *Given* N_p, *Residue composition* x_w, *Reboil Ratio B calculate Reflux Ratio R,* P_{split} (x_DDis/Wx_w), W/Dis

Given the residue composition and reboil ratio, the operating line of the stripping section can be constructed in the combination graph. The intersection point of the operating lines can be identified by letting the feed line intersect with the stripping line. The three remaining variables are W/Dis, distillate composition. By trial and error by a initial guess of reflux ratio R, the other parameters are calculated. The N_p calculated is compared with the given N_p.

XVI. *Given* N_p, *Reflux Ratio R,* W/Dis *calculate Reboil Ratio, B,* P_{split} (x_DDis/Wx_w), *Residue composition* x_w

Guess the distillate composition. Given the reflux ratio the operating line of the rectifying section can be constructed in the graph. The intersection between this line and that of the feed line is identified. Given W/Dis and the distillate composition the residue composition can be calculated from Eq. (4.26). The operating line for the stripping section is then obtained by connecting the residue point on the $y = x$ line and the intersection operating point. The number of ideal stages can be stepped off and then compared with the given N_p. The procedure is repeated till the two values come within numerical error of each other.

XVII. *Given* N_p, *Residue composition* x_w, *Reflux Ratio R, calculate* P_{split} (x_DDis/Wx_w), *Reboil Ratio B,* W/Dis.

Guess the reboil ratio. Given the residue composition and reboil ratio, the operating line of the stripping section is constructed. From the feed line intersection with the stripping line the intersection point is obtained. From the intersection point and the reflux ratio, the operating line of the rectifying section is constructed. The distillate composition is calculated from the distillate point on the $y = x$ diagonal. The procedure is repeated until the calculated N_p is within numerical error of the given N_p.

XVIII. *Given* N_p, *Reflux Ratio R,* P_{split} (x_DDis / Wx_w), *calculate Reboil Ratio B* W/Dis, *Residue composition* x_w

Guess the distillate composition. Given the reflux ratio, the operating line of the rectifying section can be constructed in the graph. The intersection between this line and that of the feed line is identified. Given product split and the distillate composition the residue composition can be calculated from Eq. (4.27).

The operating line for the stripping section is then obtained by connecting the residue point on the $y = x$ line and the intersection operating point. The number of ideal stages can be stepped off and then compared with the given N_p. The procedure is repeated till the two values come within numerical error of each other.

XIX. *Given* N_p, *W*/Dis, P_{split}(x_DDis/Wx_w) *calculate Reflux Ratio R calculate, Residue composition* x_w *Reboil Ratio, B*

Given *W*/Dis and product split, a relation between x_D and x_W can be established. This can be solved simultaneously for two unknowns with the Eq. (4.27). Guess the reboil ratio. Construct the operating line of the stripping section. This intersects with the feed line. The line connecting the intersection point and the distillate point is the operating line of the rectifying section. The reflux ratio is calculated. The N_p stepped from the graph is compared with the given value and the guessed reboil ratio is changed until they match.

XX. *Given* N_p, *W*/Dis, *Residue composition,* x_w *calculate Reboil Ratio B, Reflux Ratio R, Product Split* P_{split} (x_DDis/$x_w W$)

Given the residue composition and *W*/Dis from Eq. (4.26) the distillate composition can be calculated. The product split is then calculated. Guess the reboil ratio. Construct the operating line of the stripping section. This intersects with the feed line. The line connecting the intersection point and the distillate point is the operating line of the rectifying section. The reflux ratio is calculated. The N_p stepped from the graph is compared with the given value and the guess of the reboil ratio changed until they match.

4.2.6 Equilibrium Line

As can be seen from the discussions above and the *worked example 4.1*, the construct of the equilibrium line is an important aspect to the design procedure due to the method by McCabe and Thiele. Use of stages was made. Here a refresher is provided on the understanding of equilibrium between the vapour and liquid phases, VLE. This would provide the means to obtain the equilibrium line and later the real line limited by diffusion resistances.

Every pure liquid exerts an equilibrium pressure, the vapour pressure in the gas phase. The extent of the vapour pressure depends on the temperature. A typical P-T diagram for a pure substance is shown in Figure 4.7. *T* is the tripe point of the substance where all the three phases co-exist at one temperature and one pressure. The *phase rule* gives the degrees of freedom in terms of the number of components and phases and is,

$$F_f = N_c - \phi + 2 \tag{4.28}$$

The number of degrees of freedom F_f is given by subtracting the number of phases from the number of chemical species or components present in the system and adding 2. Thus for a 1-component system, the number of degrees of freedom for a two-phase system can be seen from Eq. (4.28) to be 1. Thus if the pressure is specified, the temperature can be calculated. When the number of phases is 3, the degrees of freedom become zero. This means that the temperature and pressure at which the three phases co-exist in equilibrium cannot be varied independently but are determinable values. This is also called the triple point of the pure substance and is shown as T in Figure 4.7. *Equilibrium* is the state of the system when after sufficient contact, the compositions of the system and other parameters of the system such as temperature and pressure do not change with time. It is at steady state. The curve *TB* in Figure 4.7 can be seen to be the *vapour pressure curve* of the system. This gives the temperature and pressure relationship of the 1 component, two-phase system in equilibrium. The points in the curve TB represent the *saturated vapour* and *saturated liquid*. The region sectored by CTB is pure liquid. The region sectored by CTA is solid. This is when the Pressure, P is high and temperature T is low. The region bordered with ATB is pure vapour and the region beyond the point B is gas. These are regions of high temperature, T and low pressure P. B is said to be the critical point and the corresponding temperature T_c the *critical temperature* and pressure P_c the *critical pressure*. Beyond point B, the liquid and vapour phases are indistinguishable from each other. State of the system beyond this point is the gas. Gas cannot be liquefied by pressure increase beyond this critical point. The temperature corresponding to 1 atm pressure in the curve TB is called the normal boiling point of the system. The path EF in Figure 4.7 is an isobaric process where the phase changes from liquid to vapour upon supplying heat. The *latent heat of vaporisation* at constant temperature is the amount of heat per mole supplied to effect the phase change.

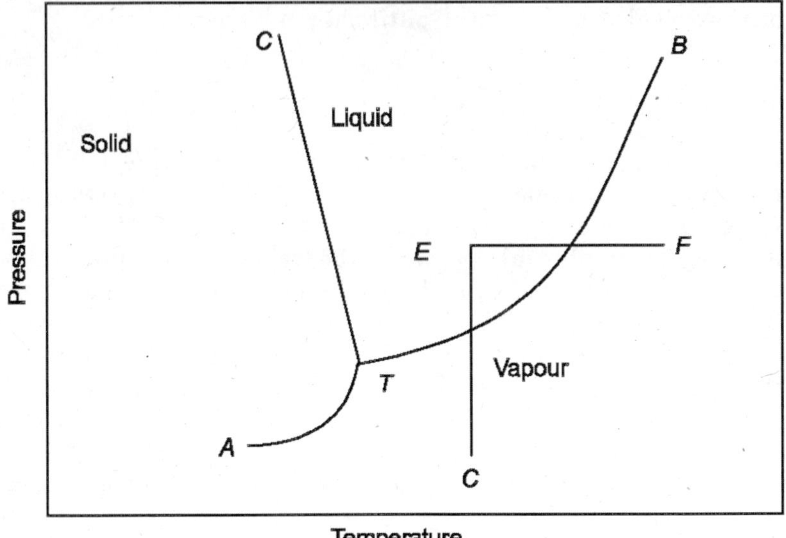

FIGURE 4.7 Phase diagram of a pure substance.

Vapour pressure data is scarce. The vapour pressure temperature curve TB can be established by interpolation or extrapolation of available data using the *Clausius Clapeyron equation*. The Clausius Clapeyron equation is,

$$\frac{dP}{dT} = \frac{\lambda_{\text{vap}}}{T(v_{\text{v}} - v_{\text{L}})} \tag{4.29}$$

where λ_{vap} is the latent heat of vaporisation. The liquid molar volume can be neglected and the vapour molar volume can be written assuming ideal gas law as:

$$\frac{dP}{dT} = \frac{P\lambda_{\text{vap}}}{RT^2} \tag{4.30}$$

Integrating both sides of the equation,

$$\ln(P) = -\frac{\lambda_{\text{vap}}R}{T} + c_1 \tag{4.31}$$

For a two-component system with two phases, the number of degrees of freedom can be derived from Eq. (4.28) to be 2. Thus if both the temperature and pressure of the system are known, then the compositions of the two-phases, are fixed and can be determined. This can be represented as a Txy *phase diagram* for a vapour-liquid mixture. This would be important in the equilibrium calculations during distillation as shown in the above sections. The VLE for a binary system of species A and B where B is the more volatile is shown in Figure 4.8. The temperature vs. y^* is given by the *dew line* and the temperature vs. x is given by the *bubble line*. The horizontal tie lines shown join the equilibrium mixtures of liquid and vapour phases. *tvu* is a *tie line*. The compositions become fixed as the pressure and temperature are known. There can be infinite tie line. The relative amounts of the two phases, liquid and vapour in this example can be calculated as:

$$\frac{\text{Moles (Liquid)}}{\text{Moles (Vapour)}} = \frac{vu}{vt} \tag{4.32}$$

The VLE data can be represented as a *distribution diagram* as shown in section B of Figure 4.8. The greater the distance between the equilibrium line and the $y = x$ diagonal, greater is the differences in composition between the vapour and liquid phases. A *separation factor* is or a *relative volatility* can be defined as follows in order to quantitatively monitor the nature of the equilibrium line:

$$\alpha = \frac{(y^*(1 - x))}{(x(1 - y^*))} \tag{4.33}$$

When α is 1, separation is not possible. Some examples of the distribution diagrams are given in Figures 4.8 and 4.9. Some of the systems shown exhibit the condition of *azeotropy* where the two phases behave like one phase and boil at a constant temperature.

Ideal solutions are first considered as a model for the liquid phase in the distillation operations. It is good for several systems but may not be for all the systems. Later, the special provisions can be made for dealing with liquid phase systems that

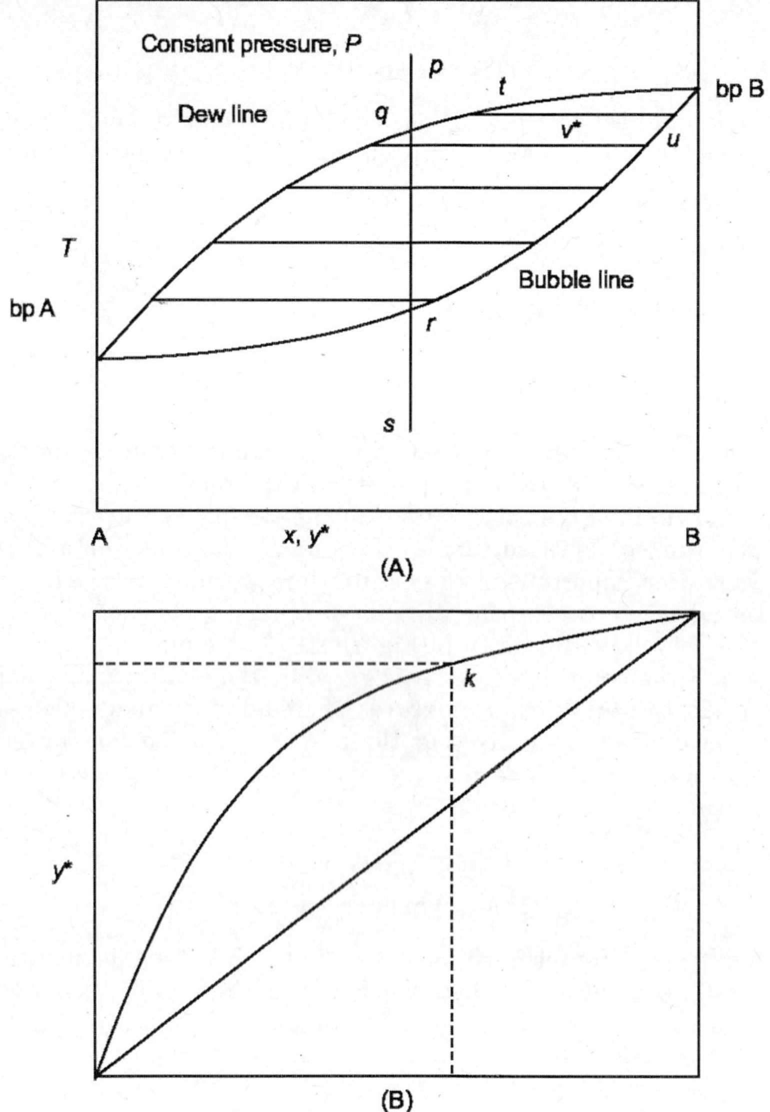

FIGURE 4.8 (A, B) VLE phase diagram of two component system point *k* represents the tie line, *tvu*. *B* is the more volatile species.

deviate from the ideal solutions. For ideal solutions, the vapour phase partial pressure can be calculated from the knowledge of the composition of the liquid phase alone *a priori* to any experimentation. This can be useful for systems where data is not readily available. Four characteristics that can be mentioned for ideal solutions are:

- The average intermolecular forces of attraction and repulsion in the solution are unchanged on mixing of the constituents.

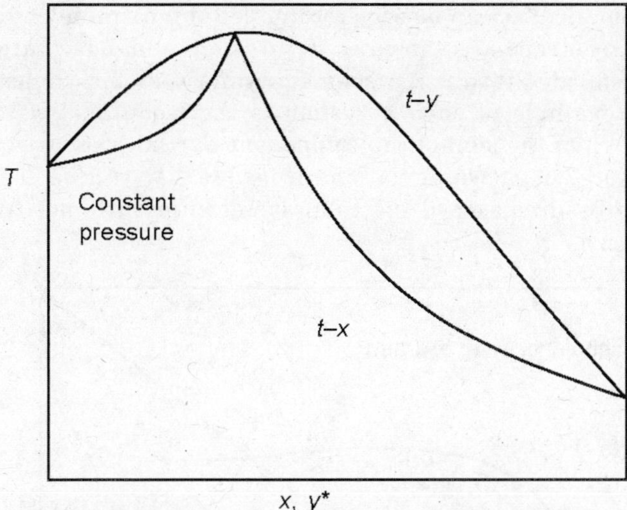

FIGURE 4.9 Maximum boiling azeotrope.

- Volume of solution varies linearly with composition.
- The mixing is adiabatic.
- The total vapour pressure of the system varies linearly with the composition expressed as mole fraction.

Real systems only approach the ideal solution in limits. Search can be made from among isomers, or compounds from a series with similar structure differing only with a methyl group. The Raoults law states that:

$$p_A = p_A^* x \tag{4.34}$$

The partial pressure of the vapour phase in equilibrium with the ideal solution is given by the product of the liquid phase composition and the pure component vapour pressure of the component at that temperature. For dilute non-ideal solutions, the Henry's law states that:

$$y^* = \frac{p_A}{P} = mx \tag{4.35}$$

where m is the Henry's law constant.

Other models for VLE in addition to the Raoult's law are available in the literature. These models are derived using the concept of fugacity and at equilibrium, the fugacities of the component in different phases are equal. Activity coefficients are defined for liquids. Van Laar, Margules, Wilson, NRTL and UNIFAC are some of the other models that can be used to interpolate and extrapolate VLE data. The resulting information will only be good for conditions where the data was measured and for those circumstances where the assumptions of the model are reasonably good. When the sum of the partial pressures of the components calculated exceed the total system pressure deviations from ideality can be seen. Examples of maximum boiling azeotrope and minimum boiling azeotrope are shown in Figures 4.9 and 4.11. Large positive or negative deviations from ideality can be seen for some systems. Some systems are partially miscible. An example of such a system is the isobutanol-water system as shown in Figure 4.10. When the sum of the component partial pressures are less than the total system pressure, negative deviations from ideality can be seen. Similarly, when the component pressures exceed the total system pressure, positive deviations from ideality can be seen.

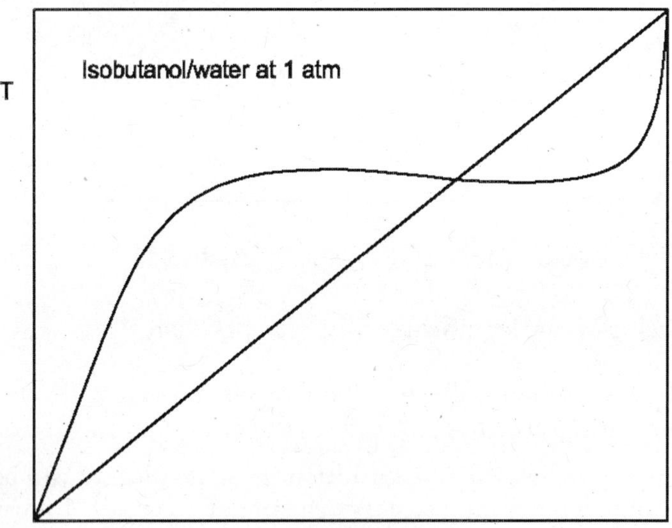

FIGURE 4.10 Distribution diagram of isobutanol-water partially miscible systems.

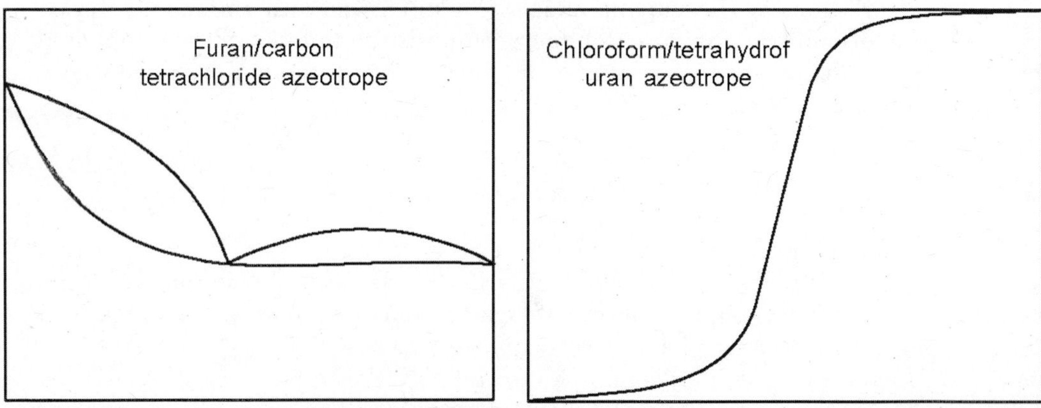

FIGURE 4.11 Minimum boiling azeotrope in a constant pressure Txy diagram and distribution diagram.

Worked Example 4.2 *Separation of n-Pentane from a Pentane/Heptane System.* The VLE data for a binary system of n-pentane/n-heptane is given below:

x	y
0	0
0.059	0.271
0.145	0.521
0.254	0.701
0.398	0.836
0.594	0.925
0.867	0.984
1	1

This data was input into the MS Excel spreadsheet in a Pentium IV, 3.0 GHz personal computer. How many (a) ideal stages are required to distill 97.33% of n-pentane as an overhead product and collect the bottoms at a concentration of 16% n-pentane when the reflux is operated at 1.5 times the minimum. The feed is 52.67% liquid with 29% pentane. (b) A second feed is injected at 100% saturated vapour with 42% pentane.

The equilibrium data was plotted using the curve fitting routine in the MS excel spread sheet. The x intercept of the q line was calculated as:

$$x_{\text{intercept}} = \frac{z_{\text{F}}}{(q)} \tag{4.36}$$

From the slope of the feed line and the x-intercept using MS paint, the feed line was constructed in Figure 4.12. The distillate point and the intersection of the feed line and the equilibrium line, the pinch point, was connected and extrapolated to the

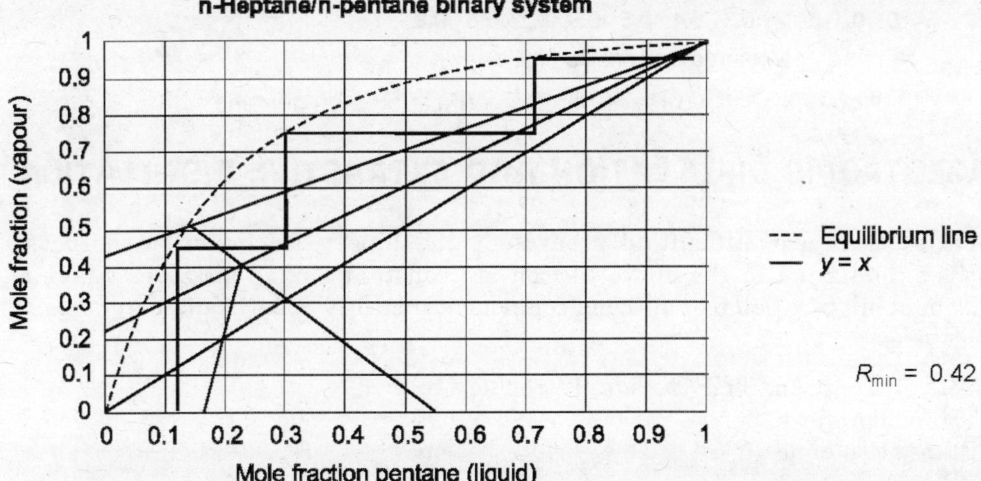

FIGURE 4.12 Refining n-pentane from a binary pentane/heptane mixture.

y axis. The y intercept is $x_D/(R_{min} + 1)$. From the y-intercept the minimum reflux ratio was calculated. The operating reflux ratio was calculated and the rectifying operating line constructed. From the intersection point the residue point was connected and the resulting line was the stripping operating line. The number of ideal stages stepped off was 3.

For the case of two feeds, the operating line from the rectifying section to the first feed is the same. For the case of the second feed, the pinch point was located and the new reflux ratio was calculated. The second intersection point was located. Thus there are three operating line segments. One rectifying, one stripping and the third intermediate with a different slope. The number of ideal stages required was 4. But the residue composition was lower at the exit of the system. The feeds can be treated independently and the stages stepped off as shown in Figure 4.13. Table 4.1 lists the number of trays that are used in the various chemical plants.

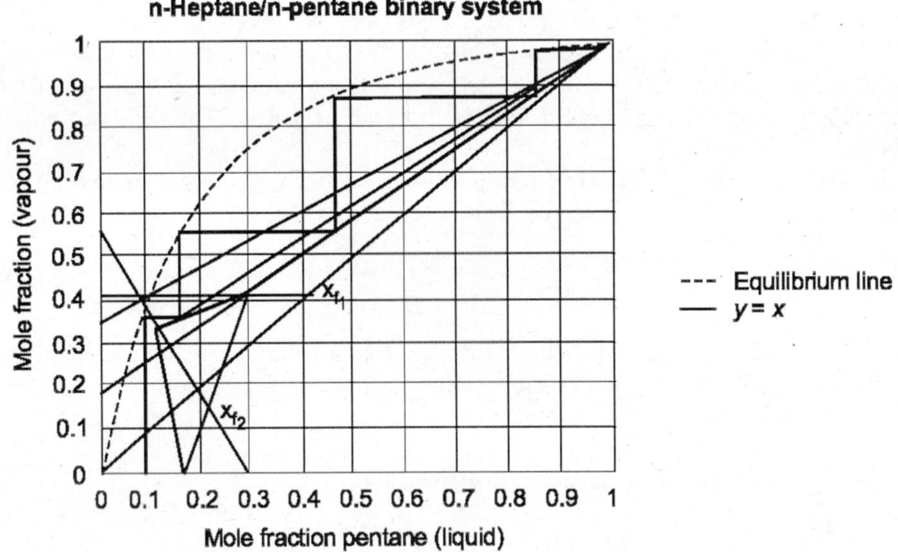

FIGURE 4.13 Two feeds during rectification of n-pentane.

4.3 AZEOTROPIC DISTILLATION AND EXTRACTIVE DISTILLATION

Azeotropic systems are difficult to separate using binary fractionation methods. Azeotropes as indicated in the above section are constant boiling mixtures where a two-component mixture behaves as a one-component mixture. Some examples of binary azeotropes are:

(1) Tetrahydrofuran, THF/Carbon Tetrachloride
(2) THF/Chloroform
(3) Ethanol/Toluene
(4) CS_2/acetone
(5) Acetone/Chloroform

(6) Acetic Acid/Water
(7) MMA/Methanol
(8) Ethanol/Water
(9) Acetone/Methanol
(10) Acrylic Acid/Water
(11) Ethyl Acetate/Ethanol
(12) Hydrazine/Water
(13) Methanol/Ethylacetate
(14) n-Heptane/Ethanol
(15) Methanol/Benzene
(16) Ethylene Glycol/Water
(17) Chloroform/Methanol
(18) Ethanol/Benzene

In order to separate such systems, a third component called the *entrainer* is added to the system. The entrainer is chosen such that the volatility is in a fashion that a new azeotrope is formed which is lower boiling and can be easily separated. For example, benzene is used as an entrainer in the ethanol/water azeotropic system. The normal boiling points of ethanol is 78.3°C, benzene 80.1°C and that of water is 100°C. Benzene is added as a entrainer to the ethanol/water azeotropic system. At the right pressure, the lower boiling binary azeotrope of ethanol/benzene is formed and boiled off. This system ethanol/water/benzene is reported to form a ternary azeotrope. One method of separating the mixture is shown in Figure 4.14. The water

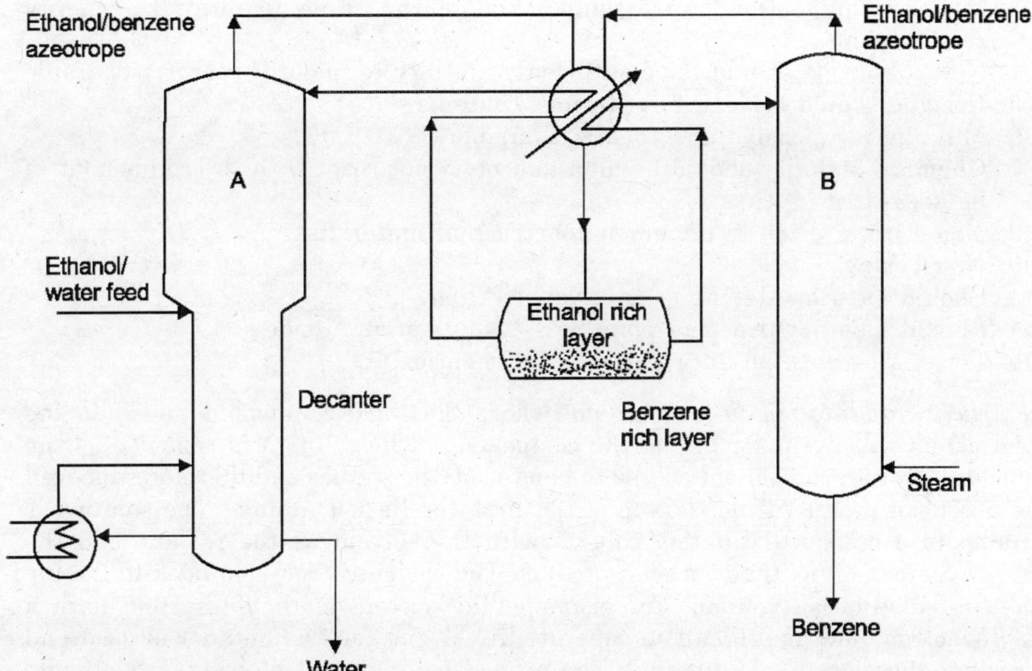

FIGURE 4.14 Azeotropic distillation of ethanol/Water using benzene as entrainer.

is collected as residue. The reboiler needs to be designed accordingly. The vapour from the first column is condensed and fed into a second column. In the second column, the benzene is collected as a residue. The benzene/ethanol azeotrope is boiled off from the top of the column 2. Another method of separating the mixture is by tapping into the formation of ternary azeotrope. This boils at 64.9°C. As this is lower than the normal boiling point of ethanol, ethanol can be collected as a residue.

Some other examples of entrainers that can be used are as follows:

1. Butyl Acetate—Acetic Acid/Water
2. Hexane—MMA/methanol
3. Toluene—Hydrazine/water

In addition to toluene, aniline, benzene, pyridine, xylene, propylalchohols, cresol, glycol, hexyl amine and glycol ethers can be used as entrainer to separate the hydrazine/ water azeotropic system. The normal boiling point of the acetic acid is 118.1°C and that of water is 100°C. In the first column, glacial acetic acid is collected as the residue and the butyl acetate and water heteroazeotrope is boiled off at 90.2°C. In the second column, water is collected as residue and the ester-water azeotrope is collected as an overhead product. A decanter can be used in between to let the ester rich and water rich layer separate.

Some of the criteria for the choice of the selection of entrainer:

1. Choice of entrainer is the most important consideration
2. Added compound should form a low boiling azeotrope with only one of the components of the binary mixture
3. It has to have the lowest composition of the three to minimize energy requirements
4. New azeotrope should have sufficient volatility to make it readily separable
5. Residue should be lean in entrainer contents
6. Must be inexpensive and readily available
7. Chemical stability should be high and must not react with the component to be separated
8. Non-Corrosive toward common construction materials
9. Non-Toxic
10. Should have low latent heat of vaporisation
11. Should have low freezing point and easily stored outside
12. Low viscosity to produce high tray efficiencies

Extractive distillation is a technique where the third component added to the system acts as a solvent for one of the components. The solution is removed as the residue as the other component will be the more volatile species and therefore removed as an overhead product from the top of the first distillation column. The solution is separated in a second distillation column with the solvent as the residue and the dissolved component as the overhead product. The solvent is recycled back to the top of the first distillation column. For example, the acetone/methanol system form a binary azeotrope and is difficult to separate. A solvent such as butanol is used and the acetone dissolves in the butanol. The normal boiling point of acetone, methanol and butanol are 56.4°C, 64.7°C and 117.8°C respectively. As shown in Figure 4.15,

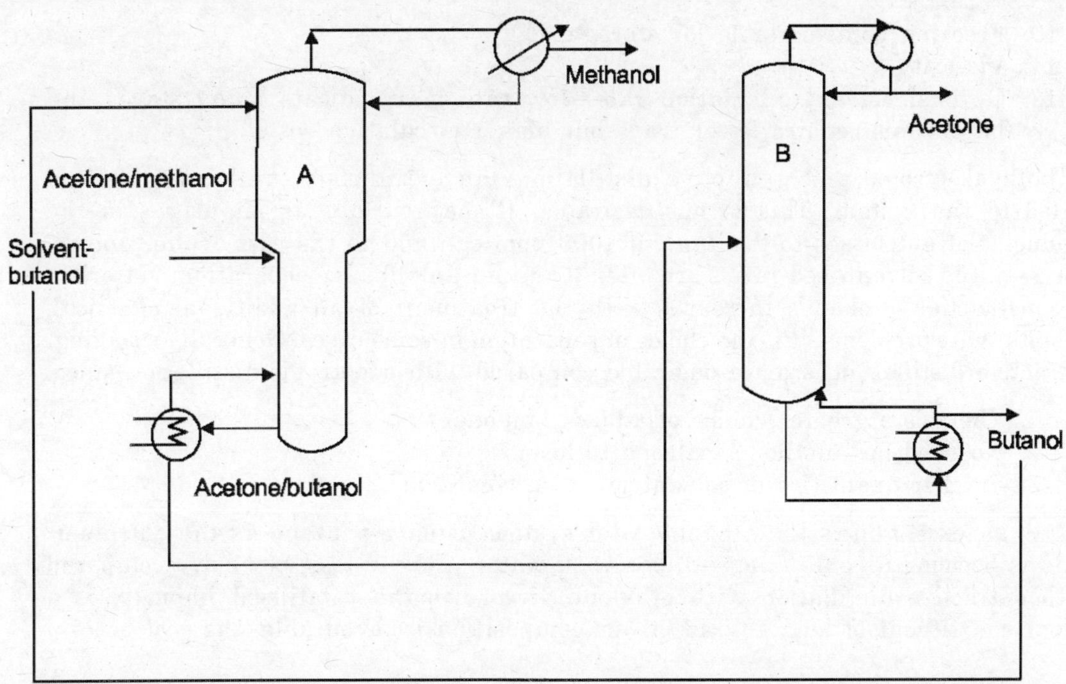

FIGURE 4.15 Extractive distillation of acetone/methanol with butanol as solvent.

as the acetone dissolves in butanol, the VLE changes such that the azeotrope is no longer present. Methanol although less volatile than acetone is removed as an overhead product from column A and the butanol is collected as the residue from column B. The residue from column A, the solution of acetone in butanol is fed into column B and the acetone is collected as an overhead distillation product from column B. Thus butanol serves as an extractive solvent. Another example of an extractive solvent is acetone or furfural to aid in the separation of butene-2/n-butane mixture. Phenol is used as a solvent in the separation of toluene/isooctane mixture. The second column is a auxiliary column. The solvent changes the relative volatility of the binary system. The solvent itself a low boiler does not enter into the rectifying section much.

The requirements of an extractive distillation solvent are as follows:

1. High selectivity—ability to change VLE of original mixture such that easy separation is permitted
2. High capacity to dissolve the components of the system
3. Low volatility—must not enter the overhead condensers
4. Separability—solvent must be readily separated from the mixture to which it is added
5. Solvent must not form azeotrope
6. Low cost
7. Less toxic
8. Non-corrosive
9. Good chemical stability

10. Freezing point suitable for storage
11. Viscosity
12. Optimal solvent circulation rate—low rate may indicate many stages and large rates require fewer trays but higher circulation costs.

In both azeotropic and extractive distillations an extraneous third component is added to the system. This is not desirable. It may surface as impurities in the product. Solvent losses to the tune of 1000 ppm can add to the cost of the process. The required solvent/feed ratios are often 3 or 4 for an effective separation. Materials of construction problems increases with the treatment of an additional chemical. Despite these problems, it is the choice of separation in some cases. Generally speaking, extractive distillation is more desirable compared with azeotropic distillation since:

1. There is a greater choice of added component
2. No need for another azeotrope to form
3. Smaller quantities of solvent must be volatilised

As an exception is the ethanol/water system using n-pentane as the entrainer. This is because n-pentane azeotropes with water. This is less expensive compared with extractive distillation with ethylene glycol. As the volatilised impurity is a minor constituent of feed and azeotrope composition is favourable, the cost is less.

4.4 FLASH VAPORISATION

Flash vaporisation is a single stage operation. Energy is supplied to the liquid mixture that needs to be separated. Sufficient time is allowed for equilibrium to be attained between the vapour and the liquid. The vapour is removed as an overhead product and the remaining liquid is removed as a residue from the bottom. The energy input to the process is provided by a tubular heat exchanger. Vapour formation is facilitated by a reduction in vacuum pressure. A centrifugal separator splits the mixture into vapour and liquid (Figure 4.16). The overhead product may be passed through an overhead condenser.

For a binary system, the distillate composition and the residue composition can be calculated from a overall and component mass balance.

$$F = \text{Dis} + W \tag{4.37}$$

$$x_F F = x_D \text{Dis} + x_w W \tag{4.38}$$

An energy balance gives,

$$FH_F + Q = \text{Dis}\, H_D + W\, H_w \tag{4.39}$$

Combining Eqs. (4.37 and 4.38),

$$\text{Dis}(x_D - x_F) = W(x_F - x_w) \tag{4.40}$$

or

$$-\frac{W}{\text{Dis}} = \frac{x_D - x_F}{x_W - x_F} \tag{4.41}$$

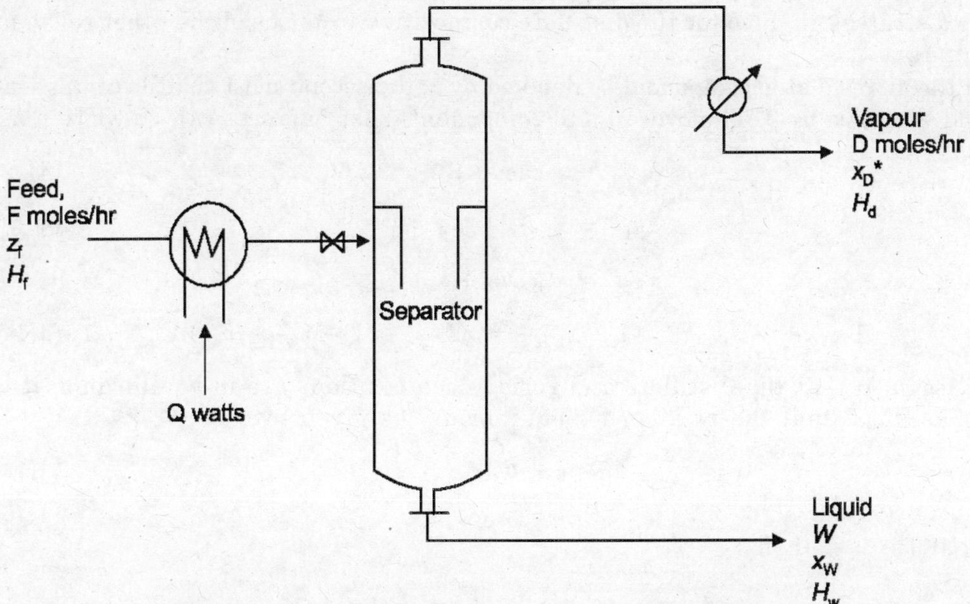

FIGURE 4.16 Continuous equilibrium flash vaporisation.

The operating line for the flash vaporisation is shown in Figure 4.17. The distillate composition can be as high as the equilibrium concentration. Often times the operating time is far from equilibrium.

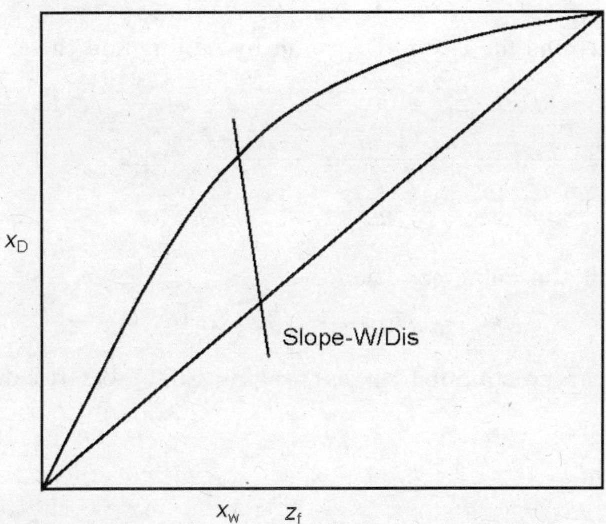

FIGURE 4.17 Operating line during flash vaporisation.

Worked Example 4.3 *Equilibrium Flash Vaporisation of a Ternary System.* Consider a three-component system that is subjected to flash vaporisation. Assuming that the distillate compositions are in equilibrium with the residue compositions, develop a strategy to solve for the distillate composition in terms of the other relevant parameters.

Let the more volatile component be denoted by 1, the second most volatile component by 2 and the third by 3. The overall and component mass balances can be written as:

$$F = \text{Dis} + W \tag{4.42}$$

$$x_{F1}F = x_{D1}\text{Dis} + x_{w1}W \tag{4.43}$$

$$x_{F2}F = x_{D2}\text{Dis} + x_{w2}W$$

$$(1 - x_{F1} - x_{F2})F = (1 - x_{D1} - x_{D2})\text{Dis} + (1 - x_{w1} - x_{w2})W \tag{4.44}$$

As assumed that the distillate and residue compositions are in equilibrium, it is further assumed that the relation for equilibrium be given by:

$$x_{D1} = m_1 x_{w1} \tag{4.45}$$

$$x_{D2} = m_2 x_{w2} \tag{4.46}$$

Let $W/\text{Dis} = f$, then

$$x_{F1}(1 + f) = x_{D1}\left(1 + \frac{f}{m_1}\right) \tag{4.47}$$

$$x_{D1} = \frac{x_{F1}(1+f)}{\left(1 + \dfrac{f}{m_1}\right)} \tag{4.48}$$

Writing similar equations for x_{D2} and x_{D3} and by adding the three compositions, the total needs to be 1.

$$\frac{x_{F1}(1+f)}{\left(1+\dfrac{f}{m_1}\right)} + \frac{x_{F2}(1+f)}{\left(1+\dfrac{f}{m_2}\right)} + \frac{x_{F3}(1+f)}{\left(1+\dfrac{f}{m_3}\right)} = 1.0 \tag{4.49}$$

f can be solved for, from the cubic equation:

$$a_0 f^3 + a_1 f^2 + a_2 f + a_3 = 0 \tag{4.50}$$

where a_0, a_1, a_2 and a_3 can be obtained by rearranging Eq. (4.49). It can be seen that

$$a_0 = 1 - m_1 x_{F1} - m_2 x_{F2} - m_3 x_{F3}$$

$$a_1 = m_1(1 - x_{F1}) + m_2(1 - x_{F2}) + m_3(1 - x_{F3}) - x_{F1}(m_1 m_2 + m_2 m_3) - x_{F2}(m_1 m_2 + m_2 m_3) - x_{F3}(m_3 m_1 + m_3 m_2)$$

$$a_2 = m_2 m_1(1 - x_{F1}) + m_3 m_2\,(1 - x_{F2}) + m_1 m_3(1 - x_{F3}) - x_{F1} m_1 m_2) - x_{F2} m_1 m_2 - x_{F3} m_3 m_2$$

since $a_3 = 0$ the cubic reduces to a quadratic equation in f when $f \neq 0$.

Thus for equilibrium ternary flash vaporisation, when the feed is specified and the distribution coefficients known as the W/Dis, residue to distillate ratio can be calculated. From W/Dis = f, the distillate composition can be calculated from Eq. (4.48) and the same for the other two components can be obtained similarly. From the distribution coefficients and the distillate compositions, the residue compositions can be calculated. The methods of matrix inversion, numerical solution of three simultaneous equations and three unknowns and solution of a quartic polynomial may also be used to solve for the distillate compositions.

4.5 METHOD OF RAYLEIGH FOR DIFFERENTIAL DISTILLATION

Batch distillation of binary mixtures are performed in stills that are large scale versions of a laboratory distillation flask and a condenser. The distillate is collected as different batches or cuts. The vapour as it is formed is removed rapidly from the flask into a condenser. This way the vapour is always in equilibrium with the liquid assuming rapid establishment of equilibrium. During the course of the distillation, there are W moles of liquid of the still of composition of x mole fraction. When d Dis moles of distillate are vapourised with the mole fraction of the more volatile that is in equilibrium with the liquid at a composition of y^*. Thus a component and overall mass balance would yield:

$$-y^* d\text{Dis} = W dx + x dW \tag{4.51}$$

$$-d\text{Dis} = dW \tag{4.52}$$

or

$$dW\,(y^* - x) = W dx \tag{4.53}$$

or

$$\frac{dx}{(y^* - x)} = \frac{dW}{W} \tag{4.54}$$

Integrating both sides of the equation;

$$\ln\left(\frac{W}{F}\right) = \int_{x_F}^{x_w} \frac{dx}{(y^* - x)} \tag{4.55}$$

Eq. (4.55) is called as *Rayleigh equation* named after Lord Rayleigh. Lord Rayleigh was a recipient of the Nobel Prize in physics and was one of the very few members of higher nobility who won name as an outstanding scientist. He derived the equation for batch distillation. He had a fine sense of literary style; every paper he wrote, even on the most abstruse subject, is a model of clearness and simplicity of diction. The 446 papers reprinted in his collected works clearly show his capacity for understanding everything just a little more deeply than anyone else. The composited distillate composition $\langle x_D \rangle$ can be given by:

$$F x_F = \text{Dis}\,\langle x_D \rangle + W x_w \tag{4.56}$$

For differential condensation,

$$\ln\left(\frac{F}{\text{Dis}}\right) = \int_{x_F}^{x_D} \frac{dy}{(y - x^*)} \tag{4.57}$$

Worked Example 4.4 *Generalized Rayleigh Equation*

In the derivation of the Rayleigh equation for differential distillation, Dis dy^* was neglected. As the vapours are removed rapidly the change in the equilibrium concentration in the overhead vapour space in the batch still was assumed to be constant. With variation in x, y^* also varies. During the operation, x changes in composition and moves from the feed composition more toward the residue composition. As the x changes y^* that is in equilibrium with x also changes. Assuming establishment of rapid equilibrium at the interface how will you account for the variation in y^*?

The exact component balance equation can be written as:

$$-d(Wx) = d(y^*\text{Dis}) \tag{4.58}$$

$$Wdx + xdW = -y^*d\,\text{Dis} - \text{Dis}\,dy^* \tag{4.59}$$

Plugging Eq. (4.52) into Eq. (4.59),

$$Wdx + xdW = y^*dW - \text{Dis}\,dy^* \tag{4.60}$$

$$\frac{\partial x}{(y^* - x)} + \partial y^* \frac{\text{Dis/W}}{(y^* - x)} = \frac{\partial W}{W} \tag{4.61}$$

Dis/W can be seen to be from Eq. (4.56),

$$\frac{\text{Dis}}{W} = \frac{(x_F - x_w)}{(<x_D> - x_F)} \tag{4.62}$$

The average distillate composition can also be calculated as:

$$<x_D> = \frac{1}{(x_F - x_w)} \int_{x_w}^{x_F} y^* dx \tag{4.63}$$

Eq. (4.61) can now be written after plugging Eq. (4.63) and (4.62) into it as:

$$\frac{dx}{(y^* - x)} + \frac{dy^*(x_F - x_w)}{(<x_D> - x_F)(y^* - x)} = \frac{dW}{W} \tag{4.64}$$

Integrating both sides of the equation,

$$\int_{x_w}^{x_F} \frac{dx}{(y^* - x)} + \frac{(x_F - x_w)}{<x_D> - x_F} \int_{y_w^*}^{y_F^*} \frac{dy^*}{(y^* - x)} = \ln\left(\frac{F}{W}\right) \tag{4.65}$$

4.6 STEAM DISTILLATION

Often times one of the components in the liquid mixture that is separated by multi-stage binary distillation column is water. When water is the residue product, the reboiler is dispensed away and steam is injected at the bottom plate of the column (Figure 4.18).

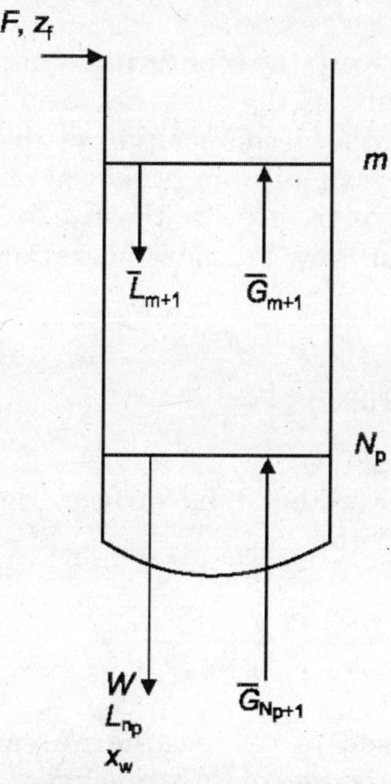

FIGURE 4.18 Steam distillation.

Overall and component balances can be written for the scheme in Figure 4.18 as:

$$x_m \bar{L}_m = x_W W + y_{m+1} \bar{G}_{m+1} \tag{4.66}$$

or

$$\frac{x_m \bar{L}_m}{\bar{G}_{m+1}} - \frac{x_W W}{\bar{G}_{m+1}} = y_{m+1}$$

$$\bar{G}_{m+1} + W = \bar{L}_m + G_{N_{p+1}} \tag{4.67}$$

$$\frac{\bar{G}_{m+1}}{W} + 1 = \frac{\bar{L}_m}{W} + \frac{G_{N_{p+1}}}{W}$$

or

$$y_{m+1} = \frac{x_m W}{G_{N_{p+1}}} - \frac{x_W W}{G_{N_{p+1}}} \tag{4.68}$$

where G_{n_p+1} is the steam flow rate. Equimolal flow rate is assumed for the liquid underflow and at the bottom plate removed as the residue. The operating line of the stripping section is given by Eq. (4.68) and the number of stages can be stepped off from the combined graph.

4.7 MOLECULAR DISTILLATION

During distillation of some products, the operating temperatures and the time-temperature history of the exposure of the material cannot be high. Higher time-temperature exposure will lead to the decomposition of the material. Examples are the separation of vitamins from fish oils, organic substances, natural products, plasticizers, certain type of copolymer, etc. In these cases the distillation can be performed at low pressures. For very low pressure operation, the Langmuir equation can be used,

$$N = 1006\, p_A \left(\frac{1}{2\pi M_A RT} \right)^{0.5} \tag{4.69}$$

where N is the rate of evaporation from the liquid surface. The ratio of the compositions in the distillate product can be written as:

$$\frac{N_A}{N_B} = \frac{p_A / M_A^{0.5}}{p_B / M_B^{0.5}} \tag{4.70}$$

The wiped film devolatiliser is used to remove the unreacted monomers from the polymer in continuous mass polymerisation of SAN copolymer. The pressure of operation is a few torr. The film spreads thin by a centrifugal operation at a certain RPM. The polymer strand is extruded and the monomers are devolatilised.

4.8 KREMSER EQUATION FOR NUMBER OF IDEAL STAGES WHEN EQUILIBRIUM LINE IS LINEAR

Sometimes the equilibrium line is linear. In such cases the number of ideal stages required to perform a desired degree of separation can be calculated from an analytical expression. This expression can be derived as follows. A component mass balance between the overhead product and stage n can be written as:

$$Lx_0 + Gy_{n+1} = Lx_n + Gy_1 \tag{4.71}$$

or

$$y_{n+1} = \left(\frac{L}{G}\right)x_n - \left(\left(\frac{L}{G}\right)x_0 - y_1\right) \tag{4.72}$$

The equilibrium from Henry's law for dilute solutions can be written as:

$$y_n^* = mx_n + b \quad (4.73)$$

or

$$x_n = \frac{(y_n^* - b)}{m} \quad (4.74)$$

$$y_{n+1} = \frac{L/G(y_n^* - b)}{m} + \frac{(y_1 - (y_0^* - b)L)}{mG} \quad (4.75)$$

Let the absorption factor be denoted by $A_f = L/mG$

Then,

$$y_{n+1} = y_1 - A_f y_0 + A_f y_n^* \quad (4.76)$$

y_0 is the hypothetical vapour concentration in equilibrium with x_0. The $A_f b$ cancels out. In a similar fashion the Eq. (4.76) can be written for each stage,

$$y_2 = (1 + A_f)y_1 - Ay_0 \quad (4.77)$$

$$y_3 = y_1 - A_f y_0 + A_f y_2 = y_1 - A_f y_0 + A_f(1 + A_f)y_1 - A_f^2 y_0$$

$$= (1 + A_f + A_f^2)y_1 - (A_f + A_f^2)y_0 \quad (4.78)$$

Thus for the staircase for all stages,

$$y_{N+1} = (1 + A_f + A_f^2 + \ldots A_f^N)y_1 - y_0(A_f + A_f^2 + \ldots + A_f^N) \quad (4.79)$$

Using the sum of geometric series,

$$y_{N+1} = \left(\frac{1 - A_f^{N+1}}{(1 - A_f)}\right)y_1 - y_0\left(A_f \frac{1 - A_f^N}{1 - A_f}\right) \quad (4.80)$$

where, $A_f = L/mG = \dfrac{y_{N+1} - y_1}{y_N - y_0}$ (4.81)

Combining Eqs. (4.81, 4.80) and solving for N,

$$N = \frac{\ln [(y_{N+1} - y_N)/(y_1 - y_0)]}{\ln [(y_{N+1} - y_1)/(y_N - y_0)]} \quad (4.82)$$

$$= \frac{\ln [(y_{N+1} - y_N)/(y_1 - y_0)]}{\ln [A_f]} \quad (4.83)$$

Eqs. (4.80, 4.82, 4.83) are called Kremser equations. The number of ideal stages needed to separate a mixture can be calculated from the above expressions. This is applicable when the equilibrium line is linear. This expression is similar to the number of transfer units NTU needed to achieve a certain degree of separation. Eq. (4.83) is valid when $A_f \neq 1$.

Worked Example 4.5 *Absorption Factor* $A_f = 1$

The Kremser equation shown in Eq. (4.83) is valid when $A_f \neq 1$. What happens when the absorption factor is 1?

When the absorption factor = 1,

$$A_f = 1 = \frac{L}{mG} \quad \text{or} \quad \frac{L}{G} = m \tag{4.84}$$

The equilibrium line given by Eq. (4.73) is now written as:

$$y_n^* = \left(\frac{L}{G}\right) x_n + b \tag{4.85}$$

The operating line given by Eq. (4.72) can be written after including Eq. (4.85) as:

$$y_{n+1} = \left(\frac{L}{G}\right) x_n - \left(\frac{L}{G}\right) x_0 + y_1 = y_n^* - b - \left(\frac{L}{G}\right) x_0 + y_1 \tag{4.86}$$

$$y_n^* - b - mx_0 + y_1 = y_n^* - y_0^* + y_1 \tag{4.87}$$

Assuming that equilibrium has been reached at the lower plate,

Thus
$$y_2 = 2y_1 - y_0;\ y_3 = 3y_1 - 2y_0 \tag{4.88}$$

$$y_{N+1} = N(y_1 - y_0) + y_1 \tag{4.89}$$

or
$$N = \left(\frac{y_{N+1} - y_1}{y_1 - y_0}\right) \tag{4.90}$$

Thus when the absorption factor is 1, the number of plates can be calculated using Eq. (4.90).

4.9 FENSKE EQUATION FOR SYSTEMS WITH CONSTANT RELATIVE VOLATILITY

For binary systems that exhibit constant relative volatility, the minimum number of ideal stages required to achieve a certain degree of separation can be estimated using an analytical expression. This can be derived as follows (Fenske, 1931),

At total reflux, the operating line superposes on the $y = x$ diagonal.

$$\frac{y_w}{1 - y_w} = \alpha\left(\frac{x_W}{1 - x_W}\right) = \frac{x_{Nm}}{1 - x_{Nm}} \tag{4.91}$$

For the bottommost tray of the column,

$$\frac{y_{Nm}}{1 - y_{Nm}} = \alpha\left(\frac{x_{Nm}}{1 - x_{Nm}}\right) = \alpha^2\left(\frac{x_W}{1 - x_W}\right) \tag{4.92}$$

This can be continued until the tray at the top of the column is reached,

At the top of the column,

$$\frac{y_1}{1 - y_1} = \alpha^{Nm+1}\left(\frac{x_W}{1 - x_W}\right) = \frac{x_D}{1 - x_D} \tag{4.93}$$

Obtaining the natural logarithm,

$$N_m + 1 = \frac{\ln((1 - x_W)x_D)}{(x_W(1 - x_D))\ln \alpha} \tag{4.94}$$

Eq. (4.87) is named after its inventor, Fenske (1931). It gives the minimum number of ideal stages required to produce a distillate with composition, x_D and residue composition, x_W using a reboiler and a total condenser. This equation may also be used for systems where the relative volatility changes with the plate. In such cases a geometric average of the relative volatilities are used in place of a in Eq. (4.87).

***Worked Example** 4.6 High Purity Products*

When the desired purity level in the product composition is high, say for example, 99.99% of benzene is required from a benzene/toluene mixture. The standard combined graph procedure discussed in the above sections may not be sufficient. In such cases log paper can be utilised. The stages can be stepped off a semi-log graph. Here the higher purity levels are more readily visible. Else the Kremser equations can be used especially when the equilibrium relationship can be taken as linear.

A 99% benzene is produced from a feed containing 95% benzene and 5% toluene. A multi-staged distillation column is proposed to be used for the separation with a total condenser. The residue leaving the column has 10% toluene and 90% benzene. If the reflux ratio used is 3, what is the number of ideal stages required for the separation? The VLE data for benzene/toluene system is given in problem 14.

The VLE data given in problem 14 is used to construct the equilibrium line and the $y = x$ diagonal in Figure 4.19 using an MS excel spreadsheet in a Pentium 4, 3.0 GHz personal computer. MS paint was used to construct the operating lines for rectification and stripping and the stages were stepped off the graph. Four ideal stages are required. The purity level can go as high as 99.99% by this method.

***Worked Example** 4.7 99.99% Benzene Rectification*

For the benzene/toluene mixture whose VLE data is given in problem 14, 99.99% benzene is desired as an overhead product using a multi-stage distillation column operated at a reflux ratio of 3.0. A total condenser and reboiler are used. The feed

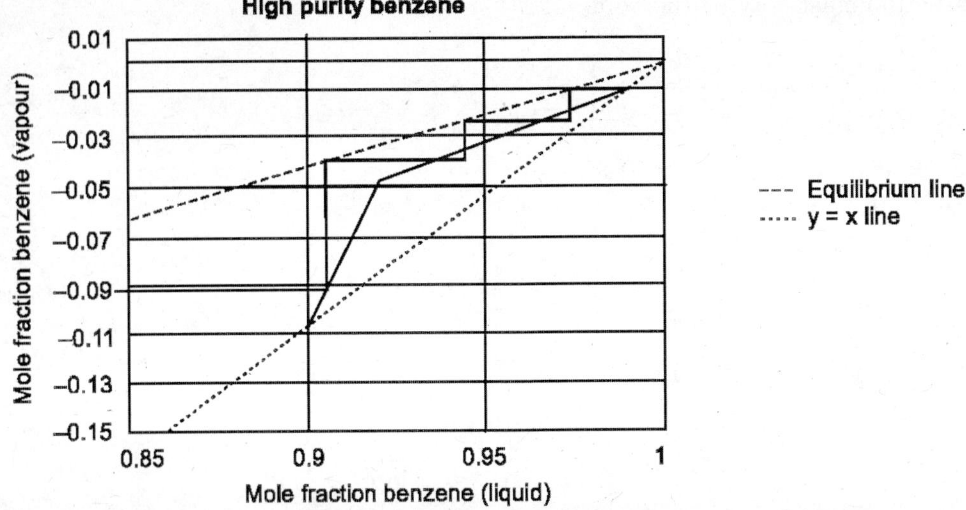

FIGURE 4.19 High purity benzene (99.0%) using log graph paper.

contains 99% benzene and the residue leaves at 90% benzene. In this region, Henry's law for dilute solutions may be applicable for the equilibrium line. Thus,

$$y_n^* = 0.58 + 0.42\, x_n \tag{4.95}$$

This is so for the more volatile benzene. Calculate the number of ideal stages required for the indicated degree of separation using the Kremser equations.

The overhead product composition, $y_1 = 0.9999$ (4.96)
The bottoms composition $x_n = 0.9$
For total condenser, $x_0 = y_1 = 0.9999$

$$Y_0 = mx_0 + b = 0.58 + 0.42(0.9999) = 0.99996 \tag{4.97}$$

Reflux Ratio, $R = 3$
The enriching operating line can be given by,

$$y_{n+1} = 0.75\, x_n + 0.9999/4 \tag{4.98}$$

From Kremser equation given by Eq. (4.82) in terms of the mole fractions,

$$N = \frac{\ln((0.925 - 0.958)/(0.9999 - .99996))}{\ln((0.925 - 0.9999)/(0.958 - .99996))}$$

$$= \frac{6.30991}{.57944} = 10.89 = 11 \tag{4.99}$$

4.10 METHOD OF PONCHON AND SAVARIT FOR MULTISTAGE BINARY DISTILLATION

This method uses enthalpy information and comes close in prediction of the number of ideal stages required to the McCabe and Thiele method. Here the energy balances are used in addition to the mass balances. Some trial and error is associated with the procedure and is more tedious and makes it less attractive compared with the McCabe and Thiele method.

In this method an internal reflux ratio is defined in addition to the external reflux ratio discussed in section 4.2. This way the equimolal overflow and underflow need not be the only criterion to effect the separation. For purposes of analysis Figure 4.3 may be used.

$$\text{Internal Reflux Ratio} = \frac{\bar{L}_n}{\bar{G}_{n+1}} = \frac{L_m}{G_{m+1}} \tag{4.100}$$

$$\text{External Reflux Ratio} = \frac{L_0}{\text{Dis}} \tag{4.101}$$

In addition to the mass balances,

$$G_{n+1} = L_n + \text{Dis} \quad \text{or} \quad \text{Dis} = G_{n+1} - L_n \tag{4.102}$$

$$y_{n+1}G_{n+1} = x_nL_n + z_D\,\text{Dis} \quad \text{or} \quad y_{n+1}G_{n+1} - L_nx_n = \text{Dis}\,z_D \tag{4.103}$$

where z_D can either be y_D when a partial condenser is used and x_D when a total condenser is used. The RHS of Eq. (4.103) is the net flow of the more volatile material upward and is permanently withdrawn at top. An Energy or Enthalpy balance is also written as:

$$Q_c + L_nH_{Ln} + H_D\text{Dis} = H_{Gn+1}G_{n+1} \tag{4.104}$$

Let the heat removed in the condenser and the permanently removed distillate, per mole of distillate be denoted by Q'.

$$Q' = \frac{(Q_c + DH_D)}{D} = \frac{Q_c}{\text{Dis}} + H_D \tag{4.105}$$

Combining Eqs. (4.105 and 4.104)

$$\text{Dis}Q' + L_nH_{Ln} = H_{Gn+1}G_{n+1} \quad \text{or} \quad \text{Dis}Q' = H_{Gn+1}G_{n+1} - L_nH_{Ln} \tag{4.106}$$

The RHS of Eq. (4.106) is the heat permanently taken out at the top from the distillate at the condenser. Eliminating Dis between Eqs. (4.102, 4.103) and re-arranging an expression for the internal reflux ratio in terms of the mole fractions can be obtained as:

$$\frac{L_n}{G_{n+1}} = \frac{z_D - y_{n+1}}{z_D - x_n} \tag{4.107}$$

Eliminating Dis between Eqs. (4.101) and Eq. (4.106) and re-arranging to obtain an expression for internal reflux ratio in terms of the enthalpies as:

$$\frac{L_n}{G_{n+1}} = \frac{Q' - H_{Gn+1}}{Q' - H_{Ln}} \tag{4.108}$$

The difference point is located in the Hxy diagram as shown in Figure 4.20. The coordinates of the difference point is (Q', z_D). It is a fictitious stream and is used to obtain the number of ideal stages required to achieve a desired degree of separation. Lines can be constructed that connect the difference point, the L_n and G_{n+1}. The slope of the line can be found to be L_n/G_{n+1}. The difference point can be located using the reflux ratio R.

$$R = \frac{L_0}{D} = \frac{Q' - H_{G1}}{Q' - H_{L0}} = \frac{\text{line } \Delta_D G_1}{\text{line } G_1 L_0} \tag{4.109}$$

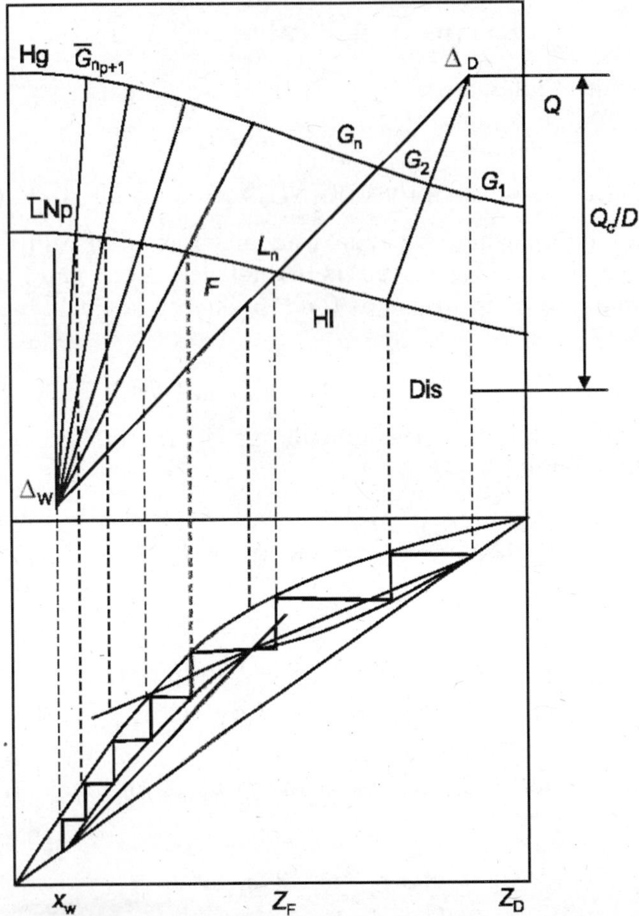

FIGURE 4.20 Enthalpy concentration and distribution diagrams in the method of ponchon and savarit for multi-stage binary distillations.

In a similar fashion the stripping section can also be analysed. The difference point for the stripping section would be Δ_w with coordinates of (x_W, Q'') in the Hxy diagram. Thus the number of ideal stages required can be calculated using the Ponchon Savarit method. The feed $q = (H_G - H_F)/(H_G - H_L)$.

4.11 METHOD OF THIELE AND GEDDES FOR MULTI-COMPONENT DISTILLATION

The principles established for multi-stage distillation of binary systems can be extended often times without much modification to multi-component systems. Some new design problems are encountered. As a matter of rule n–1 distillation columns may be required to separate n components. When the procedure is established, the VLE data for multicomponent systems may not be readily available. The solution procedure requires trial and error calculations. The method of Thiele and Geddes establishes when for a given system the feed, number of trays, position of the feed tray, liquid/vapour ratio, temperature for each tray are known at the start of the procedure, the resulting distillate and residue compositions can be calculated. There is another procedure by Lewis and Matheson.

The specifications limitations discussed in section 4.2.5 is applicable for multicomponent systems. Given the specified quantities only 3 quantities need be supplied. The rest can be calculated. This gives some idea of the degrees of freedom in the design variables. The key components of the feed mixture need be listed in the order of their relative volatility. The more volatile components are called *light* and the less volatile component is called *heavy*. The light key component is present in the residue in the minority and the heavy key component is a small portion of the distillate. The relative volatilities are compared with respect to the heavy key,

$$\alpha_j = \frac{m_j}{m_{hk}} \tag{4.110}$$

where j is any component and h_k is heavy key

The relative volatility is less than 1 for heavier components and greater than 1 for lighter components. Shiras (1950) showed that at the pinch point,

$$\frac{x_{j,D}\,\text{Dis}}{Fz_{j,F}} = \frac{x_{lkD}\,\text{Dis}}{Fz_{lkf}}\,\frac{(\alpha_j - 1)}{(\alpha_{l,k} - 1)} + \frac{x_{hkD}\,\text{Dis}}{Fz_{lhf}}\,\frac{(\alpha_{l,k} - \alpha_j)}{(\alpha_{l,k} - 1)} \tag{4.111}$$

When the LHS of Eq. (4.111) is less than 0.01 or greater than 1.01, the component j will not distribute. For values of LHS between 0.01 and 0.99 the component j will distribute. Underwood equations (1949) can be used for obtaining the condition of minimum reflux. For constant relative volatilities, L/G = constant and

$$F(1 - q) = \frac{\Sigma \alpha_j z_{jF} F}{\alpha_j - \phi} \tag{4.112}$$

$$\text{Dis}(R_m + 1) = \frac{\Sigma\alpha_j x_{jD}\text{Dis}}{\alpha_j - \phi} \tag{4.113}$$

Equation (4.112) is written for all j components and ϕ values are solved for. These lie between αs and for each ϕ, R_m is solved for simultaneously with x_{jD}Dis values. The relative volatilities are computed at the average of the distillate dew point and the residue bubble point which may require some trial and error solutions. The Fenske equation, derived for binary systems and yields the minimum number of ideal stages required to achieve the desired degree of separation, can be extended to multi-component systems,

$$N_m + 1 = \ln\frac{\left(\dfrac{x_{lkD}x_{hkw}}{x_{hkD}x_{lkw}}\right)}{(<\alpha_{lk}>)} \tag{4.114}$$

N_m+1 includes the reboiler and partial condenser. At total reflux the product split of other components can be calculated:

$$\frac{x_{j,D}\ \text{Dis}}{x_{jw}W} = <\alpha_j>^{N_m+1}\left(\frac{x_{hkD}\ \text{Dis}}{x_{hkw}W}\right) \tag{4.115}$$

The average relative volatility can be taken as the geometric mean of the values at the distillate due point and residue bubble point which may require a few iterations. Thus the minimum number of ideal stages N_m, the minimum reflux ratio R_m and the product compositions can be calculated. The operating line of the enrichment of a key component and the operating line of the stripper intersect at the optimal feed location.

The method of Thiele and Geddes is as follows:

For a total condenser, $x_0 = z_D = x_D$,

$$\frac{G_1 y_1}{\text{Dis}\ z_D} = 1 + \frac{L_0 x_0}{\text{Dis}\ z_D} = R + 1 = A_{f0} + 1 \tag{4.116}$$

For a partial condenser, $z_D = y_D$, $m_0 = y_D/x_0$. This behaves like an ideal stage.

$$\frac{G_1 y_1}{\text{Dis}\ z_D} = \frac{R}{m_0} + 1 = A_{f0} + 1 \tag{4.117}$$

Thus, A_{f0} is R or R/m_0 depending upon the type of condenser.

For tray 1,

$$A_{f1} = \frac{L_1 x_1}{G_1 y_1} = \frac{L_1}{G_1 m_1} \tag{4.118}$$

where A_{f1} is the absorption factor for tray 1. Then $L_1 x_1 = A_{f1} G_1 y_1$

For tray 2,

$$\frac{G_2 y_2}{\text{Dis } z_D} = \frac{L_1 x_1}{\text{Dis } z_D} + 1 = 1 + \frac{A_{f1} G_1 y_1}{\text{Dis } z_D} = 1 + A_{f1} + A_{f0} A_{f1} \tag{4.119}$$

Generally for the nth tray,

$$\frac{G_n y_n}{\text{Dis } z_D} = \frac{L_{n-1} x_{n-1}}{\text{Dis } z_D} + 1 = 1 + \frac{A_{n-1} G_{n-1} y_{n-1}}{\text{Dis } z_D} = 1 + A_{fn-1} + \ldots + A_{f0} A_{f1} A_{f2} \ldots A_{fn-1} \tag{4.120}$$

For the feed tray,

$$\frac{G_F y_F}{\text{Dis } z_D} = 1 + A_{fF-1} + \ldots + A_{fF0} A_{f1} A_{fF2} \ldots A_{fFn-1} \tag{4.121}$$

For the stripping section using kettle type reboiler,

$$S_W = \frac{\bar{G}_{Np+1} y_{Np+1}}{W x_W} = \frac{G_{Np+1} m_W}{W} \tag{4.122}$$

where S_W is the reboiler stripping factor. For a thermosiphon reboiler, $y_{Np+1} = x_w$ and $S_w = \bar{G}_{Np+1}/W$. For the bottom tray,

$$S_W + 1 = 1 + \frac{\bar{G}_{Np+1} y_{Np+1}}{W x_W} = \frac{\bar{L}_{Np} x_{Np}}{W x_w} \tag{4.123}$$

$$S_{Np} = \frac{\bar{G}_{Np} y_{Np}}{\bar{L}_{Np} x_{Np}} \tag{4.124}$$

For any tray,

$$\frac{\bar{L}_m x_m}{W x_w} = \frac{S_{m+1} \bar{L}_{m+1} x_{m+1}}{W x_w} = 1 + S_{m+1} + \ldots + S_{m+1} \ldots S_{Np-1} \tag{4.125}$$

For the feed tray,

$$\frac{\bar{L}_F x_F}{W x_w} = 1 + S_{F+1} + \ldots + S_{F+1} \ldots S_{Np} \tag{4.126}$$

Edmister (1957) provides a short cut to calculate an effective absorption factor A_f or stripping factor S_w.

$$A_f = \frac{\bar{L}_f}{G_f m_f} \tag{4.127}$$

$$\frac{x_W W}{\text{Dis } z_D} = \frac{A_f G_f y_f}{(D z_D)(\bar{L}_F x_F / W x_W)} \tag{4.128}$$

$$F z_F = W x_w + \text{Dis } z_D \tag{4.129}$$

$$\text{Dis } z_D = \frac{F z_F}{\left(1 + \frac{x_w}{z_D}\left(\frac{W}{\text{Dis}}\right)\right)} \tag{4.130}$$

The data consistency can be checked by verifying that the mole fractions of the different components in the feed, distillate and residue, each should add to 1. The L/G values, x_m and y_n values can be checked for validity. Especially in multicomponent distillation, the specification limitations are a salient consideration.

4.12 USE OF RICATTI DIFFERENCE EQUATION

The calculus of finite differences can be used to establish an analytical procedure to solve for the relevant design variables during multi-stage binary distillation. There is interest in this approach especially with the availability of computer resources. A component mass balance between the distillate and the tray n (Figure 4.3) can be written as:

$$G_{n+1} y_{n+1} - L_n x_n - \text{Dis } x_D = 0 \tag{4.131}$$

Dividing Eq. (4.131) by G_{n+1}

$$y_{n+1} - x_n\left(\frac{L_n}{G_{n+1}}\right) - \left(\frac{\text{Dis}}{G_{n+1}}\right) x_D = 0 \tag{4.132}$$

The definition of relative volatility, Eq. (4.33) is used to express the equilibrium composition in the nth plate as assuming that equilibrium has been established:

$$\alpha_n x_n - y_n + x_n y_n(\alpha_n - 1) = 0 \tag{4.133}$$

Further,

$$\alpha_{n+1} x_{n+1} - y_{n+1}(1 + x_{n+1}(\alpha_{n+1} - 1)) = 0 \tag{4.134}$$

Combining Eq. (4.132) and Eq. (4.134);

$$\alpha_{n+1} x_{n+1} - \left(\frac{L_n x_n}{G_{n+1}} + \frac{\text{Dis } x_D}{G_{n+1}}\right)(1 + x_{n+1}(\alpha_{n+1} - 1)) = 0 \tag{4.135}$$

Assuming constant overflow and under flow, $L_1 = L_2 = ... = L_n = L$ and $G_1 = G_2 = ... = G_{n+1} = G$ and constant relative volatility, $\alpha_1 = \alpha_2 = ... = \alpha_n = \alpha$.

$$\alpha x_{n+1} - \left[\left(\frac{L}{G}\right) x_n + \left(\frac{\text{Dis}}{G}\right) x_D\right] (1 + x_{n+1}(\alpha - 1)) = 0 \tag{4.136}$$

Re-arranging the terms,

$$x_n x_{n+1} + a_1 x_{n+1} + b_1 x_n + c_1 = 0 \tag{4.137}$$

where,

$a_1 = (\text{Dis}\ x_D(\alpha - 1) - \alpha G)/L(\alpha - 1)$
$b_1 = 1/(\alpha - 1)$
$c_1 = \text{Dis}\ x_D/L(\alpha - 1)$

The solution of Eq. (4.137) which is a Riccati difference can be made using calculus of finite differences. Eq. (4.137) is a non-linear difference equation. It can be linearised by the method of transformation of variable.

Let, $x_n = u_n + \delta$

Eq. (4.137] can be written as:

$$(u_n + \delta)(u_{n+1} + \delta) + a_1 (u_{n+1} + \delta) + c_1 = 0 \tag{4.138}$$

or

$$u_n u_{n+1} + \delta^2 + \delta(u_n + u_{n+1}) + a_1 u_{n+1} + a_1\delta + b_1 u_n + b_1\delta + c_1 \tag{4.139}$$

δ can be chosen such that, $\delta^2 + (a_1 + b_1)\delta + c_1 = 0$ (4.140)

Then,

$$u_n u_{n+1} + (a_1 + \delta)\, u_{n+1} + (b_1 + \delta) u_n = 0 \tag{4.141}$$

Dividing Eq. (4.141) by $u_n u_{n+1}$ and let $V_n = 1/u_n$

$$1 + V_n(b_1 + \delta) + V_{n+1}\,(a_1 + \delta) = 0 \tag{4.142}$$

Eq. (4.142) is a linear difference equation with constant coefficients.

4.13 HEIGHT OF TRANSFER UNITS AND NUMBER OF TRANSFER UNITS

Packed towers may be used instead of the tray towers to affect the desired degree of separation by distillation. This is used in low pressure distillation, treatment of heat sensitive materials. As opposed to stepwise changes in composition in the case of packed towers the change in composition is continuous. The schematic is essentially similar to that in Figure 4.3. For equimolar counter transfer,

$$N''_A = -N''_B \tag{4.143}$$

$$N''_A = k'_y(y_{A1} - y_{A2}) = k'_x\,(x_{A1} - x_{A2}) \tag{4.144}$$

Consider a thin slice of differential volume ΔZ in the packed bed of cross-sectional area A. A mass balance can then be written as:

$$N''_A = \frac{d(Gy)}{\rho adZ} = k'_y(y_i - y) = \frac{dx}{\rho adz} = k'_x(x - x_i) \tag{4.145}$$

For constant, G and L,

$$\frac{G}{\rho a}\frac{dy}{dZ} = k'_y(y_i - y) = \frac{Ldx}{\rho adZ} = k'_x(x - x_i) \tag{4.146}$$

$$\int_{y_1}^{y_2} \frac{dy}{(y_i - y)} = \int_0^{z_e} \frac{\rho k'_y adz}{G} = \int_{x_1}^{x_2} \frac{dx}{(x - x_i)} = \int_0^{z_1} \frac{\rho k'_x a}{L} \tag{4.147}$$

A similar expression can be written for the stripping section of the tower. The slope of the operating line of the rectifying section can be seen to be $-k'_x/k'_y$. The height of the transfer unit can be written as:

$$H_{tG} = \frac{G}{k'_y a\rho} \tag{4.148}$$

$$H_{tL} = \frac{L}{k'_x a\rho} \tag{4.149}$$

$$Z_e = H_{tG}\int_{y_a}^{y_2} \frac{dy}{(y_i - y)} = H_{tL}\int_{x_a}^{x_2} \frac{dx}{(x - x_i)} \tag{4.150}$$

The number of transfer units and the height of the transfer unit can be written as:

$$Z_e = H_{tG}N_{tG} = H_{tL}N_{tL} \tag{4.151}$$

For dilute solutions, Henry's law is applicable and $y^* = mx$. Then,

$$Z_e = H_{tG}\int_{y_a}^{y_2} \frac{mdx_i}{(mx_i - y)} = H_{tL}\int_{x_a}^{x_2} \frac{dx}{(x - x_i)} \tag{4.152}$$

The overall transfer coefficients can be written as:

$$\frac{1}{K'_y} = \frac{1}{k'_y} = \frac{m}{k'_x} \tag{4.153}$$

$$\frac{1}{K'_x} = \frac{1}{k'_x} = \frac{1}{mk'_y} \tag{4.154}$$

$$H_{tOG} = H_{tG} + \frac{mG}{L} H_{tL} \tag{4.155}$$

$$H_{tOL} = H_{tL} + \frac{L}{mG} H_{Tg} \tag{4.156}$$

4.14 MURPHREE STAGE EFFICIENCY

In the above analysis and solutions, the vapour phase was assumed to have reached the equilibrium. Often times in the real world due to diffusional mass transfer limitations, the equilibrium would not have been reached. The vapour phase composition as a function of residence time at stage n is showed in Figure 4.21. The time taken to reach equilibrium or steady state depends on the contacting pattern, nature of the diffusivities and the geometry of the plate. In addition, finite speed diffusion effects can become significant especially for shallow beds and systems with high relaxation times (Sharma, 2005).

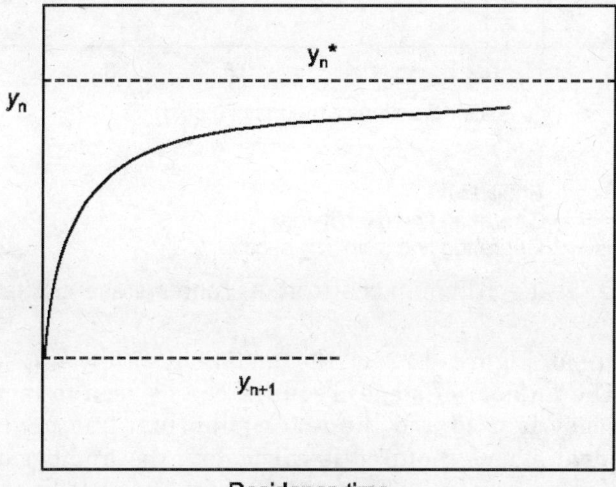

FIGURE 4.21 Approach to equilibrium of vapour phase at stage *n*.

Most design approaches given for stage-wise contacting assume that the streams leave each stage at equilibrium. The vapour composition y_n and the liquid composition x_n at the nth stage are in equilibrium with each other. In order to account for the discrepancy between the composition values achieved in practice and that assumed an efficiency, η is defined.

$$\eta = \frac{(y_n - y_{n+1})}{(y_n^* - y_{n+1})} \tag{4.157}$$

The definition in Eq. (4.143) is called the Murphree stage efficiency. It provides a measure of how close the equilibrium in the gas phase can be achieved. A real line can be constructed using the Murphree stage efficiency as shown in Figure 4.22.

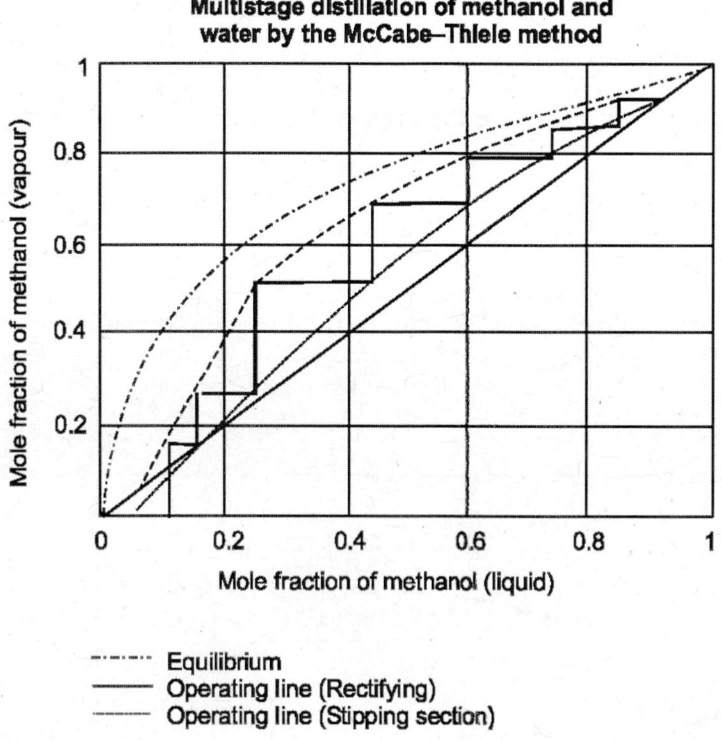

FIGURE 4.22 Real equilibrium line from murphree stage efficiency.

The system shown in Figure 4.22 is the methanol/water system discussed in *worked example 4.1*. The number of stages required can be seen to increase when the Murphree stage efficiency is used and the real equilibrium line used instead of the ideal line. After the ideal stages required is calculated, the number of actual plates required can be stepped as shown in Figure 4.22 from the real line or by using an overall efficiency of the distillation column.

Once the shape of the transient composition profile in the vapour phase is known the law of marginal returns can be used. Thus for a certain residence time and above the additional capital cost associated with a larger residence time and the incremental advantage in the number of plates required can be calculated. Thus to optimise the total cost, the column may be operated at a certain residence time where the sharp increase in composition profile tapers off and approaches the asymptotic limit of equilibrium composition at infinite time.

Although a elaborate treatment of transient composition profile would be beneficial here, it is left as further reading (Sharma, 2005) for space constraints. There the finite speed diffusion and relaxation effects will also play a pivotal role in determining the vapour phase composition profile. In a typical sieve tray in a distillation column, the vapour phase bubbles up the liquid that flows down the column. The more volatile substance transfers from the bulk of the liquid to the bubble liquid interface and from the interface to the bulk of the bubble. The size of the bubble may be assumed to stay the same for equimolar overflow and underflow or it may assumed to vary in size during the contacting time. Good models for transient diffusion and relaxation are needed to more accurately determine the Murphree stage efficiency for different systems and different contacting patterns.

For a well mixed system a relation between the Murphree stage efficiency and the mass transfer coefficient may be derived as follows:

A mass balance on the vapour phase can be written as:

(mass in) – (mass out) + mass transfer from liquid = accumulation

$$G''A(y_{n+1} - y_n) + \frac{K_y alA}{\rho(y_n^* - y_n)} = 0 \tag{4.158}$$

where K_y is the overall mass transfer coefficient, (mol/m^2/s) and a is the interfacial area of contact (m^2/g), l the depth of the liquid vapour froth and ρ the density of the vapour, A the cross-sectional area (m^2) of the column. G'' is the vapour flow rate per unit area (mol/m^2/s). The overall mass transfer coefficient can be written in terms of the individual mass transfer coefficients assuming Henry's law for the equilibrium line as:

$$\frac{1}{K_y} = \frac{RT}{pk_p} + \frac{m}{ck} \tag{4.159}$$

where c is the total molar concentration in the liquid and p/Rt is the total molar concentration in the vapour and m is the Henry's law constant. Adding and subtracting y_{n+1} in Eq. [4.158);

$$G''A(y_{n+1} - y_n) + \frac{K_y alA}{\rho}(y_n^* - y_{n+1} - (y_n - y_{n+1})) = 0$$

or
$$(y_n - y_{n+1})\left(\frac{K_y alA}{\rho} + G''A\right) = \left(\frac{K_y alA}{\rho}\right)(y_n^* - y_{n+1}) \tag{4.160}$$

Combining Eq. (4.160) and Eq. (4.157),

$$\eta = \frac{1}{\left(1 + \frac{\rho G''}{K_y al}\right)} \tag{4.161}$$

It can be seen that at very high interfacial area the Murphree efficiency will tend to 1.

SUMMARY

The method of McCabe and Thiele to design a multi-stage distillation column for separation of binary systems was presented in detail. The 8 different components of a typical distillation column include central feed, vertical cascade of stages, enriching section, overhead condenser, reflux, distillate, stripping section and reboiler. The main steps in the design procedure are the construction of operating lines derived from mass balances, equilibrium line from VLE data, a real line construction from diffusion and mass transfer model and estimate of the Murphree plate efficiency and stepping of the number of stages required. The operating line for the enriching and stripping sections were derived as:

$$y_{n+1} = x_n\left(\frac{R}{R+1}\right) + x_D\left(\frac{1}{R+1}\right)$$

$$y_{n+1} = x_n\left(\frac{B}{B-1}\right) - x_W\left(\frac{x_W}{B-1}\right)$$

where R and B are the reflux and reboil ratio respectively. The derivation of the minimum number of stages required, minimum reflux ratio required was shown with the introduction of a pinch point. The optimal feed location was discussed. There can be 6 different feed types. The intersection point of the operating line from the rectifying section, operating line from the stripping section and the feed line was identified. The different types of reboilers were discussed. A worked example was shown to illustrate the method to separate methanol from a methanol/water mixture.

The design variables involved in the calculations and the degrees of freedom available were derived. There are 20 different problem types. With the feed specified, and 3 inputs all other variables can be calculated. A set of problems called converse problems require trial and error solution when number of stages is given as an input parameter. The information needed for the construct of the equilibrium line was discussed starting with the phase rule, Clausius Clapeyron equation, Antoine equation for vapour pressure temperature relationship, Raoult's law, fugacity and VLE model from Margules, Wilson equations for activity coefficient, etc. Azeotropes with examples were discussed. Henry's law for dilute systems were also mentioned. A worked example on Pentane/Heptane system was discussed. Two feeds were introduced and seen to increase the number of stages required highlighting the optimal feed location. The typical number of trays used in the industry was given as a table. The methods for azeotropic distillation with the use of entrainer was discussed and extractive distillation to separate systems was elaborated. The characteristics of a good entrainer and solvent were enumerated. The operating line for flash vaporisation of a binary system was derived. As a worked example different methods to treat the equilibrium flash vaporisation of a ternary system were derived. The method of Rayleigh for differential distillation was derived. A generalized Rayleigh equation was also derived that included the vapour phase changes of equilibrium concentration.

The derived equation was:

$$\int_{x_W}^{x_F} \frac{dx}{(y^* - x)} + \frac{x_F - x_W}{<x_D> - x_F} \int_{y_W^*}^{y_F^*} \frac{dy^*}{(y^* - x)} = \ln\left(\frac{F}{W}\right)$$

When steam replaces the reboiler the modifications to the construct of the stripping operating line were derived. At low pressures when the mean free path of the molecule is greater in dimension compared with the critical dimension of the apparatus the Langmuir equation may be more applicable.

The Kremser equation for the number of stages required when the equilibrium relationship is linear was derived and given as:

$$N = \frac{\ln\left(\dfrac{y_{N+1} - y_N}{y_1 - y_0}\right)}{\ln\left(\dfrac{y_{N+1} - y_1}{y_N - y_0}\right)}$$

When the absorption factor was 1, an alternate equation was also derived as:

$$N = \left(\frac{y_{N+1} - y_1}{y_1 - y_0}\right)$$

The Fenkse equation for systems with constant relative volatility was derived and it provides the minimum number of stages required to affect the desired degree of separation and given by:

$$N_m + 1 = \frac{\ln (1 - x_W)x_D}{x_W(1 - x_D)\ln (a)}$$

The use of log scale was shown to obtain high purity products with compositions of 99.99%, and so on.

The use of enthalpy data and the method of Ponchon and Savarit were discussed. The method of Thiele Geddes for multi-component distillation was elaborated. The number of transfer units and height of the transfer units was derived for a packed tower. Use of Riccati equation was shown to obtain analytical solution for the distillation problem. The recurrence relation was found to be:

$$1 + V_n(b_1 + d) + V_{n+1}(a_1 + d) = 0$$

The Murphree plate efficiency was derived for a well mixed system without accumulation from a mass balance and representation of a overall mass transfer coefficient. The efficiency can be given as:

$$\eta = \frac{y_n - y_{n+1}}{y_n^* - y_{n+1}}$$

REVIEW QUESTIONS

1. Is there a critical point in the solid-liquid equilibrium line for a pure substance. Why?
2. A professional inadvertently wrote y_n instead of y_{n+1} in the derivation of the operating line of the rectifying section. Does it make any difference to the final result. Why?
3. Given the product split, W/Dis and residue composition the feed composition can be calculated and cannot be specified as it is in the beginning of the problem. How can the rule for specifications be restated to include this observation?
4. Can the operating line from the rectifying section and the operating line *f* from the stripping section intersect at a point lower than the $y = x$ diagonal. Why?
5. Rewrite the operating line for the stripping section,

$$y_{n+1} = x_n \frac{B}{B-1} - \frac{x_W}{B-1}$$

showing the x intercept, i.e. show that,

$$x_n = \frac{B-1}{B} y_{n+1} + \frac{x_W}{B}$$

The x intercept is 1/*B*. Can the x intercept be negative?

6. What happens when the reboil ratio is 1 to the operating line of the stripping section. What do you do?
7. What is the meaning of a zero x intercept or a infinite reboil ratio. What are the implications?
8. Can the multi-stage distillation tower be horizontal? What provisions have to be made to segregate the liquid and vapour phases in either direction of the column?
9. One of the distillation columns with 80 plates was built at an incline of 85° with the horizontal. Will this affect any of the results derived by the McCabe-Thiele method. Is more energy or less energy needed to complete the operation. Discuss.
10. Often times the equilibrium is not reached in real world industrial operations at the different stages. What happens to the efficiency when the stage is built wider compared with narrower?
11. A distillation column was built with a taper at an apex angle of 16°. What are the implications. What changes are observed in the results?
12. What happens to the feed line given by Eq. (4.14) for superheated vapour? How do you get the positive slope as shown in Figure 4.4.

13. What is the difference between distillation and evaporation when one of the binary components is water and is the more volatile substance?
14. Derive the number of stages using Kremser equation for the stripping section. What is the desorption factor?
15. At low pressure distillation, the relaxation time is expected to be large Eq. (1.49). How will you modify Langmuir equation in order to account for the relaxation phenomena?
16. Can the Kremser equation be used for multi-component systems?
17. When the distribution diagram was attempted to be constructed using the De Priester K value charts, negative values rose for the mole fractions. What does this mean?
18. In McCabe-Thiele method, using the generalised Fick's law of diffusion say an overall efficiency was calculated. The real equilibrium line can be constructed. A new pinch point is identified. The appropriate reflux ratio is used and the operating lines of the rectifying and stripping sections are constructed. The minimum reflux ratio needed is now different. After the construction of the number of stages required, the overall efficiency can be recalculated again. This may lead to a different pinch point. So a iterative procedure is required to achieve the number of real stages required for separation. How do you improve this method?
19. Using the McCabe-Thiele method, can a column be designed to produce a product of 100% purity. Why?

PROBLEMS

1. *Residue Composition*

The VLE data for methyl acetate and methanol system is provided here for 1 atm pressure from the Chemical Engineers Handbook by Perry and Green, 1997 and is as follows:

T°C	57.8	55.5	55.04	53.88	53.82	53.90	54.5	56.9
x_A	0.173	0.321	0.380	0.595	0.643	0.710	0.849	1.0
y_A	0.342	0.477	0.516	0.629	0.657	0.691	0.783	1.0

It can be seen that beyond 0.65 mole fraction of methyl acetate in liquid the vapour is richer in methanol and not in methyl acetate. In order to distil methyl acetate the distillate composition has to be maintained at less than 0.657 mole fraction to have a favourable condition for separation. Given the distillate composition of 0.63, and a feed of 0.25 mole fraction methyl acetate all liquid; (a) show that the minimum reflux ratio needed is 2.5. Given a reflux of 1.5 times the minimum, (b) Show that the residue composition for a set of 4 ideal stages used would be 0.14. (c) What are the temperatures in each stage for atmospheric distillation?

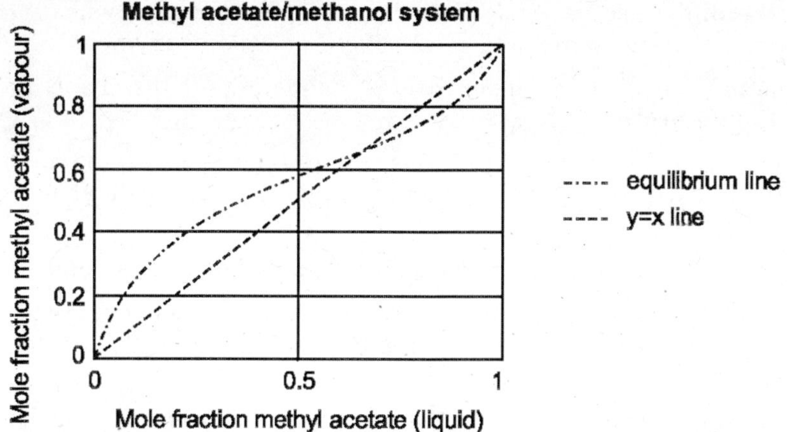

FIGURE 4.23 Residue composition estimate in MA/Methanol system.

2. *Kremser Equation*
   Dimethylene dichloride, present in saline water at 35%, is fed into a 7-stage distillation column from the top. The vapour mole fraction entering the bottom of the column is 0.85. Total condenser is used in the system. Take the m value of the equilibrium line to be 42 and the b value as 0 and find the absorption factor. (**Ans:** 0.6)
3. *Pure Benzene*
   Calculate the number of ideal stages required to produce 99.4% benzene and 0.6% toluene. A total condenser is used. The liquid leaving the column has 85% benzene. The m and b values for the equilibrium line may be taken as 0.56 and 0.42 respectively. (**Ans:** 3 stages)
4. *Partial Condenser*
   During multi-stage binary distillation, the procedure outlined in section 4.2 is for a total condenser. A partial condenser can sometimes be used. The vapour product is in equilibrium with the liquid reflux after sufficient contact time. If the condensate is removed rapidly as it forms a differential condensation may occur, show that for a partial condenser,

$$y_{n+1} = \left(\frac{L}{G}\right)x_n + \left(y_1 - \frac{L}{Gx_0}\right)$$

   Verify that partial condenser figures as an extra stage at the staircase graph.
5. *Reboil Ratio*
   The VLE data for acetone-water system is given below in a table and shown in Figure. Acetone is the more volatile substance and forms the component A.

x_A	y_A
0.05	0.6381
0.1	0.7301
0.15	0.7716
0.2	0.7916
0.3	0.8124
0.4	0.8269
0.5	0.8387
0.6	0.8532
0.7	0.8712
0.8	0.8950
0.9	0.9335
0.95	0.9627
1.0	1.0

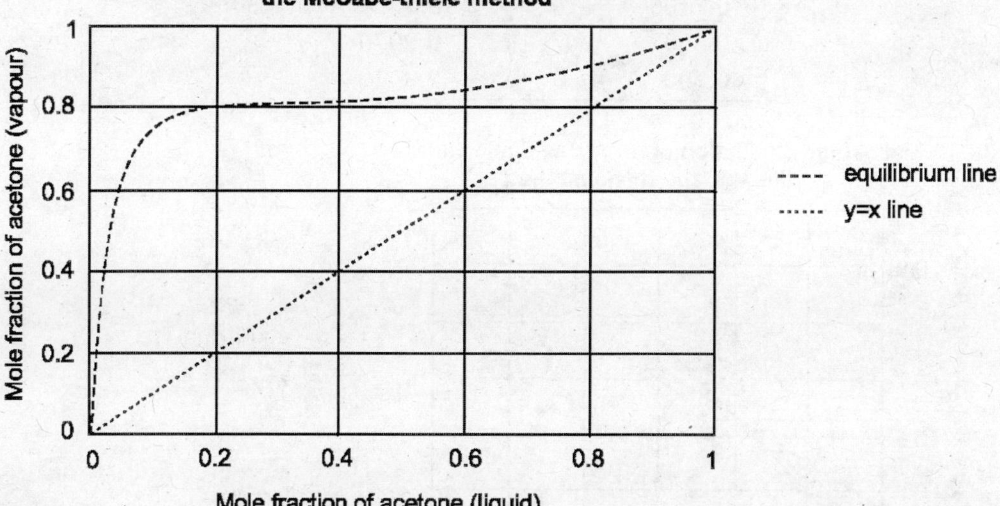

FIGURE 4.24 Reboil ratio estimate in acetone/water system.

A *W*/Dis ratio of 0.63 is used in a multi-stage binary distillation column with a total condenser and a reboiler. A product split of 146 is used and a reflux ratio of 4 is used in the rectifying section. Show that the number of ideal stages to perform the distillation at a distillate composition of 90% needed is 4 and calculate the reboil ration *B*. What is the residue composition?

6. *Reflux Ratio*

The VLE data for water and formic acid binary system at atmospheric pressure is given from Perry and Green (1997) as follows. Water is more volatile compared with formic acid and forms the component A. A product split of 1.5 is sought and a reboil ratio of 4 is used. A residue composition of 42% formic acid is collected from the multi-stage distillation column. Show that the ideal number of stages required to perform the distillation is 10. Confirm that the distillate composition in 90.84%, and the reflux ratio is 0.9747. The feed is all liquid at a composition of 38% formic acid.

x_A	y_A
0.0405	0.0245
0.1550	0.102
0.2180	0.1620
0.3210	0.2790
0.4090	0.4020
0.4110	0.4050
0.4640	0.4820
0.5220	0.5670
0.6320	0.7180
0.7400	0.8360
0.820	0.9070
0.900	0.9510

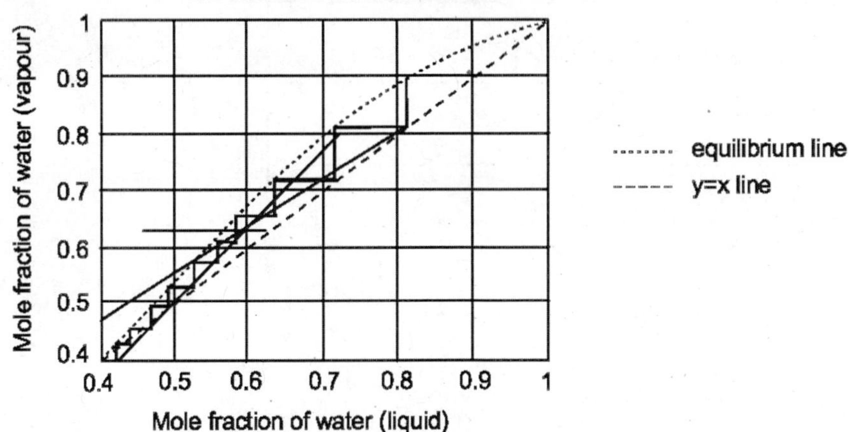

FIGURE 4.25 Estimation of reflux ratio in formic acid/water system.

7. *Ratio of Residue to Product*
Use the VLE data given for the formic acid and water in problem 6 and the same feed conditions (Figure 4.25). Given the product split of 4.5 reboil ratio of 4 and reflux ratio of 14, calculate the residue composition, the number of ideal stages required for the given degree of separation and the W/Dis ratio.
(**Ans:** 10 stages; x_D = 0.9; x_w = 0.42; W/Dis = 0.4762)

8. *Product Split*
Use the VLE data for the acetone water system given in problem 5 (Figure 4.24). Given a reflux ratio of 6 and a reboil ratio of 4 and bottoms composition of 5% calculate the product split, number of ideal stages required to effect a distillate product of 90%, the desired separation, and feed composition.
(**Ans:** 3 stages; $x_F = 0.4$)

Converse Problem
For problems 9, 10, 11, 12 and 13 use the VLE data given in problem, 6 for the formic acid water system. The same feed can be used or modified if needed.

9. Given that the distillation column contains 9 stages, and used a reflux ratio of 4 and a reboil ratio of 3.5, calculate the residue composition, product split and W/Dis ratio.
10. Given a residue composition of 49%, product split of 4 and a distillation column that contain 6 stages, calculate the reflux ratio, W/Dis ratio and reboil ratio used.
11. Given that the number of ideal stages required is 12, and the reboil ratio used in 3.5 and W/Dis ratio used is 0.5, calculate the residue composition x_W, reflux ratio R, and product split P_{split} (x_DDis/Wx_W).
12. Given that the number of ideal stages required is 10, the reflux ratio used is 2.2 and the W/Dis ratio the distillation column was operated with was 0.75, calculate the reboil Ratio B, product split, m P_{split} (x_DDis/Wx_W) and residue composition x_W for the formic acid water system.
13. Given that the number of ideal stages required to effect the separation of water from the formic acid water solution is 9, the reflux ratio used is and the product split achieved is 1.46 calculate the reboil ratio B, W/Dis ratio and the residue composition x_W.
14. *Steam Distillation*
A benzene-toluene mixture needs to be separated by distillation. The feed contains 63% toluene and 37% benzene. The overhead product is 97.3% benzene and a bottoms residue of 90% toluene. The feed rate is 4.5 kmole/s and a reflux ratio of 3 is used. (a) Calculate the mass of product and residue per unit time. (b) What are the number of ideal stages required and the location of the feed provided the feed is 53% liquid at the bubble point. (c) What is the steam flow rate required should the reboiler be replaced with steam as shown in Figure 4.18. What is the composition of the liquid in the 4th plate. The VLE data for the binary benzene/toluene system with benzene as the more volatile phase is given below:

x_A	0.1	0.2	0.3	0.4	0.5	0.6	0.7	0.8	0.9
y_A	0.22	0.38	0.51	0.63	0.7	0.78	0.85	0.91	0.96

(**Ans:** 9 stages; 4th plate – 0.65; steam = 1.65 kmole/s)

15. *Reboiler Type*

Derive the Kremser equations for the kettle type reboiler and for the Thermosiphon reboiler. Compare the two results for the number of ideal stages required. Show that for the kettle type reboiler,

$$N_{p-m+1} = \ln\left(\frac{(1 - A_f)\left(m\dfrac{x_m}{x_w} - 1\right)}{(m - 1)} + A_f\right)$$

Show that for the thermosiphon reboiler,

$$N_p - m = \ln\frac{\left(\dfrac{x_m}{x_w}(1 - A_f) + A_f\right)}{\ln\left(\dfrac{1}{A_f}\right)}$$

16. *Distilled Water from the Seas*

Show that 139.1 mole/h of steam at 101.33 kPa will be required to separate 1000 mole/h of sea water at an NaCl concentration of 3.5% at 25°C into an overhead product from a total condenser at 906.6 mole/h of distillate at 99.99% purity and a bottoms residue with a 15% brine solution at 232.5 mole/h. Use the Kremser equation and find the number of plates required when the reflux ratio is 2. The equilibrium line may be obtained from Henry's law that is applicable for dilute concentrations and is:

$$y^* = 0.1x + 0.9$$

If there is a 15 plate distillation column available what would be the bottoms composition. (**Ans:** 30%)

17. *Separation of Ethylbenzene from Styrene*

In commercial manufacture of polystyrene, HIPS, high impact polystyrene, SAN, styrene acrylonitrile copolymer and ABS, acrylonitrile, butadiene and styrene thermoplastic using the continuous polymerisation process ethyl benzene is used as a solvent. After the polymerisation is completed the unreacted monomers are separated from the product and recycled back to the feed tank. One of the critical steps in the separation is the separation of ethylbenzene and styrene. The normal boiling points of ethylbenzene and styrene are 136.2°C and 1465.2°C respectively. The system may be assumed to have constant relative volatility at 1.37. For an overhead distillate product of 97.33% composition and a residue composition of 0.04. For a feed composition of 42% ethylbenzene and a reflux ratio of 8.7 show that 20 ideal stages are needed in a multi-stage binary distillation column. The feed is 91% liquid.

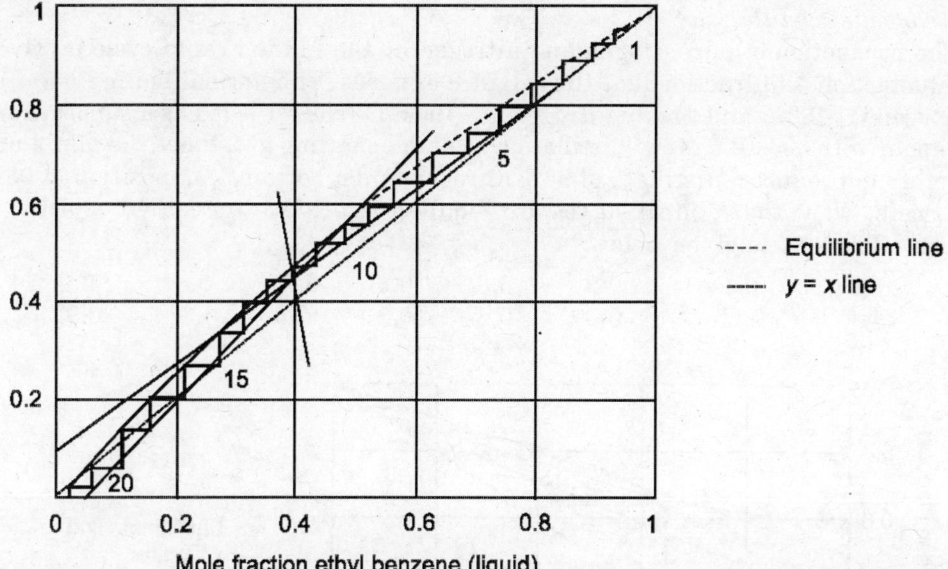

18. In problem 7, should the Murphree plate efficiency be 80% what is the number of real stages required to achieve the indicated product and residue compositions for the formic acid water system?

(**Ans:** Favorability for distillation changes to formic acid in vapour phase)

19. *Methyl Acetate as Residue*

In problem 1 the VLE data for methylacetate/methanol system shows that at compositions of the methyl acetate greater than 65% in the vapour phase, the methanol becomes the more volatile species. Construct a second stripping column with the feed composition at 0.63 mole fraction methyl acetate. For a methyl acetate composition in the residue of 98% and an all liquid feed at 0.72 mole fraction methyl acetate show that 6 ideal stages are required by the stripping section. What is the reboil ratio? (**Ans:** B = 6)

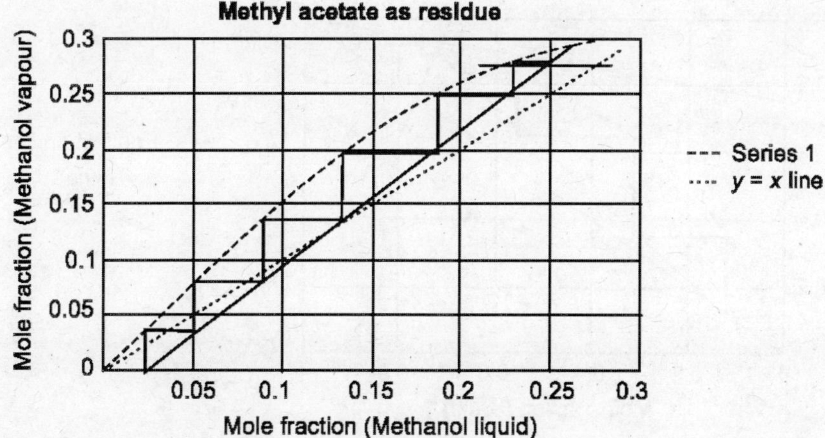

20. *Cryogenic Distillation*

The production of pure oxygen and nitrogen by the Linde Frankl cycle involves liquefaction and fractional of the mixture of gases. The normal boiling point of oxygen is –183°C and that of nitrogen is – 195.8°C. The air is fed as a superheated vapour with $q = -0.5$ into a distillation column operating at 2 times the minimum reflux to produce a distillate of 96% nitrogen and a bottoms composition of 98% oxygen. Show that 6 ideal stages are required to achieve the stated objectives. The VLE data is given below:

x_A	0.1	0.2	0.3	0.5	0.7	0.8	0.9
y_A	0.26	0.5	0.6	0.8	0.85	0.86	0.92

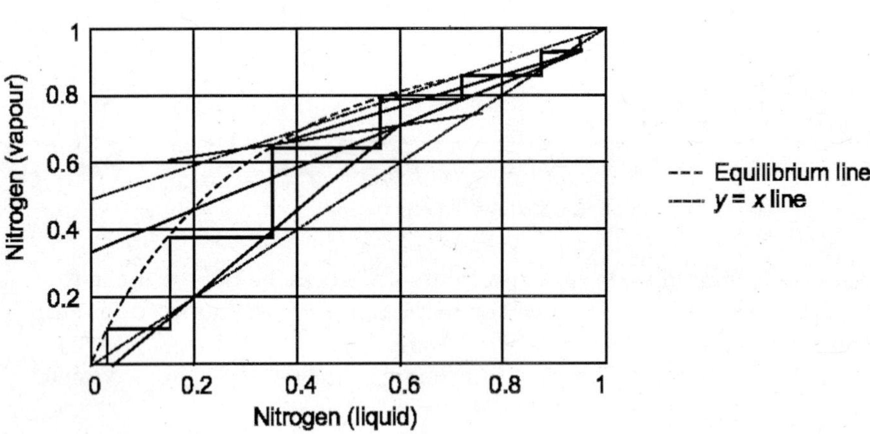

21. *Side Stream*

During fractionation of hydrocarbon mixtures, sometimes in addition to the distillate another stream is withdrawn from the top section of the column. For a system with a relative volatility of 4 and an overhead distillate composition of 97.33%, reflux ratio of 2, and a residue composition of 0.1, construct the

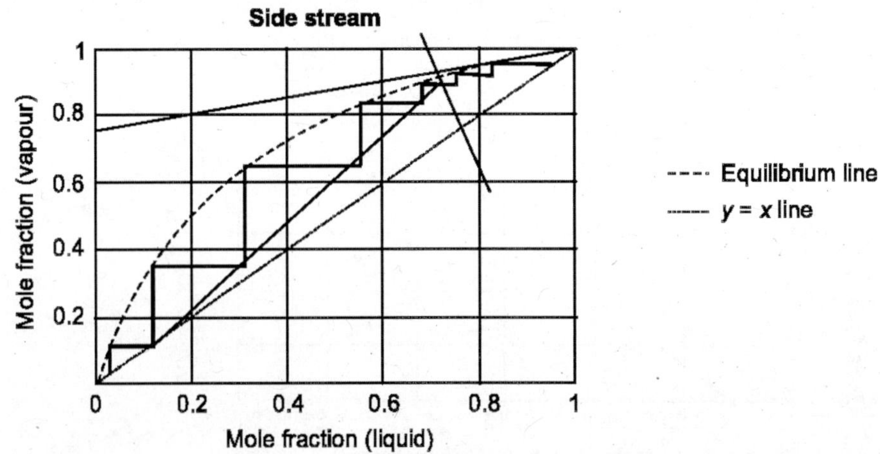

staircase graph and show that the number of ideal stages required is 7 to meet the desired objectives. A second stream at a composition of the more volatile species at 90% is withdrawn at twice the rate of the distillate product removal rate. The feed is 72% liquid at a composition of 70%. What is the reboil ratio used for a kettle type reboiler? (**Ans:** 4)

REFERENCES

Cussler, E.L., 1997, *Diffusion—Mass Transfer in Fluid Systems,* Cambridge University Press, UK.

De Priester, C.I., 1953, *Chem. Eng. Prog. Symp. Ser.*, **49,** 7, 1.

Edmister, W.C., 1957, *AIChE J*, **3**, 165.

Fenske, M.R., 1931, *Ind. Eng. Chem.*, **24**, 482.

Kister, H.Z., 1992, *Distillation Design*, McGraw Hill, New York.

Lewis, W.K., and Matheson, G.L., 1932, *Ind. Eng. Chem.*, **24,** 494.

McCabe, W.L., J.C. Smith and P. Harriott, 1993, *Unit Operations of Chemical Engineering*, 5th Edition, McGraw Hill, New York.

McCabe, W.L. and Thiele, E.W., 1925, *Ind. Eng. Chem.*, **17**, 605.

Murphree, E.V., 1925, *Ind. Eng. Chem.*, **17**, 747.

Perry, R.H. Green, D.W., 1997, *Perry's Chemical Engineers Handbook*, McGraw Hill, New York.

Ponchon, M, 1921, *Tech. Mod.,* **13**, 20, 55.

Prausnitz, J.M., 1969, *Molecular Thermodynamics of Fluid Phase Equilibria*, Prentice Hall, NJ.

Savarit, R., 1922, *Arts Metiers*, **65**, 142, 178, 241, 266, 307.

Seader, J.D. and E.J. Henley, 1998, *Separation Process Principles*, John Wiley, pp. 292–4, 299.

Smith, J.M. and Van Ness, H.C., 1986, *Introduction to Chemical Engineering Thermodynamics,* McGraw Hill, New York.

Sharma, K.R., 2005, *Damped Wave Transport and Relaxation*, Elsevier, Amsterdam.

________, October 2002, *Flash Calculations for a Ternary System for Equilibrium Vaporisation—A Comparative Study of the Methods of Matrix Inversion, Analytical Solution of a Quartic Polynomial and Iterative Solution of Three Simultaneous Equations and Simultaneous Unknowns*, 52nd Southeast Regional Meeting of the ACS, SERMACS, New Orleans, LA.

________, April, 2001, *Analysis of Multistage Binary Distillation Design with Split of the Product as Input Parameter*, AIChE Spring National Meeting, Houston, TX.

Sharma, K.R., April, 2001, *Boiler Feedwater from Seawater by Cogeneration*, AIChE Spring National Meeting, Houston, TX.

_________, April, 2001, *Graphical Iteration for Separation Problems given Np as Input Parameter*, 221st ACS National Meeting, San Diego.

_________, April, 2001, *Analysis of Multistage Binary Distillation Design with W/D as Input Parameter*, 221st ACS National Meeting, San Diego.

_________, April, 2001, *Analysis of Multistage Binary Distillation Design with Bottoms Composition as Input Parameter*, 221st ACS National Meeting, San Diego, CA.

_________, April, 2001, *Overall Efficiency in Binary Distillation Design*, 221st ACS National Meeting, San Diego, CA.

_________, June, 2001, *Specification Limitations in Binary Distillation*, 34th Middle Atlantic Regional Meeting of the ACS, Baltimore, MD.

_________, September, 2001, *Modified Rayleigh Equation to Represent Differential Distillation,* 53rd Southeast Regional Meeting of the ACS, SERMACS, Savannah, GA.

_________, December, 2001, *Limitations of Specifications in Binary Distillation Design*, 38th Annual Convention of Chemists, Jodhpur.

_________, May, 2002, Generalized *Rayleigh Equation for Batch Distillation*, Middle Atlantic Regional Meeting of ACS, MARM 02, Fairfax, VA.

_________, May, 2002, *Distribution Diagram Represented with Error Function*, Middle Atlantic Regional Meeting of ACS, MARM 02, Fairfax, VA.

_________, November, 2002, *Molecular Distillation Process for Moisture Removal for Blood Storage in Army*, 54th Southeast Regional Meeting of the ACS, SERMACS, Charleston, SC.

Shiras, R., Hanson, D.R., and Gibson, D.H., 1950, *Ind. Eng. Chem.*, **42,** 871.

Thiele, E.W., and Geddes, R.L., 1933, *Ind. Eng. Chem.*, **25,** 289.

Treybal, R.E., 1980, *Mass Transfer Operations,* McGraw Hill, New York.

Underwood, A.J.V., 1949, *Chemical Engineering Progress*, **44**, 603.

Wankat, 1990, *Rate-Controlled Separations*, Chapman-Hall, New York.

CHAPTER 5

Adsorption, Chromatography and Ion Exchange

Nomenclature

a	interfacial area (m^2/gm)
a_w	characteristic dimension of the apparatus (m)
a^*	dimensionless interfacial area ($\rho a/a_w$)
A_d	adsorption factor (mL_s/S_s)
C_0	concentration of ion in feed solution (meq/lit)
C^*	equilibrium concentration of solution (meq/lit)
D	binary diffusivity (m^2/s)
$-\bar{H}$	differential heat of adsorption (kJ/kg)
K	equilibrium constant
K_y	overall mass transfer coefficient ($mol/m^2/hr$)
k_s	mass transfer coefficient in the adsorbent ($mol/m^2/hr$)
k_y	mass transfer coefficient in the bulk fluid ($mol/m^2/hr$)
l	adsorbent bed length (m)
L_s	kg solvent/m^2/hr
M	molecular weight (kg/mol)
M_r	molecular weight of the reference substance (kg/mol)
n	power exponent on the Freundlich isotherm solution concentration
N_p	number of stages
Pe_m	Peclect number mass, $U/\sqrt{D/\tau_r}$
q	concentration of solute in adsorbent (mol/m^3)
S_s	kg adsorbent/m^2/hr
Sha	Sharma number ($k\tau_r/a_w$)
t_B	breakthrough time (hr)
t_E	saturation time (hr)
U	superficial velocity (m/s)
u	dimensionless concentration (y/y_0)
X_0	kg solute/kg adsorbent before contact
X_1	kg solute/kg adsorbent after contact

y	concentration of solute in solution (mol/m^3)
Y_0	kg solute/kg solvent before contact
Y_1	kg solute/kg solvent after contact
Y^*	equilibrium concentration at the interface
Z	height of the packed tower for adsorption (m)

Greek

α	mass action law constant
λ_T	latent heat of vaporisation (J/mol)
ε	adsorbent bed voidage
ρ	density (kg/m^3)
τ	dimensionless time (t/τ_r)
τ_r	relaxation time of the material (s)

5.1 ADSORBENTS, ISOTHERMS AND SOLUTION

Adsorption is a process by which separation of the desired component is effected by contacting a liquid or vapour with a packed column and letting the contaminant or undesired species to rest in the column and the continuous vapour or liquid to pass through the column. The packed column may be stationary or moving in a counter-current fashion. The adsorbed bed can be regenerated thermally or by other means. The process of adsorption is essentially transient. If the process does not involve any chemical change, it is called *physisorption*. If during the process there is a chemical reaction, reversible or otherwise, it is referred to as *chemisorption*. Physisorption is a reversible process. It is also called *van der Waals adsorption*. When the intermolecular forces are in such a fashion that the forces between the solute and solid is greater than that between the solute and gas or between the solute molecules adsorption is favoured. In chemisorption, the adhesive force is greater than in physisorption. It is an irreversible process usually. The heat released during the process is comparable to the heat of the reaction.

Some applications of adsorption include dehumidification of air, removal of objectionable odours, contaminant removal, separate difficult to separate hydrocarbons, desulphurisation of gasoline, decolourisation of petroleum products, cleaning semi-conductor manufacturing devices, fractionation of aromatic and paraffinic hydrocarbons and protein purification.

Adsorption can be distinguished from absorption, distillation and extraction. The latter three unit operations involve two fluids flowing steadily in opposite directions at steady state. Adsorption is an unsteady process. It involves a fluid and a solid. Solids adsorb mere traces of solute. Adsorption is energy intensive. It tends to be more expensive compared with the other operations.

The basic question in adsorption is how tall a tower is needed to meet the desired objectives. This question is answered by a mass transfer analysis, including an operating line and an equilibrium line. It also includes overall and individual coefficients summarised by dimensionless correlations. Knowledge about adsorbents and isotherms is required.

The relationship between the amount adsorbed on the solid and the concentration of the fluid is called an *isotherm.* The basic behaviour of adsorption, the solute concentration eluted from the bed as a function of volume processed is summarised in the form of a plot called the *breakthrough curve*.

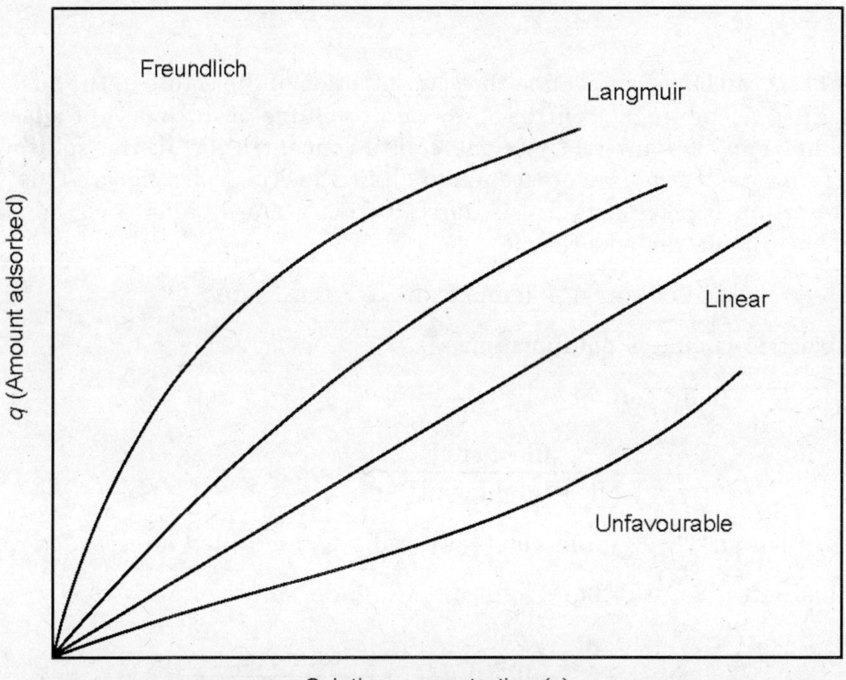

FIGURE 5.1 Typical isotherms during adsorption.

Adsorption is essentially a surface phenomenon. Molecules adsorb on any surface. Other molecules in the solid attract to the molecules in a solid from all directions. The molecules at the surface are in a stage of imbalance and are pulled inward. When the solution is contacted with the adsorbent, the molecules at the surface of the adsorbent will interact more with the available molecules, i.e. the solute molecules leading to its retention. This has given impetus to the technique of adsorption. The amount adsorbed is proportional to the surface area available. Commercial adsorbents are extremely porous with surface area typically of several hundred square meters per gram. Some specialised adsorbents have surface areas as high as 3000 m^2/gm. With the introduction of nanoporous carbon, the surface area can even be higher. Porous materials vary in the type of material and the isotherm generated from it. Adsorption and desorption operations are conducted in a cycle to achieve a continuous separation process. Sometimes they may exhibit hysteresis (Figure 5.2).

Typical isotherms are given in Figure 5.1, any isotherm with a downward curvature is considered favourable for adsorption. The ones with the upward curvature are unfavourable for adsorption. There are three different isotherms:

- Langmuir
- Freundlich
- BET—Brunauer-Emmett-Teller

The linear isotherm is assumed in most theories.

$$q = Ky \tag{5.1}$$

Different basis can be used to define the concentration of the solute in the adsorbent. Here, q is defined as solute concentration per unit volume of dry mass of adsorbent. K is the equilibrium constant and y is the solute concentration in the solution.

The *Langmuir isotherm* is more common than the linear isotherm. This has a sound theoretical basis. The isotherm is derived from a mass balance of the sites in the adsorbent available for adsorption.

$$[\text{filled sites}] + [\text{empty sites}] = \text{total sites} \tag{5.2}$$

Sites are subject to chemical equilibrium:

$$[\text{bulk solute}] + [\text{empty site}] \leftrightarrow [\text{filled site}] \tag{5.3}$$

$$\frac{[\text{filled site}]}{[\text{bulk solute}]\,[\text{empty site}]} = K'$$

$$[\text{filled site}] = K'\,[\text{bulk solute}]\,([\text{total sites}] - [\text{filled sites}]) \tag{5.4}$$

$$[\text{filled sites}]\,(1 + K'\,[\text{bulk solute}]) = K'\,[\text{bulk solute}]\,[\text{total sites}]$$

$$[\text{filled sites}] = K'[\text{bulk solute}]\left(\frac{[\text{total sites}]}{1 + K'[\text{bulk solute}]}\right) \tag{5.5}$$

$$q = \frac{q_0 y}{K + y} \tag{5.6}$$

where $K = 1/K'$ and q_0 is the total concentration of sites. Eq. (5.6) is the Langmuir isotherm. The reciprocal of q varies linearly with the reciprocal of y. The slope of such as plot is K/q_0 and its intercept is $1/q_0$.

Freundlich isotherms are represented by:

$$q = Ky^{n} \tag{5.7}$$

where K and n are both empirical constants. Sometimes, n is found to be less than 1. It is linear in logarithmic coordinates.

The solid material chosen to act as the packed column is called the *adsorbent*. The *solute* is referred to the species that gets adsorbed. The *adsorbate* refers to the continuous fluid that leaves the column. The *feed* refers to the mixture that needs the separation.

The *BET isotherm* is given by,

$$q = \frac{yq_0}{(1-y)(1+(B-1)y)} \tag{5.8}$$

Adsorbents can be classified into four categories—Carbons, Inorganic materials, Synthetic polymers, and Clay.

The adsorbent particle size may range from a few mm down to a few nanometers. Adsorbents must not provide a large pressure drop and may not be elutriated with the flow. The mechanical strength of the adsorbent needs to be sufficient so that it does not break during the operation. Adsorptive ability of the solids needs to be good and sometimes they may exhibit high selectivity. The surface area per unit weight of the adsorbent is usually very high (upto a million m^2/gm for nanoporous carbon).

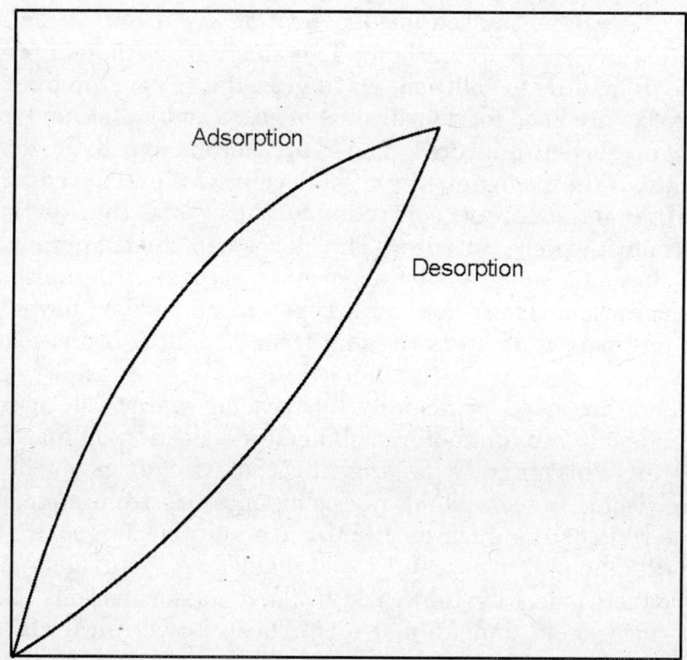

FIGURE 5.2 Hysteresis during adsorption desorption cycles.

Carbons can be further classified into decolourising carbons gas adsorbent carbons, molecular sieve carbons and activated carbons. *Decolourising carbons* are used to decolourise solutions of sugar, industrial chemicals, drugs, and dry-cleaning liquids, water purification, refining of vegetable and animal oils and recovery of gold and silver from cyanide ore-leach solutions. It is manufactured by (a) Mixing vegetable matter with calcium chloride, carbonising and leaching away the inorganic matter. (b) Mixing organic matter such as sawdust with porous substances such as pumice stone followed by heating and carbonising to deposit the carbonaceous matter throughout the porous particles. (c) Carbonising wood, sawdust followed by activation with hot air or steam. Lignite and bituminous coal are also raw materials.

Gas-adsorbent carbons are used for the recovery of solvent vapours from gas mixtures, gas masks, collection of gasoline hydrocarbons from natural gas, and the fractionation of hydrocarbon gases. It is made by carbonisation of coconut shells, fruit pits, coal, lignite and wood. Essentially a partial oxidation process must activate it by treatment with hot air or steam. *Molecular screening activated carbon* is used in the fractionation of acetylene compounds, alcohols, organic aides, ketones, aldehydes and many others. They are made with controlled pore openings from 5–5.5 A. In most activated carbons, the pore size ranges from 14–60 A. For example, pores can admit paraffin hydrocarbons, but reject isoparaffins of large molecular diameters. Bone char is used in refining of sugar made by destructive distillation of crushed, dried bones at temperatures in the range of 600–900°C.

Inorganic materials are used as adsorbents. Examples are alumina, silica gel, zeolites and molecular sieves. *Alumina* is used as desiccant for gases and liquids. Aluminium oxide that is hydrated and heated and thus activated is available as granules or powders. *Silica gel* is used for dehydration of air and other gases, in gas masks and for fractionating of hydrocarbons. It is made from the gel precipitated by acid treatment of sodium silicate solution. Hard granular porous product is obtained. *Zeolites, molecular sieves* are used for dehydration of gases and liquids and for separation of gas and liquid hydrocarbon mixtures. These are porous synthetic zeolite crystals metal aluminosilicates. The cages of the crystal cells can entrap adsorbed matter, and the diameter of the passageways controlled by the crystal composition, regulates the size of the molecules, which can enter. The sieves can thus separate according to molecular size but they also separate by adsorption according to molecular polarity and degree of unsaturation. There are nine types of molecular sieves industrially available with nominal pore diameters ranging from 3 to 10 A in the form of pellets, beads and powders.

Synthetic polymers are used principally by treating water solutions. There are porous spherical beads 500 mm diameter each bead a collection of microspheres with 100 nm in diameter. Polymers from unsaturated aromatics such as styrene, divinylbenzene, are useful for adsorbing nonpolar organics from aqueous solutions. Those made from acrylic esters such as PMMA are suitable for polar solutes.

Clay such as fuller's earth are useful in decolourising, neutralising, and drying such petroleum products such as lubricating oils, transformer oils, kerosene and gasoline as well as vegetable and animal oils. These are natural clays, American varieties coming largely from Florida and Georgia. These are magnesium silicates in the form of minerals attapulgite and montmorillonite. Clay is heated and dried during which it develops a pore structure, ground and screened. The commercially available sizes range from coarse granules to fine powders.

Activated clays are useful for decolourising petroleum products. These are bentonite, which have no adsorptive property unless treated with sulphuric acid or hydrochloric acid. Following such treatment the clay is washed, dried and ground to fine powder.

The concentration of the adsorbed gas will decrease with an increase in temperature at a given equilibrium pressure. A plot is obtained with respect to a reference substance such as acetone vapour on activated carbon. Points of constant adsorbate concentration are called *isosteres.* They are frequently straight lines in a plot of equilibrium partial pressures of acetone adsorbate vs the vapour pressure of pure acetone. The adsorption potential is given by RT ln(p/p^*). A master curve can be developed for all values obtained at several temperatures as shown in Figure 5.3.

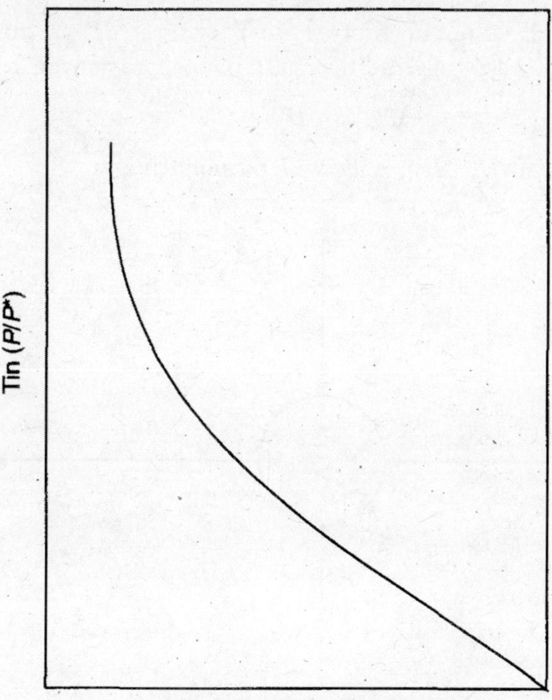

FIGURE 5.3 Master curve for adsorption potential for values obtained at several temperatures.

The *differential heat of adsorption* ($-\bar{H}$) is defined as the heat liberated at constant temperature when unit quantity of vapour is adsorbed upon a large quantity of solid already containing adsorbate so that the adsorbate concentration is unchanged. Integral heat of adsorption at any concentration X of adsorbate upon the solid is defined as the enthalpy of adsorbate—adsorbent combination—the sum of the enthalpies of unit weight of pure solid adsorbent and sufficient pure adsorbed substance to provide required concentration X, all at the same temperature. These are functions of temperature and adsorbate concentration for any system. The slope of the isostere is,

$$\frac{d\ln p^*}{d\ln p} = \frac{(-\bar{H})M}{\lambda_r M_r} \tag{5.9}$$

where M, M_r are the molecular weights of the vapour and reference substance respectively and λ_r is the latent heat of vaporisation of the reference substance. If the differential heat of adsorption is calculated at constant temperature for each isostere, the integral heat of adsorption can be calculated from:

$$\Delta H'_A = \int_0^Y \bar{H}\, dY \tag{5.10}$$

This is the area under the curve of $\bar{H}$ vs Y, where $\Delta H'_A$ is the energy per mass of adsorbate free solid and Y is the adsorbate concentration on a solute free basis.

$$\Delta H_A = \Delta H'_A + \lambda X \tag{5.11}$$

When heat is evolved the H, ΔH_A, $\Delta H'_A$ will be negative.

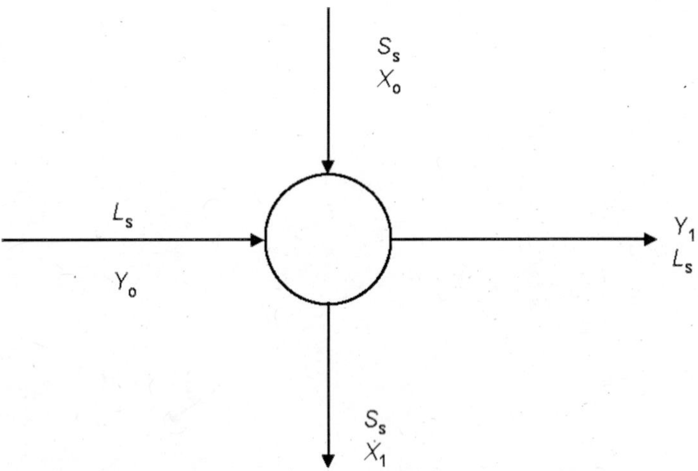

FIGURE 5.4 Single stage adsorption operation.

5.2 STAGED ADSORPTION OPERATIONS

The method of contact between the adsorbent and solution is extremely important in determining the efficiency of separation and the cost and residence time, etc. Compared to stirred tank, which represents a completely mixed system, contact of solution with a packed bed will lead to better removal of the contaminant in the feed stream. This is so because in a stirred tank, the equilibrium adsorbent concentration is with the exit concentration of solution. This would be expected to be in low quantity and hence the corresponding equilibrium concentration in the adsorbent is also low. In a packed bed, the equilibrium is with the feed solute concentration. This gives a higher q, and hence a higher concentration of solute in adsorbent leading to greater capacity for adsorption. Thus contact with a packed bed would be superior to contact, which is completely mixed from the standpoint of adsorptive capacity.

In a similar fashion, the method of contact such as cross-current and counter-current can play a pivotal role in the design of the adsorption system. The co-current system will require more number of stages to achieve the same degree of separation as down the stages, the adsorbent will have attained a concentration of solute which may be more than the equilibrium value permitted by the lower solute concentration in the solute downstream. The details of the cross-current, co-current and counter-current schemes are discussed below. The overall and component mass balances determine the operating line. The isotherm information may be used in the operating lines to obtain analytical solutions. Else graphical methods may be employed to step of the number of stages required to achieve a certain level of separation using adsorption.

Consider L_s kg of solvent that comes in contact with S_s kg of adsorbent as shown in Figure 5.4. The solute concentrations are defined in a dry basis. Thus the solute concentration in the feed is Y_0 and is expressed as kg of solute/kg of solvent and Y_1 is the solute concentration on a solvent basis after the completion of the adsorption operation. The solute concentration in the adsorbent on an adsorbent basis is given by X_0 and X_1 respectively for the cases before and after the adsorption.

An overall mass balance yields,

$$S_s + L_s = G_s + L_s \tag{5.12}$$

A component mass balance gives,

$$Y_0 L_s + X_0 S_s = L_s Y_1 + S_s X_1 \tag{5.13}$$

$$L_s(Y_0 - Y_1) = S_s (X_1 - X_0) \tag{5.14}$$

$$-\frac{S_s}{L_s} = \frac{(Y_2 - Y_0)}{(X_1 - X_0)} \tag{5.15}$$

The output from the first stage can form the input of a second stage and

$$-\frac{S_s}{L_s} = \frac{(Y_2 - Y_1)}{(X_2 - X_0)} \tag{5.16}$$

When the isotherm has a favourable distribution of solute towards the adsorbent, the adsorption operation is a powerful tool. The operating lines are shown in the distribution diagram in Figure 5.5. For systems that obey the Freundlich isotherm and where the fresh adsorbent is used, Eq. (5.15) can be written as:

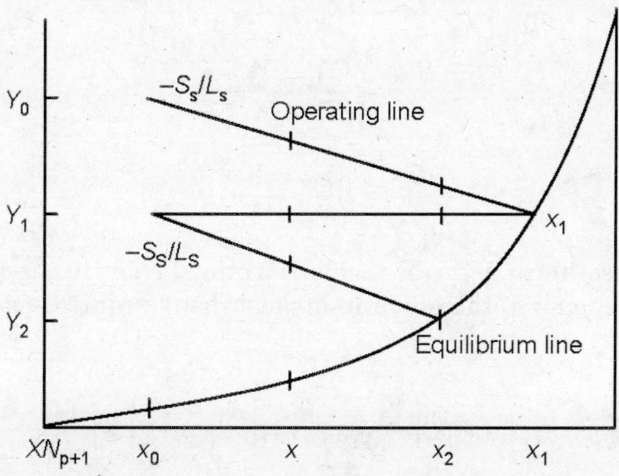

FIGURE 5.5 Operating line, equilibrium for staged adsorption operations.

$$-\frac{S_s}{L_s} = \frac{(Y_1 - Y_0)}{(Y_1/m)^{1/n}} \tag{5.17}$$

Eq. (5.17) permits analytical calculation of adsorbent/solvent ratio, for a given change in solution concentration from Y_0 to Y_1.

5.2.1 Multi-stage Cross-current Adsorption Operations

Cross-current operations are usually conducted in the batch mode and continuous operation is possible. Filtration is used in between the stages.

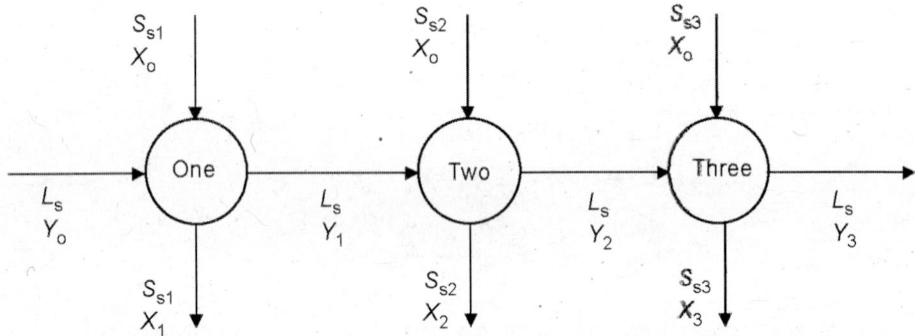

FIGURE 5.6 Multi-stage cross-current adsorption operation.

When the costs are tallied, the total cost including the filtration costs in between stages and the actual carbon used as a adsorbent, it is found that usually two stages are economical. From the mass balances as is shown in Eqs. (5.15 and 5.16),

$$-\frac{S_{s1}}{L_s} = \frac{(Y_0 - Y_1)}{(X_0 - X_1)} \tag{5.18}$$

$$-\frac{S_{s2}}{L_s} = \frac{(Y_1 - Y_2)}{(X_0 - X_2)} \tag{5.19}$$

$$-\frac{S_{s3}}{L_s} = \frac{(Y_2 - Y_3)}{(X_0 - X_3)} \tag{5.20}$$

When Freundlich isotherm describes the adsorption behaviour and fresh adsorbent is used ($X_0 = 0$). The least total amount of adsorbent required can be calculated from:

$$\left(\frac{S_{s1} + S_{s2} + S_{s3}}{L_s}\right) = m^{1/n}\left(\frac{Y_1 - Y_0}{Y_1^{1/n}}\right) + \left(\frac{Y_1 - Y_2}{Y_2^{1/n}}\right) + \left(\frac{Y_3 - Y_2}{Y_3^{1/n}}\right) \tag{5.21}$$

For minimum use of adsorbent,

$$\frac{\partial(S_{s1} + S_{s2} + S_{s3})}{\partial Y_1 L_s} = 0 \tag{5.22}$$

$$\frac{\partial(S_{s1} + S_{s2} + S_{s3})}{\partial Y_2 L_s} = 0 \tag{5.23}$$

where m, n, Y_0, Y_3 are constants,

$$0 = \left(1 - \frac{1}{n}\right) + \frac{1}{n}\left(\frac{Y_0}{Y_1}\right) + \left(\frac{Y_1}{Y_2}\right)^{1/n} \tag{5.24}$$

$$\frac{1}{n}\left(1 - \frac{Y_0}{Y_1}\right) = 1 + \left(\frac{Y_1}{Y_2}\right)^{1/n} \tag{5.25}$$

Similarly, Y_2 can be calculated for the adsorbent used to be a minimum. The minima can be confirmed by verifying that the second derivative is positive at that value of intermediate composition.

$$-\frac{1}{n}\left(\frac{Y_1}{Y_2}\right) - \left(1 - \frac{1}{n}\right) = \left(\frac{Y_2}{Y_3}\right)^{1/n} \tag{5.26}$$

Thus for the optimal case the intermediate compositions and the adsorbent needed can be calculated.

5.2.2 Multi-stage Counter-current Adsorption Operations

Multi-stage counter-current adsorption operations are superior to both batch and cross-current operations. It is operated at steady state in a continuous manner. For example, the simultaneous dissolution of gold and silver from finely ground ore by cyanide solution and adsorption of the dissolved metal upon granular activated carbon in a 3-stage operation in one testimonial to the benefit of counter-current contact adsorption operations. Coarse screens between agitated vessels separate the large carbon particles from the pulped-ore-liquid mixture. A mass balance of the solute about N_p stages can be performed as:

$$L_s(Y_0 - Y_{Np}) = S_s(X_1 - X_{Np+1}) \tag{5.27}$$

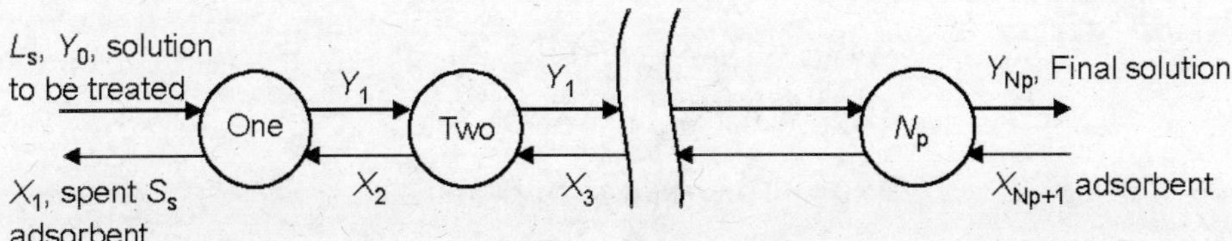

FIGURE 5.7 Multi-stage counter-current adsorption operations.

Eq. (5.27) denotes the operating line and passes through the points whose coordinates are (X_1, Y_0), (X_{Np+1}, Y_{Np}) and has a slope of S_s/L_s. The number of stages is obtained by stepping of the combination graph by staircase construction. The adsorbent/solvent ratio for a predetermined number of stages can be found by trial and error location of the operating line. When the operation is desorption the operating line is located below the isotherm. The pinch point can be located with the operating line with the largest slope, which requires infinite number stages for separation (Figure 5.8). If the equilibrium line is concave downward the pinch point is located at the point of tangency. When filtration is used in between it has been found that it is rarely economical to operate more than two stages. If the solvent is a gas, G_s can be used instead of L_s.

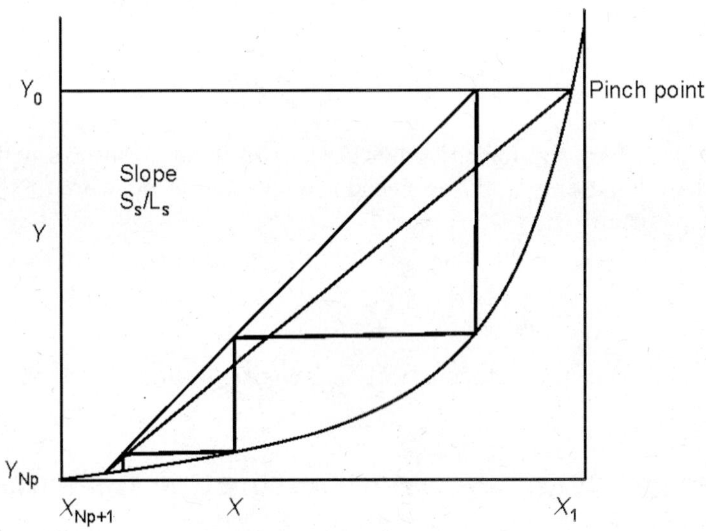

FIGURE 5.8 Number of ideal stages required for counter-current multi-stage adsorption.

When the isotherm is linear, the Kremser equations may be applicable. The expressions for the intermediate compositions can be written as:

$$mL_s/S_s(Y_0 - Y_{Np}) = (Y_1 - mX_{Np+1}) \tag{5.28}$$

Let the adsorption factor be defined as, $A_d = mL_s/S_s$ and the equilibrium relation be given by, $Y^* = mX$.

$$A_dX_0 - A_dX_{Np} = X_1 - X_{Np+1} \tag{5.29}$$

$$X_{Np+1} = A_d(X_{Np} - X_0) + X_1 \tag{5.30}$$

$$X_{Np} = A_d(X_{Np-1} - X_0) + X_1 \tag{5.31}$$

$$X_{Np-1} = A_d(X_{Np-2} - X_0) + X_1$$

Adding and subtracting X_1 on the RHS of Eq. (5.30):

$$X_{Np+1} - X_1 = A_d(X_{Np} - X_0) = A_d(X_{Np} - X_1 + X_1 - X_0) \tag{5.32}$$

$$= A_d(A_d(X_{Np-1} - X_0) + X_1 - X_0) \tag{5.33}$$

$$= A_d^2(X_{Np-1} - X_1) + (X_1 - X_0)\ (A_d + A_d^2) \tag{5.34}$$

$$= (X_1 - X_0)(A_d + A_d^2 + \dots + A_d^{Np}) \tag{5.35}$$

Sum of the geometric series:

$$S_n = A_d + A_d^2 + \dots + A_d^{Np}$$

$$S_n = \left(\frac{A_d^{Np+1} - A_d}{A_d - 1}\right) \tag{5.36}$$

Combining Eq. (5.36) with Eq. (5.35),

$$\left(\frac{X_{Np+1} - X_1}{X_1 - X_0}\right) = \left(\frac{A_d^{Np+1} - A_d}{A_d - 1}\right) \tag{5.37}$$

$$\frac{\ln\left((A_d - 1)\dfrac{(X_{Np+1} - X_1)}{(X_1 - X_0)} + A_d\right)}{\ln (A_d)} - 1 = N_p \tag{5.38}$$

The number of stages required can be calculated as a function of the adsorption factor and the inlet solution concentration and the outlet solution concentration, the input adsorbent solute concentration and the linear isotherm's slope m. Eq. (5.38) is applicable when $A_d \neq 1$, i.e., the adsorption factor is not equal to one.

Worked Example 5.1 *Special Case When Adsorption Factor is a (A_d = 1)*
Calculate the number of ideal stages required for counter-current multi-stage adsorption when the adsorption factor is 1 ($A_d = 1$).

When the isotherm is linear, the Kremser equations may be applicable. The expressions for the intermediate compositions when $A_d = 1$, can be written as:

$$Y_0 - Y_{Np} = Y_1 - mX_{Np+1} \tag{5.39}$$

Realising that the equilibrium relationship is linear, $y^* = mx$, q. (5.39) becomes:

$$X_0 - X_{Np} = X_1 - X_{Np+1} \tag{5.40}$$

$$X_{Np+1} = (X_1 - X_0) + X_{Np} \tag{5.41}$$

$$X_{Np} = (X_1 - X_0) + X_{Np-1} \tag{5.42}$$

Adding and subtracting X_1 to Eqs. (5.41, 5.42).

$$X_{Np+1} - X_1 = (X_1 - X_0) + X_{Np} - X_1 \tag{5.43}$$

$$X_{Np} - X_1 = (X_1 - X_0) + X_{Np-1} - X_1 \tag{5.44}$$

Substituting Eq. (5.44) into Eq. (5.43),

$$X_{Np+1} - X_1 = 2(X_1 - X_0) + X_{Np-1} - X_1 \tag{5.45}$$

Repeating this N_p+1 times,

$$X_{Np+1} - X_1 = (N_p + 1)\,(X_1 - X_0) + X_0 - X_1 \tag{5.46}$$

or

$$N_p = \frac{X_{Np+1} - X_1}{X_1 - Y_0/m} \tag{5.47}$$

5.2.3 Multi-stage Co-current Adsorption Operations

Multi-stage co-current adsorption operations are easier to analyse and are used sometimes. Counter-current contact is superior to co-current adsorption operations. As the adsorbent goes through the N_p stages, the concentration of the solute increases in concentration and the solution concentration decreases in concentration. As equilibrium is allowed to be established, the adsorption capacity of the operation will be lesser compared with a counter-current operation where fresh adsorbent meets the exit solution and the spent adsorbent encounters the solution concentration in the feed. The concentration profile when plotted as a function of distance travelled will never cross each other in a counter-current operation and may cross in a co-current operation. They cannot be allowed to cross by design as beyond that point there will be no transfer. Further as the number of stages is increased in a counter-current operation, the efficiency of separation will increase. In a co-current operation depending on the equilibrium characteristics it could well be that after p stages, the concentration in the solution is low enough to cause a low equilibrium concentration in the adsorbent, lower than the adsorbent concentration attained by then thus precluding further transfer of mass from the solution to the adsorbent. In some cases, co-current operation is used when the control of the operation is of paramount significance or when efficiency of separation takes a back seat to control and simplifies the cleaning operations.

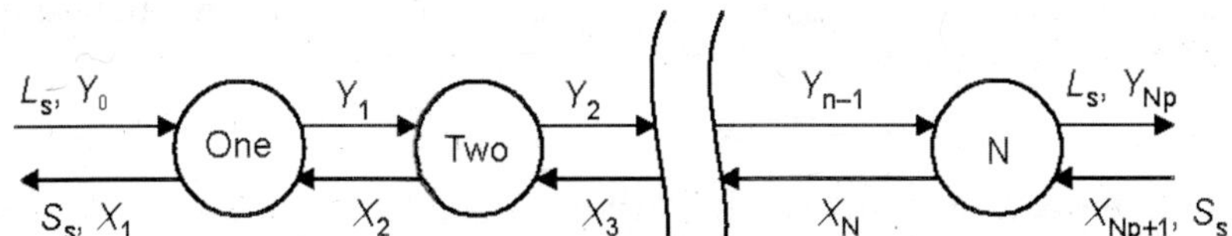

FIGURE 5.9 Multi-stage co-current adsorption operations.

A component mass balance can be written from Figure 5.9,

$$L_s(Y_0 - Y_{Np}) = S_s(X_{Np+1} - X_1) \tag{5.48}$$

or

$$-\frac{S_s}{L_s} = \frac{Y_0 - Y_{Np}}{X_{Np+1} - X_1} \tag{5.49}$$

The operating lines are shown in Figure 5.10. The lines have slope $-S_s/L_s$ as shown in Eq. (5.49) and the stage can be operated until equilibrium is reached and then the adsorbent and solution proceed to the next stage. It can be easier to control multi-component operations by co-current compared with counter-current operations. In order to achieve the results shown in Figure 5.10, fresh adsorbent needs to be added after each stage. This is because the equilibrium concentration in the adsorbent after the first stage of operation is greater than the desired outlet concentration of the solution after the subsequent stage. This can be facilitated only by addition of fresh adsorbent. If addition of fresh adsorbent is not desired then the maximum removal of solute from the solution would be to Y_1, which is in equilibrium with X_{1e}.

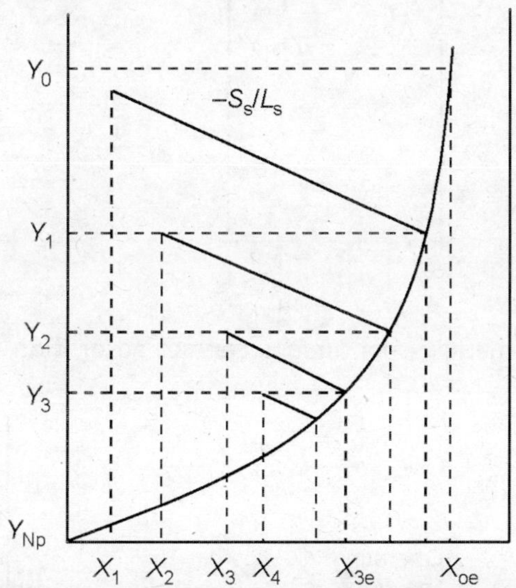

FIGURE 5.10 Operating lines and stages in continuous co-current adsorption operations.

5.2.4 Continuous Contact Adsorption Operations

The complete adsorption operation can be conducted in a packed tower where the adsorbent and solution are in intimate contact with each other throughout the entire apparatus. There are no periodic separation or stages. Moving bed adsorbers are used in the industry in the agitated form or as a fluidised bed contacting arrangement. *Hyperadsorber* is used for fractionating hydrocarbon gases with activated gas adsorbent carbon in the counter-current contact. The solid motion downward is essential in plug flow. In large bed operation, attrition losses from solid breakage have been found. *Higgins Contactor* is used to contact solid and liquid in an hourglass type configuration in a counter-current fashion. The conditions where fluidisation may occur are avoided.

A schematic of continuous counter-current contact adsorption operations is shown in Figure 5.11. Figure 5.12 depicts a design diagram of continuous contact adsorption operation. A component mass balance on the input and output streams can yield:

$$S_sX_2 + L_sY_1 = L_sY_2 + S_sX_1 \tag{5.50}$$

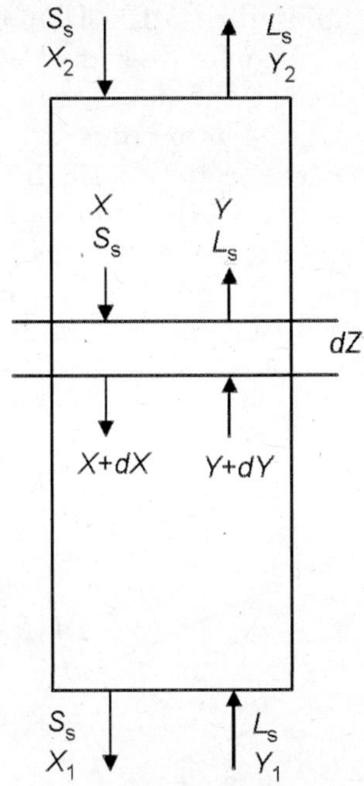

FIGURE 5.11 Schematic of continuous contact adsorption operations.

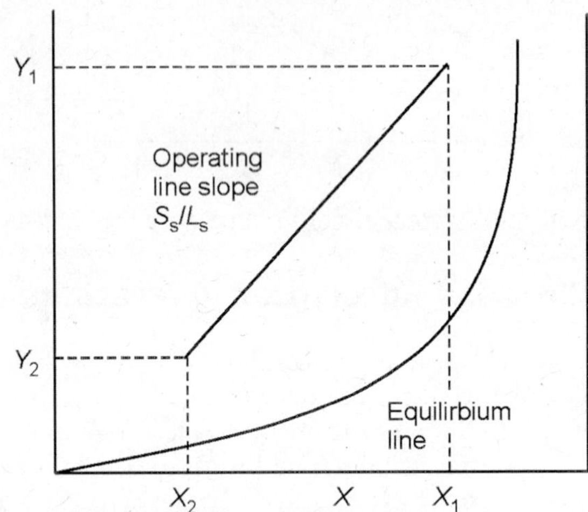

FIGURE 5.12 Design diagram of continuous contact adsorption operations in a counter current fashion.

or

$$\frac{S_s}{L_s} = \frac{Y_1 - Y_2}{X_1 - X_2} \tag{5.51}$$

A component mass balance on a thin slice of thickness dZ in the direction of the cross-section of the tower as showed in Figure 5.11 at steady state gives,

$$S_s dX = L_s dY = K_y a\rho(Y - Y^*)\, dZ \tag{5.52}$$

The number of transfer units and height of the transfer unit needed for performing the desired separation is as follows:

$$N_{tOG} = \int_{Y_2}^{Y_1} \frac{dY}{Y - Y^*} = \frac{K_y a\rho}{L_s} \int_{O}^{Z} dZ = \frac{Z}{H_{tOG}}$$

where the height of the transfer unit, $H_{tOG} = L_s/K_y a$ and K_y is the overall mass transfer coefficient and is the sum of the k_y and k_s the mass transfer coefficients in the solution and adsorbent respectively and m is the slope of the equilibrium line. The mass transfer correlations for moving bed adsorbers can be used for the mass transfer coefficients in the adsorbent phase. Further:

$$\frac{L_s}{K_y a} = \frac{L_s}{k_y a} + \frac{mL_s}{S_s}\frac{S_s}{k_s a} \tag{5.53}$$

$$H_{tOG} = H_{tG} + A_d H_{ts} \tag{5.54}$$

5.3 BREAKTHROUGH CURVE, SATURATION AND EXHAUSTION TIMES

The characteristics of the adsorption operation can be studied using the breakthrough curve. The breakthrough curve is a plot of the adsorbate concentration as a function of time. As shown in Figure 5.13, the concentration of solute in the exit stream of a fixed bed adsorber starts lean in the solute concentration. The first few layers of the adsorbent are involved in the separation of the solute from the feed stream leaving a zone of adsorbent that is untouched by the solute.

With the passage of time, the adsorbent bed gets completely saturated. When the adsorbent bed gets saturated, the adsorbate concentration at the exit of the bed begins to increase. This point is called the *breakthrough point*. This is shown as (t_B, C_c) in Figure 5.13. Here, t_B is the breakthrough time and denotes the time taken for the volume of effluent to register an increase in solute concentration post adsorption. After the adsorbent has reached saturation, the exit stream will have the same concentration as the feed stream. This point is referred to as the *exhaustion point*. This is shown as (t_E, C_d) in Figure 5.13.

Thus,

$$t \le t_B,\ y = 0 \tag{5.55}$$

$$t_B \le t_E,\ y = y_0\left(\frac{t - t_B}{t_E - t_B}\right) \tag{5.56}$$

$$t > t_E,\ y = y_0 \tag{5.57}$$

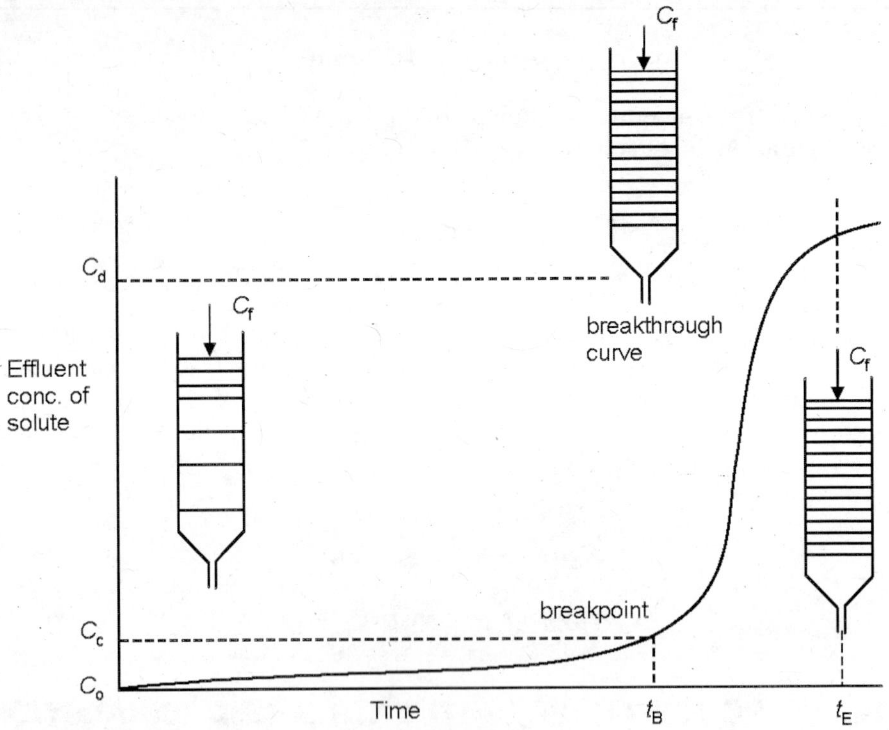

FIGURE 5.13 Breakthrough curve, breakthrough time, exhaustion time.

At the break point,

$$z \leq 1_{sat}, \; y = y_0 \tag{5.58}$$

$$1_{sat} < z \leq 1, \; y = \frac{y_0(1 - z)}{(1 - 1_{sat})} \tag{5.59}$$

$$z > 1, \; y = 0$$

where,

$$1_{sat} = (t_E - t_B)\frac{(y_0 U)}{q_0(1 - \varepsilon)} \tag{5.60}$$

where U is the superficial velocity of the solution (m/s) and ε is the bed voidage. Thus the adsorbent bed's saturation level can be estimated using Eq. (5.59).

The effect of diffusion resistance can be seen in the breakthrough curves. This can be expected in a transient operation. From the breakpoint to the exhaustion point, the saturation zone advances with a velocity U_{sat}. This can be related with the feed velocity U and time taken to saturation as:

Time taken to establish a saturation zone $t_{sat} = q_0(1 - \varepsilon)/ka\rho(y_0 - 0)$ (5.61)

The interfacial concentration of the solute in the solution is assumed to be zero indicating the potential for complete removal. In the time in excess of the saturation time, the volume of the adsorbent zone up to where the saturation zone has advanced is given by $U_{sat}(t - t_{sat})A$, where A is the area of the cross-section of the bed.

$$Uay_0(t - t_{sat}) = q_0(1 - \varepsilon)U_{sat}A(t - t_{sat}) \tag{5.62}$$

Eq. (5.62) is written from a mass balance from the amount that would have been gained by the adsorbent in the excess time with the amount that would have been removed from the feed solution. A steady state is assumed and any accumulation in the bed is neglected. Rewriting Eq. (5.62),

$$U_{sat} = U\left(\frac{y_0}{q_o(1 - \varepsilon)}\right) \tag{5.63}$$

The length of the saturation zone can be written as:

$$l_{sat} = U_{sat}(t - t_{sat}) \tag{5.64}$$

$$= \frac{Uy_o t}{q_o(1 - \varepsilon)} - \frac{U}{ka\rho} \tag{5.65}$$

In the adsorption zone at steady state assuming no accumulation of the solute,

$$ka\rho(y - 0) = -\frac{U\partial y}{\partial z} \quad \text{or} \quad \frac{dy}{y} = -\frac{ka\rho}{Udz} \tag{5.66}$$

$$y = y_0 \text{ at } z = 1_{sat} \tag{5.67}$$

Integrating Eq. (5.67) and solving for the integration constant using the boundary condition given in Eq. (5.67):

$$\frac{y}{y_0} = \exp\left(-\frac{ka\rho}{U}\right)(z - 1_{sat}) \tag{5.68}$$

5.4 CHROMATOGRAPHY

Chromatography is a special case of adsorption. In chromatography a pulse of feed solution is used at the top of the adsorbent bed. The column is washed with solvent repeatedly until the separation is effected. Chromatography is a technique used to identify unknown chemical species. By calibration or theory the elution time of every chemical is paired up with its unique elution time. Based upon the elution time of the unknown sample, the chemical structure is identified. This technique is preferred in expert witness court testimonies and detective agencies with good reason. It is a powerful technique. The human genome project was completed ahead of time in part due to good chromatographic techniques to identify the sequence distribution of the polynucleotide. When a pulse is injected, it is called *elution chromatography*. When a step change in concentration is injected, it is called *frontal chromatography*.

There are different kinds of chromatography. The classification is shown in Figure 5.14. Gas chromatography was introduced in the market by Burrel corp. and is the most frequently used technique. Each chromatographic system shown in Figure 5.14 can be operated by three different methods of operation. The three principle methods of chromatography are *frontal, displacement and elution*. Adsorption of mobile phase onto the stationary phase is allowed to happen in the method of frontal chromatography. *Frontal analysis* involves the study of the breakthrough curve, the saturation and exhaustion times, etc. In *displacement chromatography,* a small sample is displaced by a more strongly held mobile phase so that the sample is gradually pushed out of the column as the mobile phase advances. Bands develop and are used in environmental analysis. In *elution chromatography,* the solute is not allowed to stay in the adsorption column as in the cases of frontal and displacement chromatography.

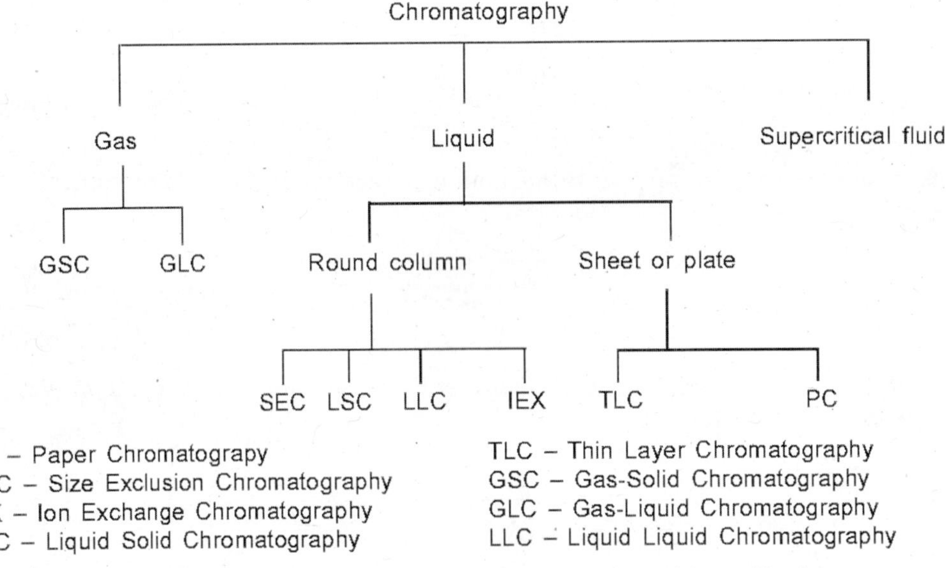

FIGURE 5.14 Classifications of chromatographic methods.

The packing in the column used can have an important role in the separation. This is so because of the interaction between the sample and the surface and pore structure of the column. Gas-solid chromatography has some advantages over liquid-solid chromatography. Column has greater thermal stability. The problems of evaporation and loss of liquid are absent. The demerit of the system is the strong interaction between the sample and the column.

Molecular sieve is the most common type of solid phase packing in gas-solid chromatography. Molecular sieves are zeolites or carbon sieves that have a regular pore structure. It can be used to separate small molecules such as O_2, N_2, CO gases and low carbon number hydrocarbons. Isomers can be separated by this method. Porous polymers can also be used as column material provided high temperature of operation is not needed. Silica, alumina and metal oxides and salts have been used as packing material. Reproducibility is a limiting criterion for ubiquitous use of these materials. The column tubing can also play a part in the separation. For sensitive compounds such as pharmaceuticals like steroids and pesticides, glass tubes are used. The glass surface is inert compared with metals. Transparency also allows for visual inspection that comes in handy to detect column contamination and degradation. Glass is fragile and not as durable as metal. Compared to glass, materials such as stainless steel, nickel, copper or aluminium are less likely to get damaged during handling. On account of less corrosion the stainless steel alloy is preferred among this class of materials. Columns made out of Teflon, PTFE, polytetrafluoroethylene are difficult to pack and connections without leaks are difficult to achieve. For capillary columns, fused silica is the material of choice for packing. Data from this packing is reproducible. A polyimide layer coating is given over the fused silica columns. The fused silica column is an inherently straight wire of material and is straight at its rest state. The process of making a fused silica column is complex and requires complex equipment such as high temperature furnace of operating temperature of about 2000°C and a laser-based system for determining the true diameter of the ultimate product. Prior to placing in the oven, the material must be wound onto a frame that secures its coiled configuration. Au contraire, the glass columns can be prepared in standard laboratories by using inexpensive drawing machines.

PLOT and SCOT columns are certain types of glass capillary columns. PLOT is the porous layer open tubular column. The packing in the capillary column is etched and pores created. SCOT columns are support coated open tubular, and have a coating on the wall of the column.

The mobile phase selection is based on the detector used. Gases with higher molecular weight are selected. Examples are N_2, CO_2 or Ar. If high velocities are not permitted, a lower molecular weight gas such as H_2 or He may be selected. The purity of the gas is an important consideration. Gases used are at least 99.995% pure. More than 50 ppm impurity in the solution phase can obviously interfere with the detection of the desired species, which is also present in small quantity. Further the nature of the contaminant is also a consideration. Water and air as contaminants can affect the stability for the liquid phase in a packed column. Sometimes a compromise has to be made in the selection for meeting a balance of requirements.

The role of a GC detector is to identify the presence of a component of the sample at the exit of the column. *Selectivity* is the property that allows the detector to discriminate between constituents. Some detectors respond well to a certain component type and not to other chemical species. Sensitivity and signal to noise ratio are salient considerations in the selection of a detector. The TCD, thermal conductivity detector gets used a lot of the time. Hot body loses heat at a rate that depends on the composition of the material. It can be used for quantitative analysis. Accurate analysis requires response analysis or a reliable transient conduction mathematical model. FID, the flame ionisation detector, taps into the current generated from ions formed when organic materials are burned in a small oxygen-hydrogen flame at the end of the column. In order to ease the detection of the current, a voltage is applied across the region and a responsive electrometer is used for measurement of the current. ECD is the electron capture device which is very responsive and can be used to detect highly electronegative compounds such as chlorinated hydrocarbons, pesticides and polychlorinated biphenyls. A radioactive source emits electrons which is interacted with by the sample. At the detector the loss of the electrons are registered. Nickel-63 and tritium are used in the detectors. FPD, flame photometric detector is selective for organic compounds containing phosphorous and sulphur detecting chemiluminescent species formed in a flame from these materials. The chemiluminescence is detected through a filter by a photomultiplier. The photometric response is linear in concentration for phosphorous but it is second order in concentration for sulphur. The minimum detectable level for phosphorus is 1 pico gram and for sulphur it is about 50 pica grams. AFID, the alkali flame ionisation detector, detects the current that arises from ions formed in excess of thermal ionic formation when a phosphorous or nitrogen containing organic material is placed in contact with an alkali salt above a flame. Such a detector at the end of a column then reports on the elution of these compounds.

In *liquid chromatography,* the mobile phase is a liquid. Thin layer chromatography, paper chromatography and high pressure liquid chromatography are all members of this class of processes. Liquid chromatography originated in the late 60s and 70s. Use of pressurised liquid is made of in the technique of HPLC. Materials that cannot be handled in the gas chromatography such as non-volatile compounds are handled in the liquid chromatography. Many thermally unstable compounds such as natural products, pharmaceuticals and biomacromolecules are separable by partitioning between a liquid mobile phase and a stationary phase. The composition of the liquid phase can be changed during the operation in order to improve the efficiency of the separation. In classical chromatography, the usual system consists of a polar adsorbent, and a non-polar solvent such as a hydrocarbon. In practice, the situation is often reversed in which case the technique is known as reversed-phase liquid chromatography.

Paper chromatography originated in the 40s the 50s. A chamber is used to isolate the column which is a piece of filter paper and a glass plate coated with an adsorbent such as silica gel. The chromatogram is developed by allowing a mobile phase to creep through the column carrying with it materials soluble to various extents. After this process has proceeded sufficiently, the column is removed from the solvent tank and the mobile phase evaporated. The separated components are visualised elsewhere, for example under an ultraviolet lamp in which various fluorescent bands indicate how fluorescent materials are separated by the movement of the solvent. Paper chromatography is a special type of liquid-liquid chromatography. The paper inherently contains bound water that acts as a stationary liquid phase. In TLC, the usual mechanism for separation is partitioning resulting from adsorption on the stationary. Paper and TLC may be further classified as either 1 or 2 dimensional, ascending, descending, analytical or preparative.

Affinity chromatography is a technique that involves the use of a bio selective stationary phase placed in contact with the materials to be purified, the ligate. Because of its rather selective interaction, sometimes called a lock-and-key mechanism, this method is more selective than other LC systems based on differential solubility. Affinity chromatography is sometimes called *bio selective adsorption*.

Chiral chromatography is used for the analysis of enantiomers and is most useful for the separations of pharmaceuticals and biochemical compounds. Chiral adsorbents that use attractive interactions are metal ligands, inclusion complexes and protein complexes. The separation of optical isomers has important ramifications especially in biochemistry and pharmaceutical chemistry. One form of a compound may be bioactive and the other inactive, inhibitory or toxic.

In *Ion-exchange chromatography*, the column contains a stationary phase having ionic groups such as sulphonate or carboxylate. The charge of these groups is compensated by counterions such as sodium or potassium. The mobile phase is usually an ionic solution, e.g., NaCl having a pH and salt concentrations that acts as the separation variables, having ions similar to the counterions. Ionic samples are introduced into the mobile phase and retardation in movement results from ion exchange with the stationary phase of the form:

$$M^+ + NaSO_3 \leftrightarrow Na^+ + MSO_3 \tag{5.69}$$

where M^+ is the ion in the sample to be analyzed and sulphonate groups are assumed to be a part of the stationary phase. The more the ion interacts with the exchanger, the more strongly it is retained. For cation-exchange chromatography, positively charged ions are separated as shown in the equation. In anion-exchange chromatography, negatively charged ions in the sample interact with and bind to cationic stationary phases. In IC, ion-chromatography, a weak ion-exchange column is used for separation. Detection is usually done conductimetrically. After passing through the weak ion-exchange column, the eluent passes through a subsequent column called a stripper column in which the stream usually made acidic or basic in the ion-exchange column is neutralized.

In SEC, *size exclusion chromatography* or GPC, *gel permeation chromatography*, the material with which the column is packed has pores in a certain range of sizes. Molecules or solvent-molecule complexes too large to pass through these pores pass rapidly through the column, whereas molecules or complexes of sufficiently small size are retained and are the last to exit the column. Molecules of intermediate size are partially retained and elute from the chromatographic column at intermediate times. SEC is extremely useful as a tool for characterisation of polymer materials because the retention mechanism is reproducible enough to give good comparative data. SEC can also give valuable information about the distribution of sizes of molecules in a sample.

5.5 ION EXCHANGE

Ion exchange is a special case of adsorption. Metathetical reactions take place between the ion exchange resin and the solution treated. Ion exchange materials first used were porous, natural or synthetic minerals containing silica, zeolites, such as the mineral $Na_2O.Al_2O_3.4SiO_2.2H_2O$. Positively charged ions of a solution diffuse through the pores will exchange with the positive ions of such a mineral which is referred to as a *cation exchanger*. For example,

$$Na^+ + RSO_3H \rightarrow RSO_3Na + H^+$$

$$H^+ + Cl^- \rightarrow HCl$$

where R represents the residual mineral of the zeolite. Thus sea water can be desalinated using two ion exchange beds. First the sea water is passed through a cation exchange bed. The sodium ions are exchanged with the sulphonic acid resin. Subsequently the water with acid is passed through the anion exchanger.

$$Cl^- + RNH_3OH \rightarrow RNH_3Cl + OH^- \tag{5.70}$$

$$H^+ + OH^- \rightarrow H_2O \tag{5.71}$$

The anions are exchanged and the water is now free of sodium chloride. The ion exchange beds are subsequently recharged with treatment with strong acid and alkali respectively. Similarly, hard water can be softened by exchanging the calcium Ca^{2+} ions, with the less objectionable Na^+ ions. The zeolite exchange resin can be regenerated by contacting with a solution of salt. The ion exchange resin is in the form of fine, granular solids or beads.

Ion exchange operations can be batchwise or continuous, multi-stage or continuous contact, counter-current or co-current or cross current, fluidised bed and packed bed. The design procedures for the ion-exchange technique are similar to that of adsorption. The equilibrium line can be obtained from various empirical equations such as the Freundlich isotherm. The equations of mass action type can be used to represent the equilibrium reactions. For example in the softening of hard water,

$$Ca^{2+} + Na_2R \rightarrow CaR + Na^+ \tag{5.72}$$

The zeolite is regenerated by washing with salt later. The harness causing calcium ion is immobilised in the ion exchange resin and is replaced by sodium in the water. The mass action law constant may be written as:

$$\alpha = \frac{(CaR)\,(H^+)}{(Na_2R)\,(Ca^{2+})} \tag{5.73}$$

α is the relative adsorptivity and the quantities within the brackets in Eq. (5.73) are concentrations. Eq. (5.73) can be written for any general ion exchange resin and system used for treatment as:

$$\alpha = \left(\frac{X}{X_0 - X}\right)\left(\frac{C_0 - C^*}{C^*}\right)$$

$$= \left(\frac{X/X_0}{1 - X/X_0}\right)\left(\frac{1 - C^*/C_0}{C^*/C_0}\right) \tag{5.74}$$

where, X is the equilibrium cation concentration after the exchange, X_0 is the concentration if all of the available ions in the exchanger were exchanged by the cation of interest, C_0 is the initial concentration of the cation in solution and C^* is the equilibrium concentration of the cation after the exchange. In some cases, the relative adsorptivity was found to be constant through the entire range of the concentration.

Worked Example 5.2 *Sea Water Desalination by Ion Exchange*
An ion-exchange column with sulphonic acid cation exchange resin is used to remove Na^+ ions from sea water. The solution is percolated through the bed and is at a feed concentration of 0.62 meq/litre. The resin contains 8 meq Na^+/litre resin at saturation. The solution flow rate is 4.5 lit/s. The relative adsorptivity α can be taken as 13.5 for this system and assumed to be constant for this system at the concentration range of operation. For a counter-current continuous operation, operate the ion-exchange column at twice the minimum ion-exchange resin needed for a final purity of 100 ppm of water at the exit. What is the concentration of NaCl of the spent resin if fresh ion-exchange resin is used at the beginning of the operation?

The equilibrium line can be constructed in Figure 5.15 by re-arranging Eq. (5.74) as follows:

$$\alpha\left(\frac{X_0}{X} - 1\right) + 1 = \frac{C_0}{C^*} \tag{5.75}$$

The operating line can be constructed as:

$$L_s(C_0 - C_1) = S_s\,(X_2 - 0)$$

The minimum S_s/L_s needed is found from Figure 5.15 by identifying the pinch point and is 0.075. The operation is twice the minimum needed. So,

$$\frac{S_s}{L_s} = 0.150 = \left(\frac{0.62 - 0.0017}{X_2}\right)$$

or

$$X_2 = 4.12$$

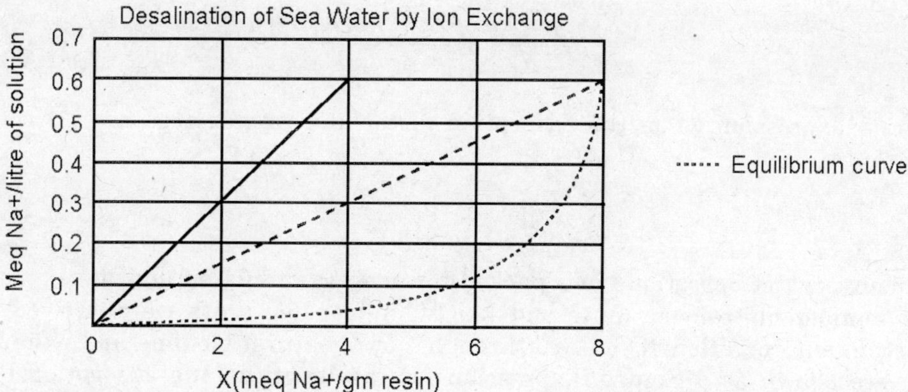

FIGURE 5.15 Counter-current continuous exchange of Na^+ from sea water by sulphonic acid resin.

SUMMARY

Adsorption is a process by which separation of the desired component is effected by contacting a liquid or vapour with a packed column. Different kinds of adsorbents such as carbons, synthetic polymers, activated clay and inorganic zeolite materials can be used. The commonly found isotherms are linear isotherm, Langmuir isotherm, Freundlich isotherm and BET isotherm.

$q = Ky$ (linear isotherm); $q = q_0 y/(K + y)$ (Langmuir isotherm)

$q = Ky^n$ (Freundlich isotherm); $q = yq_0/\{(1 - y)(1 + (B - 1)y\}$ (BET isotherm)

The operating line for a single-stage contact adsorption operation was derived and shown in a schematic along with the equilibrium line,

$$-\frac{S_s}{L_s} = \frac{Y_2 - Y_1}{X_2 - X_0}$$

A multi-stage cross-current adsorption operation was discussed. Three stages were selected and the minimum adsorbent weight required was derived. The intermediate compositions were calculated using calculus. A counter-current multi-stage operation is developed as the most efficient of contacting methods. The operating line forms a parallel to the equilibrium line and is given as:

$$L_s(Y_0 - Y_{Np}) = S_s(X_1 - X_{Np+1})$$

The Kremser equations were derived for a linear isotherm to give an expression for the number of ideal stages required as a function of the adsorption factor, compositions of adsorbent and solution compositions and given as:

$$\frac{X_{Np+1} - X_1}{X_1 - X_0} = \frac{A_d^{Np+1} - A_d}{A_d - 1}$$

A separate expression when the adsorption factor is 1 was derived as:

$$N_p = \frac{X_{Np+1} - X_1}{X_1 - Y_0/m}$$

Continuous contact operations in a packed tower were discussed and the expression for the number of transfer units and height of transfer units were derived for a counter-current operation. The breakthrough curve, saturation time and exhaustion times were shown as a Figure. Expression for the length of the adsorption bed to attain saturation was derived from mass balance on a slice of packed bed. The different kinds of chromatography were outlined. The column packing type, detector type and applications were discussed. Ion exchange as a special case of adsorption was discussed. The law of mass action was used to obtain the equilibrium relationship. Adsorptivity was defined. An example with sea water desalination and the use of cation exchange resin and anion exchange resin in sequence were elaborated.

EXERCISES

1. What is the difference between dehumidification and dehydration?
2. Which would fare between in the contacting pattern for adsorption—fluidised bed or packed bed and why?
3. Why is there no recycle in the multi-stage adsorption operations?
4. A scrubber is proposed to be placed after the fuel tank in the automobile. The objective of the scrubber is to reduce the sulphur emissions to less than 40 ppm according to the California phase II specifications. Can a fluidised bed be used as a scrubber? What are the advantages and limitations?
5. The conditions of flooding, coning, weeping and dumping in a fixed bed chemisorber can be avoided using a PI, proportional-integral flow controller. The variables selected were liquid flow rate and liquid solid ratio. What other variables will be of interest for the purpose of control?
6. What is the difference between desorption and elution?
7. Why is co-current contact not as attractive as counter current-contact in multi-stage adsorption operations.
8. When is an isotherm considered favourable for adsorption and Why?
9. Show that the Langmuir isotherm is essentially linear at low solution concentrations.
10. Show that when the exit concentration of the n–1 stage in a cross-current multi-stage operation is the same as the inlet concentration of solute in the adsorbent of the nth stage for all stages, the contact scheme becomes a co-current scheme. Thus co-current is a special case of cross-current continuous multi-stage adsorption operations.
11. Can the cross-current operations be performed in a continuous contact tower? Why?
12. At what concentrations of the solution and adsorbent would the use of a log paper be recommended?
13. How will a figure similar to that of Figure 5.13 look like should the concentration at a fixed point in the adsorbent bed be plotted? Why?
14. In multi-stage cross-current contact operations should the adsorbent weight used in each stage be made equal. Why?
15. Can the Y values used in adsorption problems become greater than 1? Explain.
16. Can the Kremser equation shown in section 5.2.2 for counter-current multi-stage adsorption operations be derived for a Langmuir isotherm? Given reasons.

PROBLEMS

1. *Contaminant Purification in Cabin of Airplane*
The contaminant concentration in the cabin of an airplane is 1 wt%. Design a cross-current adsorption multi-stage operation with three stages. What are the intermediate concentrations for the solution with a target output at 1000 ppm. The spent adsorbent concentration is 0.005. Fresh adsorbent is used. The isotherm obeyed is the Freundlich isotherm with K and n values of 0.413 and 0.86 respectively. What is the ratio of the adsorbent used to the solution feed rate processed? (**Ans:** 0.0052;0.0024; $-S_s/L_s = 0.88$)

2. *Desulphurisation of Gasoline using Adsorption onto Nickel*
The EPA mandate for engines from 2006 and beyond, and the California phase II specification require that the sulphur content of gasoline be less than 40 ppm. The thiopenes, mercaptans, sulphides and disulphides present in the gasoline pump prior to operation in an internal combustion engine can be treated in a chemisorption bed of Nickel.

$$Ni + CH_3SH \longrightarrow NiS + CH_4$$

For a feed solution of 5% sulphide concentration, design a counter-current multi-stage contact operation. The spent adsorbent concentration is 1%. How many ideal stages are required, should the isotherm obeyed is of the Langmuir type with a K value of 0.08 and q_0 is 0.25? What is the adsorbent to solution flow ratio recommended? (**Ans:** 10 stages; $S_s/L_s = 0.5$)

3. *Desalination of Sea Water by Molecular Sieve*
Sea water at a salt concentration of 3.6 wt% is adsorbed onto nanoporous carbon. The outlet concentration of potable water is 413 ppm. Assume that the adsorbent solution system follows the BET isotherm with a B value of 0.75. Fresh adsorbent is contacted in a counter-current fashion in a packed tower. What is the height and number of transfer units needed to effect the desired separation. The interfacial area available for liquid-solid contact is 1 million m^2/gm. Select a suitable correlation for the mass transfer coefficient from chapter 3. What is the solution to adsorbent flow ratio?

4. *Cleaning of Semi-Conductor Manufacturing Device by Adsorption onto Teflon*
Semi-conductor chip manufacturers have recognised the deleterious effects of oxide deposits on the reaction chamber walls in which the various chemical reactions and deposition processes take place during chip manufacture. As impurities build up on the reaction chamber surfaces such as interior chamber walls, the risk increases that such impurities may be co-deposited on target work piece surfaces such as the computer chips. Therefore such chambers must be periodically cleaned during down cycles in the chip manufacturing process. One way to clean the unwanted deposits from the interior reaction chamber walls is to produce F, fluorine in the reaction chamber under sub-atmospheric conditions to remove unwanted silicon containing oxide deposits. Adsorption is carried out in an arrangement of 2 or more adsorption beds operated in parallel

and operated out of phase so that at least one bed is undergoing adsorption while another bed is being regenerated.

$$2NF_3 \longrightarrow N_2 + 3F_2$$

Fluorine gas with contaminant is passed through an adsorbent bed and the waste gets adsorbed. The fluorine is recycled. The adsorbent is selected in a fashion that it does not react with the fluorine. PTFE, polytetrafluoroethylene, Teflon is a suitable candidate. Pores are formed by reacting the adsorbent with isopropanol and extruding the mixture and then removing the isoproponal after extrusion. Pores with less than 1 micron diameter are formed.

For a contaminant concentration of 2% and an exit concentration of 1000 ppm using fresh adsorbent and spent adsorbent at a concentration of .003 on the basis of dry adsorbent volume, how many ideal stages are required by counter-current contact. What is the ratio of the solution feed rate and adsorbent flow rate? The isotherm obeyed is of the Langmuir type at a K value of 0.3 and q_0 of 0.1. (**Ans:** 3.5 stages; S_s/L_s = 6.67)

5. *Ethane/Ethylene Separation*
   Design a multi-stage co-current operation to separate ethane from ethylene by adsorption onto montmorillonite clay. After three stages what is the adsorbent concentration at a L_s/S_s ratio of 1.5. The inlet concentration of the ethane is 12 wt% and the outlet composition of the ethane is 500 ppm. The fresh adsorbent is brought in contact with the feed solution. Assume a Fruendlich isotherm with K and n values of 0.413 and 1.4 respectively. (**Ans:** 0.0025)

6. *Flue Gas Desulphurisation and SO_2 Recovery using Styrenic Polymer*
   Flue gas is brought in contact with adsorbent consisting of calcium oxide. The sulphur-dioxide adsorbs onto the bed by a method of chemisorption.

$$2CaO + 2SO_2 + O_2 \longrightarrow 2CaSO_4$$

   The flue gas concentration is 2% and the target exit composition is 42 ppm. The continuous contact is in a packed bed tower in a counter-current fashion. What is the height and number of transfer units needed to meet the desired objectives. Fresh adsorbent is used to begin with and the spent adsorbent composition is 1/10th of 1%. Which mass transfer correlation did you select. The porosity of the bed is 0.43. Assume a linear isotherm with an m value of 10. The flow rates of flow of the flue gas and adsorbent are 3 kg/s and 4 kg/s respectively.

7. *Capture of Mercury Vapour from Coal Fired Power Plant*
   Calculate the number of ideal stages required to adsorb mercury vapour at 2.6% feed solution and reduce the stream content of mercury to 56 ppm by counter-current adsorption. Fresh adsorbent is used and a spent adsorbent concentration is 0.002 on the basis of dry adsorbent volume. The linear isotherm has an m value of 0.3. What is the ratio of the feed rate of the solution and that of the adsorbent? (**Ans:** 3 stages; S_s/L_s = 12)

8. *Separation of p-xylene from a mixture of xylenes*
Isomers can be separated using the technique of adsorption. p-xylene can be adsorbed onto zeolite from a mixture of p-xylene, o-xylene and m-xylene. Assuming a counter-current multi-stage contact of a feed solution at 8% concentration of p-xylene and a spent adsorbent concentration of 2000 ppm, what is the ratio of adsorbent flow rate to the solution flow rate for a spent adsorbent concentration of 2.5%. Show that three ideal stages are required to effect the separation. The isotherm is of the Langmuir type as shown in Fig. 5.16. (**Ans:** S_s/L_s = 2.61)

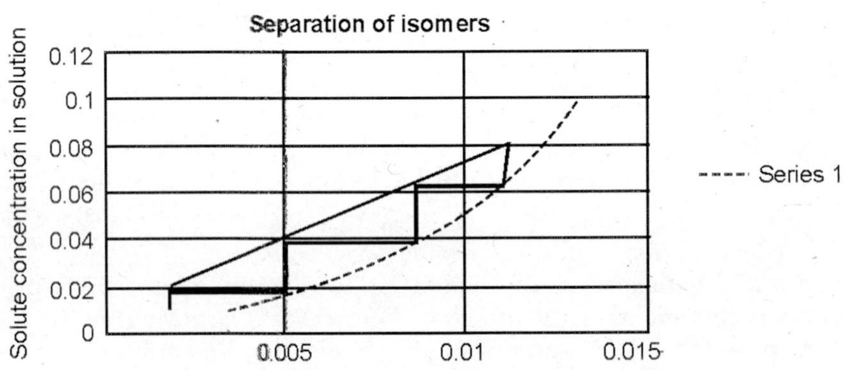

FIGURE 5.16 P-xylene separation from xylenes.

9. *Kremser Equation for Multi-stage Co-current Continuous Contact*
Derive an expression for the number of stages required to achieve the desired purity level as shown in Figure 5.9 using Kremser equation. Assume that the equilibrium relationship between the solute in solution and adsorbent is linear. Show that

$$N_p = \frac{\ln\left(\dfrac{1 - N_{Np+1}(1 - A_d)}{X_0 A_d}\right)}{\ln A_d}$$

What happens when adsorption factor becomes 1?

10. *Colour Removal*
Two thousand kg of a tinted solution must be clarified by adsorbing 97.3% of the tint on adsorbent carbon. Given that the tint concentration is 0.95 and that the equilibrium is a Freundlich's isotherm with m = 0.275 and n = 0.6. (a) How much fresh adsorbent is needed for batch adsorption? (b) How much is needed for a packed bed? (**Ans:** (a) 61.0 kg; (b) 6.9 kg)

11. *Steroid Recovery*
Chemically modified steroid hormone is suggested to be used like hydrocortisone for reducing allergic reactions. The breakthrough curve is shown in Figure 5.17. The superficial velocity is 1 cm/min through a packed bed of 310 μm spheres

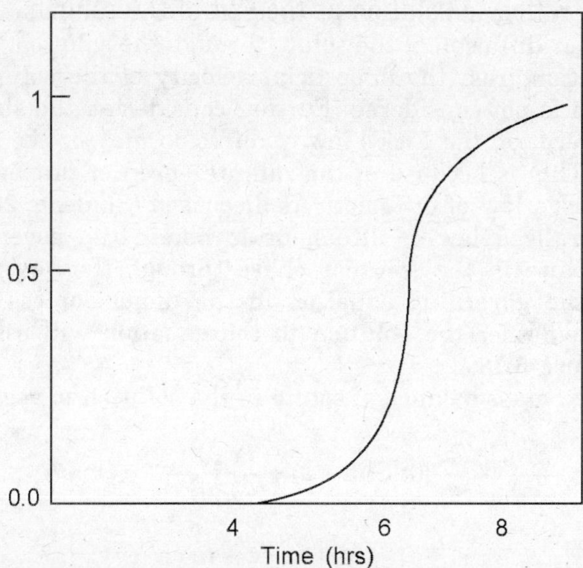

FIGURE 5.17 Breakthrough curve of steroid hormone.

with a void fraction of 0.4. The saturation concentration in the adsorbent is 300 times the feed concentration in the solution. What is $ka\rho$?

(**Ans:** 0.047 s^{-1})

12. *Decolourising Sugar*
During the extraction of sugar from sugarcane water is removed in a triple effect evaporator. Prior to that, the colour of the sugar needs to be removed before the product is acceptable. A 1 m packed bed is used as adsorbent and the solution is fed into the column at a superficial velocity of 2.2 cm/s. The bed has a void fraction ε of 0.36. The mass transfer coefficient is 0.26 s^{-1}. A linear isotherm may be assumed with an m value of 0.413. Find the concentration at the break point. What is the length of the adsorption zone? What is the breakthrough time in a 3 m bed at the same breakthrough concentration?

13. *Desalination of Sea Water by Co-Current Ion Exchange*
Repeat worked example 5.2 for a co-current operation of ion-exchange. What is the exit concentration of ion-exchange resin for a ratio of S_s/L_s of 2.5. Sea water contains 3.6% NaCl and the exit water needs to be less than 100 ppm concentration. Fresh ion exchange resin needs to be used at the beginning of the operation. (**Ans:** 0.0144)

14. *Accumulation in the Solution Phase*
One possible explanation for the nature of the breakthrough curve shown in Figure 5.13 is the diffusion in the solution phase. During the adsorption operation up to saturation, the solute gets adsorbed leaving the solution phase lean in solute throughout the column of the adsorption bed. When saturation point is reached the adsorption of solute ceases from solution to the adsorbent. At the inlet of the adsorption column the solute concentration is higher than the

solute concentration in solution at the exit of the column. This will provide a driving force for diffusion of the solute through the solution. This is in addition to the convection from the superficial velocity of the solution. So, often the axial diffusion is not considered. Further considering the short times involved during the operation the Fick's law of diffusion may not be adequate to model this system. This is because of the infinite speed of propagation of the mass implied by Fick's law of diffusion. As discussed (Sharma, 2005) in Chapter 2. use the generalised law of diffusions to obtain the governing equation for transfer of solute in the solution phase through the voids of the adsorbent bed. Obtain the governing equation in the dimensionless form and discuss qualitatively whether the solution to the equation will result in a curve as shown in Figure 5.13.

Show that a mass balance of solute in the solution phase can be written as:

$$\frac{\varepsilon \partial^2 u}{\partial \tau^2} + \frac{\partial u}{\partial \tau}(a^*\text{Sha} + \varepsilon) + \frac{Pe_m \partial^2 u}{\partial Z \partial \tau} + a^*\text{Sha}\, u$$

$$= \frac{\varepsilon \partial^2 u}{\partial Z^2} - \frac{Pe_m \partial u}{\partial Z}$$

where Sharma number, Sha $= k\tau_r/a_w$; dimensionless interfacial area, $a^* = k\rho a$

$$\text{Peclect number (mass)} = \frac{U}{v_m} = \frac{U}{(D/\tau_r)^{1/2}};\ u = \frac{y}{y_0};\ \tau = \frac{t}{\tau_r};\ Z = \frac{z}{(D\tau_r)^{1/2}}$$

$Au = W \exp(-n\tau)$ substitution will reduce the governing equation for $n = -(1/2 + a^*\,\text{Sha}/\varepsilon)$ to:

$$\frac{\varepsilon \partial^2 W}{\partial \tau^2} + \frac{Pe_m \partial^2 W}{\partial Z \partial \tau} - \frac{W}{4\varepsilon(a^*\text{Sha} - \varepsilon)^2} = \frac{\partial^2 W}{\partial Z^2} + \frac{Pe_m}{\varepsilon \partial W/\partial Z}\left(\frac{a^*\text{Sha}}{2} - \frac{1}{2}\right)$$

It can be shown that for some conditions Pe_m. Sha, a^* the concentration will exhibit subcritical damped oscillations assuming a finite length of the adsorption bed. Further for the point shown in Figure 5.13, a time lag can be derived using an infinite depth of the adsorption bed. This is the mass inertia regime. This occurs at the first zero of the Bessel function. From the stated law that concentration can never be less than zero a region of no concentration change can be seen. This is followed by two rising regimes given by a Bessel composite function and modified Bessel composite function of space and time.

15. *Crystalline Magnesium Chloride from Sea Water*
Sea water contains 1300 mg/litre of magnesium. Design a counter-current adsorption bed made of montmorillonite clay. Given that the system obeys a Langmuir isotherm of K value of 0.051 and q_0 value of 0.1, what is the minimum adsorbent flow rate required for counter-current multi-stage operation for a sea water throughput of 8 litres/s. Use a log paper if necessary and calculate the number of ideal stages required to operate at 1.8 times the minimum adsorbent solution ratio. What is the spent adsorbent concentration of adsorbent?

(**Ans:** S_{smin} = 4.4 lit/s; 4 stages; 2e–4)

16. *Vanaspati and Decolourisation*

Hydrogenation in well stirred reactors is used in the fat and oil industry to remove the unsaturation in oils and obtain a raise in the melting point of the fat and improve its resistance to rancid oxidation. The major end product in India is Vanaspati, a solidified household oil used in vegetarian cooking. Other products are vegetable ghee, hardened industrial oils and partially hydrogeneted liquid oil. The hydrogenated oil is deodourised in a vacuum tower. The next step is decolourisation using fuller's earth as adsorbent. The isotherm obeyed by the system can be assumed to be of the Langmuir type with a K value 5 and q_0 value of 11 in the colour units. The original solution has a colour concentration of 20 and it is desired to reduce the colour to 1.5% of its original value. What would be the concentration of the spent adsorbent when the operation is conducted at 1.5 times the minimum ratio of adsorbent to solution flow rate is needed. How many ideal stages are required for the operation. Use fresh adsorbent to begin with. **(Ans:** 2.8 stages; 0.1)

17. *Desorption of Heavy Water*

Electric power generation in India is mostly from coal-based thermal and hydroelectric plants. The installed capacity of nuclear power plants is expected to be increased to 10 GW in the next decade. Plants have been commissioned at Kalpakkam and Koodankulum. The nuclear programme in India has been designed to make good use of the limited uranium reserves and the five times higher thorium reserves. Apart from the first two boiling water reactors, using enriched uranium all the other reactors are pressurised heavy water reactors. PHWRs, pressurised heavy water reactors use enriched uranium as fuel and heavy water as moderator and coolant. Based originally on the CANDU design, the first two reactors were setup with Canadian collaboration in Rajasthan. The design of these reactors has been continuously improved and two PHRW units each has been set up at Kalpakkam near Madras and at Narora in UP and another unit at Kakrapar in Gujarat. Construction of 500 MW PHWR is proceeding at Tarapur. The heavy water, deuterium oxide, D_2O is found in low

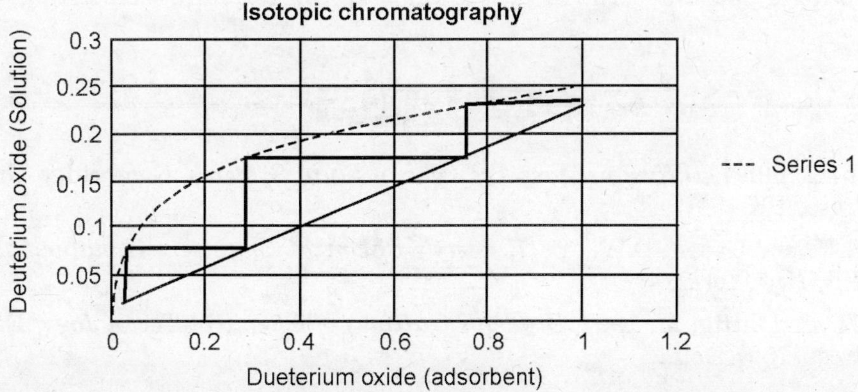

FIGURE 5.18 Desorption of D_2O from a spent silica gel adsorbent bed.

concentration in sea in the ppm range. Using isotopic chromatography, the D_2O is adsorbed onto a chromatographic column such as silica gel. The adsorbed bed can be desorbed to obtain D_2O. Deuterium oxide is produced in India in quantities that they can be exported.

Starting with an adsorbed column with a concentration of 1.0 kg/kg of dry adsorbent and a product solution of 0.24 on a solvent basis is brought in contact in a counter-current multi-stage desorption apparatus as shown in Figure 5.18. The feed solution has a concentration of 1% of the solute and a desorbed bed concentration of 0.02 on a dry adsorbent basis. Show that 3 ideal stages are needed to complete the operation. What is the ratio of the desorbent flow rate and the solution flow rate? (**Ans:** 0.24)

18. *Minimum Adsorbent Weight for Counter-Current Adsorption Operations*
In counter-current multi-stage contact adsorption operations with 3 stages, derive the minimum adsorbent flow rate required assuming Freundlich isotherm.

19. *Minimum Adsorbent Weight for Co-Current Adsorption Operations*
In co-current multi-stage contact adsorption operations with 3 stages, derive the minimum adsorbent flow rate required assuming Freundlich isotherm.

20. *Kremser Equation for Co-Current Adsorption Operations*
In co-current multi-stage adsorption operations when the isotherm is linear, derive the relation between the numbers of stages required and the adsorption ratio, spent adsorbent concentration, feed concentrations of the solution and adsorbent for a desired level of purification.

21. *Riccati Equation*
For ion-exchange resin systems with a constant adsorptivity, use Riccati difference equation and using calculus of finite differences, derive a recurrence relation for the composition at each stage for a multi-stage counter-current contact operation. Can you calculate the number of stages?

22. *Desorption and Kremser Equation*
Using the approach shown in 5.2.2, derive an expression for the number of stages that are required to desorb a spent adsorbent bed assuming a linear isotherm and a desorption factor. Indicate the relevant compositions needed.

REFERENCES

Cussler, E.L., 1997, *Diffusion Mass Transfer in Fluid Systems,* Cambridge University Press, UK.

Perry, R.H. and Green, D.W., 1997, *Perry's Chemical Engineers Handbook,* McGraw Hill, New York.

Rao, G.M. and Sittig, M, 1997, *Dryden's Outlines of Chemical Technology,* East West Press, New Delhi.

Sharma, K.R., 2005, *Damped Wave Transport and Relaxation*, Elsevier, Amsterdam.

________, October 2002, *Two Stage Cross Current Chemisorber for Desulfurizing Automobile Gasoline*, Rocky Meeting Regional Meeting of the ACS, RMRM 02, Albuquerue, NM, USA.

________, October 2002, *Hyperbolic Wave Propagative Equation Solution to Chemisorption of Organic Sulfides onto Palladium*, Rocky Meeting Regional Meeting of the ACS, RMRM 02, Albuquerue, NM, USA.

________, October 2002, *Fluidized Bed Chemisorber for Desulfurization of Automobile Gasoline,* Rocky Meeting Regional Meeting of the ACS, RMRM 02, Albuquerue, NM, USA.

________, September 2003, *Removal of Arsenic from Drinking Water by Molecular Sieve Adsorption*, 226th ACS National Meeting, New York.

________, October 2003, *Transient Fixed Bed Adsorption Examination using Hyperbolic Partial Differential Equations,* 52nd Southeast Regional Meeting of the ACS, SERMACS, New Orleans, LA.

________, October 2000, *Sea Water to Boiler Feed Water—Comparative Study of the Methods of Distillation, Adsorption, Reverse Osmosis, Extraction, Ion Exchange and Ultrafiltration*, 52nd Southeast Regional Meeting of the ACS, SERMACS, New Orleans, LA, USA.

Treybal, R.E., 1980, *Mass-Transfer Operations*, McGraw Hill, New York.

CHAPTER 6

Reverse Osmosis, Molecular Sieves and Electrodialysis

Nomenclature

a	solute radius (m)
A_p	cross-sectional area of the pore in capillary (m^2)
A_N	Avagadro number (6.023 E23 molecules/mol)
C_s	concentration of solute (mol/m^3)
C_{sf}	concentration of solute in feed (mol/m^3)
C_{sp}	concentration of solute in product (mol/m^3)
f_w	pure component fugacity of solvent, (water)
f_s	fugacity of solution
J	volumetric fluid flow rate across the membrane (m^3/s)
L_p	hydraulic conductance (m^2s/kg)
M_w	molecular weight (gm/mol)
N	number of solutes
r	capillary radius (m)
t_m	capillary wall thickness (m)
S_a	sieving coefficient
T	temperature (K)
P^w	pressure of solvent, (water) (N/m^2)
P^s	pressure of solution (N/m^2)
R	universal gas constant (J/mol/K)
S	cross-sectional area of the membrane (m^2)
S_c	circumferential area of a given capillary (m^2)
V_w^L	molar volume of water (m^3/mol)
x_w	mole fraction water
x_s	mole fraction of solute

Greek

γ_W	activity coefficient
π	osmotic pressure (N/m^2)
μ	viscosity of fluid (kg/ms)
ε	porosity of the capillary wall
ρ	density of solute (kg/m^3)

6.1 OVERVIEW

Reverse osmosis, ultrafiltration, molecular sieves and electrodialysis can be used to separate liquid solutions. Reverse osmosis and ultrafiltration differ in the solute size rejected. The conventional filtration rejects, particles of the size of 10 microns and above, *ultrafiltration* rejects solute sizes of 10 nm to 10 microns and reverse osmosis rejects solute sizes in the ionic range of 1 pico meter to 0.1 nm. The driving force for RO, ultrafiltration and molecular sieve operations come from a pressure difference. No filter cake is allowed to form. The driving force in *electrodialysis* is an induced electric field and polarisation of the ions to the anode and cathode. The compartments are made from alternating anode and cathode and the ions get segregated from the diluate water in alternating compartments. It is used to desalinate brackish water and is found to be the most economical method at low salt concentration in the feed at little over 1 wt%. Any material that can exclude molecular species by size is referred to as *molecular sieve*. Inorganic materials that possess uniform pores have diameters in either less than micro- range 2 nm, or meso range 2–20 nm.

In *reverse osmosis* the solvent from the solution is pumped across a semi-permeable membrane opposing the osmotic pressure difference with the solute largely rejected by the membrane. (Figure 6.1). For example in sea water desalination by reverse osmosis, the osmotic pressure will cause a flow from the region of low solute

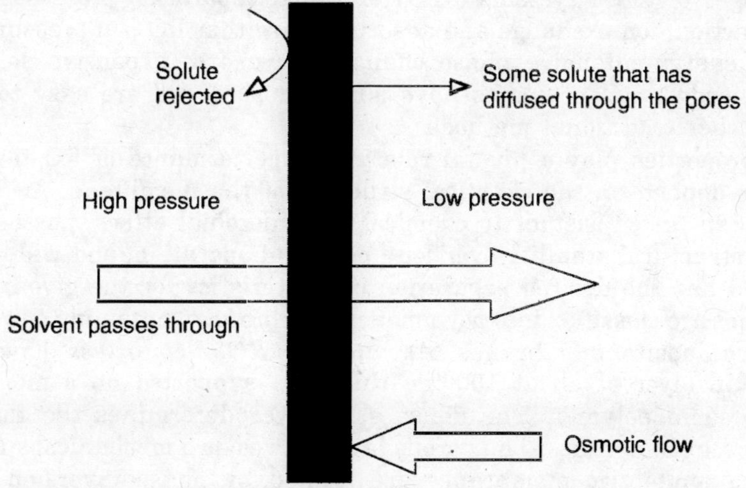

FIGURE 6.1 Transport processes during reverse osmosis.

concentration to a region of higher solute concentration. Pressure is supplied in order to overcome this, so that the solvent from the sea water flows across a semi-permeable membrane that permits only the solvent and not the solute. Although the membrane rejects the solute, some will diffuse across the pores.

6.2 INTRODUCTION TO REVERSE OSMOSIS

The drinking water needs of major cities around the world can be met from the reverse osmosis method for desalination as the drought conditions in major cities have become an important concern. In the Middle-East gulf region where the energy is cheap and river water scarce, sea water reverse osmosis, SWRO plants are increasingly relied upon to obtain potable drinking water. Reverse osmosis is also used in the food industry to concentrate fruit juices. Reverse osmosis is dominated by desalination that is by removing salt from water. It is used in the treatment of sea water, brackish water and reclamation of municipal waste water. It is used in boiler feed water pretreatment in order to avoid scale formation later on. The market for ultra pure water is growing and is used in injectable pharmaceuticals and semi-conductors, domestic use, sweetener concentration, and fruit juice concentration and fermentation product recovery.

In the 1950s, it was first shown that cellulose acetate RO membranes were capable of separating salt from water. But the water fluxes obtained were small to be feasible in practice. Since then the RO membrane technology has improved a lot. Recently Du Pont patented a polyamide membrane made from interfacial polymerisation and has a high solvent flux and turbulence to reduce the concentration polarisation layer thickness formed during the operation of the RO during desalination. Advances in thin-film composite membranes and polymer materials have widened the applications of RO from desalination to the treatment of hazardous wastes, material recovery in electroplating industries, production of ultra pure water, water softening and food processing, dairy and semi-conductor industries.

The advantages of these systems over traditional separation processes such as distillation, extraction, ion exchange and adsorption are that RO is a pressure-driven process and no energy intensive phase changes, or use of expensive solvents or adsorbents are needed. RO processes have simple design and are easy to operate compared with other traditional methods.

Membrane properties play a pivotal role in the performance of RO technology. These properties depend on the chemical structure of the membrane. An ideal RO membrane is low in cost, resistant to chemical and microbial attack, possesses high mechanical and structural stability over long period of operation, and wide range of temperature, and has the desired separation characteristics for the given system.

RO membranes are classified into asymmetric membranes (containing one polymer) and thin film composite membranes. *Asymmetric RO membranes* have a thin permselective skin layer of about 100 nm thickness supported on a more porous sublayer of the same polymer. The dense skin layer determines the fluxes and selectivities of these membranes. The porous layer serves as a mechanical support for the skin layer. Asymmetric membranes are formed by phase inversion polymer precipitation process. In this process, a polymer solution is precipitated into a polymer-rich solid phase that forms the membrane and a polymer-poor liquid phase that forms the membranes or void spaces. *Composite RO and ultrafiltration membranes* are thin films consisting of a thin polymer barrier layer formed on one or more porous support layers which is a different polymer compared with the barrier layer. The surface layer determines the flux and separation characteristics of the membrane. The porous backing is intended largely for the support of the barrier layer. The barrier layer is extremely thin thus allowing high water fluxes. The most important thin-film composite membranes are made by interfacial polymerisation, a process in which a highly porous membrane such as polysulphone is coated with an aqueous solution of a polymer or monomer and then reacted with a cross-linking agent in a water-immiscible solvent.

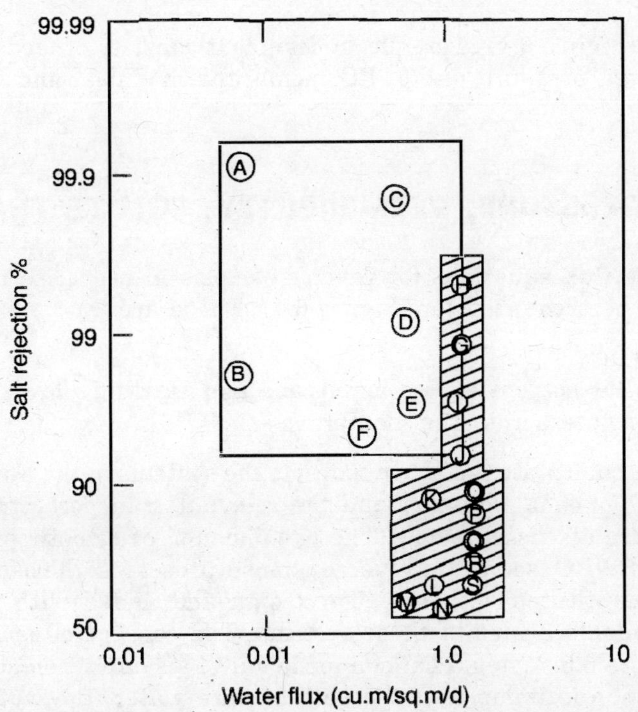

(A-F—Sea water membranes-operate at 5.5 Mpa and 25°C. H-J Brackish water membranes—operate at 1500 mg/L NaCl feed, 1.5 Mpa and 25°C. K-S—nanofiltration membranes—operate at 500 mg/L NaCl feed, 0.74 Mpa and 25°C.

A—cellulose acetate/triacetate
B—linear aromatic polyamide
C—cross-linked polyether
D—cross-linked fully aromatic polyamide
I—Du Pont A-15
J,K, L—Nitto-Denko, NTR-739HF
NTR—729HF, NTR-7250
O—Toray, UTC-40HF
E—other thin film composites
F—asymmetric membranes
G—FilmTec, BW 30
H—Toray, SU –700
Q—Toray, UTC 60
R—Toray, UTC 20HF
S—Filmtec, NF50
P—Filmtec, NF70

FIGURE 6.2 Water flux and salt rejection for various SWRO semi-permeable membranes.

The most important RO membrane materials are linear and cross-linked aromatic polyamide, aryl-alkyl polyetherurea and cellulose acetate/triacetate. Cellulose acetate materials hydrolyze and exhibit compaction upon application of higher pressure thus reducing the water flux and have a narrow temperature and pH application range. Polyamide and polyurea composite membranes have higher water fluxes and solute rejections (Figure 6.2) and operate at higher temperature and pH range and better resistance to biological attack, and mechanical compaction. These membrane are less resistant to chlorine attack.

The membrane module needs to offer mechanical support to the RO membrane upto 8 Mpa, offer low pressure drop, and be resistant to fouling and concentration polarisation. The commercially available modules are plate and frame, tubular, spiral wound and hollow fibre elements. For large scale desalination plants, spiral wound modules are used.

The RO process (Figure 6.4) is simple in design. It consists of feed water source, fed pretreatment, high pressure pump, RO membrane modules, and in some cases post-treatment steps.

6.3 OSMOTIC PRESSURE, PERMEABILITY, SOLUTE TRANSPORT

In order to write the flux equations for reverse osmosis in a precise manner, three phenomena need to be accounted for (Figure 6.1). These are:

1. Osmotic pressure
2. Permeability for solvent across membrane and starling's law
3. Diffusion of solute across the membrane

In order to better quantitate the above concepts the systems in the human anatomy, the solute transport, plasma filtration and the relevant transport processes across the cell membrane are discussed below. The phenomenon of osmotic pressure needs to be understood. In SWRO, sea water reverse osmosis processes, the osmotic pressure drop needs to be quantitated with the correct sign and added with the hydraulic pressure drop for reliable design of these systems. The concept of osmotic pressure is illustrated in Figure 6.3. When a balloon made out of a semi-permeable membrane filled with salt solution is immersed in a bath of pure water, the water will travel from the jar into the balloon and the size of the balloon will increase until equilibrium is reached in terms of the chemical potential on both sides of the membrane. The semi-permeable membrane can be chosen in a manner that it permits only water and not the solute in large measure. The flow of water is an example of the concept of osmotic pressure. *Osmosis* is the flow of solvent from a region of low solute concentration to a region of high solute concentration. The pressure difference that causes this flow is called as *osmotic pressure*. This pressure is caused by the presence of solutes. Hence it is called *colloid osmotic pressure*. For the human plasma in the blood, the colloid oncotic pressure is about 28 mm Hg. The colloid osmotic pressure is small compared with the osmotic pressure developed when a human cell is placed in pure water. The total osmotic pressure of the intracellular fluid would be 5450 mm Hg at 37°C.

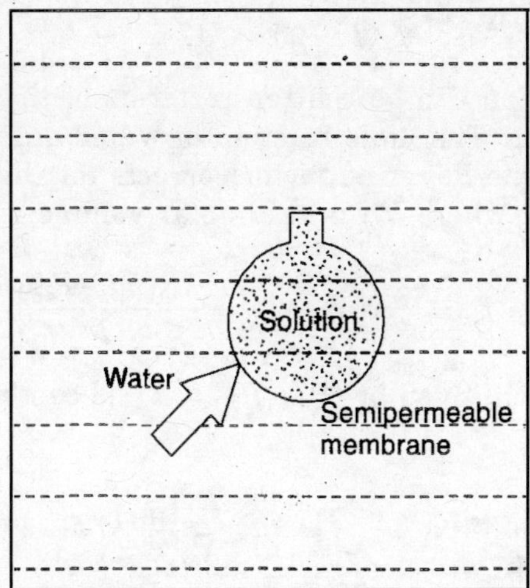

FIGURE 6.3 Concept of osmotic pressure.

This is caused by the presence of solutes such as K^+ ions, phosphocreatine, Mg^{++}, Cl^-, HCO_3^-, HPO_4^{2-} ions, carnosine, amino acids, creatine, lactate, Na^+ ions, urea, ATP, hexose monophosphate, and others. A number of solute molecules contribute to the osmotic pressure. Some of them are dissociating. For example, NaCl dissociates to the Na^+ and Cl^- ions. Each ion particle exerts its own osmotic pressure. The charge of the ion has no bearing on the osmotic pressure. Substances such as glucose do not dissociate and the osmotic pressure exerted is based upon its concentration. The term *osmole* is introduced to account for the effect of a dissociating solute. One osmole is therefore defined as one mole of a nondissociating substance. One mole of dissociating NaCl is equivalent to two osmoles.

Osmolarity defines the number of osmoles per litre of solution. For physiological solutions the units used for convenience is mOsM. If a cell is placed within a solution that has a lower concentration of solutes or osmolarity, then the cell is in a *hypotonic solution* and establishment of osmotic equilibrium requires the osmotic flow of water into the cell. The influx of water into the cell results in swelling of the cell and a subsequent decrease in its osmolarity. On the other hand, if the cell is placed in a solution with a higher concentration of solutes or osmolarity, i.e., *hypertonic solution*, then osmotic equilibrium requires osmotic flow of water out of the cell. An *isotonic solution* is a fluid that has the same osmolarity as that of the cell. When cells are placed in an isotonic solution there is neither swelling not shrinkage of the cells. Examples of isotonic solutions are 0.9% by weight NaCl in water solution and 5% by weight of glucose solutions with respect to a human cell.

The ***vant Hoff's law*** to determine the osmotic pressure in terms of the concentration of the solution can be derived from the concept that fugacity of different phases needs to be equal at equilibrium. Thus at equilibrium, in Figure 6.3 the temperature and fugacity of the water and that of the solution must be equal. Fugacity is a measure of the chemical potential of the system. As discussed above, the difference in the chemical potential between the solution and solvent causes the osmotic flow from a region of low solute concentration to a region of high solute concentration.

$$f_w(T, P_w) = f_s(T, P_s) \tag{6.1}$$

The fugacity of solution can be written in terms of the pure component fugacity using the Poynting factor. The mole fraction of water and the activity coefficient of water are also needed. The Poynting factor corrects for the effect of pressure on the pure component fugacity where V_w is the molar volume.

$$f_w = \gamma_w x_w f_w \exp\left(-\frac{V_w(P_w - P_s)}{RT}\right) \tag{6.2}$$

The osmotic pressure is given by $(P_s - P_w) = \pi$ and can be solved for from Eq. (6.2) as:

$$\pi = (P_w - P_s) = -\frac{RT}{V_w} \ln(\gamma_w x_w) \tag{6.3}$$

For an ideal solution, the activity coefficient may be taken as 1. For dilute systems using Taylor series expansion, the logarithmic functionality can be approximated as:

$$\ln(x_w) = \ln(1 - x_s) \approx -(x_s) \tag{6.4}$$

Eq. (6.3) becomes on substituting for Eq. (6.4) in Eq. (6.3);

$$\pi = \frac{RTx_s}{V_w} = RTC_s \tag{6.5}$$

Eq. (6.5) is called the van't Hoff's law to determine the osmotic pressure. Eq. (6.3) may be used where activity coefficient information is available. If the solution contains N ideal solutes, then the osmotic pressure can be obtained as a sum of the contributions from each solute,

$$\pi = \frac{RTx_s}{V_w} = RT\sum_{1}^{N} C_s \tag{6.6}$$

The number of molecules and not the absolute weight of the solute determine the osmotic pressure of the solution. The combined effect of osmotic pressure and hydrostatic pressure can be seen in *Starling's Law* which gives the relation between the flow of fluid across the capillary wall or a porous membrane and the pressure difference across the capillary. The volumetric fluid transfer rate, J, across the capillary membrane is given as:

$$\frac{J}{L_p S} = \Delta P_h - \Delta\pi = \overline{\Delta P_h} \tag{6.7}$$

where the effective pressure ΔP_h is the resultant of the hydrodynamic pressure drop ΔP_h, and osmotic pressure difference $\Delta\pi$. L_p is the hydraulic conductance and S is the effective cross-sectional area through which the fluid flows. The hydraulic conductance is often determined by experiment. The hydraulic conductance varies from 1E–9 m^2s/kg in capillaries in the kidney's glomeruli to 1E–14 m^2s/kg for endothelial cells found in the capillaries of the rabbit brain. One use of Eq. (6.7) is to better understand the flow of plasma across the capillary wall in human anatomy. This can be used in the SWRO systems used to desalinate sea water to drinking potable water. The hydraulic conductance can also be derived from the properties of the system. Thus should the membrane be viewed as a series of parallel cylindrical pores,

$$L_p = \left(\frac{r^2 A_p}{8 \mu t_m S_c} \right) \tag{6.8}$$

where A_p is the cross-sectional area of the pore of radius r in the capillary wall, S_c is the circumferential area, t_m is the wall thickness. The ratio A_p/S_c is the porosity of the capillary wall, ε.

Often times the solvent moving across the membrane will carry with it the solute molecules. Some molecules will be filtered on account of their large size. Even when the membrane is semi-permeable, some solute diffusion will take place. The solute separation on account of the size can be accounted for by the introduction of the sieving coefficient S_a. The *sieving coefficient* is defined as the ratio of the solute concentration in the filtrate C_{sp} to the solute concentration of the feed solution C_{sf}. Theoretical expressions based on the motion of a spherical solute moving through a cylindrical pore have been developed to estimate the value of the sieving coefficient (Deen, 1987).

$$S_a = \frac{C_p}{C_f} = (1 - \lambda)^2 (2 - (1 - \lambda)^2) \left(1 - \frac{2}{3} \lambda^2 - 0.163 \lambda^3 \right) \tag{6.9}$$

where λ is the ratio of the solute radius a, to the capillary pore radius, r. *Concentration polarisation* refers to the formation of a coat of retained solutes on the feed side of the membrane.

The solute diffusion across the membrane can be treated with the Fick's laws of diffusion at steady state as discussed in chapter 1 and the generalised Fick's laws of diffusion in chapter 2 for transient applications. Different diffusivities such as pore diffusion, diffusion in polymeric systems and convective effect can be added together. The size of the solute can be estimated from the Stokes Einstein relation presented in chapter 1. (Eq. 1.39). The diffusivity of the solute in the liquid needs to be known. If the diffusivity is not known, the solute size can first be estimated from the following equation that is written assuming that the solute of molecular weight M_w is a sphere with a density of approximately 1 gm/cm^3 and is the same as that of the solute in the solid phase.

$$a = \left(\frac{3 M_w}{4 \pi \rho A_N} \right)^{1/3} \tag{6.10}$$

Steric exclusion and hindered diffusion can also be accounted for in the expressions for diffusion of the solute depending on the problem at hand. *Steric exclusion* refers to the problem that only the volume in the pores other than that of the solute is available for diffusion. The hydrodynamic drag experienced by the solute is referred to as the hindered diffusion. The Renkin equation (1954) gives the ratio of the pore diffusivity to that of the bulk diffusivity. The solute flux can be written as:

$$J_s'' = \frac{DH}{1}(C_{sf} - C_{sp}) \tag{6.11}$$

$$J_s = P_m S(C_{sf} - C_{sp}) \tag{6.12}$$

where P_m is the permeability of the solute. The combined effect of hydraulic pressure and osmotic pressure and the solute flux can be written from the application of irreversible thermodynamics (Kedem and Katchalsky, 1958). The cross coefficients that are from secondary effects are equal according to the Onsager relationships. The relative flow between the solvent and solute is capable of providing a separation of the solute and solvent. The sieving mechanism is called *ultrafiltration* and when the solute permeability is low, it is referred to as *reverse osmosis*.

$$J = SL_p(\Delta P - \sigma RT\Delta C) \tag{6.13}$$

$$J_s = SL_p\left(-\sigma\Delta P + \frac{L_s}{L_p}RT\Delta C\right) \tag{6.14}$$

The parameter $\sigma = -L_{sp}/L_p$ is called the *Staverman reflection coefficient*. If the membrane is permeable to solvent and not to the solute, $\sigma = 1$. When $\sigma = 0$, the membrane is equally permeable to both solvent and solute. Eq. (6.13) gives the flux of solvent across the semi-permeable membrane and Eq. (6.14) gives the flux of solute.

Worked Example 6.1 *SWRO Plant*

Prepare a preliminary estimate of a SWRO plant where the inlet NaCl composition is 3.6% by weight and the outlet composition is 100 ppm. A typical RO process schematic is given in Figure 6.4. What ought to be the size of the pump. The RO membrane used is a linear aromatic polyamide. The expected water supply for the township is 63 lakh litres per day. Spiral wound modules are used with interfacial area of 5 m^2/gm. How much membrane is needed? The operating temperature is 25°C.

Volumetric flow of water = 63E–5 × 1E–3/24/3600 = 0.07292 cu.m/s

Water flux for linear aromatic polyamide
(From Figure 6.2, Salt Rejection vs. Water Flux Diagram) = 0.08 cu.m/m^2/d

= 9.2593E–7 cu.m/m^2/s

Amount of membrane needed = 0.07292/9.2593E–7/5000 = 1575 gm

= 605.77 cm^3

Radius of the cylindrical pores in the membrane module, $r = 2V_m/a$ = 605.8 cm^3

= 24.2 μm.

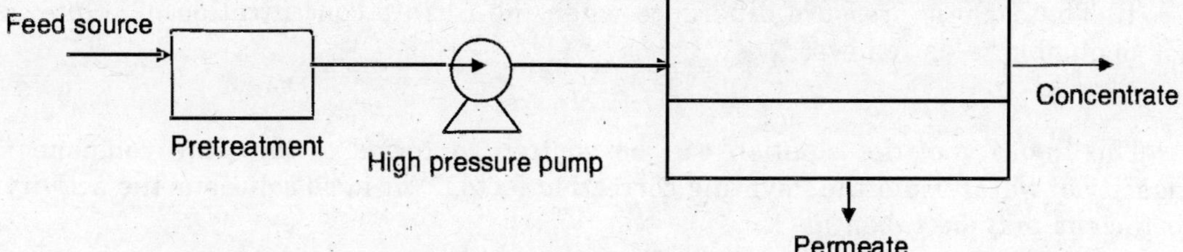

FIGURE 6.4 Schematic of a SWRO semi-permeable membrane process.

The effective length of the membrane module, $L = 605.77/\text{Pi}/r^2 = 328.415$ km.

The hydraulic conductance $SL_p = \left(\dfrac{r^2 A_p}{8\mu t_m}\right) = 8.89\text{E}{-6}\ \text{m}^4\text{s/kg}$

Hydraulic Pressure Drop at each stage = $J/L_pS + \Delta\pi = \Delta P_{h1}$

$$0.104 + 2477.6*(615.4 - 1.71) = 15.0 \text{ atm}$$

Pressure at the high pressure pump = 16.0 atm.

Worked Example 6.2 *Effect of Solute Concentration*

van't Hoff's law as given by Eq. (6.5) was derived assuming that the solute concentration was small. (a) For not so dilute systems, what would be the expression for osmotic pressure. What effect does it have in the solution in the worked example 6.1. (b) Use the $\ln(x_w)$ in the van't Hoff's law instead of the Taylor series approximation. What effect does it have?

(a) In the Taylor series expansion of $\ln(1 - x_s)$ another two terms are taken,

$$\ln(x_w) = \ln(1 - x_s) \approx -(x_s) + \frac{x_s^2}{2} - \frac{x_s^3}{6} \tag{6.15}$$

Eq. (6.3) becomes on substituting for Eq. (6.15) in Eq. (6.3):

$$\pi = \text{RT}x_s \frac{(1 - x_s/2)}{V_w} = \text{RTC}_s\left(1 - \left(\frac{C_s}{2}\right)V_w + \frac{(C_sV_w)^3}{6}\right) \tag{6.16}$$

In worked example 6.1,

$$\pi = 2477.6 * 615.4\left(1 - \frac{(0.0111)^2}{2} - \frac{(C_sV_w)^3}{6}\right)$$

$$= 2477.6 * 615.4\ (1 - 2.27\text{E}{-7}) = 15.0 \text{ atm.}$$

For the solute concentration of 3.6 wt% NaCl, Eq. (6.5) is a reasonable approximation of the logarithmic quantity.

(b) The osmotic pressure difference when the filtrate concentration of solute is not negligible is as follows:

$$f_{sf} = f_{sp} \tag{6.17}$$

The fugacity of the solution can be written in terms of the pure component fugacity of water using the Poynting correction factor. For ideal solutions the activity coefficient may be taken as 1.

$$X_{wf} f_w \exp\left(-V_w \frac{P_{sf} - P_w}{RT}\right) = x_{wp} f_w \exp\left(-V_w \frac{P_{sp} - P_w}{RT}\right) \tag{6.18}$$

or

$$-\frac{RT}{V_w} \ln\left(\frac{x_{wf}}{x_{wp}}\right) = P_{sp} - P_{sf} = \pi \tag{6.19}$$

In the worked example 6.1 if the Taylor series was not used to approximate for the logarithmic function of the mole fraction of the solute, Eq. (6.19) will be applicable.

Concentration of solute on the product side = 100 ppm = 3.156E–5

$$\pi = \frac{2477.6}{V_w} \ln\left(\frac{1 - x_{sf}}{1 - x_{sp}}\right) = 15.48 \text{ atm}$$

The error involved in using the Taylor series approximation is 3.2% in the osmotic pressure estimation.

Worked Example 6.3 *Multi-stage Sea Water Reverse Osmosis Operation*

For the problem discussed in worked example 6.1, for some practical reason say due to mechanical compaction of the membrane under high pressure water from the pump, the same objectives can be met by a multi-stage operation. At each stage 10% of the solute is rejected. How many stages are needed? What are the intermediate compositions? What is the net pressure drop required?

Equation (6.19) is used to obtain the pressure drop required to overcome the osmotic pressure at each stage. An MS excel spreadsheet is prepared on a Compaq X 86 desktop personal computer. At each stage 10% of the solute is rejected. At stage 9, 19.99% of the solute is rejected.

	X_{wf}	x_{wp}	Pi	atm
1	0.9886	0.989776	163630.98	1.614833
2	0.989776	0.990912	157887.6	1.558153
3	0.990912	0.992048	157706.7	1.556367
4	0.992048	0.993184	157526.21	1.554586
5	0.993184	0.99432	157346.14	1.552809
6	0.99432	0.995456	157166.47	1.551036
7	0.995456	0.996592	156987.22	1.549267
8	0.996592	0.997728	156808.37	1.547502
9	0.997728	0.999886	297444.64	2.935406
		0.000369	369.24 ppm	15.41996

It can be seen that the outlet from the 9th stage is 369.2 ppm. The osmotic pressure difference that is needed to be overcome is given in column *Pi* in N/m^2 and as atmospheric pressure in the column atm. The pressure drop required at each stage is 1.6 atm except at the 9th stage at 2.9 atm. The total pressure drop is 0.4% lower than from a direct 1 stage operation. 9 pumps of a factor of 10 lower than that required by a single stage operation can be used in this design.

Worked Example 6.4 *Use of Reflection Coefficient*

A brackish water body is desalinated using the reverse osmosis method. The concentration of the NaCl feed is 15 gm/litre. A Filmtec NF40HF semi-permeable membrane is used. How much water flux can be handled when a spiral module with interfacial area of 8 m^2/gm is used and an exit water at a concentration of 100 ppm is produced. The operating temperature is 37°C. How much membrane is needed. What is the pressure needed at the high pressure pump?

From Figure 6.2, for Filmtech NF40HF resin, the water flux can be seen to be 1.0 m^3/m^2.day and the salt rejection % is 51%.

The sieving coefficient and the Staverman reflection coefficient is then 0.51.

$$\text{The solute radius of NaCl, } a = \left(\frac{3M_w}{4\pi\rho A_N}\right)^{1/3} = 3*58.5/4/3.14)/2/6.023E23)^{0.3333}$$

$$= 22.7 \text{ nm}$$

By using Eq. (6.9), which can be used as a relation between, σ and a/r, the pore radius r can be found by trial and error to be 50 nm.

The volumetric flow of water in cu.m/s, Water flux for Filmtech NF40HF resin in 1 cu.m/m^2/d, amount of membrane needed in gms and cm^3, radius of the cylindrical pores in the membrane module in nm and ($r = 2V_m/a$), and the effective length of the membrane module L for a 1 mm thick membrane are all related. These quantities were derived for the worked example 6.1 in MS excel spreadsheet. In this problem the volumetric flow rate was changed until the radius of the pore of the membrane was equal to 50 nm by trial and error. Here are the set of values for which this is the case:

Volumetric flux = 42 lakh/lit/day

= 1.15741 cu.m/s

Amount of membrane needed = 525 gm, 202 cc at 8 m^2/gm

$$\frac{J}{L_p S} = 0.9$$

The water flux is, $J = SL_p(\Delta P - \sigma RT\Delta C)$

and,

$$\Delta P = 0.9 + 0.51*2577.34*(256 - 1.7)$$

$$= 0.9 + 6.48 = 7.38 \text{ atm}$$

The pressure at the pump = 8.4 atm.

6.4 OPTIMAL REVERSE OSMOSIS DESIGN

Many mechanistic and mathematical models have been proposed to describe reverse osmosis membranes. Reliable models reduce the number of experimental data needed. The mathematical models that describe the RO process can be classified into three categories: (a) models from irreversible thermodynamics (Kadem and Katchalsky), (b) solute diffusion model and (c) pore models.

Nanofiltration membranes can be explained using charged RO membrane theories. The principle of Donnan exclusion and extended Nernst—Planck equation include the electrostatic effects. Most models focus on the thin asymmetric layer that determines the water fluxes and selectivities. The homogeneous models and porous membrane models differ in the assumption of the porosity of the membrane. One assumes a pore size and certain geometry of pores, and the other assumes that there are no pores.

The use of stages can be made of once the equilibrium characteristics of the liquid-liquid systems on the feed and filtrate side is better quantitated. Some relations have been developed for the estimation of sieve coefficient. Some of the stages at lower level of solute concentration can be substituted with other methods for a minimum total cost. For instance RO and electrodialysis can be used in sequence in desalination of sea water to make potable drinking water at a minimum total cost.

The feed characteristics of membrane polymers and organics solute interactions, fundamental transport mechanisms, hazards, flux drop phenomena, concentration polarisation layer formation need to be better understood for an optimal design of RO-based separations. Some experimental studies have shown that substituted phenolic compounds can cause reduction in solvent flux.

Liquid chromatography can be used to measure the membrane solute equilibrium characteristics. Equilibrium distribution coefficients, Gibbs free energy and surface energy for a number of different polymers can be measured. Quantum chemistry calculations, force field computer molecular simulations, and IR spectroscopy have been used to calculate the interactions. Small hydrocarbons occupy the void spaces in the membranes and reduce the water flux.

The ability of RO membranes to remove both inorganic and organic compounds have made these attractive to treat drinking water supplies. RO processes can remove the hardness in water, various inorganics such as bromate, colour, various kinds of bacteria and viruses, and organic contaminants such as agricultural, chemicals and trihalomethane precursors. Natural organic matter, NOM, and disinfection byproducts, DBP can also be removed using RO technology. The application of RO technology to treat municipal wastes has also been successful. Dissolved solids that cannot be separated by other processes can be removed and Colour, nitrate and organic levels can be reduced using RO membranes. The water factory 21 site in Orange County, California, USA, is a large-scale municipal wastewater treatment site. The membrane throughput is 190 lakh litres/day. The process reduces TDS and organics to levels that allows the effluent to be injected into ground water aquifers used for water supplies. TDS solids of various slats rejection of more than 99% and TOC rejections as high as 90% have been found to be feasible. Losses in water flux over a period of time have been offset by periodic cleaning.

From the mid 90s, the primary use of RO technology has been in desalination of sea water and brackish water. The industry is driven by the demand for potable water in some regions of the world. Desalination involves reduction of the TDS concentration to less than 200 ppm. RO technology was first applied to brackish water which has a 1 wt% salt concentration compared with 3.6 wt% in sea water. This difference in concentration as shown in the worked examples translates to large differences in pressure needed at the pump. RO membranes that can withstand 103 atm pressure were developed. The current desalination plants in operation around the world treat both brackish water and sea water.

The brackish water plant in Yuma, Arizona, has a capacity of 27.5 crore litres/day with a TDS intake of 0.3 wt%. Spiral-wound cellulose acetate membranes produce permeate water having 200 ppm of solids. Sea water desalination requires a much higher pressure to achieve a given separation than brackish water. The SWRO desalination plant in Jeddah, Saudi Arabia has a capacity of 56.8 crore litres/day. The Bhaba Atomic Research Center, BARC, at Trombay in Mumbai has set up a large desalination plant at the atomic power plant at Kalpakkam about 50 km from Chennai. This plant has two sections. One section produces 18 lakh litres of potable water a day from sea water using the reverse osmosis method and the other section 45 lakh litres a day using the thermal method. Another desalination plant with two units at 50,000 litres per hour was inaugurated at Koodankukulum in Tirunelveli, Tamil Nadu using reverse osmosis technology. Some desalination plants combine distillation with RO to produce both power and water. MSF, multi-stage flash processes, are used to produce both power and distilled water.

Costs for RO technology include both capital and recurring costs. Equipment costs in terms of capacity corresponding to 1000 litres/day range from Rs. 18,000 for brackish water plants and Rs. 40,500 per litre. Operating costs include energy, labour, chemicals and membrane replacement and is 0.70 paise per litre for brackish water plants and 2.250 paise per litre for SWRO plants. Costs may vary depending on the pre-treatment and post treatment needed. RO membrane modules are available from Du Pont, Dow/FilmTech, UOP, Millipore, Nitto-Denko and Toray.

Worked Example 6.5 *Use of Recycle to Reduce the Size of the High Pressure Pump* In worked example 6.1, 63 lakh litres of water at a salt concentration of 100 ppm was generated from a feed with a salt concentration of 3.6 wt% NaCl. The size of the pump needed was 16 atm. In order to reduce the load of the high pressure pump, consider recycling the treated water with the feed water. The actual feed into the RO module will be lower in solute concentration, which will decrease the osmotic pressure need to be overcome (Figure 6.5). What is the pressure at the pump for a recycle ratio of 3.3? The residue to product ratio is 0.5. What is the % reduction in pump size needed?

The product flow rate of water, P = 63 lakh litres/day

Feed flow rate, $F = P + W$

where W is the concentrate flow rate, R' is the rate of return to the feed tank

Let R be the recycle ratio of treated water, $R = R/P$ (6.20)

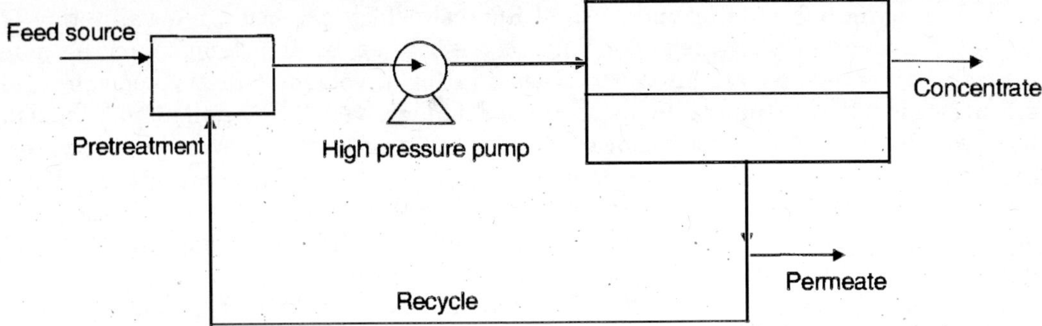

FIGURE 6.5 Use of recycle to reduce the load on the high pressure pump.

Then, the concentration at the throat of the membrane module

$$= \left(\frac{x_f F + R' x_{sp}}{F + R'}\right) = \left(\frac{0.0114*94.5 + 3.11E\text{-}5*207.9}{207.9 + 94.5}\right)$$

$$x_f = 0.0114$$

$$x_{sp} = 3.11\ E\text{-}5$$

$$R = 0.8$$

$$x_f' = 0.0036$$

The new osmotic pressure needed to be overcome, Eq. (6.19)

$$-\frac{RT}{V_w}\ln\left(\frac{1 - x_f'}{1 - x_{sp}}\right) = 2.67 \text{ atm} \tag{6.21}$$

Pressure needed at the pump = 3.7 atm.

Thus 82% reduction in the pump size is needed. However, at this recycle ratio some fresh water is called for. Large recycle storage tanks and metering devices are needed. But the pressure at the pump can be lowered in this fashion.

6.5 MOLECULAR SIEVES

Molecular sieve adsorbents can be used to perform difficult separations including solutes from solutions, liquids from liquids and gases from gases. They are supplied as pellets, granules or beads and occasionally as powders. Molecular sieve adsorbents can be used for selective removal of one component from a mixture. The concept it is based upon is that molecular species differ in their sizes. Contaminants in feed streams can be reduced to quantities low enough to reach undetectable levels using standard detection apparatus.

Any material that can reject molecular species by virtue of their size can be called a molecular sieve. Materials that have uniform pore size with less than 2 nm are *microporous* molecular sieves, and those that have pore sizes between 2–20 nm are referred to as *mesoporous* molecular sieves. Zeolites are the most important class of molecular sieves that are of technological importance. Zeolites were first discovered by Swedish mineralogist Cronstedt in 1756. Zeolites possess a regular, crystalline structure with channels of diameters less than 1.2 nm made of aluminosilicate framework. The general formula is written as $M_nOal_2O_dySiO_2.wH_2O$. Several structures with aluminium replaced by boron, gallium or iron and silicon replaced by germanium have been prepared and studied. The family of microporous framework structures have been expanded by the replacement of silicon with phosphorous and synthesis of aluminophosphate and silicoaluminophosphate. Some have a zeolite analogue and some others are unique in crystallographic structure. Metalloaluminophopshates and metallosilicoaluminophopshates have been introduced.

The synthesis effort has been driven by achieving pore diameters within the mesoporous range. The 14-member phosphate based ALPO-8, VPI-5, cloverite 20 MR have been made with pore diameters within the 0.8–1.3 nm. 1.4 nm channels are present in cacoxenite, natural ferroaluminophosphate.

The application areas of the molecular sieves are heterogeneous catalysis and adsorption processes. Over *200 synthetic zeolite types* and *50 natural zeolites* are known. Zeolite minerals are found all round the globe including the bottom of the sea, volcanic rocks, alkaline lake bed and underlain land surface. Percolation of surface water through appropriate sediments is a traditional method of zeolite formation. By and large, high grade zeolite ore is mined and then processed by crushing, drying, powdering and screening. Depending on its application, zeolite may be chemically modified by ion exchange, acid extraction and calcined at 350–550°C. Some of the zeolite compositions are given in Table 6.1. Of the 50 known zeolite minerals, chabazite, erionite, mordenite and clinoptilolite occur in sufficient quantity to allow their use in commercial products.

Of the 120 known framework aluminosilicates, 50 occur naturally, the rest being synthetic. There are 56 structural types known. The topology of zeolites can be studied by the following three aspects:

(1) spatial arrangement of the basic structural units
(2) location of charge balancing cations
(3) channel filling material

The void filling property of zeolites makes it suitable for applications in ion-exchange, adsorption, catalysis, etc. The current beliefs of zeolite structure were developed by nobel laureate Pauling in 1930. Modern characterisation tools such as x-ray diffraction, electron/neutron diffraction, infrared spectroscopy and Raman spectroscopy, NMR and electron microscopy have provided a very detailed description of many structures. Computer-aided techniques such as FTIR are used for characterisation of zeolites. There exists an internal pore system in zeolites comprising interconnected cage-like voids. This comes in handy in determining diffusion type in catalysis and adsorption.

The primary structural unit is a tetrahedral and they are assembled into secondary building blocks which may be simple polyhedra such as cubes, hexagonal prisms, truncated octahedral, etc. The final framework structure is an assemblage of the secondary units (Figure 6.6).

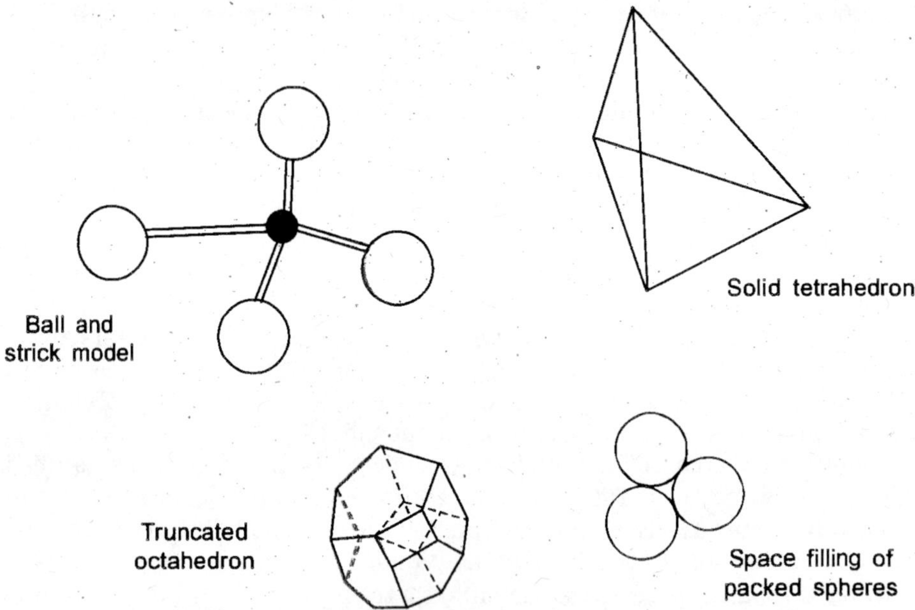

FIGURE 6.6 Methods of representing SiO_4 and AlO_4 tetrahedra.

TABLE 6.1 Zeolite compositions

Chabazite	$Ca_2((AlO_2)_4(SiO_2)_8).13H_2O$
Mordernite	$Na_8((AlO_2)_8(SiO_2)_{40}).24H_2O$
Erionite	$(Ca, Mg, Na_2K_2)_{4.5}((AlO_2)_9(SiO_2)_{27}).27H_2O$
Faujasite	$(Ca, Mg, Na_2K_2)_{29.5}((AlO_2)_{59}(SiO_2)_{133}).235H_2O$
Clinoptilolite	$Na_6((AlO_2)_6(SiO_2)_{30}).24H_2O$
Philipsite	$(0.5Ca,Na,K)_3((AlO_2)_3(SiO_2)_5).6H_2O$
Zeolite A	$Na_{12}((AlO_2)_{12}(SiO_2)_{12}).27H_2O$
Zeolite X	$Na_{86}((AlO_2)_{86}(SiO_2)_{106}).264H_2O$
Zeolite Y	$Na_{56}((AlO_2)_8(SiO_2)_{27}).250H_2O$
Zeolite L	$K_9((AlO_2)_9(SiO_2)_{27}).22H_2O$
Zeolite Omega	$Na_{6,8}(TMA(AlO_2)_8(SiO_2)_{28}).24H_2O$
ZSM-5	$(Na,TPA)_3((AlO_2)_3(SiO_2)_{93}).16H_2O$

Chabazite and mordenite have crystal structures. The structure of chabazite is hexagonal and the framework consists of double six-membered rings of $(Si_y)AlO_4$ tetrahedra arranged in parallel layers in AABBCC sequence. These tetrahedrals are cross-linked by four-membered rings resulting in cavities of 0.67×1.0 nm. Exchangeable metal ions such as calcium, occupy positions within or near the double six-membered rings. Modernite has channels of width 0.6 × 0.7 nm. The framework itself is built from chains of 5-membered rings which are cross-linked. Several properties of zeolite minerals are studied including adsorption and ion-exchange. It was used in water softening as ion-exchangers. Molecular sieve behaviour was first discovered in 1920s when the gas adsorption properties of dehydrated natural zeolite crystals was detected. These materials can reject molecules based upon their molecular size. They are microporous with pore size that range from 0.3–0.8 nm. Many crystalline zeolites have been synthesized. The formula for zeolite is $M_{x/n}((AlO_2)_x(SiO_2)_y).wH_2O$ where n is the valence of cation M, w is the number of water molecules per unit cell x and y are the number of AlO_4 and SiO_4 tetrahedra per unit cell respectively, and y/x usually has values of 1–5. Examples of important synthetic zeolites are shown in Table 6.1. The secondary structure in zeolites A, X and Y is truncated octahedron. The internal cavity is 1.1 nm in zeolite A and is entered by 6 circular apertures of 0.42 nm in size. The interlinked cavities form 3-dimensional unduloid channels with a 0.42 nm minimum diameter. The faujasite structure consists of a tetrahedral arrangement of the truncated octahedral by joining 6-membered rings. The resulting cages are 1.3 nm in size and are entered by a 12-membered ring, 0.74 nm in diameter. This is the largest known pore size in zeolites. EMC-3 has a slightly different stacking of the truncated cubooctahedrahegaonal prisms combination resulting in a hexagonal structure. It has one straight channel formed by a 12-membered rings of 0.71 nm crystallographic diameter. These channels are connected in the a direction by 12-membered rings of 0.74 × 0.65 nm size. Several defects can occur in the structure.

The structure of zeolites lets them have some unique properties. These properties are:

(a) Adsorption
(b) Catalytic properties
(c) Ion exchange

Of the three, adsorption and ion exchange are of interest here. Zeolites are high capacity, selective adsorbents capable of separating molecules based on the size and shape of the molecule relative to the size and geometry of the main apertures of the structure. The adsorbed molecules, in particular those with a permanent dipole moment show other interaction effects, with a selectivity that are not found in other solid adsorbents. The topology of the zeolite determines the pore size and void volume of the voids. Activated diffusion phenomena occur at higher temperatures. The pore size of the zeolite has been compared with the molecular sizes from Lennard-Jones potential function (Figure 6.7).

There are several different types of separations that can be performed on zeolite adsorbents. One is the recovery of normal paraffins from hydrocarbon feedstocks. Paraffin isomers and cyclic hydrocarbons are too large to be adsorbed and are excluded. The recovered n-paraffin is used in the manufacture of biodegradable detergents. The affinity of zeolites for water and other polar molecules decreases with rising SiO_2/Al_2O_3. Therefore low silica zeolites can be used for drying gases and liquids. In order to avoid unwanted side reactions potassium exchanged form of zeolite may

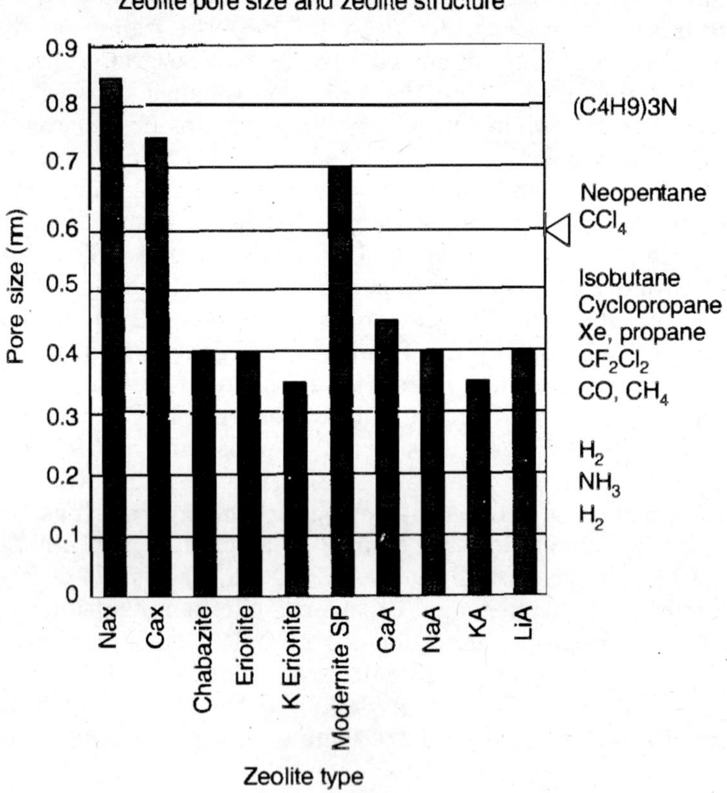

FIGURE 6.7 Molecular dimension and zeolite pore sizes.

take place. Refrigerants can be dried using the molecular sieve effect of zeolite. The effective pore size of 3A excludes all hydrocarbons including ethane. Another type of separation is based on the differences in selectivity of two or more co-adsorbed gases or vapours. Sulphur compounds are separated from natural gas in this manner. The structure of zeolite can be tailor-made. Given the structure property relationships, properties such as adsorption capacities can be dialed in. During adsorption in zeolites the adsorbates migrate into the zeolite crystals. Transport occurs between crystals contained in a pellet and this is followed by diffusion within the crystals. Diffusion can be measured by obtaining the adsorption rates, jump times using NMR. The factors such as channel geometry, molecular size, shape, polarity, zeolite cation distribution, charge, temperature, adsorbate composition, impurity molecules, crystal surface defects are important in determining the kinetics and diffusion.

The nature of the cation in the zeolite determines the exchange behaviour of nonframework cations. Properties such as selectivity and degree of exchange, thermal stability, adsorption behaviour and catalytic activity are examples of the behaviour of zeolites. The ion-exchange isotherms and relative adsorptivity was discussed in section 5.6. Here the selectivity of the cations are discussed and derived from the metathetical reactions that take place during the ion-exchange process. A typical ion exchange process is represented by:

$$Z_A B^{ZB}(\text{zeolite}) + Z_B A^{ZA}(\text{solution}) \leftrightarrow Z_B A^{ZA}(\text{zeolite}) + Z_A B^{ZB}(\text{solution}) \quad (6.22)$$

where Z_A and Z_B are the ionic charge cations A and B. The mole fraction cations in the zeolite and solution phases, when plotted give the ion-exchange isotherm. Thus A (zeolite) is plotted against A (solution). The separation factor, α_{AB} and rational selectivity coefficient, K_{AB} can be written as:

$$\alpha_{AB} = \frac{\text{A(zeolite) B(solution)}}{\text{B(zeolite) A(solution)}} \quad (6.23)$$

$$K_{AB} = \frac{A^{ZB}\text{(zeolite)}\, B^{ZA}\text{(solution)}}{B^{ZA}\text{(zeolite)}\, A^{AB}\text{(solution)}} \quad [6.24)$$

The exchange capacities of typical ion exchange isotherms are given in Table 6.2.

TABLE 6.2 Ion exchange capacity of various zeolites

Zeolite	*Si/Al Ratio*	*meq/g*
Chabazite	2	5
Mordenite	5	2.6
Erionite	3	3.8
Clinoptilolite	4.5	2.6
Zeolite A	1	7.0
Zeolite X	1.25	6.4
Zeolite Y	2	5.0
ZSM-5	2	0.78

The kinetic of ion-exchange is controlled by ion diffusion within the zeolite structure. The ion diffusion coefficients, particle radii and temperature are important parameters of interest. It has been found for example, that diffusion of sodium ions occurred by migration from type II sites in the 6-membered rings into the large cages followed by diffusion to the surface through the large channels. Cation sieving in zeolites is credited to solvation and charge distribution in the topology of the zeolite.

The manufacture of zeolites involves crystallization from aqueous systems containing various types of reactants. Many synthetic zeolites have been made with a natural mineral counterpart. The starting materials for the preparation of zeolites are amorphous solids, co-precipitated gels, strong bases and the operating conditions are hydrothermal with a low autogeneous pressure and a high degree of supersaturation of the gel components leading to nucleation of a large number of crystals. The temperature range is from room temperature to 200°C and time required for crystallization varies from a few hours to several days. Gel preparation and crystallization are the important steps in the process.

$$NaOH + NaAl(OH)_4 + Na_2SiO_3 \xrightarrow{25°C} Na_a(AlO_2)_b(SiO_2)_c.NaOH.H_2O(gel)$$

$$\xrightarrow{25\text{–}175°C} Na_x(AlO_2)_x(SiO_2)_y.mH_2O \quad (6.25)$$

The crystallisation of zeolites consist of three stages: (a) formation of precursors; (b) nucleation, (c) crystal growth. x-ray diffraction can monitor the progress of crystal growth. The first two steps are the induction period. Nucleation without appreciable crystal growth can be accomplished by ageing at low temperatures. Ageing can be used to prevent crystallisation of impurity. Seeding can induce crystal growth. In this period the induction period step is skipped and crystal growth starts immediately. The two major processes to prepare zeolites are hydrogel, and clay conversion, besides other processes.

Zeolites are used in adsorption. Adsorption can be used for purification as well as for bulk separations. The dehydration of water vapor can also be accomplished by adsorption. The adsorptive separations are classified as shown in Figure 6.8.

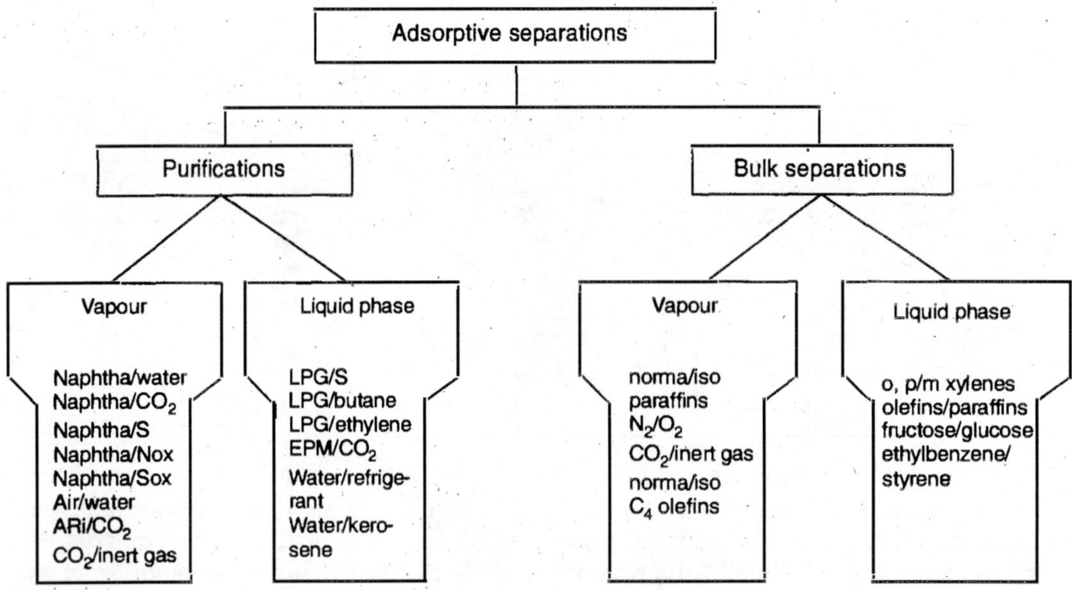

FIGURE 6.8 Classification of adsorptive separations.

The unique features of molecular sieves are demonstrated in the removal of water from natural gas streams containing high percentages of acid gases such as hydrogen sulphide and carbon dioxide. They can dehydrate gas streams containing corrosive components such as chlorine, sulphur dioxide and hydrogen chloride. Water can be removed from liquid streams as well using molecular sieve adsorbent zeolite. Molecular sieve is designed for the lifetime of a unit. Refrigerant drying cartridge is one example. Molecular sieves can be sued to purify gas streams containing CO_2 in cryogenic applications where free-out of CO_2 would cause fouling of low temperature equipment. Sulphur compounds can be removed from air streams using molecular sieves. Trace water and hydrogen sulphide can be removed using zeolites. LPG sweetening is another large application of zeolite with plant capacities of over 8000 m^3/d.

Normal and isoparaffins can be separated using molecular sieve zeolite. Type 5A zeolite can be used to reject the branched and cyclic hydrocarbons from n-paraffins.

There are 7 commercial processes in operation. The Universal Oil products operates in the liquid phase. The purex process is used to separate p-xylene from xylene/ ethylbenzene mixture. Olefins are separated from isobutenes at Union Carbide. PSA, pressure swing adsorption is used to prepare oxygen from air.

6.6 ELECTRODIALYSIS

Dialysis is a membrane separation technique to remove toxic metabolites from blood in patients suffering from kidney failure. The first artificial kidney was developed in 1940 and was based on cellophane. Since the 1990s, most artificial kidneys are based on hollow-fibre modules having a membrane area of 1 m^2. Cellulose fibres are replaced with polycarbonate, polysulphone and other polymers, which have higher fluxes and are less damaging to the blood. Blood is circulated through the centre of the fibre, while isotonic saline, dialysate urea, creatinine and other low molecular weight metabolites in the blood diffuse across the fibre wall and are removed with the saline solution. The process is slow, requiring several hours to remove the required low molecular weight metabolites from the patient and must be repeated one to two times per week. Over 10,000 patients in hospitals use these devices on a regular basis. The largest application of membranes is the artificial kidneys. Similar hollow fibre devices are being explored for medical uses, including an artificial pancreas, in which islets of Langerhans supply insulin to diabetic patients or an artificial liver in which adsorbent materials remove bilirubin and other toxins. In *carrier facilitated transport*, the membrane used to perform the separation contains a carrier which preferentially reacts with one of the components to be transported across the membrane. Liquids containing complexing agent are used. Membranes are formed by holding the liquids by capillary action in the pores of a microporous film. The *carrier agent* reacts with the solute on the feed side of the membrane and then diffuses across the membrane to release the solute on the product side of the membrane. The carrier agent is then reformed and diffuses back to the feed side of the membrane. The carrier acts as a shuttle to transport one component of the feed to the other side. Metal ions can be transported selectively across a membrane driven by the *s* pumped counter currently around the outside of the fibres. Figure 6.9 shows the flow of hydrogen or hydroxyl ions in the other direction. High membrane selectivities can be achieved using facilitated transport. There are no commercial processes still using this method. This is due to the instability of the membrane and the carrier agent. Dialysis is the earliest molecularly separative membrane process discovered. The Ficks law of diffusion and the generalised laws of diffusion are applicable to describe the transport of solute molecules across the membrane to the other side. A multi-stage dialysis separation procedure can be envisioned for desalinating sea water.

Depending on the ratio of the pore size of the membrane and the solute radius, the salt concentration on the other side of the first stage can be estimated using the sieving coefficient expressions discussed in the above sections. Upto one-half the feed concentration of the feed of solute can be obtained on the permeate side of the membrane. Thus at every stage a maximum of halving of concentration takes place.

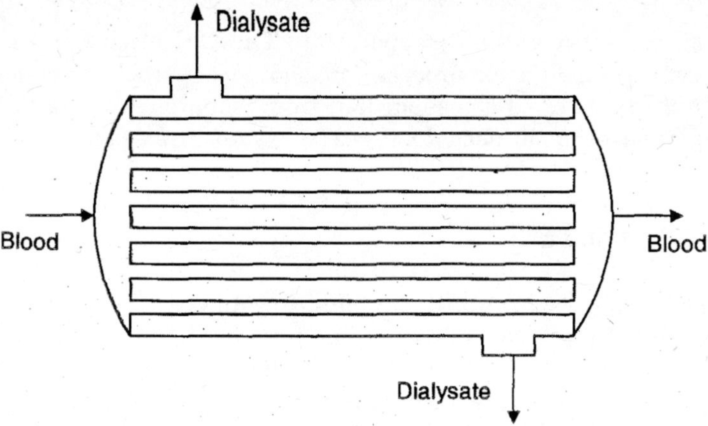

FIGURE 6.9 Schematic of a hollow fiber mass exchanger used as an artificial kidney dialyser used to remove urea and toxic metabolites from blood.

Starting with 3.6 wt% NaCl of sea water a suitable dialysis membrane can in after *n* stages reduce the concentration of the NaCl in both sides of the membrane down to 100 ppm or 0.01% (Figure 6.10). After the first stage if sufficient time allowed for equilibrium to be attained by diffusion both sides of the dialysis membrane will be at 1.8 wt%. This repeated over *n* stages would be 1.8, 0.9, 0.45, 0.225, 0.113, 0.055, 0.028, 0.014, 0.0007 respectively. Thus after 9 stages, the concentration of sea water will reach potable water allowable limits down to less than 100 ppm.

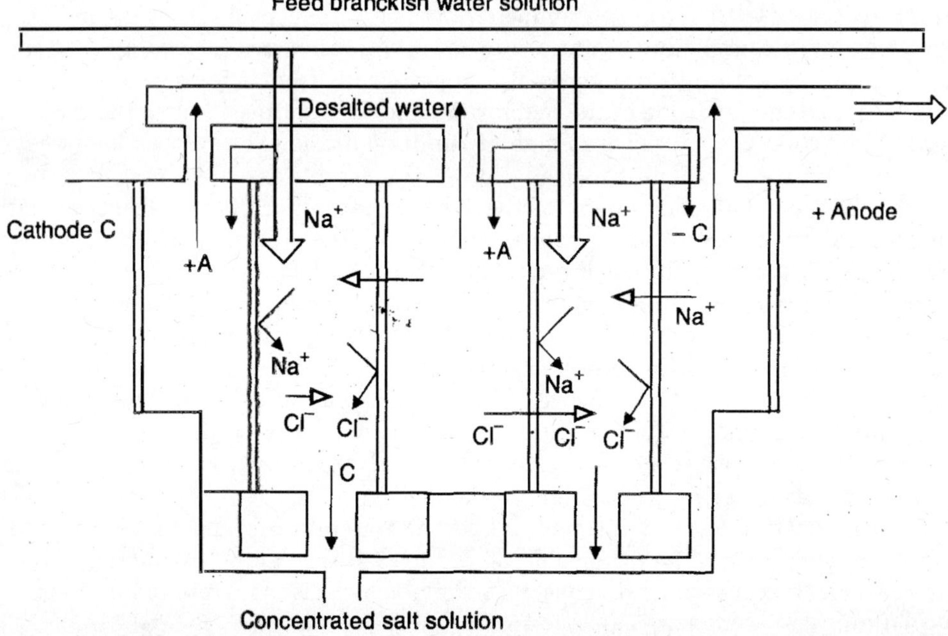

FIGURE 6.10 Schematic of electrodialysis apparatus with alternating anode and cathode upto 100 Cell pairs.

The recovery of caustic from hemicellulose in the rayon process was well established in the 30s and is used in modern times in the paper pulp industry. *Isobaric dialysis* as a unit operation is emerging and is used in the removal of alcohol from beverages and in the production of products derived from biotechnology. By the end of 1992, 40 key beer breweries had installed worldwide industrial dialysis plants with an annual capacity of 18.9 crore litres. An example industrial schematically dialysis process is shown in Figure 6.11.

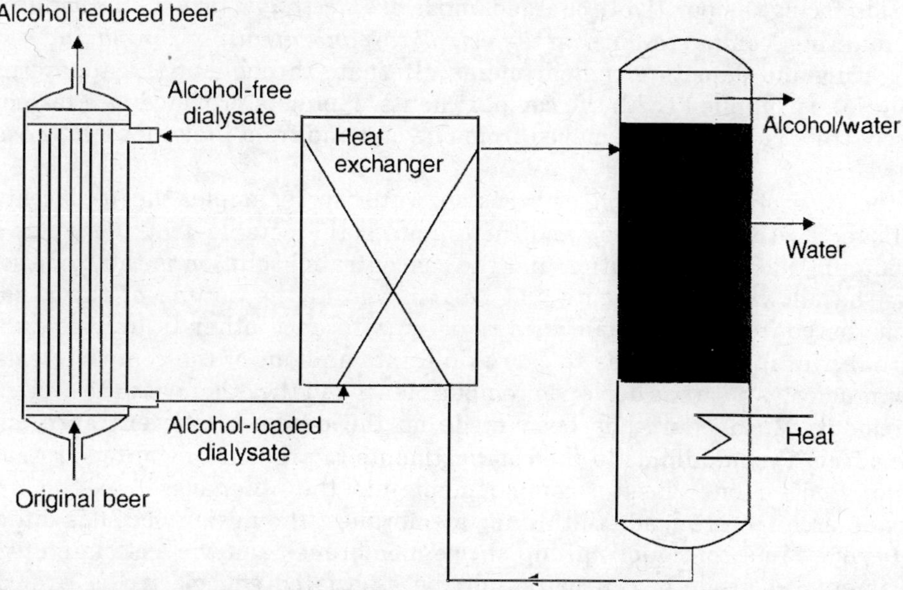

FIGURE 6.11 Countercurrent dialysis and distillation to separate alchohol from beverages.

Alcohol is removed from beer by dialysis and it is distilled from the dialysate. The raffinate is recycled as a distillate stream. The combination of dialysis and distillation preserves the flavour of the product. Dialysis is an isothermal operation and it is an important parameter in the biotechnology industry. It facilitates the removal of salts from heat sensitive or mechanically *labile* compounds such as vaccines, hormones, enzymes, and other bioactive cell secretions. Dialysis is combined with ultrafiltration to offer *diafiltration* with higher process efficiency. The media and extracellular environment in bioreactors can be controlled using diafiltration. Novel bioreactor designs are possible using dialysers. The extraluminal region of a hollow fibre dialyser provides an excellent growth environment for mammalian cells when the lumen is perfused with oxygen and nutrients. In the production of monoclonal antibodies for example, a benchtop bioreactor can readily equal the antibody production of several thousand mice. This technology is in the development stage. Attempts have been made to separate biological fluids using dialysis membrane. Examples are removal of buffer from a protein solution, or concentrating polypeptide and hyperosmotic dialysate. *Microdialysis* is a specialised application of the technique. A U-shaped dialysis capillary is surgically implanted into the tissue of a living animal. Isotonic dialysate is pumped through the tubing at a flow rate low enough to allow equilibration with small solutes in the host's extracellular portion of the tissue. It helps in sampling tissues. Perfusate rate is low at 1 nanolitre per minute.

Electrodialysis or ED technique has been found to be the most economical to desalinate brackish water at a feed NaCl concentration of a little over 1 wt%. The apparatus for ED as shown in Figure 6.10 is basically an array of anode and cathode membranes terminated by electrodes. The membranes are separated from each other by gaskets which form fluid compartments. Compartments that have the A membrane on the side facing the positively charged anode are *electrolyte depletion compartments*. The remaining compartments are *electrolyte concentrating compartments*. The concentrating and depleting compartments alternate throughout the apparatus. The feed solution is supplied to all the compartments. Piping is provided in a fashion that the concentrated solution is removed from one end and from the other end the diluate is removed.

In the case of desalination of brackish water for example, the feed contains a little of over 1 wt% salt solution and the diluate is the potable drinking water at less than 100 ppm NaCl concentration and the concentrated solution is the brine solution that can be allowed to segregate and be removed at the bottom of the apparatus. Holes in the gaskets and membranes register with each other to provide two pairs of internal hydraulic manifolds to carry fluid into and out of the compartments. One pair communicates with the depletion compartments and the other with the concentrated compartment. Much effort has been made on the design of the entrance and exit channels from the manifolds to the compartments to prevent unwanted cross-leak of fluid intended for one class of compartment into the other class. As the trend in membrane architecture leads to thinner membranes, the design becomes difficult. A *cell pair* refers to a contiguous group of two membranes and the associated two fluid compartments. A group of cell pairs and the associated end electrodes are called a *stack* or *pack*. Generally *100–600 cell pairs* are arranged in a single stack. The choice depends on the capacity of ED, the uniformity of flow distribution achieved among the several compartments of the same class in stack and the maximum total direct current potential desired. One or more stacks may be arranged in a filter press configuration designed to compress the membranes and gaskets against the force of fluid flowing through the compartments thereby preventing fluid leaks to the outside and internal cross-leaks between compartments. Hydraulic rams are used for large presses and rods provide the compression for small presses. Commercial membranes have thickness of 150–500 microns. The compartments between the membranes have typical thicknesses of 0.5–2 mm. The thickness of a cell pair is therefore in the range of 1.3–5.0 mm. 100 cell pairs have a combined thickness of 30 cm. The effective area of a cell pair for current conduction is generally in the order of 0.2–2 m^2. Electric current applied to the stack is limited by economic considerations. The power consumption is I^2R. In relatively dilute electrolyte, the electric current that can be applied is diffusion limited. This is the ability of ions to diffuse through the membranes.

The membranes used in the ED apparatus are ion-selective. The stack of membranes is prepared on a plate and frame concept. Anion-exchange membranes contain fixed positively charged entities such as quaternary ammonium groups, fixed to the polymer backbone. These membranes may permit negatively charged ions and exclude positive ions. Cation exchange membranes contain fixed negatively charged groups such as sulphonic acid groups. They permit positively charged ions to move through them.

SUMMARY

In *reverse osmosis* the solvent from the solution is pumped across a semi-permeable membrane opposing the osmotic pressure difference with the solute largely rejected by the membrane. A chart that contains water flux per unit interfacial area per day vs. salt rejection rate for RO processes was presented. Data from 19 different membranes were included in the chart. The membranes in the chart are available in the market and can be purchased off the shelf. In order to write equations that describe RO processes, the principles of osmotic pressure, permeability of solvent across the membrane and diffusion of solutes across the membrane needs to be described. An expression for the osmotic pressure across a hydrophilic membrane was developed from equilibrium thermodynamics and concept of fugacity, the van't Hoff's law. The combined effect of osmotic pressure and hydraulic permeability is given by the Starling's law. The hydraulic conductance in terms of the characteristics of the pores in the membrane was presented. The sieve coefficient developed by Deen gives the solute concentration in the permeate side. The solute diffusion across the membrane can be written in terms of Ficks laws of diffusion and generalised Ficks laws of diffusion. The Staverman reflection coefficient can be obtained from the solute rejection. The pore radius information about the membrane and other characteristics are needed. The equations that describe the RO processes can be written as:

$$\pi = (P_w - P_s) = -RT/V_w \ln(\gamma_w x_w) \qquad \text{(osmotic pressure)}$$

$$J/L_pS = \Delta P_h - \Delta\pi = \overline{\Delta P}_h \qquad \text{(permeability of solvent)}$$

$$J_s = SL_p\left(-\sigma\Delta P + \left(\frac{L_s}{L_p}\right)RT\Delta C\right) \qquad \text{(solute diffusion)}$$

As a worked example a SWRO plant with the same capacity as that of the desalination plant at Kalpakkam was presented. The pressure at the pump was calculated. The effect of stages was illustrated as a worked example. The optimal design of the RO technology was discussed. The cost of generation of drinking water using RO was presented.

The use of recycle was shown to cut the pressure needed at the pump on the feed side of the RO membrane as a worked example.

Molecular sieves can be used in difficult separation operations. Zeolites both natural and synthetic are good examples of molecular sieves. The pore size is a few angstroms. There are 200 synthetic zeolite types. The zeolite structure can be understood by looking at the spatial arrangement of the basic structural units, locating the charge balancing cations, and channel filling material. The zeolite compositions are presented as a Table. The zeolite pore size and zeolite structure was shown along with molecular dimensions as a figure.

The separation factor and selectivity coefficient were introduced. The ion exchange capacity vs. the Si/Al ratio was presented as a Table. The manufacturing method for zeolite was presented. The adsorptive separations were classified into liquid and gas operations, purification and bulk.

Dialysis was introduced as a membrane separation process to let solute diffuse across the membrane and form the permeate and retentate at desired concentrations. A mass exchanger is used in hospitals to treat patients with kidney disease by removing the toxic metabolites and urea. Industrial dialysis and reclamation of alcohol from beer was shown with examples from commercial operations worldwide. The electrodialysis apparatus was shown with schematic as a method to let salt concentrate and water become dilute in alternate compartments of anode and cathode during desalination for example.

EXERCISES

1. Can Staveman reflection coefficient be calculated from the % salt rejected.
2. Can the osmotic pressure be used to cause flow that can operate a turbine and generate electricity? Why?
3. Why is RO a single-stage operation?
4. What are the differences between hollow fibre module and spiral wound Module?
5. What is the reason for the higher water flux in the membranes developed later such as the polyamide membrane compared with the cellulose acetate membrane first discovered by Loeb and Sourirajan?
6. How much pressure does the membrane need to withstand for RO application in sea water desalination?
7. What is mechanical compaction?
8. If there is a temperature difference across the feed side and filtrate side, what will happen to the van't Hoff's law? Why?
9. Can the osmotic pressure concept be extended to gaseous solutions? Give reasons.
10. Is there any special reason behind making the shape of the membrane into a cuboid in Figure 6.1.
11. Consider a solution with a layered solute profile. A higher concentration of the solute is found at the bottom and a lower concentration of the solute is found at the top of the jar. Will there be solvent flow from the top to the bottom of the jar on account of osmotic pressure profile, higher on the top and lower at the bottom?
12. Explain why are the membranes used in SWRO made thin.
13. What is the energy transfer during osmotic flow?
14. What is the physical significance of Eq. (6.7) becoming zero?
15. Why don't you find a lot of trees near the beach or coastlines?
16. Can you have 100% separation using semi-permeable membrane RO technology?

PROBLEMS

1. *Hydraulic Conductance in Human Body*
Calculate the normal rate of net filtration for the human body. Assume the capillaries have a total surface area of 413 m^2 and that the slit-pore surface area is 3/1000 of the total capillary surface area. Assume that the porous structure of the capillary wall is a series of parallel cylindrical pores with a diameter of 7 nm. Plasma filtrate may be considered as Newtonian fluid with a viscosity of 1 cp. The mean net filtration pressure for the capillary was just calculated to be 0.3 mm Hg. The capillary characteristics are as follows:

Inside diameter — 10 μm	Length L — 0.1 cm
Wall thickness, t_m — 0.5 mm	Average blood velocity — 0.05 cm/s
Pore fraction — 0.001	Wall pore diameter — 6–7 nm
Inlet pressure — 30 mm Hg	Outlet pressure — 10 mm Hg
Mean pressure — 17.3 mm Hg	Colloid osmotic pressure — 28 mm Hg

Interstitial fluid pressure — –3 mm Hg
Interstitial fluid colloid osmotic pressure — 8 mm Hg

2. *Orange Juice Concentration by Osmosis*
During concentration of orange juice, the water needs to be removed. A plastic bag containing orange juice at 1% by weight sucrose concentration is dropped into a brine solution at 35% NaCl by weight. Calculate the osmotic pressure developed that will concentrate the juice. **(Ans:** 250 atm)

3. *Blood Storage in Army*
For transfusion in Army, human blood donated is stored for a month. There is interest in improved storage procedure such as concentration of red cells, white cells, platelets, vitamins, proteins, sugars, minerals, hormones and enzymes by water removal. Ultrafiltration devices are sought by the army to remove water with 50% by volume water to levels low enough to effect significant volume reduction during blood storage. The temperatures have to be kept low during separation to prevent hemolysis. Use an ultrafiltration membrane and for a solute rejection of 80%, find the pore radius of the membrane from the expression for sieving coefficient. The solute radius can be taken as an effective radius of the different ions present in the blood.

For the given pore size of the membrane and for a membrane thickness of 1 mm, what is the hydraulic conductance during flow of water across the membrane. The interfacial area of the membrane, amount of water that can be treated, volume and weight of the membrane needed can be related to the information in Figure 6.2. Choose a membrane for the given solute rejection and by trial and error in an MS excel spreadsheet, obtain the throughput that can be handled during ultrafiltration of blood, the pressure needed at the high pressure pump.

4. *Reverse Osmosis to Separate Acrylonitrile from Water*
In the manufacture of ABS engineering thermoplastic using continuous polymerisation process for every pound of product manufactured there is a

little over a pound of water generated that contains acrylonitrile ($CH_2 = CHCN$). A 5% AN solution of water needs to be separated by Reverse Osmosis and the product needs to have AN less than 1 ppm. A membrane made of cross-linked polyether resin at an interfacial area of 10 m^2/gm and 1 mm thickness is used. How much membrane is needed for a solute rejection of 99.9%. What is the pore size of the membrane. What is the throughput of water it can handle? What is the pressure at the high pressure pump on the feed side.

(**Ans:** 7.2 atm; 20 lakh litres/day; 300 gm; 115 cc; 23 nm)

5. *Effect of Concentration Polarization Layer*
During the operation of the reverse osmosis in sea water desalination, the salt gets rejected by the semi-permeable membrane and gets accumulated near the feed side of the membrane (Figure 6.1). Perform a mass balance of the solvent in the region of the concentration polarisation layer and the membrane and show that the flux will decrease with time, i.e.,

$$\frac{J}{L_p S} = \frac{\left(\Delta P - \dfrac{RTC_{sf0}}{\delta S}\right)}{\left(\dfrac{1}{L_p S} + \dfrac{RTC_{sf0}t}{\delta S}\right)}$$

where δ is the polarization layer, and t is the time of operation.

6. *Hydraulic Conductance*
For water that contains PCBs, polychlorinated biphenyls and TCE, tetrachlorinated ethylenes at 4.5% and 2.0% by weight respectively, a reverse osmosis membrane of 0.5 mm thickness is used. The membrane is Toray, SU-700. What is the rejection rate for SU-700? What is the flux rate the Toray membrane can handle per day when an interfacial area is 13 m^2/gm. How much membrane is needed to produce a filtrate with a concentration of 1 ppb? For a throughput of 22,500 litres/day, what is the osmotic pressure? What is the hydraulic conductance (L_pS) of the system. What should the pressure at the pump be?

7. *Starch Removal*
A new membrane on the market was tested and found to have permeability, L_pS of 1 E-4 m^4s/kg under a pressure difference of 10 atm. The membrane handles 4.8% of partially hydrolyzed starch (MW 17,000) as feed and puts out a product at 175 ppm. What is the Staverman reflection coefficient. What is the throughput of water the membrane can handle? What is the solute rejection rate of the membrane? Can you provide the pore diameter and length of the membrane for a thickness of the membrane of 500 microns when 100 cc of the membrane is used.

8. *Tallest Tree in the World*
What is the limit on the height of a tree. Include the Bernoulli law as well as the osmotic pressure drop. Assume that the leaves on the tree top have a starch concentration of 10 wt%. (**Ans:** 35 ft)

9. *Porous Membrane*

A copolymer', with a high acrylonitrile content of SAN, styrene acrylonitrile is tested for use as a reverse osmosis membrane. The pore radius is 200 nm. For a solution of 1% by weight of polyethylene glycol of 18,000 molecular weight in water, what is the sieving coefficient? Taking this to be the Staverman reflection coefficient what is the pressure at the pressure pump to reduce the water content to less than 1 ppb in the filtrate side. What is the hydraulic conductance? Show that the effect of the molecular weight has reduced the pressure needed at the pump in such a fashion that the solvent filtration pressure drop is the limiting factor compared with most RO processes where the osmotic pressure is the limiting factor. Show that the water flux rate this membrane can handle is 26,000 litres/day for a 1 mm thick membrane and effective volume of 1 cc.

10. *Effect of Molecular Weight of the Solute*

In some applications of RO technology, such as the desalination of sea water, the cost limiting step is the osmotic pressure that needs to be overcome in order to achieve the desired degree of separation. This leads to a large size of the pump. As the molecular weight of the solute increases, for the same solute concentration by weight, say 3.6 wt% what is the molecular weight of the solute when the pressure drop needed for the hydraulic motion of the solvent alone is greater than or equal to the osmotic pressure from the solute? Make suitable assumptions about the Staverman reflection coefficient.

11. *Effect of Operating Temperature*

In the worked example 6.1, what happens to the pressure needed at the high pressure pump during winter? Say an operating temperature at 4°C and a inlet salt concentration of 4.4 wt% and a product expected salt concentration is of 42 ppm. For the same membrane used in worked example, what is reduction in pressure needed at the pump?

(**Ans:** Pressure drop needed is reduced by 10.5%)

12. *Salt Precipitation by Freezing*

Based on the results in worked example 6.1, estimate the energy needed to pump the sea water using a high pressure pump to achieve the desired separation objectives. Make a comparative study of the decrease in solubility of NaCl with temperature. As you compress the sea water, at what point does it become ice, or at what point does it become favourable for salt precipitation. In which route is the energy required less?

REFERENCES

Applegate, L., 1984, *Chem. Eng.*, **91**, 64.

Colton, C.K., 1091, *AIChE J*, **17**, 772–780.

Daniekl, F.K., "*Dialysis and Electrodialysis*", in Encyclopedia of Chemical Technology, **5**, 1–20, USA.

Deen, W.M, 1987, *Hindered Transport of Large Molecules in Liquid-Filled Pores, AIChE J*, **33**, 1409–1425.

Ferguson, P., 1980, *Desalination*, **32**, 5.

Fournier, R.L., 1999, *Basic Transport Phenomena in Biomedical Engineering*, Taylor & Francis, Philadelphia.

Ho, W.S. and Sirkar, K.K., 1992, *Membrane Handbook*, Van Nostrand-Reinhold, New York.

Katchalsky, A. and Curan, P.F., 1967, *Non-Equilibrium Thermodynamics in Biophysics*, Harvard university Press, Cambridge, MA.

Kedem, O. and Katchalsky, A., 1958, *Thermodynamic Analysis of the Permeability of Biological Membranes to non-Electrolytes*, Biochmica et Biophysica Acta, **27**, 229.

Lonsdale, H., 1982, *J. Membrane Sci.*, **10**, 81.

Noble, R.D. and Stern, S.A., 1993, *Handbook of Membrane Separations*, Marcel Dekker, New York.

Renkin, E.M., 1954, *Filtration, Diffusion and Molecular Sieving through Porous Cellulose Membranes,* J. Gen. Physiol., **38**, 225–243.

Sharma, K.R., 2005, *Damped Wave Transport and Relaxation,* Elsevier, Amsterdam.

________, November 2002, *Concentration Polarization Layer in Reverse Osmosis*, 94th AIChE Annual Meeting, Indianapolis.

________, November 2002, *Reverse Osmosis Unit for Water Removal for Blood Storage in the Army*, 54th Southeast Regional Meeting of the ACS, SERMACS, Charleston, SC.

________, November 2002, *Ultrafiltration Device for Concentration of Transfusion Blood during Combat*, 54th Southeast Regional Meeting of the ACS, SERMACS, Charleston, SC.

________, December 2003, *Water Processing using Polyamide Membrane by Reverse Osmosis and with Composite Resin in Ion Exchange Technology*, Chemistry Preprint Archive, **2003**, 12, 98–103.

________, October 2001, *Process Design for a Polyamide Membrane used in Water Purification*, 17th National Convention of Chemical Engineers, Kochi.

________, March 2003, *Molecular Sieve Design for Reuse of Army Vehicle Washeries under Mild Conditions*, AIChE Spring National Meeting, New Orleans.

________, September 2003, *Removal of Arsenic from Drinking Water by Molecular Sieve Adsorption*, 226th ACS National Meeting, New York.

Stein, W.D., 1984, *The Movement of Molecules across Cell Membranes*, Marcel Dekker, New York.

CHAPTER 7

Gas Humidification and Solid Drying

Nomenclature

A	wet surface area (m^2)
C_p	heat capacity (J/K/mol)
C_s	humid heat (J/K/mol)
C_A	heat capacity of component A (J/K/mol)
G'_s	gas flow rate (kg dry gas/s/m^2)
L'	liquid flow rate (kg dry gas/s/m^2)
h_G	heat transfer coefficient (w/m^2K)
H	enthalpy (J/mole)
H'_1	enthalpy at 1 (J/mol)
H'_2	enthalpy at 2 (J/mol)
H_L	enthalpy of liquid (J/mol)
k_Y	mass transfer coefficient (mol/m^2/s)
K_Y	overall mass transfer coefficient (mol/m^2/s)
M_A	molecular weight of component A, (kg/mol)
M_B	molecular weight of component B, (kg/mol)
N	rate of drying (kg moisture/m^2/hr)
p_A	partial pressure of component A (N/m^2)
p_B	partial pressure of component B (N/m^2)
p_t	total pressure of mixture (N/m^2)
P	pressure (N/m^2)
S_s	mass of dry solid (kg)
t_{G1}	dry bulb temperature (°C)
t_L	temperature (°C)
t_w	wet bulb temperature (°C)
U	internal energy (J/mol/K)
U_0	overall heat transfer coefficient (w/m^2/K)
V	molar volume (m^3/mol)

v_H	humid volume (m^3/kg)
X	moisture content of solid (kg of water/kg dry solid)
X^*	moisture content of the substance that is in equilibrium with the vapour
y_A	mole fraction of component A
y_B	mole fraction of component B
Y'_1	molal absolute humidity at 1 (moles of H_2O/moles of dry air)
Y'_2	molal absolute humidity at 2 (moles of H_2O/moles of dry air)
Y_s	saturation absolute humidity (kg water/kg dry air)
Y'_s	saturation molal humidity (mol water/mol water)

Greek

λ_w	latent heat of vaporisation (J/kg)
θ	time of drying (hr)

7.1 INTRODUCTION

Gas humidification operations refer to those operations where transfer of mass occurs through a gas-liquid interface with the supply of necessary energy. The mass that transfers is usually water. Cooling towers are the largest mass transfer devices in common use. Hot water sprayed into the top of the tower packed with inert material such as wooden slats evaporates as it goes. Air enters from bottom of the tower and rises up through the packing. In small towers, the air is pumped up by using fans. In larger ones, it is allowed to rise by natural convection. The *drying operations* refer to those operations where moisture is transferred from a solid to vapour phase. *Evaporation* refers to the transfer of mass from liquid to vapour and usually the mass is water. During *dehumidification operations* the moisture is removed from the vapour phase or usually from air. Another operation where the liquid transfers from the vapour-gas mixture to a bone-dry solid can be introduced as *wetting* or *moistening* process.

The VLE of water is of interest during the design of these operations. As discussed in Chapter 4, the phase diagram of water will contain a triple point, a curve to indicate the vapour-liquid equilibrium, vapour-solid equilibrium and solid liquid equilibrium. The Clausius Clapeyron, Eq. (4.29) gives the vapour pressure and temperature relationship. The phase rule is discussed and the degree of freedoms can be calculated for a 2-component, two-phase system as 2. The critical pressure has also been mentioned. The enthalpy of any substance can be written as:

$$H = U + PV \tag{7.1}$$

The enthalpy change between two conditions can be given by,

$$\Delta H = H_1 - H_2 = C_p(t_1 - t_2) \tag{7.2}$$

where C_p is the heat capacity and t_1 and t_2 are the two temperatures at the two conditions. The absolute humidity is a measure of moisture in the vapour-liquid mixture. It gives the mass of liquid vapour to the mass of moisture free dry gas, Y'. If the quantities are expressed in moles, it is called molal absolute humidity, Y'.

$$Y = \frac{y_A}{y_B} = \frac{p_A}{p_B} = \frac{p_A}{(p_t - p_B)} \tag{7.3}$$

$$Y' = Y\left(\frac{M_A}{M_B}\right)$$

For air-water system,

$$Y' = \left(\frac{18}{28.84}\right)Y = 0.6241\,Y \tag{7.4}$$

The *absolute humidity*, based on mass was introduced by Grosvenor and is called Grosvenor humidity. The *relative humidity* expressed as a percentage is the ratio of the partial pressure of the water vapour to the equilibrium vapour pressure of water vapour at the dry bulb temperature, i.e. RH = 100 (p_{AG}/p_{AG}^*). For any system such as air-water mixture, the lines of constant relative humidity are shown in Figure 7.1. In the figure, the humidity can be read from the right hand side ordinate and the absolute humidity of the mixture from the left hand side ordinate.

When an insoluble dry gas is brought in contact with a liquid A, the liquid will enter the gas phase as vapour until equilibrium is reached. At equilibrium, the partial pressure of A will reach its saturation value at the prevailing temperature. The *absolute humidity of the vapour-gas mixture at saturation* can be written as:

$$Y_s = \left(\frac{p_A^*}{p_t - p_A^*}\right) \tag{7.5}$$

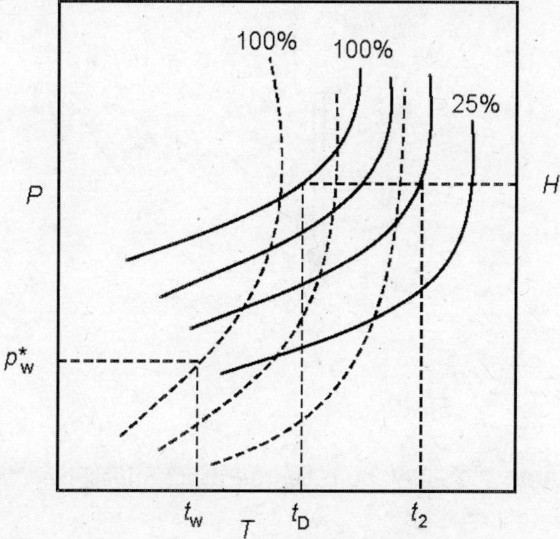

FIGURE 7.1 Pressure, temperature, humidity chart for vapour-gas mixture.

For a water-vapour air mixture the molal humidity can be written as:

$$Y_s' = 0.6241\, Y_s \tag{7.6}$$

The *percentage saturation* or percentage absolute humidity is represented by 100 Y/Y_s and 100 Y'/Y_s'. The lines of constant relative saturation as shown in the figure need not be identical with the lines of constant relative humidity. The percentage unsaturation will be equal to the relative humidity, $RH^*(p_t - p_{AG}^*)/(p_t - p_{AG})$. At the boiling temperature this becomes infinity. The *dew point* is the temperature at which a vapour-gas mixture becomes saturated when cooled at constant total pressure. In Figure 7.1, this is given by t_D. The path of the process is followed from t_2 to t_D. All mixtures with the same humidity will have the same dew point. If the temperature is reduced, an infinitesimal amount below t_D the vapour will condense at liquid dew.

The *dry-bulb temperature* is the temperature of the vapour-gas mixture as determined by immersion of a thermometer in the mixture. The *wet bulb temperature* is the temperature reached by a drop of liquid evaporating into a large amount of unsaturated vapour-gas mixture. It can be used to measure the humidity of the mixture. The wet bulb temperature is lower than the dry bulb temperature. The heat lost from the vapour gas mixture is gained by the drop of liquid.

The equations that describe the use of wet bulb temperature are a good example of an application in simultaneous heat and mass transfer. The wet bulb thermometer is a convenient way to measure the relative humidity. Two thermometers, one that measures the air's dry bulb temperature and the other that is clad in cloth wick wet with water. The cloth clad wet-bulb temperature measures colder temperature caused by evaporation of water (Figure 7.2). The measured temperature difference can be used to estimate the relative humidity of the vapour gas mixture.

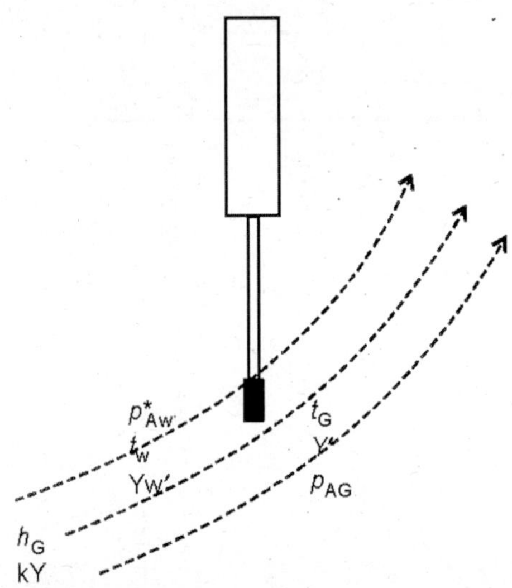

FIGURE 7.2 Wet bulb temperature measures.

An energy balance on the evaporating vapour can be written as:

$$h_G (t_G - t_w) - \lambda_w M_A k_G (p^*_{Aw} - p_{AG}) = 0 \tag{7.7}$$

where h_G is the heat transfer coefficient, and k_G is the mass transfer coefficient. The wet bulb depression temperature is then written after expressing the mass transfer rate in terms of mole fractions instead of the partial pressures as:

$$(t_G - t_w) = \lambda_w \left(\frac{Y'_w - Y'}{h_G/k_Y} \right) \tag{7.8}$$

The denominator of Eq. (7.8) is the psychrometric ratio and is given by (h_G/k_Y). The psychrometric ratio can be calculated from Lewis relations for a given geometry and flow pattern. The *Lewis number* is the ratio of the thermal diffusivity to binary diffusivity. The Lewis relation for the air water system, $(h_G/k_Y C_s) = \text{Le}^{0.567}$ correlated well for 18 vapour-gas systems for flow of gases past cylinders. Lewis number for air-water system is nearly 1. Therefore, the psychrometric ratio is equal to the humid heat capacity, C_{s1}. When the psychrometric ratio is taken as 1, the adiabatic saturation temperature and wet bulb temperature become equal.

The *humid volume* v_H of a vapour-gas mixture is the volume of unit mass of dry gas and its accompanying vapour at the said temperature and pressure. Assuming ideal gas behaviour

$$v_H = 8.315 \left(\frac{1}{M_B} + \frac{Y'}{M_A} \right) \left(\frac{t_G + 273}{p_t} \right) \tag{7.9}$$

The *humid heat,* C_s is the heat required to raise the temperature of unit mass of gas and its accompanying vapour by one degree at constant pressure. For a mixture of absolute humidity, Y'. This can be calculated as follows:

If Y' is the ratio of the moles of component A to the moles of component B, (Eq. (7.3)), then the ratio of the moles of component A to the total moles of the vapour-gas mixture can be given by $1 + Y'$. Now the humid heat per mole of the vapour gas mixture can be written as:

$$\frac{C_A Y'}{1 + Y'} + \frac{C_B}{1 + Y'} = C_s \tag{7.10}$$

For small Y' this can be approximated as $C_B + C_A Y' = C_s$. The enthalpy of the vapour-gas mixture per unit volume of dry gas can be written with respect to a reference condition o. For an unsaturated air-vapour mixture, it is said to be *superheated state*.

$$H' = C_B(t_G - t_0) + Y'(C_A(t_G - t_D) + \lambda_D + C_{AL}(t_D - t_0) \tag{7.11}$$

or

$$= C_s(t - t_0) + Y'\lambda_0 \tag{7.12}$$

where λ_D is the latent heat of vaporisation of the vapour at prevailing temperature and λ_0 is the latent heat of vaporisation at the reference temperature. The psychrometric chart for air-water system can be prepared depending on the problem at hand. The chart in SI units is given in Appendix B.

7.2 ADIABATIC SATURATION AND OPERATIONS

The vapour gas mixture is contacted with fresh liquid as shown in Figure 7.3. It can be contacted as a spray or by some other method. The exit vapour gas mixture leaves at a temperature and humidity different from the feed values. The operation is conducted in an *adiabatic* manner. A mass balance on the vapour component A yields,

$$G'_s\,(Y'_2 - Y'_1) = L' \tag{7.13}$$

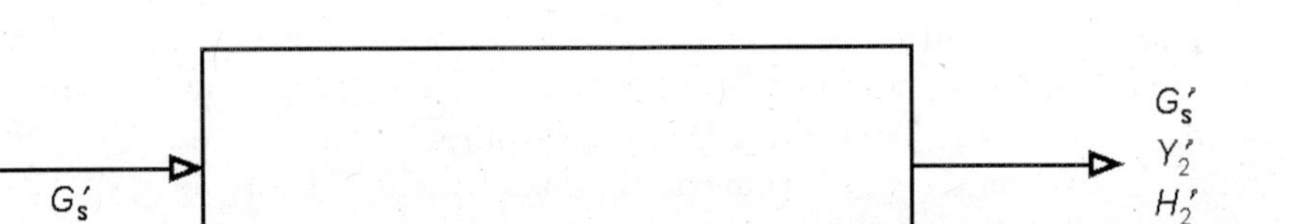

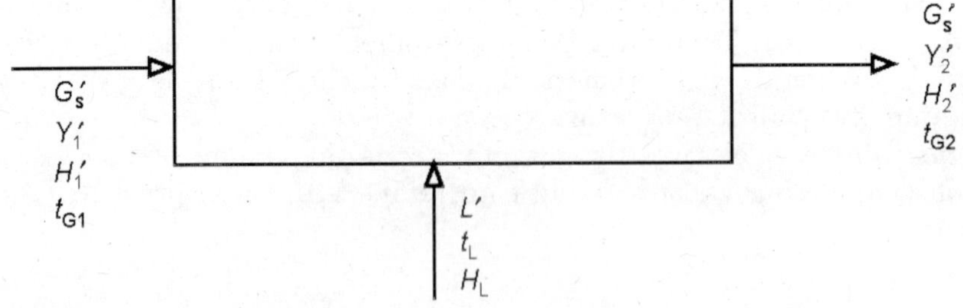

FIGURE 7.3 Adiabatic saturation.

An enthalpy balance is written as:

$$G'_sH'_1 + L'H_L = G'_sH'_2 \tag{7.14}$$

Combining Eq. (7.13) with Eq. (7.14),

$$H'_1 + (Y'_2 - Y'_1)H_L = H'_2 \tag{7.15}$$

Substituting Eq. (7.12) in Eq. (7.15)

$$C_{s1}(t_{G1} - t_0) + Y'_1\,\lambda_0 + (Y'_2 - Y'_1)C_L(t_L - t_0) = C_{s2}(t_{G2} - t_0) + Y'_2\lambda_0 \tag{7.16}$$

In special case when the conditions at the exit stream are saturated and the liquid enters at saturation temperature, t_s

$$C_{s1}(t_{G1} - t_0) + Y'_1\,\lambda_0 + (Y'_2 - Y'_1)C_L(t_s - t_0) = C_{s2}(t_s - t_0) + Y'_s\lambda_0 \tag{7.16}$$

For small values of humidity, the humid heat can be written as $C_B + C_AY' = C_s$. Writing this into Eq. (7.16),

$$(C_B + C_AY'_1)\,(t_{G1} - t_0) + Y'_1\,\lambda_0 + (Y'_s - Y'_1)\,C_L(t_s - t_0)$$
$$= (C_B + C_AY'_s)\,(t_s - t_0) + Y'_s\lambda_0$$

say $\lambda_0 - C_L(t_s - t_0) + C_A(t_s - t_0) = \lambda_s$

or,
$$(t_{G1} - t_s) = (Y'_s - Y'_1)\lambda_{as}/C_{s1} \tag{7.17}$$

This is an adiabatic saturation curve.

Worked Example 7.1: *Psychrometric Chart of Acetone-Air Mixture*

Construct a psychrometric chart for the mixture of acetone-air at a pressure of 760 mm Hg over the temperature range 0–100°C, Y' = 0–3 kg vapour/dry gas. Include 25%, 50%, 80% and 100% relative humidity curves and adiabatic saturation curves at different saturation temperatures.

Molecular weight of acetone = 58.1 gm/mole
Normal boiling point of acetone = 56°C

Using the Clausius Clapeyron equation given in Eq. (4.29) the equilibrium vapour pressure of acetone vs. temperature is calculated using an MS excel spreadsheet on a Pentium IV desktop personal computer. The reference value is the normal boiling point at 1 atm pressure of 56°C. The corresponding saturation humidity values are calculated using Eq. (7.5). This would be the 100% RH line in the psychrometric chart given in Figure 7.4. The 80% line is obtained by calculating the vapour pressure needed for 80% saturation and calculating the absolute humidity. The 100%, 80%, 50% and 25% lines are shown in Figure 7.4.

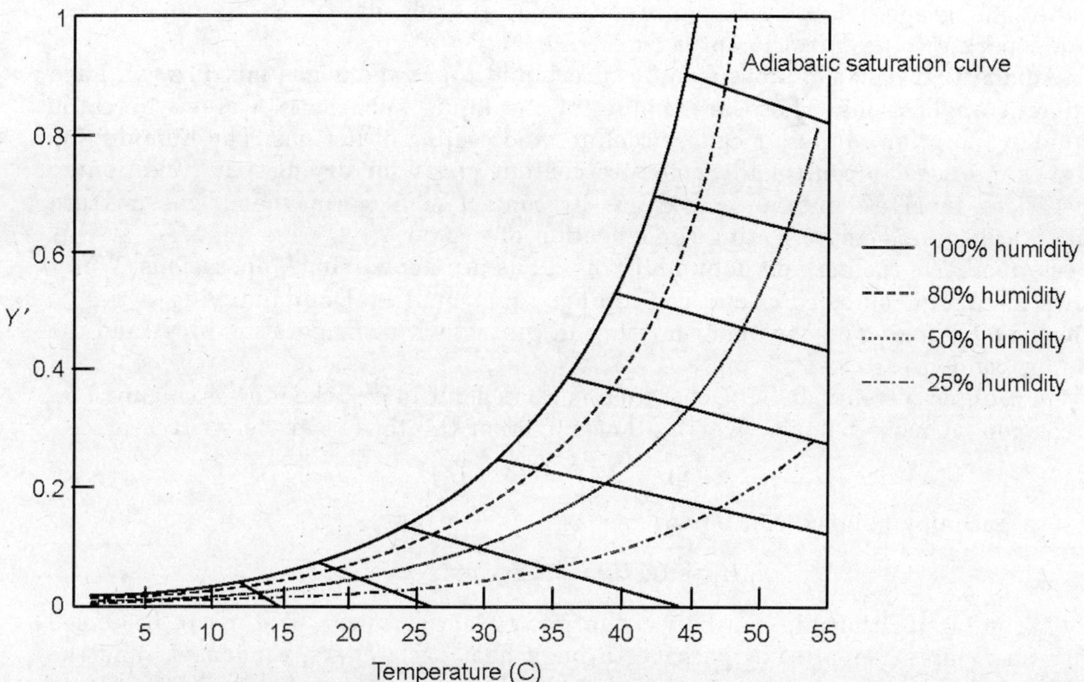

FIGURE 7.4 Psychrometric chart for acetone-air vapour-gas mixture.

			100%	80%	50%	25%
p_A^*	H	T(°C)	Y_s'	Y_s'	Y_s'	Y_s'
13.9	1	1	0.018667	0.014849	0.009229	0.004593
21.4	6	6	0.02895	0.023045	0.01428	0.007089
32.33	11	11	0.044429	0.035231	0.021732	0.010749
48.2	16	16	0.067716	0.053449	0.032749	0.016111
70.9	21	21	0.102888	0.080651	0.048927	0.023879
102.9	26	26	0.155015	0.121473	0.072613	0.035035
147.6	31	31	0.241019	0.183948	0.107549	0.05103
209.3	36	36	0.37931	0.282571	0.159686	0.073939
293.4	41	41	0.627409	0.446847	0.239198	0.106823
407	46	46	1.152747	0.74954	0.365678	0.154577
559	51	51	2.777837	1.429668	0.581686	0.225312
594.7	52	52	3.597701	1.673797	0.64271	0.243201
632.7	53	53	4.970149	1.994012	0.713062	0.262826
672.8	54	54	7.715596	2.427128	0.794145	0.284218
760	56	56				

The adiabatic saturation line given by Eq. (7.17) has a slope of C_{s1}/λ_{as}. It passes through the point (t_s, Y_s'). Few such lines are shown in Figure 7.4. The enthalpy of the vapour-gas mixture can be calculated from Eq. (7.12). In some cases this is shown in the same graph but as a second ordinate. With the advent of PCs this can be saved as a separate column in the spreadsheet file.

Adiabatic transfer of mass from a pure liquid to a vapour gas mixture can have different applications. These are cooling of the liquid such as in water where the liquid evaporation to the air causes cooling, and cooling of hot gas. The humidifying of a gas is used to control the moisture content of air for drying. Air conditioning operations involve dehumidifying a gas by contact of a warm vapour-gas mixture with a cold liquid and effecting condensation of vapour.

Evaporative cooling and dehumidifying a gas are nonadiabatic operations. Water flows as a film outside a tube containing liquid or gas. Dehumidifying a gas is effected by bringing a gas-vapour mixture in contact with refrigeration pipes and the vapour condenses upon the pipes.

A counter-current adiabatic operation is carried out in a packed tower (Figure 7.5). A component mass balance over the lower part of the tower can be written as:

$$G_s'(Y' - Y_1') = (L' - L_1') \tag{7.18}$$

An enthalpy balance can be written as:

$$L'H_L + G_s' H_1' = L_1'H_{L1} + G_s' H' \tag{7.19}$$

A similar relationship can be written for the entire tower. Water can be cooled with air. Water heated after passage through heat exchangers, condensers and the like is cooled by atmospheric air for reuse. Little vaporisation of water causes large cooling effects. The Lewis relation can be used to simplify the enthalpy balance equations for the air-water system.

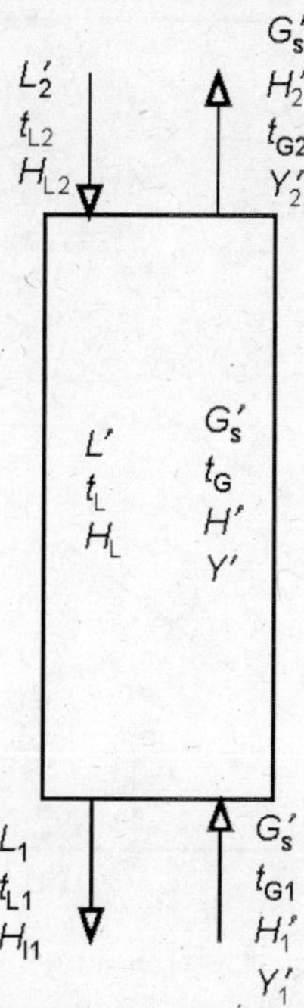

FIGURE 7.5 Continuous countercurrent adiabatic humidification operation.

$$L'C_L(t_{L2} - t_{L1}) = G'_s(H'_2 - H'_1) \tag{7.20}$$

In Figure 7.6, the enthalpy of the gas, H' is plotted against t_L the liquid temperature. Eq. (7.20) is represented by the operating line in Figure 7.6. When the $L'_2 - L'_1$ is small, the operating line is a straight line with a slope of $L'C_L/G'_s$. The equilibrium curve in Figure 7.6 is the enthalpy of saturated gas at each temperature. At small mass transfer rates performing a mass balance on the cross-section of the tower,

$$G'_s dY' = k_Y a\rho(Y' - Y')\,dZ \tag{7.21}$$

$$G'_s C_s dt_G = h_G a\rho(t_i - t_G)\,dZ \tag{7.22}$$

Combining the equations,

$$G'_s dH' = k_Y \rho a(H'_i - H')\,dZ \tag{7.23}$$

$$G'_s dH' = k_Y \rho a(H'_1 - H')\,dZ = h_L \rho a(t_L - t_i)\,dZ \tag{7.24}$$

The packed height needed can be calculated by integration,

$$\int_1^2 \frac{dH'}{(H'_1 - H')} = \frac{k_Y a\rho}{G'_s}\int_0^Z dZ = \frac{k_Y a\rho Z}{G'_s} \tag{7.25}$$

The number of gas-enthalpy transfer units N_{tG} can thus be calculated.

$$Z = H_{tG}N_{tG} \tag{7.26}$$

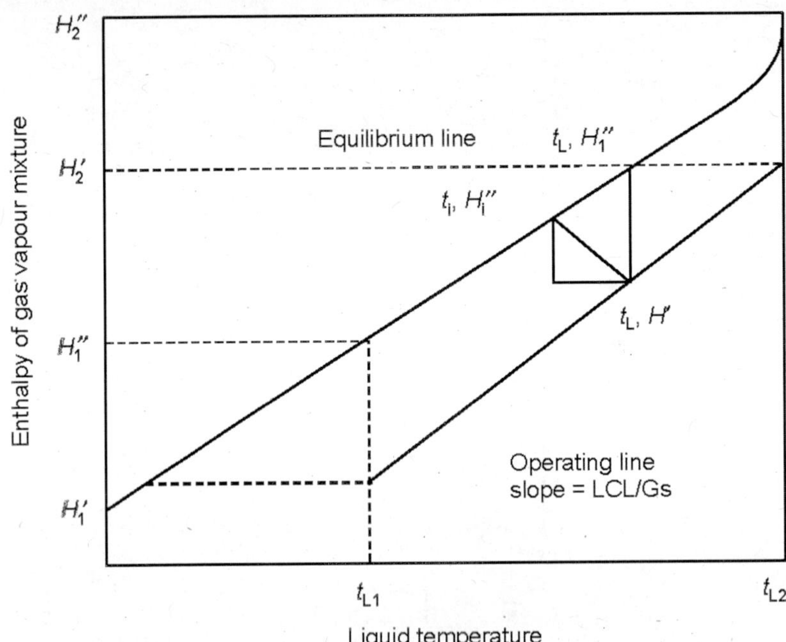

FIGURE 7.6 Operating line, equilibrium line for a water cooler.

An overall transfer coefficient can be used as discussed in the earlier chapters.

$$N_{tOG} = \int_1^2 \frac{dH'}{(H'^* - H')} = \frac{Z}{H_{tOG}} = \frac{K_Y Z \rho a}{G_s'} \tag{7.27}$$

Eq. (7.27) is applicable especially when the equilibrium line is straight.

During dehumidifying operations, the operating line will lie above the equilibrium line, the driving force in a packed tower being $H' - H'^*$. The vapour diffuses from the vapour-air mixture toward the liquid surface and the humidity of the air-vapour mixture will be reduced thus effecting dehumidification.

A gas-humidification cooling with recycle operation is shown in Figure 7.7 and the equilibrium line, operating line, operating points of a typical process are shown in Figure 7.8. The liquid enters the equipment at adiabatic saturation temperature of the entering gas. Recycle of liquid is used adiabatically. The gas will be cooled and humidified following the path of the adiabatic saturation curve on the psychrometric chart. The gas will approach equilibrium with the liquid. The height of the tower packing needed and the number of transfer units needed are calculated as follows:

$$G_s' dY = k_Y a \rho (Y_s' - Y') \, dZ \tag{7.28}$$

$$\int_1^2 \frac{dY'}{(Y_s - Y')} = \frac{K_Y a \rho}{G_s'} \int_0^Z dZ \tag{7.29}$$

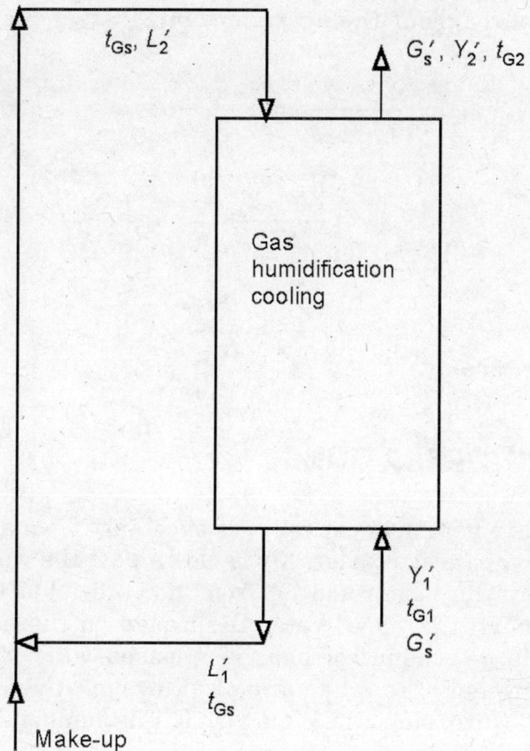

FIGURE 7.7 Gas-humidification cooling with recycle.

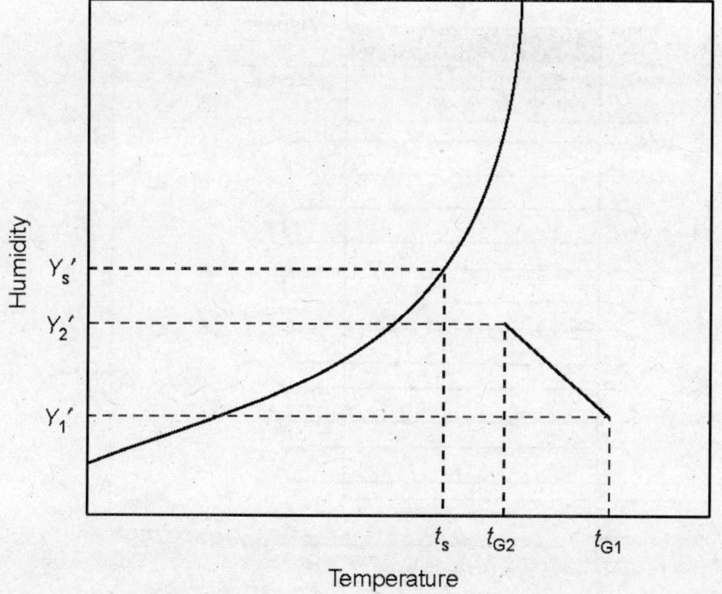

FIGURE 7.8 Equilibrium line and operating line for gas humidification cooling with recycle.

Since Y_s' is constant throughout the operation, integrating Eq. (7.29),

$$\ln\left(\frac{Y_s' - Y_1'}{Y_s' - Y_2'}\right) = \frac{k_Y a \rho Z}{G_s'} \tag{7.30}$$

$$N_{tG} = \left(\frac{Y_2' - Y_1'}{\Delta Y'}\right)_{av} \tag{7.31}$$

$$H_{tG} = \frac{G_s'}{k_Y \rho a} = \frac{Z}{N_{tG}} \tag{7.32}$$

7.3 NONADIABATIC OPERATIONS

There are some operations where heat exchange is necessary from an external source. One good example is an evaporative cooler. Air is blown past the water on the outside of the tube to carry away the heat removed from the tube-side fluid. The fluid is cooled when it flows through the tube. Water is sprayed on the outside of the tube and flows as a film. A large amount of heat is released when the sprayed water evaporates into the air stream. The schematic of an evaporative cooler is shown in Figure 7.9. The tube banks are shown in Figure 7.10. The temperature and enthalpy

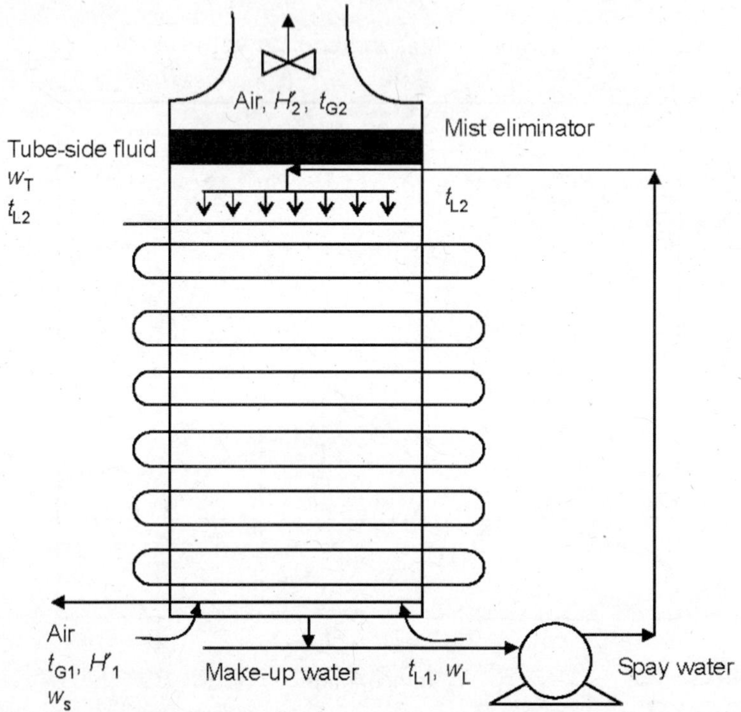

FIGURE 7.9 Schematic of an evaporative cooler.

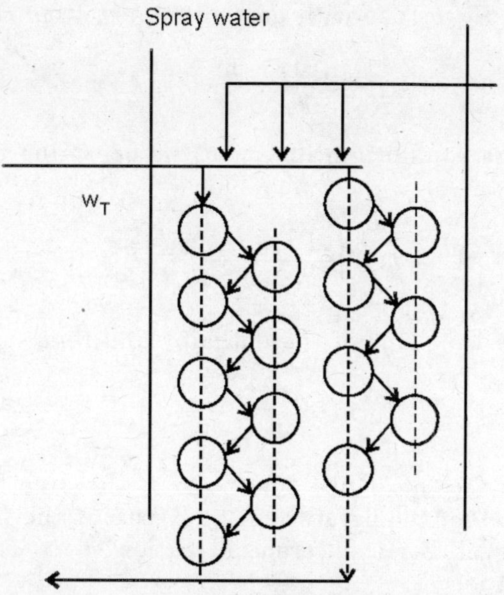

FIGURE 7.10 Schematic of tube-side fluid flow of an evaporative cooler.

profiles are shown in Figure 7.11. The bulk water at the temperature t_L corresponds to a saturation gas enthalpy, $H_1'^*$. The overall heat transfer coefficient U_0 based on the outside tube surface from tube fluid to bulk water is then given by,

$$\frac{1}{U_0} = \frac{d_0}{d_i h_T} + \frac{d_0}{d_{av}(z_m/k_m)} + \frac{1}{h_L'} \tag{7.33}$$

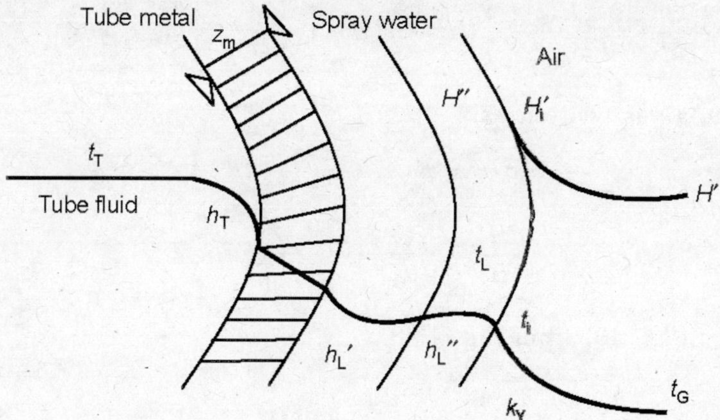

FIGURE 7.11 Temperature and enthalpy profile and resistances in evaporative cooler.

The overall coefficient K_Y for use with gas enthalpies, from $H_1'^*$ to H', is,

$$\frac{1}{K_Y} = \frac{1}{k_Y} + \frac{m}{h_L'} \tag{7.34}$$

where m is the slope of the equilibrium line near the operating region as shown in Figure 7.6.

$$m = \left(\frac{H'^* - H_1'}{t_L - t_i}\right) \tag{7.35}$$

Strictly speaking, m is the slope of the enthalpy and temperature diagram,

$$m = \frac{dH''}{dt_L} \tag{7.36}$$

If A_0 is the outside surface of all the tubes, A_0x is the area from the bottom to the level where the bulk-water temperature is t_L. Here x is the fraction of the heat transfer surface to that level. For a differential portion of the exchanger, the heat loss by the tube-side fluid is,

$$w_T C_T dt_L = U_0 A_0 dx\ (t_T - t_L) \tag{7.37}$$

$$\frac{dt_T}{dx} = \left(\frac{U_0 A_0}{w_T C_T}\right)(t_T - t_L) \tag{7.38}$$

The heat lost by the cooling water is:

$$W_{AL} C_{AL} dt_L = K_Y A_0 dx (H'^* - H') - U_0 A_0 dx (t_T - t_L) \tag{7.39}$$

$$\frac{dt_L}{dx} = \left(\frac{K_Y A_0}{w_{AL} C_{AL}}\right)(H^* - H^*) - \left(\frac{U_0 A_0}{w_{AL} C_{AL}}\right)(t_T - t_L) \tag{7.40}$$

Heat gain in the air is written as:

$$W_s dH' = K_Y A_0 dx (H'^* - H') \tag{7.41}$$

$$\frac{dH'}{dx} = \frac{K_Y A_0}{w_s}(H'^* - H') \tag{7.42}$$

Subtracting Eq. (7.40) from Eq. (7.38),

$$d\left(\frac{t_T - t_L}{dx}\right) + \alpha_1 (t_T - t_L) + \beta_1 (H'^* - H') = 0 \tag{7.43}$$

where $$\alpha_1 = -\left(\frac{U_0 A_0}{w_T CT} + \frac{U_0 A_0}{w_{AL} C_{AL}}\right)$$

$$\alpha_2 = \frac{mU_0A_0}{w_T C_T} \tag{7.44}$$

$$\beta_1 = \frac{K_Y A_0}{W_{AL} C_{AL}}$$

$$\beta_2 = -\left(\frac{mK_Y A_0}{w_{AL} C_{AL}} - \frac{K_Y A_0}{w_s}\right) \tag{7.45}$$

Eqs. (7.43) and Eq. (7.42) can be combined into a second order differential equation with constant coefficients.

$$\frac{d^2 \Delta T_{TL}}{dx^2} + \alpha_1 \frac{d\Delta t_{TL}}{dx} - \beta_1 \left(\frac{k_y A_0}{w_s}\right) m\Delta t_{TL} = 0 \tag{7.45a}$$

The auxiliary equation which is a quadratic can be examined to realise that the roots of the equation are real and so the solution will not undergo any pulsations. The solution is given in terms of the integration constants. The enthalpy is related to temperature difference by Eq. (7.35),

$$\Delta t_{TL} = c_1 \exp(r_1 x) + c_2 \exp(r_2 x) \tag{7.46}$$

$$H'^* - H' = d_1 \exp(r_1 x) + d_2 \exp(r_2 x) \tag{7.47}$$

where,

$$d_j = c_j(r_j + \alpha_1)/\beta_1 \tag{7.48}$$

The design of the evaporative cooler needs values of the transfer coefficients. Some of this data availability is scarce. Correlations for the Nusselt and Sherwood numbers for the given geometry needs to be looked up as discussed in Chapter 3.

7.4 DRYING OF SOLIDS

Solid drying refers to those operations where moisture is removed from the substance. Removal of moisture from solids is by evaporation into a gas stream. Moisture component is usually water and the gas component is usually air. The techniques discussed are equally applicable to other substances. The wet solid is brought in contact with the gas stream. The moisture evaporates from the solid as vapour into the gas steam until a dynamic equilibrium is reached. The equilibrium pressure at this condition is denoted by p_w. This value when divided by the equilibrium vapour pressure of moisture at that temperature, p_w^* will give the relative humidity. The equilibrium value is reached at by either adsorption of moisture onto the solid or by desorption of moisture from the solid to the gas stream. These two processes when repeated several times can lead to hysteresis.

Some solids may be soluble in the liquid that is transferred to the vapour phase. Hydrated crystals exhibit complex equilibrium relationships. The equilibrium moisture content of hydrated copper sulphate is shown in Figure 7.12.

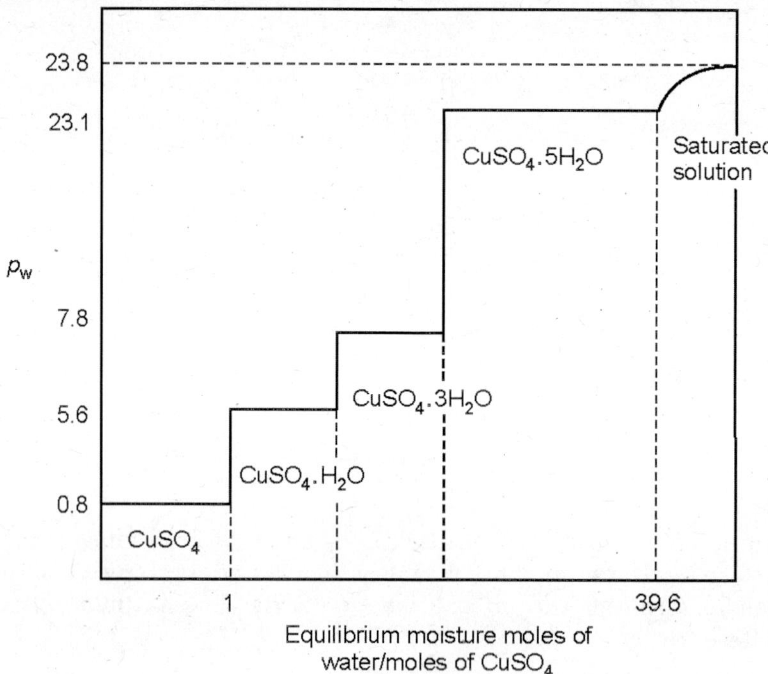

FIGURE 7.12 Equilibrium moisture content of hydrated $CuSO_4$.

In some cases when there is sufficient moisture the solids may dissolve and are said to *deliquesce*. In some cases it may *effloresce* when the moisture is least as in the formation of a monohydrate from a trihydrate. The *moisture content on a wet basis* is the weight of moisture per kg of wet solid. The *moisture content on a dry basis* is the kg moisture per kg dry solid and is denoted by X. The *equilibrium moisture* X^* is the moisture content of a substance when at equilibrium with a given partial pressure of the vapour. *Bound moisture* refers to the pressure contained by the substance which exerts equilibrium vapour pressure less than that of pure liquid at the same temperature. *Unbound moisture* refers to the moisture contained by a substance which exerts equilibrium vapour pressure equal to that of the pure liquid at the same temperature. Free moisture is contained by the substance in excess of the equilibrium moisture content: $X–X^*$. Only free moisture can be evaporated, and the free-moisture content of a solid depends upon the vapour concentration in gas. The different types of moisture such as equilibrium moisture, free moisture, bound moisture and unbound moisture introduced are shown in a schematic in Figure 7.13.

Drying operations can be operated in batch, semi-batch or continuous mode. During batch drying the wet solid is exposed to a flowing stream of gas. In continuous drying the wet solid as well as the stream of gas both move in the equipment during the exchange of mass. Different types of dryers have been developed over the years and this is shown in Figure 7.14. The equipment used in drying can be classified according to several conditions. These are the method of operation, method of supplying heat necessary for the drying operation and the nature of substance to be dried. Thus the

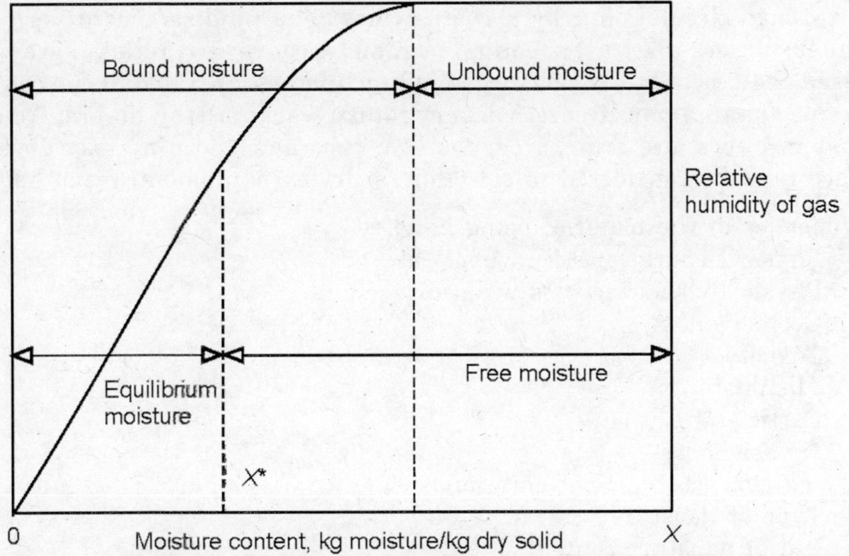

FIGURE 7.13 Different types of moisture in solid.

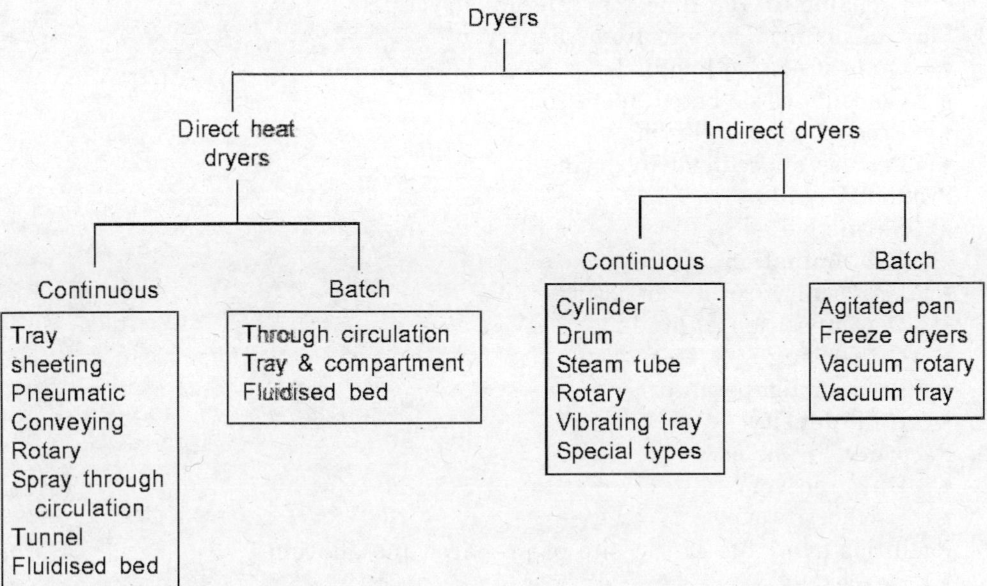

FIGURE 7.14 Classification of different types of dryers.

modus operandi can be batch, semi-batch or continuous. It can be operated at steady state or in transient mode. The heat supplied may be directly through the gas or indirectly through the walls of the container, pan, tray, tube, drum, etc. The nature of the solid substance to be dried may be rigid, flexible, polymeric, granular, crystalline, amorphous, powdered, paste or slurry. Among the direct dryers operated in the batch mode are through circulation, tray & compartment and fluidised bed dryers. Among the continuous dryers are tray, sheeting, pneumatic conveying, rotary, spray, tunnel and fluidised bed types. In the category of indirect dryers, operated in the continuous mode are the agitated pan, freeze, vacuum rotary, vacuum tray dryers. Among the indirect batch dryers, are drum, rotary, steam tube and vibrating tray dryers.

The factors to be considered in selection of drying equipment are as follows:

1. Properties of the material being handled
   - Physical characteristics when wet
   - Physical characteristics when dry
   - Corrosiveness
   - Toxicity
   - Flammability
   - Particle size
   - Abrasiveness
2. Drying characteristics of material
   - Type of moisture
   - Initial moisture content
   - Final moisture content
   - Permissible drying temperature
   - Probable drying time for different dryers
3. Flow of material to and from the dryer
   - Quantity to be handled per hour
   - Continuous or batch operation
   - Process prior to drying
   - Process subsequent to drying
4. Product Qualities
   - Shrinkage
   - Contamination
   - Uniformity of final moisture
   - Decomposition of product
   - Overdrying
   - Product temperature
   - Bulk density
5. Recovery Problems
   - Dust recovery
   - Solvent recovery
6. Facilities available at the site of prepared installation
   - Space
   - Temperature, humidity and cleanliness of air
   - Available fuels
   - Available electric power
   - Permissible noise, vibration, dust or heat losses
   - Source of wet feed
   - Exhaust gas outlets

The rate of batch drying can be analysed to determine the time taken for drying of a substance from one moisture content to another moisture content. The different parameters that can influence the time of drying are discussed. The understanding of the mechanism of drying is by no means complete. So some experimental data is needed to obtain the equilibrium characteristics using which the drying time can be calculated in a reliable manner. The rate of drying curve and the different regimes are shown in Figure 7.15. A constant rate regime and two falling rate regimes, can be seen. In the falling rate regimes, one is of constant slope and the other has some curvature.

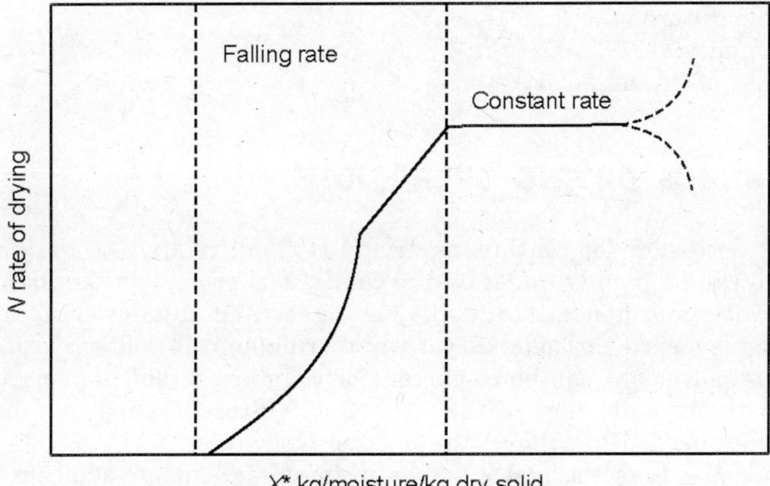

FIGURE 7.15 Rate of drying curve.

If the solid is initially wet, the film of water formed on the solid is unbound moisture and can evaporate at a constant rate as shown in Figure 7.15. When the moisture content of the solid reaches X_c, the critical moisture content the film of water removed and now water exchange mechanism changes. Dry spots appear on the surface. From here on for the drying to proceed, the moisture has to diffuse from the internal sites and come to the surface and then transfer to the vapour phase. The point X^* is the equilibrium moisture content of the solids.

The rate of drying can be defined as follows:

$$N = -\frac{S_s}{A}\left(\frac{dX}{d\theta}\right) \tag{7.49}$$

During the constant rate period, the time to dry can be estimated by integrating Eq. (7.49),

$$\theta_c = -\frac{S_s}{A}\int_{X_1}^{X_c}\frac{dX}{N_c} = S_s\left(\frac{X_1 - X_c}{AN_c}\right) \tag{7.50}$$

In the falling rate regime when the falling rate vs. moisture content of the solid is linear,

$$N = mX + b$$

$$\theta = \frac{S_s}{A}\int \frac{dX}{(mX + b)} = \frac{S_s}{mA}\ln\left(\frac{mX_c + b}{mX_E + b}\right) \tag{7.51}$$

The parameters that can affect the drying characteristics are:

1. Gas velocity
2. Gas temperature
3. Gas humidity
4. Thickness of drying solid

7.5 CONTINUOUS DRYING OPERATIONS

The equipment necessary for continuous drying is small relative to the quantity of the product and can be readily integrated to continuous process in the industry. The product uniformity, and hence the quality, is higher and cost lower in continuous dryers. The solid is moved through a dryer while bringing it in contact with a moving gas stream. The movement can be co-current or counter-current or the gas may be in cross-flow with the solid flow. Either direct or indirect heating may be used to supply the energy needed to achieve the desired objectives.

The operation may be isothermal or adiabatic. In counter-current adiabatic operation the hottest gas is in contact with driest solid and the discharged solid is therefore heated to a temperature which may approach that of the entering gas. As the last traces of moisture are the most difficult to remove, and when this is done readily by contact with the hottest gas, counter-current can be seen to be the most efficient of continuous removal operations. In parallel adiabatic operation the wet solid is contacted with the hottest gas. The solid will be heated up to the wet bulb temperature. Some commercial continuous dryers are tunnel dryers, turbo-type rotating shelf dryers, through circulation dryers, rotary dryers, drum dryers, spray dryers, fluidised bed, pneumatic dryers etc.

The material and energy balance equations for a counter-current continuous dryer can be written as follows. A schematic of the operation is shown in Figure 7.16.

$$S_sX_1 + G_sY_2 = S_sX_2 + G_sY_1 \tag{7.52}$$

$$S_s(X_1 - X_2) = G_s(Y_1 - Y_2) \tag{7.53}$$

$$H'_s = C_s(t_s - t_0) + XC_A(t_s - t_0) + \Delta H_A \tag{7.54}$$

where the integral heat of wetting as J/kg is given by ΔH_A. If the net heat loss from the drier is Q watts the enthalpy balance becomes,

$$S_sH'_{s1} + G_sH'_{G2} = S_sH'_{s2} + G_sH'_{G1} + Q \tag{7.55}$$

FIGURE 7.16 Continuous counter-current dryer.

Direct heat dryers can be operated either at low temperature or at high temperature. When the temperature of operation is higher than the boiling point of the moisture the humidity of the moisture is not a major consideration in the rate of drying. Experimental data from drying tests are necessary as a reliable mathematical model for the mechanism(s) of drying is not developed yet. This is because of the complex mechanisms involved. Three different regimes in the dryer can be recognised (Figure 7.17). The regime A is the preheat region where the solid is heated by the gas until the rate of heat transfer to the solid is balanced by the heat requirements for evaporation of moisture. Very little transfer of moisture takes place in this region. In the second regime, the unbound moisture is evaporated. It can be expected that the temperature is constant in this region. Upon attainment of the critical moisture content, the third regime of drying operations sets in. Unsaturated surface drying

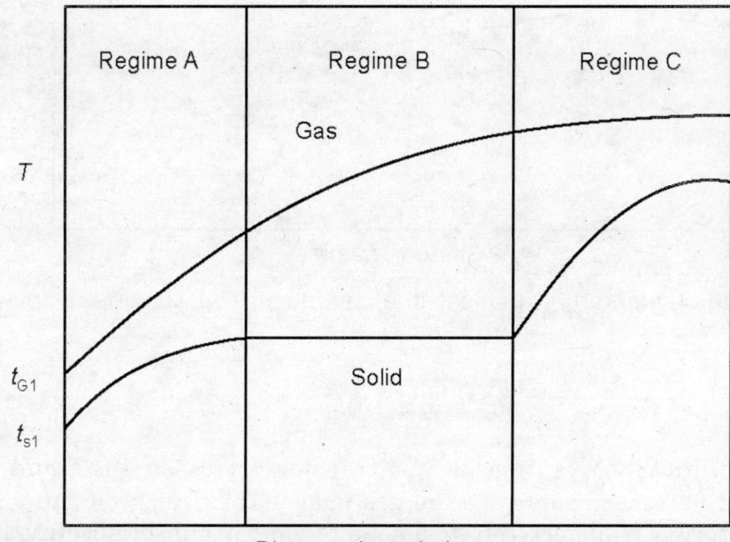

FIGURE 7.17 Temperature gradients in a continuous counter-current dryer.

and evaporation of bound moisture occur. The decreased rate of evaporation results in increased solid temperature.

Most dryers operate in regime B. Here the temperature humidity relationship is of interest (Figure 7.18). If the operation is adiabatic the adiabatic saturation line represents the temperature of the gas in the drier. The surface temperature of the solid is given by the curve with S_1 and S'_1. The number of transfer units and the height of the transfer units needed to complete the drying operations are given below:

$$dN_{tOG} = \frac{Ua\rho dZ}{G_s C_s} \tag{7.56}$$

$$N_{tOG} = \frac{Z}{H_{tOG}} \tag{7.57}$$

$$H_{tOG} = \frac{G_s C_s}{U\rho a} \tag{7.58}$$

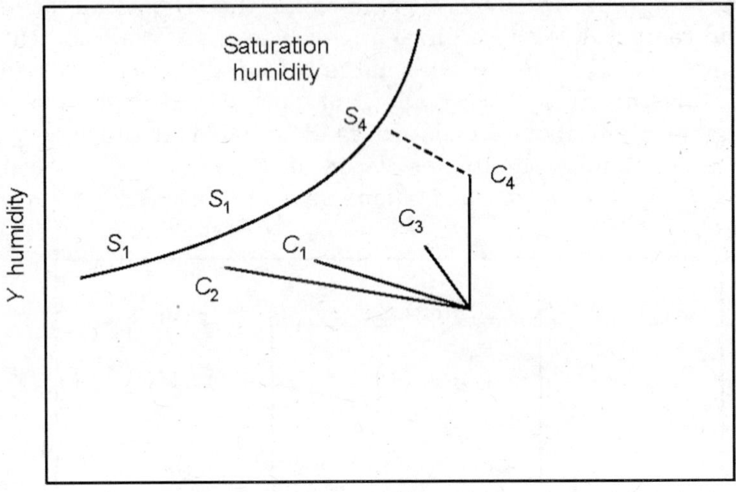

FIGURE 7.18 Temperature gradient in a continuous countercurrent dryer.

SUMMARY

During gas humdification operations water transfers across a gas-liquid interface with the supply of necessary energy to the gas phase. The *drying operations* refer to those where moisture is transferred from a solid to vapour phase. Absolute humidity, relative humidity, % saturation, dew point, dry bulb temperature are defined. The wet bulb temperature for air-water system can be taken as the adiabatic saturation temperature and can be written as:

$$(t_G - t_w) = \lambda_w \left(\frac{Y'_w - Y'}{h_G / k_Y} \right)$$

A psychrometric chart for acetone air mixture was prepared as a worked example. The operating line from mass balance, equilibrium line was shown in the operating diagram for an adiabatic countercurrent continuous operation. The operating line can be written as:

$$G'_s (Y' - Y'_1) = (L' - L'_1)$$

Use of recycle was studied in the cooling with humidification operation. The operation of an evaporative cooler was discussed in some detail as an example of a non-adiabatic operation. The enthalpy and temperature profile in an evaporative cooler was combined into a second order ODE with constant coefficients. The solution to the equation can be written as:

$$\Delta t_{TL} = c_1 \exp(r_1 x) + c_2 \exp(r_2 x) \qquad \text{(temperature)}$$

$$H^* - H' = d_1 \exp(r_1 x) + d_2 \exp(r_2 x) \qquad \text{(enthalpy)}$$

For any combination of parameters of the system the solution to the ODE can be shown to be stable.

The drying operations were distinguished between continuous, batch, semi-batch. The equilibrium relationships between the vapour and solid moisture content may change in behavior should the solid be soluble in the liquid. Different kinds of moisture such as unbound moisture, bound moisture, equilibrium moisture and free moisture are defined. The different types of driers are classified in Figure 7.14. The different factors considered for selection of drying equipment was discussed. The rate of drying for a batch of solid was derived. Three different regimes of drying were identified. One is the constant rate period, the other is a falling rate period of constant slope and a third varying falling rate period of complex nature. The time taken for drying during constant rate and falling rate periods can be written as:

$$\theta_c = S_s \frac{(X_1 - X_c)}{A N_c} \qquad \text{(constant rate)}$$

$$\theta = \frac{S_s}{mA} \ln \frac{mX_c + b}{mX_E + b} \qquad \text{(falling rate)}$$

A continuous counter current dryer was analyzed and the mass and energy balance written. Three different zones are identified in the drying equipment. The first regime is the heating period of the solid, the second regime is the constant rate and constant temperature period, and the third is the falling rate and rising temperature zone. (Figure 7.17). The height and number of transfer units needed to effect a given drying operation in a countercurrent continuous manner was derived.

EXERCISES

1. What is the difference between evaporation and humidification?
2. During drying does the mass transfer occur as a solid vapour interface or a liquid vapour interface? Give reasons.
3. What is the value of the absolute humidity and molal humidity at the boiling point of the liquid? What happens physically at this condition?
4. How many degrees of freedom exist for a vapour-gas mixture. How many triple points can it have?
5. In the adiabatic saturation operation shown in Figure 7.3, what will happen when ice is sprayed into the chamber instead of water?
6. What does $(1 + Y')$ in the denominator of Eq. 7.10 mean?
7. Can you humidify an inlet gas by contacting with steam? Discuss.
8. Can the liquid supplied for adiabatic humidification operations be sea water. Does this change any of the objectives?
9. How will the changing of cross-sectional area of the cooling tower with height affect the design equations?
10. What happens when the moisture content in the solid is at equilibrium as well as the vapour pressure in the gas of the moisture at the temperature is a saturated value?
11. Why is the last portion of the falling rate regime of drying curvilinear?
12. What does the linear relationship between the drying rate and the moisture content of the solid in the first falling rate of drying mean physically?
13. When a dry solid is contacted with a saturated vapour gas mixture and the vapour condenses and forms the moisture in the solid, can the time taken to reach the equilibrium value be calculated using the equations developed. Why?
14. When a dry solid is contacted with a vapour gas mixture at a partial pressure lower than the saturated value, and when the vapour condenses to form moisture within the otherwise dry solid, how can the time taken to moisten be calculated?
15. What will happen when a wet solid comes in contact with steam? Can the operation meet the drying objectives?
16. Sketch the schematic of a continuous counter-current wetting operation, i.e. a saturated vapour gas mixture comes in contact with dry solid that is moisture free. What are the regimes of wetting that will be expected?
17. In the continuous counter-current drying operation (Figure 7.16), should the wet solids be sprayed or dropped as a strand what difference does it make in the calculations? How?
18. In a rotary dryer, is the contact of air and wet solid co-current and counter-current?

PROBLEMS

1. *Induced Draft Cooling Tower*

The overhead condenser in distillation tower requires 15 kg/s of cooling water removing 270 watts of heat. The water will leave the condensers at 45°C. It is cooled in an induced-draft cooling tower. The entering air is at 24°C wet bulb temperature and 30°C dry bulb temperature. The water need to be cooled to within 5°C of the inlet air wet bulb temperature. An air/water-vapour ratio of 1.5 times the minimum is used. The make-up water will come from a well at 10°C, hardness 500 ppm dissolved solids. The circulating water should not exceed 200 ppm hardness. In the tower packing, $K_Y a\rho$ = 0.9 kg/m^3/s. This is for a gas rate of 2 kg/m^3/s and a liquid rate at 2.7 kg/m^2/s. Compute the dimensions of the packed section and the make-up water requirement (Fig. 7.19).

(**Ans.** H_{tOG} = 2.22 m; N_{tOG} 25)

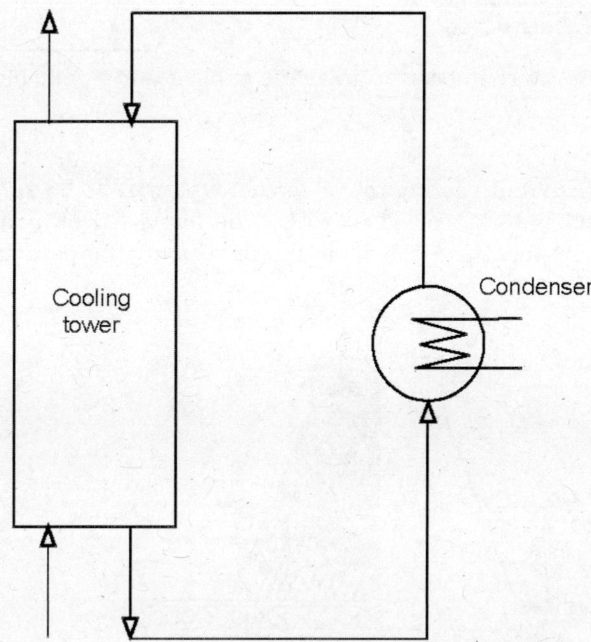

FIGURE 7.19 Induced draft cooling tower for problem 1.

2. *Adiabatic Humidification and Cooling*

A horizontal spray chamber (Figure 7.20) with recycled water is used for adiabatic humidification and cooling of air. The active part of the chamber is 2 m long and has a cross-sectional area of 2 m^2. With an air rate of 3.5 m^3/s at dry bulb temperature of 65.0°C., Y' = 0.0170 kg water/kg dry air, the air is cooled and humidified to a dry-bulb temperature of 42.0°C. If a duplicate spray chamber operated in the same manner were to be added in series with the existing chamber, what outlet conditions could be expected for the air?

(**Ans.** Y'_2 = 0.0295, t_{G2} = 34.0°C)

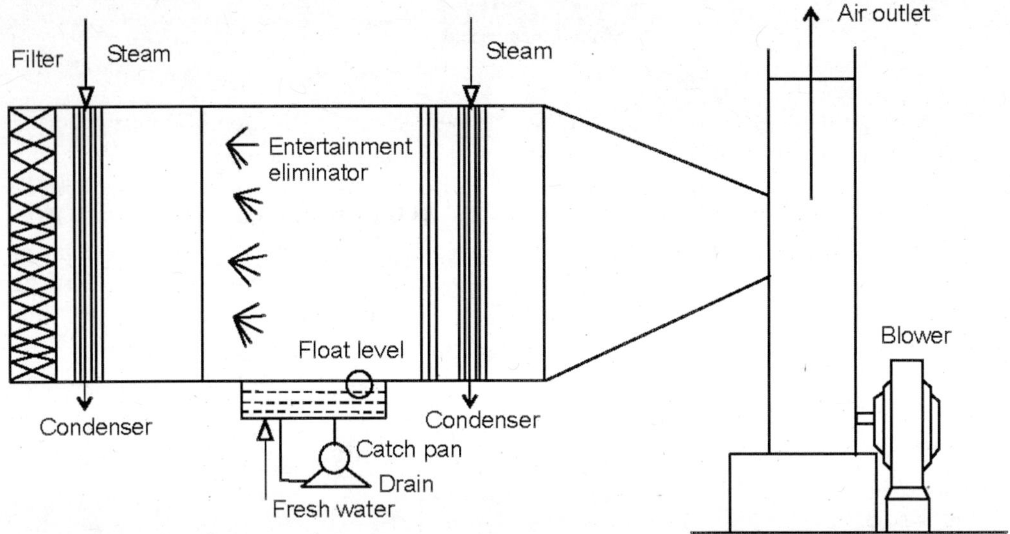

FIGURE 7.20 Spray chamber for adiabatic cooling and humidification.

3. *Cooling Tower*

Design a counter-current cooling tower to cool water at 35.83 kg/min. The water enters at 60°C and is to be cooled to 25°C. The air is fed at a rate 60 gmol/m^2/s with a dry bulb temperature of 30°C and a dew-point temperature of 10°C. The

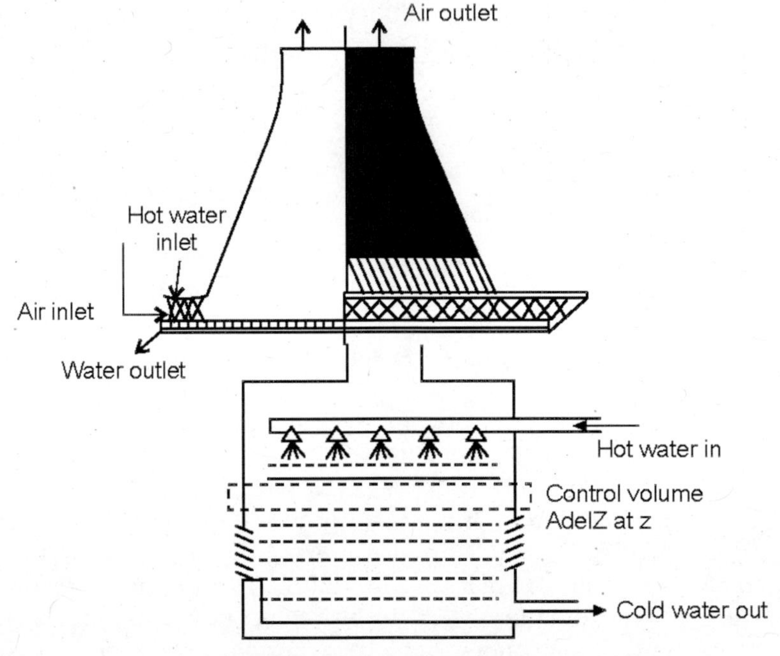

FIGURE 7.21 Cooling tower.

water flux should be 40% lower than the maximum thermodyna-mically. The tower is packed with wood slats having an area per volume of 3 m^{-1}. Find (a) the flow rate of the water per tower cross-section, (b) the tower cross section, and (c) the height of tower required.

(**Ans.** (a) 110 g-mol H_2)/m^2sec; (b) 18 m^2 and (c) 8.1 m

4. *Wet Bulb Temperature*
The wet bulb temperature and dry bulb temperature of chloroform is 23°C and 62°C respectively. Determine its concentration in air at 2 atm.

(**Ans.** 0.004 mole/lit)

5. *Psychrometric Chart*
Construct the psychrometric chart for air at 2 atm. Use the Clausius Clapeyron equation and ideal gas law if necessary.

6. *Drying at Low Temperature*
The solids dried at low temperatures also pass through different regimes as shown in Figure 7.22. Only regimes *B* and *C* need be considered as not much moisture transfers when the solid is being heated up. Show that the time taken for drying is:

$$\textbf{Ans.}\ \theta_B + \theta_C = \frac{G_s}{Ak_Y}\ln\left(\frac{Y_s - Y_1}{Y_s - Y_c}\right) + \left(\frac{G_s X_c}{S_s k_Y}\right)\left(\frac{1}{(Y_s - Y_c)G_s/X_2}\right)\ln\left(\frac{X_c(Y_s - Y_2)}{X_2(Y_s - Y_c)}\right)$$

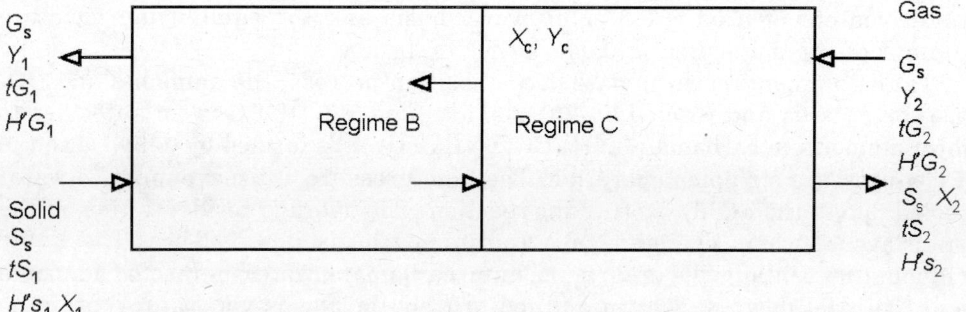

FIGURE 7.22 Continuous countercurrent low temperature drier.

7. *Prilling Tower to Granulate Urea*
Urea ($NH_2.CO.NH_2$) has a melting point of 132.7°C and has a solubility of 135 gm per 100 gm at 100°C). Urea is used as a solid fertilizer, liquid fertilizer, urea formaldehyde adhesive, melamine formaldehyde dinnerware (formica) etc. Urea is the nitrogen compound with the largest production volume in the world.

Fertilizer corp. of India manufactures 1500 tonnes/day urea granules at their Ramagundam facility. Assuming that they use the ammonium carbamate decomposition process calculate the air rate required at a dry bulb temperature of 36°C and a wet bulb temperature of 26°C when it is allowed to exit the prilling tower (shown in Figure 7.23) at a relative humidity of 97.7%. The inlet

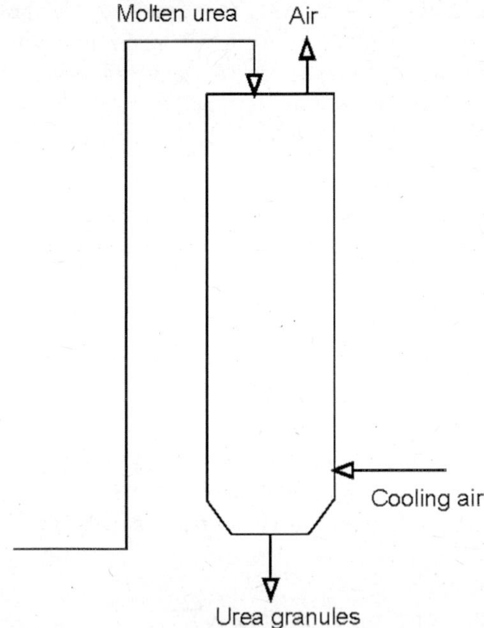

FIGURE 7.23 Prilling tower to granulate urea.

water content of urea is 8.38% on a wet basis and the equilibrium moisture content of the solid urea is 0.045 kg water/kg dry solid.

In the ammonium carbamate decomposition process, the ammonia and CO_2 are compressed and reacted at 100–200 atm and 170–190°C in an autoclave to form ammonium carbamate ($NH_4.COO.NH_2$). Urea is formed by dehydration in a low-pressure stripping operation. The high pressure autoclave must be water cooled due to the highly exothermic reaction. The average residence time of the autoclave which is operated on a continuous basis is 1.5–2 hrs. The liquid effluent that contains the urea, ammonium carbamate, water, unreacted ammonia and CO is let down at 27 atm and fed to a special flash-evaporator containing a gas-liquid separator and condenser. Unreacted ammonia CO_2 and water are recycled. An aqueous solution of carbamate urea is passed to the atmospheric flash drum where further decomposition of carbamate takes place. The off gases can be recycled at this stage. The 80% aqueous urea solution is sent to a vacuum evaporator to obtain a molten urea containing less than 1% water. Biuret formation is avoided by keeping the temperature above the melting point of the urea.

(**Ans.** 4446 tonnes/day)

8. *Rotary Drum Dryer*

Drying of large volumes of fragmented fibrous materials can be carried out in heat exchangers consisting of elongated, horizontally-oriented drums. Hot gases are caused to flow through each drum to remove moisture from the material by heat exchange between the hot gases and the fibrous product.

A burner is disposed to direct hot products of combustion directly into the inlet of the drum which also receives the moisture-bearing material to be dried. After removal of the requisite amount of moisture from the material, the dried product is directed into a collector at the outlet of the heat exchange drum. A blower is provided to provide the required rate of flow of hot gases through the drum heat exchanger. Three pass dryers have been used which include a single rotatable drum with concentric stages arranged so that the material being dried traverses the drum in a serpentine fashion.

The pass dryers are relatively expensive but have been used primarily because of the decreased product residence time necessary to obtain adequate drying, while minimizing ground space in the drying plant. 12 tonnes per hour of 70% wet solid is dried to a 10% wet solid on a wet basis. How much water is removed per hour when the inlet air is at a relative humidity of 26% and the outlet air is at a relative humidity of 97.7%. Can you show the different zones of heat transfer. What is the air supply needed at 40°C?

(**Ans.** 7.2 tonnes/hr; 200 tonnes/hr)

9. *Drying Time of Wet Toner Particles*
Chemical toner used in copiers is composed of toner particles which are applied to paper to produce an image. It is desirable that the toner particles be uniformly sized, having a narrow size distribution, to produce images with improved resolution and clarity. For example solid toner particles produced have a typical average size distribution of approximately 6 μm in diameter with most particles falling in a range of about 2–8 μm. What will the porosity for a FCC packing?

Assuming that the equilibrium moisture content is when a tenth of the pore volume is filled with water, calculate the drying time needed for a batch of 5 kg of toner material at atmospheric pressure. Calculate the initial moisture content by assuming a film of water of thickness equal to 1/4th the diameter of the toner particle each. The relative humidity of air at the drying temperature of 42°C is 100%. The critical moisture content may be taken as 5 times the equilibrium moisture content. The *m* value in the falling rate region is 1 E-7 kg/m^2/s. The drying rate during the constant rate period is 10 nanogram/sec.

(**Ans.** 0.26; 0.37 kg water/kg dry toner; θ_{tot} = 21.6 + 11.6 hr)

10. *Falling Rate Regime of Drying*
The total drying time needed for a batch of solid is 13.14 hrs. The weight of the wet solid is 160 kg and the initial and final moisture content of the solid are 25 and 6% respectively. The drying surface is 1 m^2/40 kg of dry weight. The critical moisture content in the solid is 0.2 kg moisture/kg dry solid. The rate of drying at the constant rate period is given as 0.003 kg water/m^2/s. Find the slope of the falling rate region assuming that b is zero.

(**Ans.** m = 12.97)

11. *Dry Fruits*
A critical operation in the preparation of dry grapes is the drying operation. Calculate the total time of drying needed for a batch of 5 kg of dry grapes with an equilibrium moisture content of 0.02 kg water/kg of dry solid. The drying

temperature is 35°C. Assume some unbound moisture is present in the initial condition and the moisture content of the solid is 26% on a wet basis. If the final size of each grape is 1 cm cylinder with a radius of 3 mm, calculate the drying surface area. The critical moisture content from drying tests have been found to be 0.18 kg moisture/kg dry solid. Assume that the drying takes place in a fluidised bed and there is little contact between the solids. The slope of the falling rate regime is given as 1 E-5.

(**Ans.** A = 22.21 m^2/kg; θ_{tot} = 1 + 2.75 hr)

12. *Recirculation Countercurrent Tray Dryer*
The tray dryer is essentially a hot air oven in which the material is placed in thin layers in trays. There are variations in the design depending on the source of heat used such as microwave, infrared, steam coil, and modifications such as vacuum and recirculation air is common. The air is circulated using fans. Show that for the recirculation tray dryer as shown in Figure 7.24, the operating diagram such as Figure 7.18 needs to be modified as shown in Figure 7.25. What other changes in the calculations are needed?

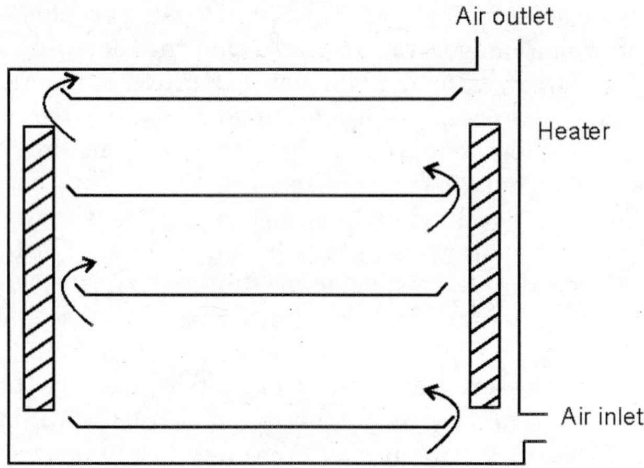

FIGURE 7.24 Recirculation tray dryer.

13. *Komline-Sanderson Spray Dryer*
An aqueous solution containing 63.5 per cent by weight of sodium 2-ethylhexanoate was atomized through a spray nozzle under 5 atm of pressure into a Komline-Sanderson Spray Dryer having an inlet temperature of about 110°C. at a feed rate of 136 cc/min. The total volume of solution fed into the spray dryer was 993 cc. The outlet temperature was recorded at about 85°C. The product, a white crystalline form of sodium 2-ethylhexanoate, was recovered from the chamber and cyclone of the spray drier, and amounted to 411 grams, a 74.9 per cent recovery. The sodium 2-ethylhexanoate so obtained contained 3.57 per cent by weight of water as determined by Karl-Fischer analysis. The surface tension of water is 0.07 N/m. What is the surface area of transfer of moisture when a jet is formed on account of the spray.

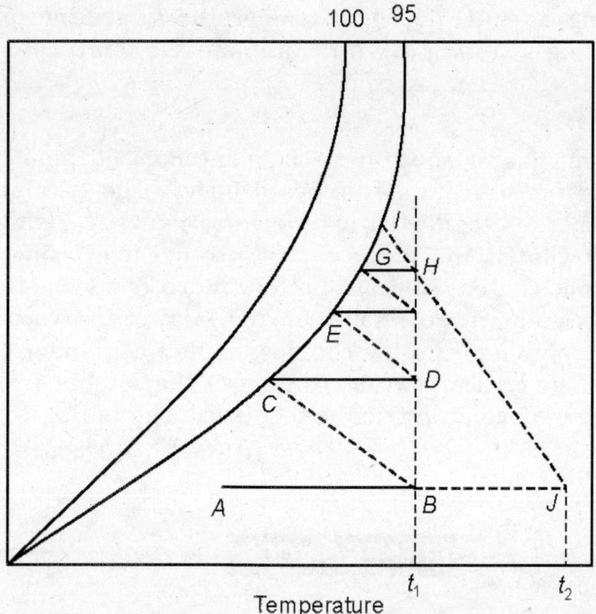

FIGURE 7.25 Temperature humidity operating diagram for recirculation tray dryer.

Spray drying typically is a process which involves the rapid dehydration of moist particles which contain solids in either the soluble or insoluble form or both. The extent of successful drying and formation of a crystalline product normally depends upon several factors, including the temperature range of the spray dryer, the concentration of the solution undergoing the drying process, and the extent of atomization of the solution which has been preimposed on the feedstock to create the moist particles. Deterrents to successful spray drying include a high viscosity of the solution being atomized, in addition to operational temperature. It is therefore important in any particular spray drying process to utilize optimum concentrations of solutions being dried, and optimum temperature ranges for the desired result. Calculate the relative humidity of stream of air at the outlet. The relative humidity of the inlet air is 11%.

(**Ans.** jet size = 86 nm/m of jet length; 50%)

14. *Evaporative Cooler*

The evaporative cooler similar to that shown in Figures 7.9 and 7.10 is used to cool oil. The oil flows at 4 kg/s in the tubes and enters the tubes at 95°C. Its density may be taken as 800 kg/m^3, viscosity as 0.005 kg/m/s, thermal conductivity as 0.1436 w/m/K and heat capacity as 2010 J/kg/K. The cooler consists of rectangular vertical shell, 0.75 m wide, fitted with 400 admiralty metal tubes, 19.05 mm od, 1.65 mm wall thickness. 3.75 m long, on equilateral-triangular centres 38 mm apart. The tubes are arranged in horizontal rows of 20 tubes through which the oil flows in parallel, in stacks 10 tubes tall. Cooling water will be recycled at the rate of 10 kg/s. The design air rate is at 2.3 kg dry air/s entering at 30°C dry bulb temperatures, at atmospheric pressure, humidity of 0.01 kg water/kg dry air. Estimate the temperature to which the oil will be cooled. (**Ans.** 47.0°C)

15. *Vacuum Drying*
The residual monomer in polystyrene is reduced to 50 ppm using a process similar to the drying operations discussed above. The devolatilization of the unreacted monomers is conducted in a vacuum chamber. Using a Henry's law constant of 0.01 what is the vacuum pressure needed. Using 50 ppm as the equilibrium styrene content in the solid how much residence time is needed to devolatilize polystyrene at 90% solids in the inlet of the vacuum chamber at 3636 kg/hr. The volume of the DV chamber is half a cylinder of 1 m diameter and 2 m height. The critical styrene content of the solid is 1 wt% styrene and the constant rate of devolatilization is 0.05 kg/s. The falling rate m value is 5. (**Ans.** 10 torr; θ_{tot} = 2.33 + 1.22 hr)

REFERENCES

Cussler, E.L., 1997, *Diffusion: Mass Transfer in Fluid Systems,* Cambridge University Press, London.

Grosvenor, W.M., 1908, *Trans. AIChE,* **1**, 184.

Mickley, H.S., 1949, *Chem. Eng. Prog.*, **45**, 739.

Sharma, K.R., June 2001, *Mass Transfer Model for Microwave Drying of Diced Apples in a Spouted Bed,* 34th Middle Atlantic Regional Meeting of the ACS, Baltimore, MD, USA.

________, 2005, *Damped Wave Transport and Relaxation,* Elsevier, Amsterdam.

Treybal, R.E., 1980, Mass Transfer Operations, McGraw Hill, New York.

CHAPTER 8

Liquid Extraction and Solid Leaching

Nomenclature

A_f	extraction factor (R/mE)
B	solvent rate (kg/hr)
C_s	co-solvent factor ($B_s m_1/Bm_2$)
E	extract rate (kg/hr)
F	feed rate (kg/hr)
R	raffinate rate (kg/hr)
M	mixture rate (kg/hr)
N	solvent concentration, solvent free basis (mass B/mass A + mass C)
N_p	number of stages
E'	solvent free basis (kg/hr) of extract
x	weight fraction C of solute
x_M	weight fraction of solute in the mixture
x'	($x/(1–x)$) wt. fraction C on a solute free basis
y	weight fraction C in extract
y'	($y/(1 - y)$) wt. fraction C on a solute free basis
y_S	wt. fraction solute in the fresh solvent
X	wt. fraction solute, on a solvent free basis in raffinate
Y	wt. fraction solute, on a solvent free basis in extract

Greek

Δ_R	difference point in the operating diagram
ρ	reflux ratio of solvent in counter-current cascade

8.1 INDUSTRIAL APPLICATIONS OF LIQUID EXTRACTION

The common denominator between the liquid extraction and solid leaching process is the solvent used. In *liquid extraction,* bulk separations are effected from liquid mixtures and in *solid leaching* the bulk separations are carried out from solid mixtures. Liquid extraction is used in areas where it is difficult to separate the components otherwise. Some *solvent loss* during a continuous process with recycle will add to the *cost of the separation*. The original liquid mixture is called the *feed* and upon extraction becomes the *raffinate*. The *extract* refers to the portion of the product that contains the desired component or solute in cases where the feed mixture was a solution. If the desired component to be separated is a gas then the operation of separation would be *stripping*.

Liquid extraction is selected after consideration has been given to distillation and absorption. Liquid-liquid extraction is central in some industries. It is used in petrochemical, pharmaceutical, metallurgical and energy industries. The general principles shown for distillation, humidification are applicable here also. Extraction is used to separate metal ions that are non-volatile. It is used in flavour and fragrance industry where the compounds that need to be handled can decompose at conditions of distillation. Solvent extraction can be used to separate azeotropes as discussed in Section 4.3. Typical solvents used and the application used for is shown in Table 8.1. Lubricants are made from specific fractions of crude oil after refining operations.

The process of liquid extraction is used in the preparation of most drugs. Synthetic drugs produced by fermentation are separated by liquid extraction. The production of bacitracin involves partitioning from the bacterial growth medium into butanol. Gelatin preparation involves the use of warm water as solvent after the skin and bone collagen is treated with lime or dilute acid. Extraction of other proteins and polypeptides involves the use of solvent on the minced animal tissue. Penicilin is produced using fermentation. The suspended microbes in beer are removed by filtration. The clarified beer is then extracted with an alkyl acetate such as amyl acetate. Pencilin undergoes four transfers before being crystallised into the final product. Supercritical CO_2 is used for extraction. One example is in coal cleaning. Liquefied coal can be further extracted into useful fractions. The different industries where liquid extraction is used are as follows:

1. **Chemical**
   - Washing of acids/bases, polar compounds from organics
2. **Pharmaceuticals**
   - Recovery of active materials from fermentation broths
   - Purification of vitamin products
3. **Effluent Treatment**
   - Recovery of phenol, DMF, DMAC
   - Recovery of acetic acid
4. **Polymer Processing**
   - Recovery of caprolactam
   - Separation of catalyst
5. **Petroleum**
   - Lube oil quality improvement
   - Separation of aromatics/aliphatics (BTX)

TABLE 8.1 Applications of Liquid Extraction in Industry

No.	*Solvent used*	*Application*	*Feed*	*Industry*
1.	Amyl acetate, Methylene chloride	Penicilin extraction	Filtered fermentation beer	Pharmaceuticals and Food
2.	Warm water	Gelatin manufacture	Skin and bone collagen	Biotechnology
3.	Propane, Hexane	Refining fats and oils	Soyabeans	Pharmaceuticals and Food
4.	Hydroxylamine in kerosene	Concentrating copper for electrowining	Acidic leach liquors	Mining and Metallurgy
5.	Tertiaryamines in kerosene	Uranium and rare earth separations	Acidic leach liquors	Nuclear Power Industry
6.	Glycols, furfural, cresol, liquid SO_2	Dewaxing lubricating oils	Crube lube oil	Petroleum
7.	Glycols, sulpholane	Higher octane aromatic fuels	Aliphatic-aromatic mixtures	Petroleum
8.	Aqueous base	Desulfurisation	Sour distillates	Petroleum
9.	Aqueous copper complex	Butadiene/Butene separation	Incompletely dehydrogenated feed	Plastic
10.	Toluene	Monomer purification	Impure caprolactum	Plastic
11.	Ethylbenzene	Continuous polymerisation of polystyrene recycle	Styrene, initiator	Plastic
12.	Methylethylketone	Continuous polymerisation of ABS–Recycle	Styrene, Acrylonitrile, Polybutadiene	Plastic
13.	Phenol	Recovery and recycle in chemical process	Polycarbonate manufacture	Plastic
14.	Ethyl acetate/ Butyl acetate	Separation of hydrogen bonded organics from water	Waste water	Water Treatment
15.	Hydrocarbon/ aqueous alcohol	Orange oil, Cinnamon oil, Peppermint oil, Lemon oil	Essential oil	Food, Fragrance industry
16.	Water	Organics	Organic/base	Organic Chemicals

6. **Petrochemicals**
    - Separation of olefins/paraffin
    - Separation of structural isomers
7. **Food Industry**
    - Decaffeination of coffee and tea
    - Separation of essential oils (flavours and fragrances)
8. **Metals Industry**
    - Copper production
    - Recovery of rare earth elements
9. **Inorganic Chemicals**
    - Purification of phosphoric acid
10. **Nuclear Industry**
    - Purification of uranium

A typical extraction process is shown in Figure 8.1. The solute is given by B instead of C and solvent denoted by C instead of B. Since most investment and operating costs are associated with the solvent recovery steps, it is very important to consider and study this aspect when designing the entire process. In a typical extraction process about 3% of the operating cost is in the extractor, with the remaining 97% in solvent recovery. Therefore, it is extremely important to consider the solvent recovery aspects early in the project since they play such an important role in the overall process economics. The feed, raffinate, extract and solvent can be recognised in the process flow diagram.

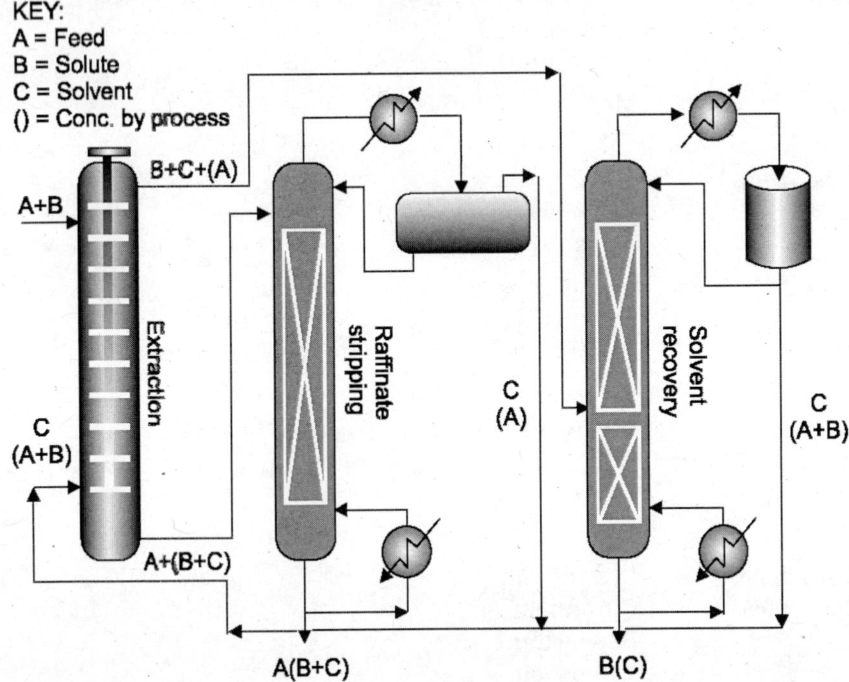

FIGURE 8.1 Typical flowsheet of a liquid extraction B-solute; C-solvent.

Figure 8.2 shows a flow chart for the recovery of hydrogen-bonded organics like formaldehyde and acetic acid from water.

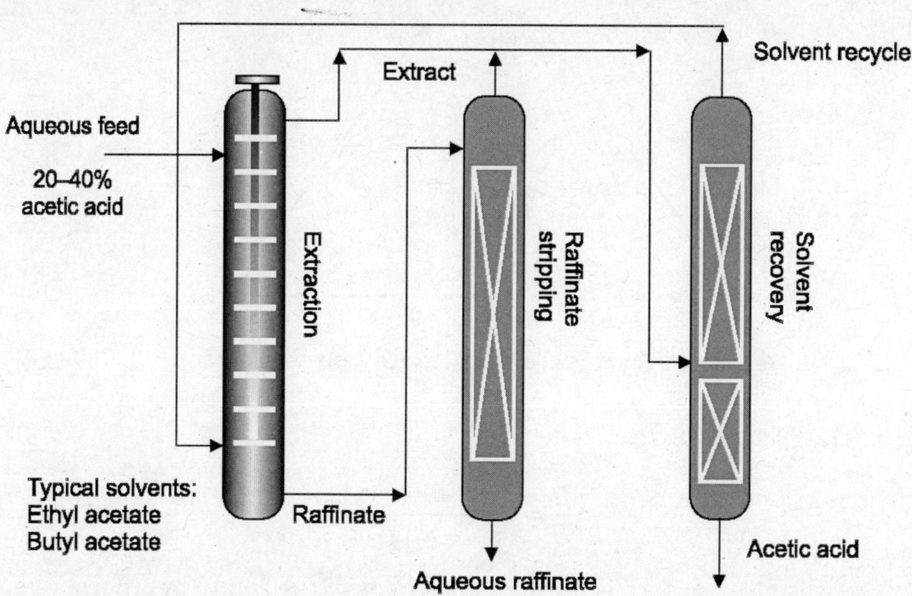

FIGURE 8.2 Recovery of tightly hydrogen-bonded organics from water such as formaldehyde, formic acid and acetic acid.

8.2 EQUILIBRIUM LINE, OPERATING LINE IN TRIANGULAR PHASE DIAGRAMS

Liquid-Liquid extraction systems are typically three-component, 2-phase systems. The number of degrees of freedom are then, 3 – 2 + 2 = 3. The desired component for separation is denoted as solute C, the solvent is denoted with B and the primary component of the feed solution by A. The equilibrium relationships and operating line for a given extraction scheme is shown in a *ternary phase diagram* such as the one shown in Figure 8.3. The equilateral triangular coordinates are used to represent concentrations for a ternary system. For a point inside the triangle, the perpendicular distance from the point to the side gives the relative proportion of the component in the mixture. If R kg of raffinate is added to E kg of extract and M is the kg of the mixture, then,

$$R + E = M \tag{8.1}$$

The compositions of the solute in the raffinate R, extract E and mixture M are represented by x_{R}, x_{E} and x_{M} respectively. A component balance of the solute can be written as:

$$x_{\mathrm{R}}R + x_{\mathrm{E}}E + x_{\mathrm{M}}M \tag{8.2}$$

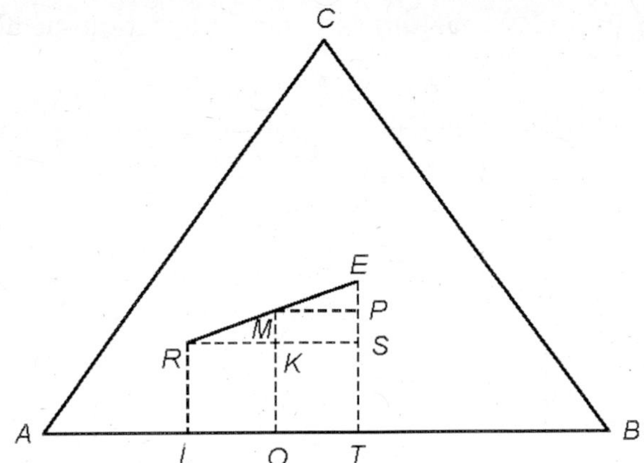

FIGURE 8.3 Mixture rule and ternary phase diagram.

Combining Eqs. (8.1 and 8.2),

$$\frac{R}{E} = \left(\frac{x_M - x_E}{x_R - x_M}\right) \tag{8.3}$$

The *R*/*E* ratio was given in terms of the compositions of the solute in different phases. The *R*/*E* ratio can also be given by the ratio of line segments in Figure 8.3 such as *EP*/*SP* and *ME*/*MR*.

Consider the system of polybutadiene, styrene and ethyl benzene. The solute is polybutadiene *C*, the solvent is ethyl benzene *B* and styrene is *A*. This is an example of a typical ternary system that is drawn upon in most extraction operation calculations. The curve shown in Figure 8.4 is isotherm. The region outside the curve *RPE* in Figure 8.4 is the one phase region and the region within the curve is the heterogeneous region. The curve *LPE* in Figure 8.4 is the binodal curve. The favour of free energy with composition expression for miscibility changes at this curve. There can exist a spinodal curve which is metastable. In this case the binodal curve is a UCST, Upper Critical Solution Temperature, i.e. the temperature identified is a value above which the system will be one phase. There can exist a LCST for some systems, the Lower Critical Solution Temperature, which is the temperature above which the system is immiscible.

Thus a mixture within the binodal solubility curve will form two insoluble saturated liquid phases of equilibrium compositions. Thus a mixture *M* will form *R* and *E* as shown in Figure 8.4. The line joining these two points is the *tie line*. Among the infinite tie lines that exist, a few are shown in Figure 8.4. The tie line is when horizontal indicates the *plait point P* and the system is said to be *solutropic*. The distribution of solute *C* is favoured in *E* compared with *R* for the Pbd-styrene-ethyl benzene system as shown in Figure 8.4. This can be plotted in the distribution diagram in the lower portion of Figure 8.4. The *distribution coefficient* is given by y^*/x. The system pressure and temperature will influence the phase behaviour. Of the 3 degrees of freedom for a 2-phase, 3-component system, the temperature and

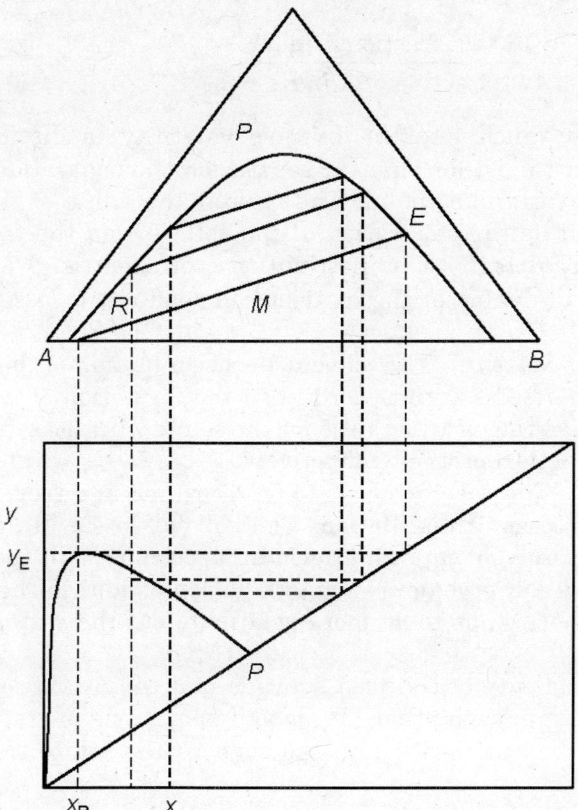

FIGURE 8.4 Ternary phase diagram of solute *C* distribution in *A* and *B* liquids.

pressure are two degrees of freedom. Once one of the compositions is known, the other can be calculated or looked up from the distribution diagram as shown in Figure 8.4.

Sometimes rectangular coordinates are preferred to triangular coordinates. In this case the abscissa is reserved for solvent-free solute fraction in the feed and the ordinate is earmarked for the solvent concentration on a solvent free basis.

8.3 CONSIDERATIONS FOR CHOOSING A SOLVENT

A number of liquids are available to be chosen from as a solvent for a given liquid extraction problem. The solvent can be chosen according to the following criteria:

1. *Selectivity*: The partitioning of the solute C into the solvent B or remain in the A phase can be quantitated by a parameter called *selectivity*.

$$\beta = \frac{(\text{wt. fraction } C \text{ in } E)(\text{wt. fraction of } A \text{ in } R)}{(\text{wt. fraction } C \text{ in } R)/(\text{wt. fraction of } A \text{ in } E)} \tag{8.4}$$

$$= \frac{(y_E^*/x)(\text{wt. fraction } A \text{ in } R)}{(\text{wt. fraction of } A \text{ in } E)} \tag{8.5}$$

The selectivity, much like the relative volatility in distillation problems, must be greater than 1 for efficient separation. No separation is possible with $\beta = 1$. It is 1 at the plait point. The greater the value of selectivity from 1, the greater will be the separation of the solute from the feed solution.

2. *Distribution Coefficient*: The distribution coefficient is given by the ratio of y_E^*/x_R. Larger the value of the distribution coefficient, lesser solvent will be required.
3. *Immiscibility if Solvent*: The solvent needs to be immiscible with the liquid contained in the feed solution. Only then the formation of two phases can be effected and when the partition ratio for the solute with the solvent is favourable the liquid extraction proceeds effectively.
4. *Recoverability*: The solvent needs to be recovered and recycled. This is often done by the process of distillation. The solvent loss will add to the cost of separation. The solvent must not form an azeotrope with the solute. For if it does, the separation problem remains! The component in the extract solution which is in minority has to be more volatile to cut the cost of the recovery of the solvent.
5. *Density*: The density of the feed solution and the extraction solution needs to be different from each other for smooth operations of stagewise contacting and continuous contacting. This is so even as the solute transfers from one phase to another.
6. *Surface Tension*: Greater the interfacial tension between the two phases during liquid extraction, greater is the coalescence but less is the dispersion. As coalescence is more a consideration than dispersion, large differences in interfacial tension is favourable for liquid extraction operations.
7. *Chemical Reactivity*: The solvent must be inert and must not react with the solute or the liquid that forms the majority in the feed solution. The solvent should be stable chemically.
8. *Viscosity, vapour pressure and freezing point*: For purposes of handling, lower these values better it is.
9. *Nontoxic, nonflammable and low cost*: The ignition point of the solvent, carcinogenity of the chemical and cost are important considerations in the choice of the solvent.

8.4 STAGEWISE CONTACTING OF LIQUID EXTRACTION OPERATIONS

The liquid extraction operations can be carried out by stagewise contacting either in a single stage or multiple stages. The method of contact can be continuous co-current, continuous cross-current or continuous counter-current operations. Batchwise and semi-batch operations are also possible when the solute is in small amount and expensive. In continuous operations, the quality is superior and the cost per pound is lowest, and often times with the use of recycle, more environmental friendly.

8.4.1 Single-stage Liquid Extraction Operation

The schematic of a single stage operation is shown in Figure 8.5. The operating line and equilibrium line of the operation is shown in Figure 8.6.

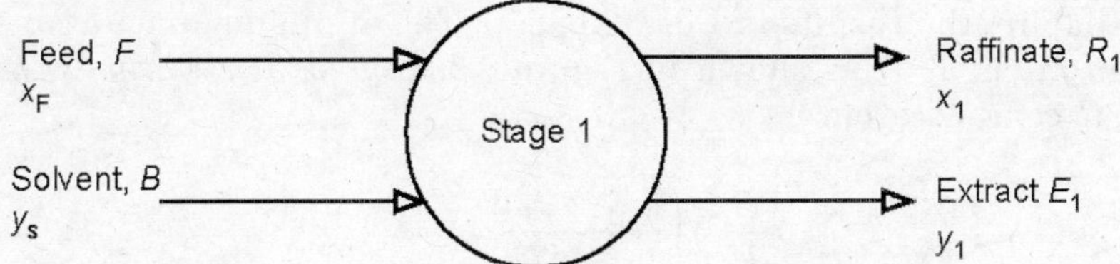

FIGURE 8.5 Schematic of a single stage liquid extraction.

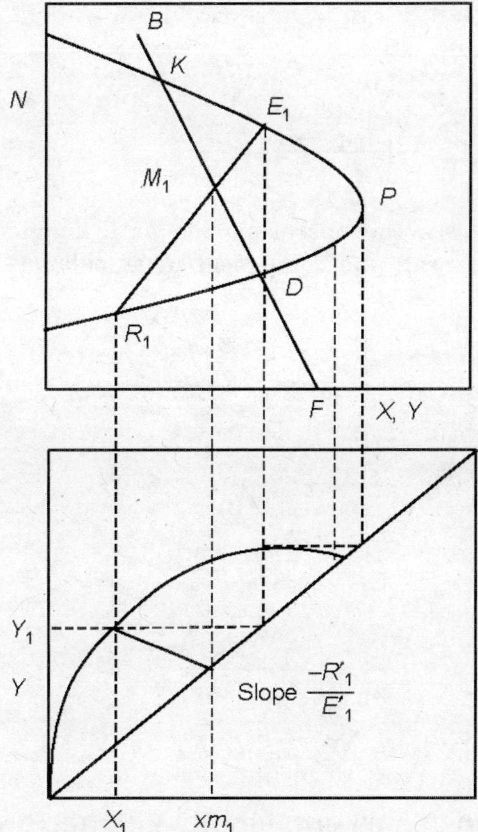

FIGURE 8.6 Operating diagram for one-stage liquid extraction.

The solvent B contains some solute to begin with. It can be introduced as a fresh solvent. For recycle operations it is good to perform the derivations with some solute in the solvent-rich B phase. This is given by y_s. A mass balance on the overall species and the solute component balance can be written as:

$$F + B = R_1 + E_1 \tag{8.6}$$

$$x_F F + y_s B = x_1 R_1 + E_1 y_1 \tag{8.7}$$

The operating line can be seen as FB in the ternary phase diagram and the rectangular distribution diagram in Figure 8.6. The mixture rule given by Eq. (8.3) can be used to simplify the results in Eqs. (8.6, 8.7). The minimum amount of solvent required is when M_1 is at D as shown in Figure 8.6. L gives the raffinate with the lowest possible C concentration.

$$\frac{B}{F} = \left(\frac{x_{m1} - x_F}{y_s - x_{m1}}\right) \tag{8.8}$$

$$M_1 x_{m1} = x_1 R_1 + E_1 y_1 \tag{8.9}$$

$$M_1 = R_1 + E_1 \tag{8.10}$$

or

$$M_1 x_{m1} = x_1(M_1 - E_1) + E_1 y_1 \tag{8.11}$$

$$\left(\frac{x_{m1} - x_1}{y_1 - x_1}\right) = \frac{E_1}{M_1} \tag{8.12}$$

In solvent-free coordinates, the operating line for a single stage operation is shown in Figure 8.7. The overall and component mass balances for the solute and solvent are then given by,

$$F' + B' = M'_1 = E'_1 + R'_1 \tag{8.13}$$

$$F'X_F + B'Y_s = M'_1 X_{M1} = E'_1 Y_1 + R'_1 X_1 \tag{8.14}$$

For solvent B,

$$X'N_s = M'_1 N_{M1} = E'_1 N_{E1} + R'N_{R1} \tag{8.15}$$

or

$$\frac{E'_1}{M'_1} = \left(\frac{X_{M1} - X_1}{Y_1 - X_1}\right) \tag{8.16}$$

$$\frac{R'_1}{E'_1} = \left(\frac{Y_1 - X_{M1}}{X_{M1} - X_1}\right) \tag{8.17}$$

$$E_1 = E'_1(1 + N_{E1});\ R_1 = R'_1(1 + N_{R1}) \tag{8.18}$$

8.4.2 Multi-stage Cross-current Liquid Extraction Operations

The schematic of a multi-stage cross-current liquid extraction operations are given in Figure 8.8.

A three-stage extraction is shown in Figure 8.8. The overall and component mass balances on the solute and solvent are as follows:

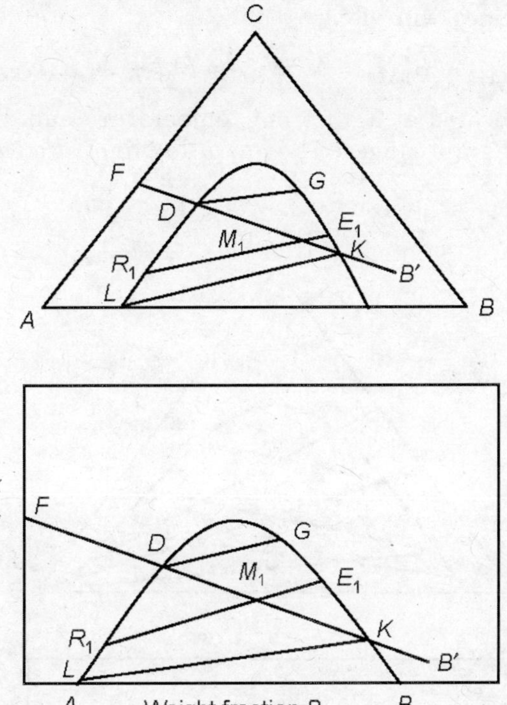

FIGURE 8.7 Single stage liquid extraction operation in solvent free coordinates.

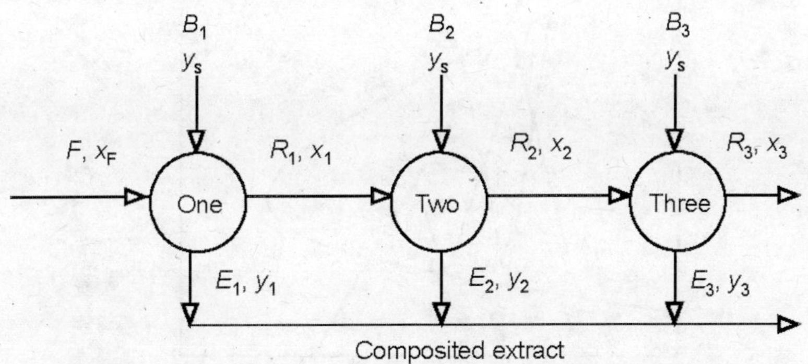

FIGURE 8.8 Schematic of multi-stage cross-current liquid extraction.

$$R_{n-1} + B_n = M_n = E_n + R_n \tag{8.19}$$

$$x_{n-1}R_{n-1} + B_n y_s = M_n x_{mn} = E_n y_n + R_n x_n \tag{8.20}$$

In the solvent free coordinates,

$$X_{n-1}R'_{n-1} + B'_n y_s = M'_n X_{mn} = E'_n Y_n + R'_n X_n \tag{8.21}$$

A solvent component balance will yield,

$$R'_{n-1}N_{Rn-1} + S'_nN_s = M'_nN_{Mn} = E'_nN_{En} + R'_nN_{Rn} \tag{8.22}$$

Each stage can be operated at a different temperature and different amounts of solvent can be used at different stages. The operating lines are shown in the ternary phase diagram in Figure 8.9.

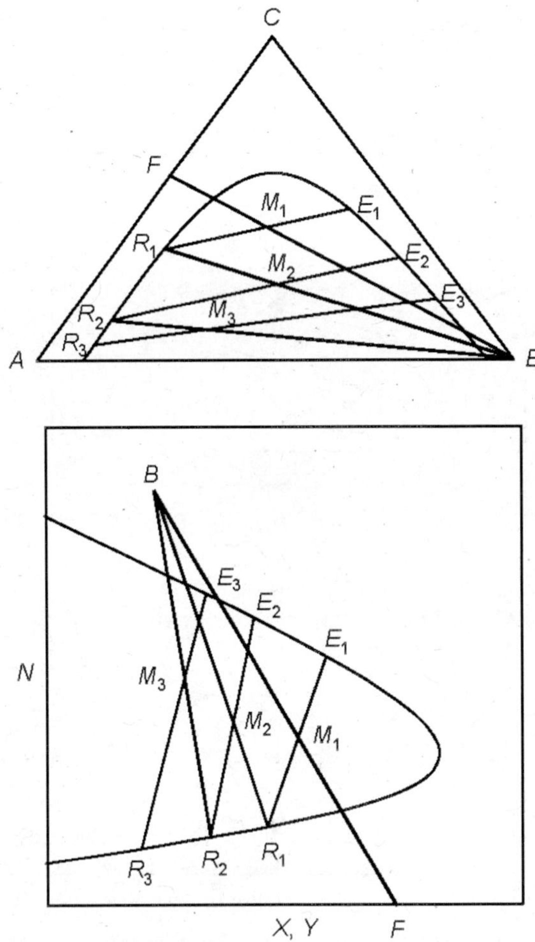

FIGURE 8.9 Operating lines in ternary phase diagram and distribution diagram during multi-stage cross-current liquid extraction.

When the solvent and the liquid that makes up the majority of the feed solution are *immiscible*, the derivations of the operating lines in multi-stage cross-current operations become simplified. Since A and B are immiscible, A is present in all raffinates. All the extract contains the solvent B. Thus the component balance simplifies to,

$$-\frac{A}{B_n} = \left(\frac{y'_s - y'_n}{x'_{n-1} - x'_n}\right) \tag{8.23}$$

The operating line equation for stage n has the slope $-A/B_n$ passing through the points, (x'_{n-1}, y'_s) and $(x'_{n-1} - x'_n)$. Each operating line at each stage intersects the equilibrium curve at the raffinate and extract compositions.

Worked Example 8.1 *Liquid Extraction of Styrene from a Mixture of Ethylbenzene and Styrene using Multi-stage Cross-current Operations*

A solution containing 50% ethylbenzene A and 50% styrene C needs to be separated at 25°C at a rate of 1000 kg/h. The extract product composition at the end of the final stage is 92% styrene. Diethylene glycol is the solvent. It is added in a cross-current fashion. The slope of $-R/E$ at each stage is maintained at constant and is equal to 1. The equilibrium data in solute free coordinates between the raffinate and extract phases are given in Figure 8.10. Rather than the composited extract as shown in Figure 8.8, the extract is removed at each stage and fed into the next stage. How many ideal stages are required to achieve 92% styrene concentration in the final extract phase. In order to make a product rich in solute, the extract from each stage is given as a feed into the next stage.

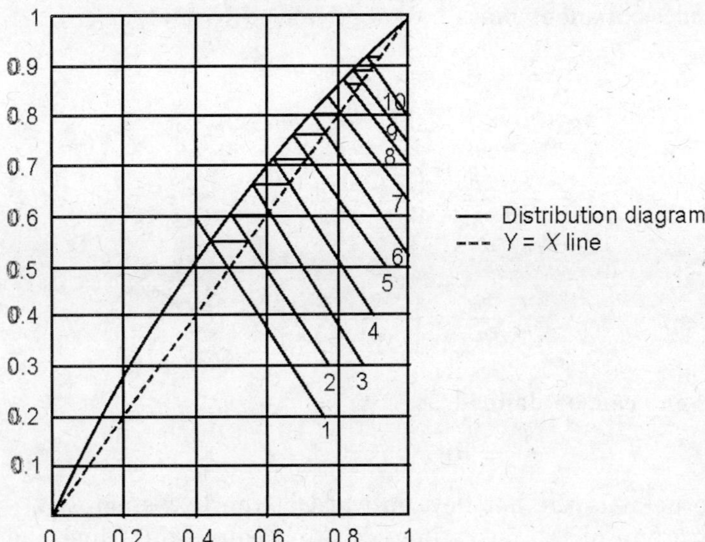

FIGURE 8.10 Multi-stage cross-current liquid extraction of styrene from a feed mixture of ethylbenzene and styrene using diethylene glycol as solvent.

Operating lines at each stage with a slope of −1 are constructed starting with the feed composition of 0.5 at $Y = X$ line. The extract from stage 1 is used as feed of stage 2 and so on and so forth. 10 ideal stages are required to achieve the target styrene composition. The composited raffinate composition can be calculated from the graph in Bancroft coordinates.

8.4.3 Multi-stage Continuous Counter-current Liquid Extraction Operations

The continuous counter-current multi-stage liquid extraction operations will result in the minimum number of stages for a given set of desired objectives such as the degree of separation, etc. The fresh solvent B, with some solute in it (y_s), meets the final raffinate thus allowing for maximum driving force for transfer of solute from the raffinate to the extract. The feed encounters the final extract unlike other operations where it encounters the fresh solvent at the very beginning of the operations.

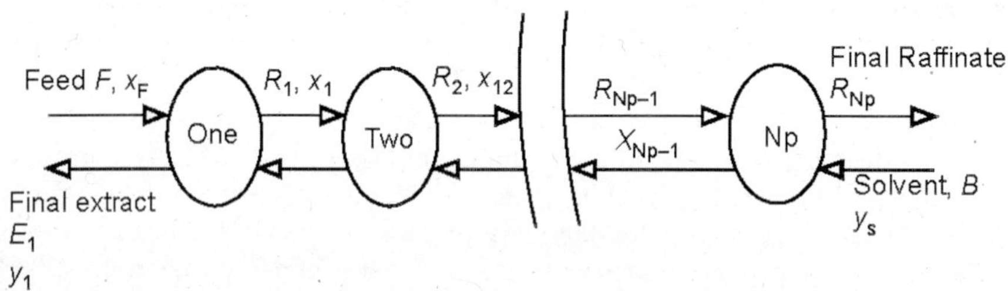

FIGURE 8.11 Schematic of continuous counter-current multi-stage liquid extraction operations.

The overall and component mass balances can be written as:

$$F + B = E_1 + R_{NP} = M \tag{8.24}$$

$$Fx_F + By_S = E_1y_1 + R_{Np}x_{NP} = Mx_M \tag{8.25}$$

$$\frac{F}{M} = \left(\frac{x_M - y_S}{x_F - y_S}\right) \tag{8.26}$$

or

$$\frac{E_1}{R_{N_p}} = \left(\frac{x_M - x_{N_p}}{y_1 - x_M}\right) \tag{8.27}$$

A difference point can be defined as,

$$\Delta_R = R_{NP} - B = F - E_1 \tag{8.28}$$

The difference point is the net flow outward at the last stage N_p.

The combined graph of the operating lines and equilibrium line along with the difference point is shown in Figure 8.12. The graphical construction involves the identification of tie lines. The feed F, solvent B, mixture M, extract from the first stage E_1 and raffinate from the last stage R_{NP} and the difference point Δ_R are first located on the combined graph. The liquid–liquid equilibrium line is constructed as shown in Figure 8.12. The region above the $R_{NP}E_1E_{Np}$ curve is a one phase region and within the curve is the heterogeneous region. Tie lines are constructed between E_1 and R_1, E_2 and R_2 and E_{Np} and R_{Np}, etc. The extract from the first stage E_1 is in equilibrium with the raffinate R_1. As more solvent is added, point M approaches

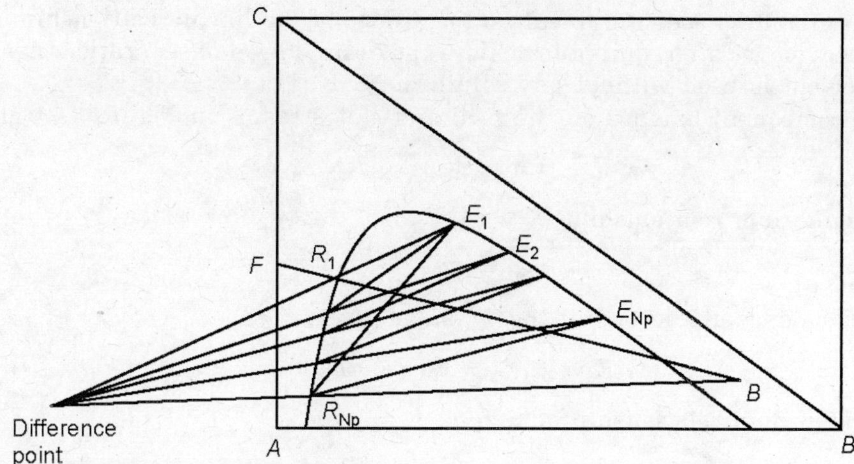

FIGURE 8.12 Operating lines and equilibrium line during continuous multi-stage counter-current liquid extraction operations.

point B. Thus the minimum solvent required can be calculated when M falls on the binodal curve. When large amounts of solvent are used, the tie lines intersect at infinity and become parallel to each other. Thus a line from the difference point connects the raffinate and extract from adjacent stages which fall on the binodal solubility curve. If a line from the difference point coincides with the tie line, infinite number of stages will be required to meet the desired objectives.

In solvent-free coordinates, the component mass balance can be written as:

$$F' + B' = E'_1 + R'_{Np} = M' \tag{8.29}$$

Since the feed is usually free of solvent, $F = F'$.

$$F'X_F + S'Y_s = M'X_M \tag{8.30}$$

The difference point in solvent free flow, out minus in at stage N_p can be written as,

$$\Delta'_R = F' - E'_1 = R'_{Np} - S' \tag{8.31}$$

The distribution diagram for a counter-current multi-stage continuous liquid extraction operation can be constructed as shown in Figure 8.13. The tie lines are used to get the operating line and equilibrium line in the resulting distribution diagram. The distribution diagram gives the relation between the solute in the extract and raffinate phases in a solute-free basis. One such typical construct is shown in Figure 8.12. The slope of the operating line at each stage is given by R'/E' at that stage.

Worked Example 8.2 *Number of Ideal Stages when Equilibrium Relationship is Linear*

In the schematic given in Figure 8.11 for a multi-stage continuous counter-current liquid extraction operation, when the equilibrium relationship of the solute in the extract and raffinate phases are linear and the slope of the linear relationship is m, how many ideal stages are required to achieve the desired level of separation. Assume that fresh solvent is used without any solute in it.

A solute component balance on the feed and exit streams in Figure 8.11 gives,

$$x_0F + 0 = y_1E_1 + x_{Np}R_{Np} \tag{8.32}$$

From the equilibrium relationship,

$$y_{Np} = mx_{Np} \tag{8.33}$$

A component balance at the end of the 1st stage gives,

$$y_1E_1 + x_1R_1 = x_0F + y_2E_2 \tag{8.34}$$

From the equilibrium relationship $x_1 = y_1/m$; $x_0 = y_0/m$

$$y_2 = y_1\left(\frac{E_1}{E_2} + \frac{R_1}{mE_2}\right) - \frac{F}{E_2}\frac{y_0}{m} \tag{8.35}$$

For the special case when $E_1 = E_2 = \ldots = E$

$$R_1 = R_2 = \ldots = R = F \tag{8.36}$$

$$y_2 = y_1\left(1 + \frac{R}{mE}\right) - y_0\left(\frac{R}{mE}\right) \tag{8.37}$$

Let $R/mE = A_f$ be the extraction factor, (8.38)

$$y_2 = y_1(1 + A_f) - y_0A_f$$

Similarly,

$$y_3 = y_2(1 + A_f) - y_1A_f = y_1\ (1 + A_f + A_f^2) - y_0(A_f + A_f^2) \tag{8.39}$$

$$y_4 = y_3(1 + A_f) - y_2A_f = y_1(1 + A_f^2)\ (1 + A_f) - A_fy_0\ (A_f^2 + A_f + 1) \tag{8.40}$$

$$= y_1(1 + A_f + A_f^2 + A_f^3) - y_0(A_f + A_f^2 + A_f^3) \tag{8.41}$$

Thus,

$$y_{N+1} = y_1(1 + A_f + A_f^2 + \ldots + A_f^N) - y_0(A_f + A_f^2 + \ldots + A_f^N) \tag{8.42}$$

The sum of a geometric series can be obtained as follows,

$$S_{N+1} = 1 + A_f + A_f^2 + \ldots + A_f^N \tag{8.43}$$

Multiplying Eq. (8.43) by A_f

$$A_fS_{N+1} = A_f + A_f^2 + \ldots + A_f^N + A_f^{N+1} \tag{8.44}$$

Subtracting Eq. (8.43) from Eq. (8.44);

$$S_{N+1}(A_f - 1) = A_f^{N+1} - 1 \quad \text{or} \quad S_{N+1} = \left(\frac{A_f^{N+1} - 1}{A_f - 1}\right) \tag{8.45}$$

Substituting Eq. (8.45) in Eq. (8.43) and that in Eq. (8.42);

$$y_{N+1} = y_1\left(\frac{A_f^{N+1} - 1}{A_f - 1}\right) - y_0 A_f\left(\frac{A_f^{N} - 1}{A_f - 1}\right) \tag{8.46}$$

Eq. (8.46) is applicable when $A_f \neq 1$.

Likewise, the raffinate compositions can be derived. For the special case when $E_1 = E_2 = \ldots = E$ and $R_1 = R_2 = \ldots = R = F$.

$$y_1 E + x_1 R = x_0 R + y_2 E \tag{8.47}$$

$$x_1 + \frac{x_1 R}{mE} = \frac{x_0 R}{mE} + x_2 \tag{8.48}$$

$$x_2 = x_1(1 + A_f) - x_0 A_f \tag{8.49}$$

Similarly,

$$x_3 = x_2(1 + A_f) - x_1 A_f = x_1(1 + A_f + A_f^2) - x_0 A_f(1 + A_f) \tag{8.50}$$

Thus,

$$x_{Np} = x_1(1 + A_f + A_f^2 + \ldots + A_f^{Np-1}) - x_0 A_f(1 + A_f + \ldots + A_f^{Np-2}) \tag{8.51}$$

$$= x_1\left(\frac{A_f^{N_p} - 1}{A_f - 1}\right) - x_0 A_f\left(\frac{A_f^{N_p\;2} - 1}{A_f - 1}\right) \tag{8.52}$$

8.4.4 Use of Reflux in Continuous Counter-current Liquid Extraction Operations

The best possible extract weight fraction is achieved using the schematic in Figure 8.11 using counter-current contacting. Here the extract is in equilibrium with the feed. In order to achieve even higher yield compared with this scheme, reflux can be used. The feed is introduced at an optimal location in the staged cascade scheme. The content of the solute in the product is increased by removing the solvent from the extract and generating a solvent-free stream. Part of the product is returned as recycle. This is not needed for the raffinate section. What would be the incentive in removing the liquid A and attempting to generate a solute stream that may not be present in sufficient measure. Liquid A is the majority component of the feed stream containing the solute that is the desired component to be separated by using a solvent to extract it. The number of stages required can be calculated by stepping off the steps in a combined graph similar to the one used for distillation. Here there is no thermal energy transfer from the reboiler. The chemical potential changes and when the fugacities of the two phases are equal, equilibrium is reached. Calculation of these stages in a distribution diagram in rectangular coordinates is preferred to triangular coordinates. Software needs to be developed to better tap into the triangular coordinates. The schematic for a counter-current cascade with reflux is shown in

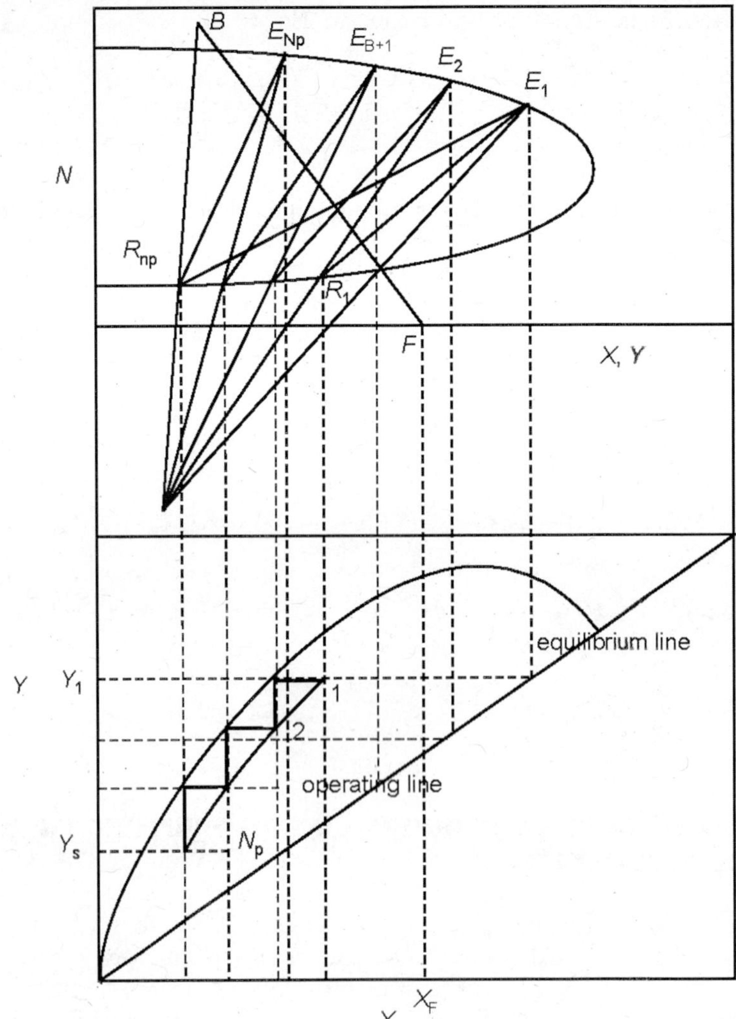

FIGURE 8.13 Multi-stage counter-current continuous liquid extraction operations: Operating curve, equilibrium line from tie lines and difference point from N-XY plot to distribution diagram.

Figure 8.14. The combined graph with the operating curves and stepping of stages in the distribution diagram is shown in Figure 8.15.

Assuming that the solvent introduced is fresh, that contains no solute, and the recovered solvent from recycle is also devoid of any solute, the overall and component mass balances in the extract enriching and raffinate stripping sections respectively are as follows:

$$F + B = P + B_R + R_{Np} \tag{8.53}$$

$$Px_p + x_{RNp}R_{Np} = x_FF \tag{8.54}$$

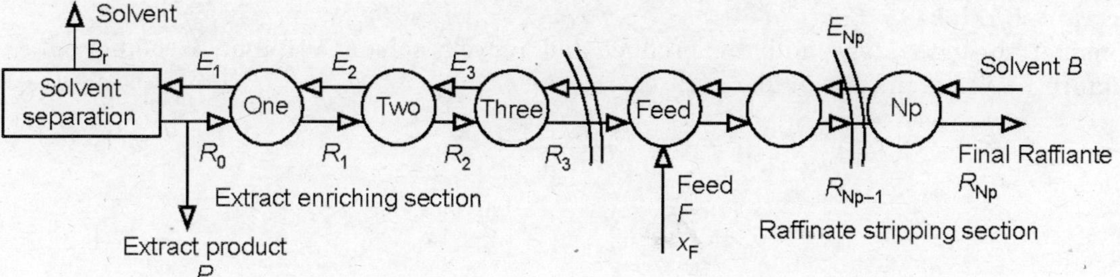

FIGURE 8.14 Multi-stage continuous counter-current liquid extraction with reflux.

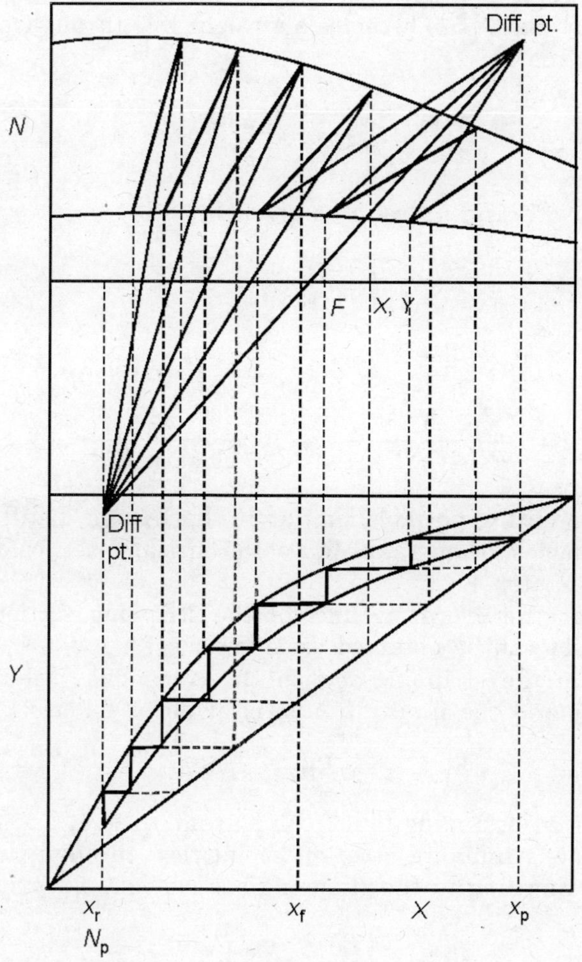

FIGURE 8.15 Counter-current cascade with reflux.

Extract Enriching Section

Consider the *m*th stage and the product and recycle solvent stream. A component balance between these would result in,

$$E_m y_m = R_{m-1} x_{m-1} + P x_p \tag{8.55}$$

or

$$y_m = \left(\frac{R_{m-1}}{E_m}\right) x_{m-1} + \left(\frac{P}{E_m}\right) x_p \tag{8.56}$$

It can be seen that the Eq. (8.51) gives the operating line for the enriching section. For a constant product solute fraction the slope of the operating curve at the *m*th stage can be seen to be R_{m-1}/E_m. For the special case when $E_1 = E_2 = E_m = E$ and $R_1 = R_2 = \ldots R_m = R$, Eq. (8.51) becomes a straight line throughout the enriching section and can be written as:

$$y_m = \left(\frac{R}{E}\right) x_{m-1} + \left(\frac{P}{E}\right) x_p \tag{8.57}$$

P is a fraction of *R*. Let the reflux ratio be given by,

$$\rho = \frac{B_r}{P} \tag{8.58}$$

Then Eq. (8.57) can be written as:

$$y_m = \left(\frac{1}{P(1+\rho)/R + 1}\right) x_{m-1} + \frac{x_P}{(1 + \rho + R/P)} \tag{8.59}$$

P/*R* or *R*/*P* is an operating variable similar to the *W*/*D* in distillation problems. This operating line resembles the operating line of the enriching section of the distillation column discussed in Chapter 4.

In a similar fashion the operating line for the stripping section shown in the schematic in Figure 8.14 can be obtained as follows:

For the kth stage in the stripping section the component solute balance with respect to the kth stage and the final raffinate is given as follows:

$$E_k y_k + x_{NP} R_{NP} = x_k R_k \tag{8.60}$$

Eq. (8.55) was derived assuming that fresh solvent without any solute in it was added at the N_p^{th} stage. Assuming constant molal flow through the cascades, the equation for the operating line in the stripping section may be written as:

$$y_k = x_k\left(\frac{R}{B}\right) - x_{N_p}\left(\frac{R}{B}\right) \tag{8.61}$$

The raffinate to solvent ratio is an important parameter of the overall design.

8.4.5 Design Specifications in Continuous Liquid Extraction Multi-stage Operations

The principal parameters of the continuous multi-stage liquid extraction operations with reflux are as follows in addition to the temperature and pressure and equilibrium relation between the extract solute phase and the raffinate solute phase. The feed conditions need to be specified completely.

These are feed solute composition, the amount of liquid *A*. The remaining principal variables are:

1. Solvent Reflux Ratio, $\rho(B_r/P)$
2. The Raffinate to Product Ratio, R/P
3. Raffinate to Solvent Ratio
4. Product extract composition, x_p
5. Raffinate extract composition, x_{NP}
6. Number of stages needed for the extraction

As discussed in Section 4.2.5, three more parameters more than the feed specifications are needed to complete the description of the problem. Once three parameters are specified, the other three can be calculated from the information available from the analysis above. Thus there are $^6C_3$ = 20 liquid extraction problem types. For 10 of them the number of stages can emerge as a result and for 10 other types, the number of ideal stages can be given as an input to the problem. The real line can be calculated using a Murphree stage efficiency as shown in Chapter 4. Given the stage efficiency an overall efficiency can be calculated. From the overall efficiency, the real line can be drawn. From the real line, the actual number of stages that may be required in the industry to achieve the desired objectives can be stepped off. The equilibrium data can be calculated from models. The fugacities in the extract and raffinate phases can be equated for the solute *C* at equilibrium. The activity coefficient models can be used to calculate the fugacities.

This approach can be extended to more than one solute present in the system. A co-solvent may also be introduced into the mix of things. This can improve the overall yield of the extract.

8.4.6 Use of Co-solvent in Counter-current Multi-stage Continuous Liquid Extraction Operations

A co-solvent can be used in addition to the solvent in order to extract more of the solute from the raffinate phase during continuous co-current multi-stage continuous liquid extraction operations. In Figure 8.16 is shown the schematic of the inclusion of the co-solvent.

Fresh solvent and co-solvent is assumed without any solute in them. For the special case when the raffinate flow, $R_{NP} = R_{Np-1} = \ldots = R_2 = R_1 = F = R$ and when the two extract phases flow rate is constant in all the stages and is equal to *B* and B_s respectively. The equilibrium relationships between the distribution of solute between the extract and raffinate phases can be assumed to be linear. So is the case when the solutions are dilute. Thus,

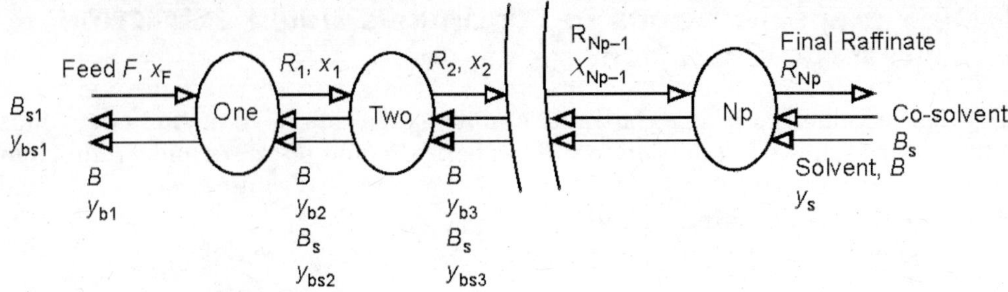

FIGURE 8.16 Use of co-solvent in continuous counter-current multi-stage liquid extraction operations.

$$y_{Bk}^{*} = m_1 x_k \tag{8.62}$$

$$y_{Bsk}^{*} = m_2 x_k \tag{8.63}$$

A solute component mass balance on stage 1 can be written as:

$$x_0 R + y_{bs2} B_s + y_{b2} B = x_1 R_1 + y_{bs1} B_s + y_{b1} B \tag{8.64}$$

Dividing throughout the Eq. (8.64) by B and using the Eqs. (8.62 and 8.63) to reduce the x_1 and x_0 values in terms of y_{b1} and y_{b0} values,

$$\frac{R y_{b0}}{B m_1} + \frac{y_{bs2} B_s}{B} + y_{b2} = \frac{R y_{b1}}{B m_1} + \frac{y_{bs1} B_s}{B} + y_{b1} \tag{8.65}$$

The extraction factor, A_f can be defined as $A_f = R/m_1 B$. Further it can be seen that,

$$y_{bs2} = m_2 x_2 = \left(\frac{m_2}{m_1}\right) y_{b2}$$

$$A_f y_{b0} + y_{b2} \frac{B_s m_2}{B m_1} + y_{b2} = A_f y_{b1} + y_{b1} \frac{B_s m_2}{B m_1} + y_{b1} \tag{8.66}$$

A co-solvent factor can be defined as,

$$C_s = \frac{B_s m_2}{B m_1}$$

$$A_f y_{b0} + C_s y_{b2} + y_{b2} = A_f y_{b1} + C_s y_{b1} + y_{b1} \tag{8.67}$$

or

$$(1 + C_s)\, y_{b2} = y_{b1}(1 + A_f + C_s) - A_f y_{b0} \tag{8.68}$$

$$y_{b2} = y_{b1}\left(\frac{A_f}{1 + C_s} + 1\right) - \frac{A_f y_{b0}}{(1 + C_s)} \tag{8.69}$$

Similarly,

$$y_3 = y_{b1}\left(1 + \frac{A_f}{1+C_s} + \left(\frac{A_f}{1+C_s}\right)^2\right) - y_0\left(\frac{A_f}{1+C_s} + \left(\frac{A_f}{1+C_s}\right)^2\right) \tag{8.70}$$

Thus,

$$y_{N+1} = y_1\left(1 + \frac{A_f}{1+C_s} + \dots + \frac{A_f^{N_p}}{(1+C_s)^{N_p}}\right) - y_0\left(\frac{A_f}{(1+C_s} + \dots + \frac{A_f^{N_p}}{(1+C_s)^{N_p}}\right) \tag{8.71}$$

$$\left(\frac{A_f}{(1+C_s)} - 1\right) y_{N_p+1} = y_1\left(\left(\frac{A_f}{1+C_s}\right)^{N_{p+1}} - 1\right) - \frac{y_0 A_f}{1+C_s}\left(\frac{A_f}{(1+C_s)^{N_p}} - 1\right) \tag{8.72}$$

Equation (8.72) is applicable when $A_f \neq 1$. In a similar fashion the raffinate compositions can be derived,

$$x_{N_p} = x_1\left(\frac{\left(\frac{A_f}{1+C_s}\right)^{N_p} - 1}{A_f - 1}\right) - x_0 A_f\left(\frac{A_f^{N_p-2} - 1}{A_f - 1}\right) \tag{8.73}$$

8.5 LIQUID EXTRACTION EQUIPMENT

A wide range of extractors are available in the industry to choose from, both static and agitated columns. The static columns include (a) sieve trays, (b) random packing and (c) structured packing. The agitated columns include the Karr column, scheibel column, RDC the rotating disc contactor and pulsed column. (KMPS, 2006). The Karr column is shown in Figure 8.17. It has the highest capacity at 30–60 M^3/M^2-hr, good efficiency, good turndown capability (25%), uniform shear mixing and best suited for systems that emulsify.

The Scheibel column is shown in Figure 8.18. It has reasonable capacity at 15–25 M^3/M^2-hr, high efficiency due to internal baffling, good turndown capability (25%) and best suited when many stages are required and is recommended for highly fouling systems or systems that tend to emulsify.

The rotating disc contactor, RDC column is shown in Figure 8.19. It has a capacity of 20–30 M^3/M^2-hr, limited efficiency due to axial backmixing and is suitable for viscous materials, fouling materials besides being sensitive to emulsions due to high shear mixing and has reasonable turndown (40%).

The pulsed column is shown in Figure 8.20. It has reasonable capacity at 20–30 M^3/M^2-hr, is best suited for nuclear applications due to lack of seal and is also suited for corrosive applications since can be constructed out or non-metals. It has limited stages due to backmixing and has limited diameter/height due to pulse energy required. One example of a static column is shown in Figure 8.21. It has a capacity

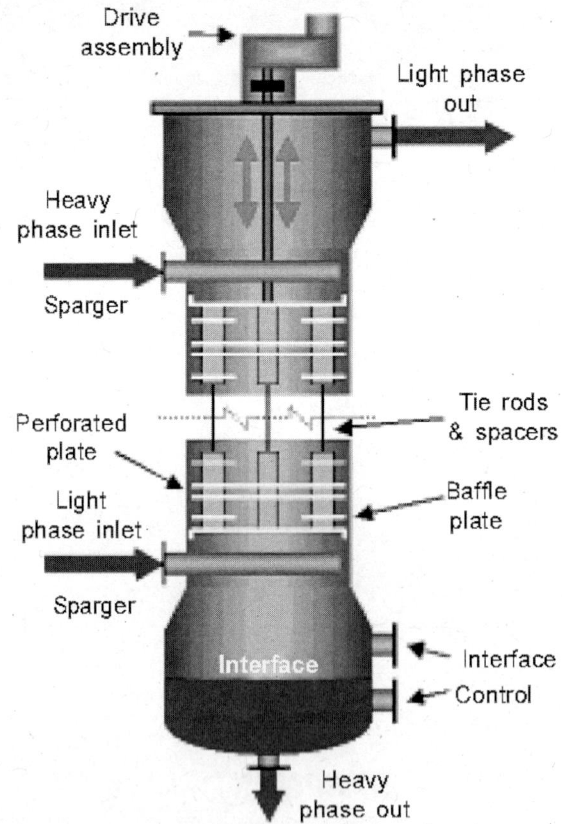

FIGURE 8.17 The Karr reciprocating agitated column.

of 30–50 M^3/M^2-h, possesses good efficiency due to minimum backmixing, is affected by changes in wetting characteristics and is limited as to which phase can be dispersed.

Either differential or staged contactors can be used to perform the liquid extraction operation. A packed column has better mass transfer characteristics but has lower capacity compared with say a spray column. The multiple stage contactors have a mixer settler arrangement. As interfacial area through which the solute transfer occurs is an important parameter of all the operations for which different designs have been developed to suit different applications, each having a different interfacial area. Spray and packed columns run the risk of flooding. Large size drops are used to minimise the risk of flooding. The spray columns have low capital and operating costs. It can handle corrosive material and it is used in petrochemical and chemical industry. It can handle density differences greater than 0.05 g/cm^3 and the flow rate of less than 100 m^3/hr. It is used in single stage operations.

Packed columns have less capacity but the mass transfer is more compared with spray columns. As mentioned it is restricted in use on account of flooding problems. It can handle density differences greater than 0.05 gm/cm^3 and flow rate less than 50 m^3/hr, and the number of stages used is less than 10. It is used in petrochemical and pharmaceutical industries.

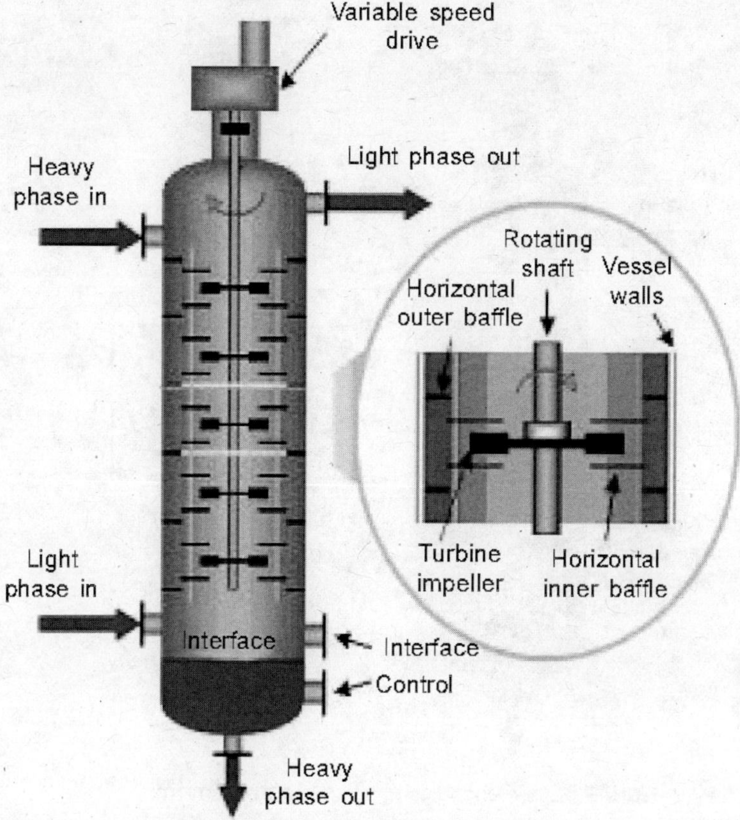

FIGURE 8.18 Schiebel agitated column.

Centrifugal extractors have high capital cost and short contact times. It is used in pharmaceutical and nuclear industries. The number of stages used is less than 5, flow rate less than 10 m^3/hr and density difference it can handle is greater than 0.01 gm/cm^3. Mixer-settlers have high capacity and flexibility and can handle materials with high viscosity. It is used in the metallurgical and petrochemical industry. The number of stages can be any value and flow rates can be greater than 250 m^3/hr and it can handle density differences greater than 0.1 gm/cm^3.

For a differential contactor, the mass balance is written on a slice, Δz in a spray tower or a packed tower. The feed rate and raffinate rate are F and R respectively and the extract rate is E.

$$\frac{Rdx}{dz} = \frac{Edy}{dz} \tag{8.74}$$

where x and y are the solute fractions in the raffinate and extract phases respectively.

Re-arranging Eq. (8.74),

$$\frac{dy}{dx} = \frac{R}{E} \tag{8.75}$$

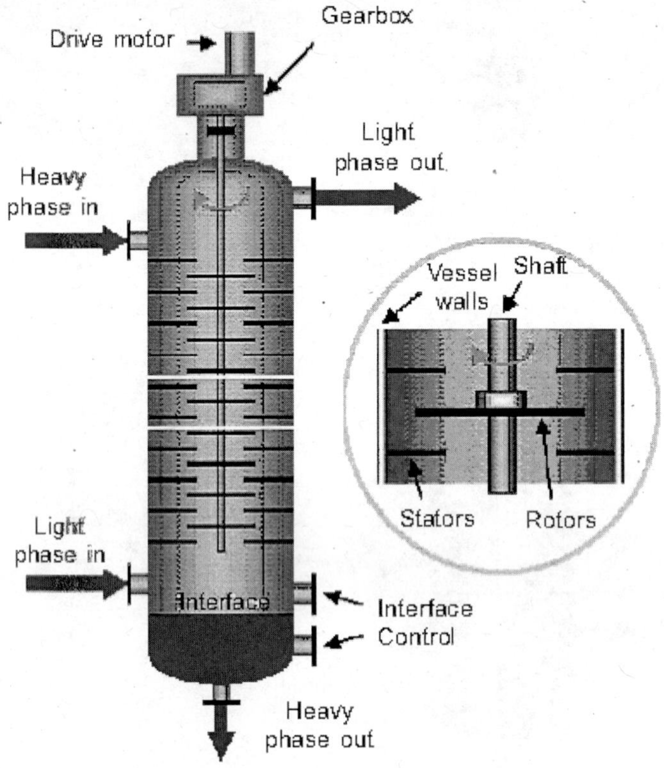

FIGURE 8.19 Rotating disc contactor column.

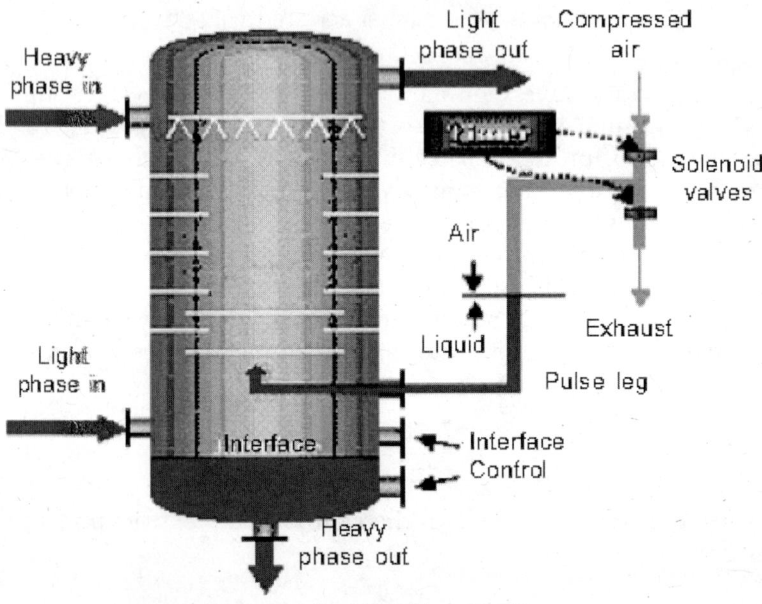

FIGURE 8.20 Pulsed liquid extraction.

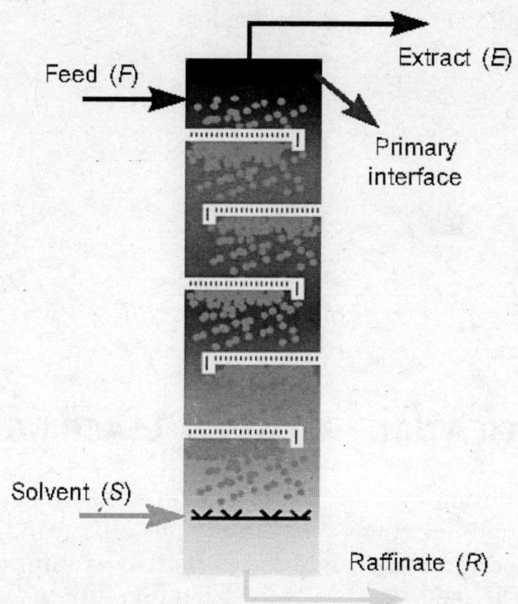

FIGURE 8.21 Sieve tray static liquid extraction column.

Eq. (8.75) is subject to the boundary conditions,

$$z = 0, \; x = x_0, \; y = y_0 \tag{8.76}$$

Integrating Eq. (8.75),

$$y = R/E \; (x + c') \tag{8.77}$$

$$y_0 = R/E \; (x_0 + c') \tag{8.78}$$

Subtracting Eq. (8.78) from Eq. (8.77),

$$(y - y_0) = R/E \; (x - x_0) \tag{8.79}$$

Eq. (8.79) is the operating line. The equilibrium line can be obtained from equating the fugacity of the two mixtures at equilibrium. A suitable model for the activity coefficient can be chosen and the equilibrium line drawn accordingly. For some dilute systems the equilibrium relationships are linear and can be written as:

$$y^* = mx \tag{8.80}$$

The diffusion resistances contributing to the efficiency was discussed in Chapter 4. Here a steady state mass balance on a differential volume $A\Delta z$ can be written as:

$$-\frac{Rdx}{dz} = K_y a \rho (x - x^*) \tag{8.81}$$

Eq. (8.81) is subject to the following boundary conditions,

$$z = 0, \; x = x_0; \; z = 1, \; x = x_1$$

Re-arranging Eq. (8.81) and integrating both sides,

$$-\frac{R}{K_y \rho a}\int_{x_0}^{x_1}\frac{dx}{(x - x^*)} = \int_0^1 dz \tag{8.82}$$

Let $x^* = m'y$ (equilibrium relationship written in terms of the raffinate phase)

or

$$l = \left(\frac{R}{K_y a \rho}\right)\ln\left(\frac{(x_0 - m'y_0)}{(x_1 - m'y_1)}\right)\left(\frac{1}{1 - m'(R/E)}\right) \tag{8.83}$$

8.6 INDUSTRIAL APPLICATIONS OF SOLID LEACHING

The process of selective dissolution of a species in a solid mixture using an external solvent is called *solid leaching*. Sometimes more than one solute can be leached. The separation of alkali from wood ashes is called *lixiviation*. It is different from liquid extraction in the sense that the feed mixture is in the solid state in solid leaching as opposed to the liquid state in liquid extraction. Should the solvent be at its boiling point, the process is referred to as *decoction*. Should the solid solute be washed off without any dissoution by the flowing liquid, it is called *elutriation*. *Solution mining* refers to operations when the solvent is percolated and circulated over and through the ore body. Heap leaching operations are used to beneficiate low-grade ores whose mineral values do not warrant the expense of crushing or grinding. The leach liquor is pumped over the ore and collected as it drains from the heap. Seven or more years are needed to reduce the copper content of heaps from 2 to 13%. Pyrite ores are in heaps containing 2.2×10^7 tons of ore using over 20,000 m^3 of leach liquor per day.

Leaching has a variety of commercial applications, including separation of metal from ore using acid and sugar from beets using hot water. Chloride can also be leached out of food. In agriculture, leaching may refer to the loss of water-soluble plant nutrients from the soil due to rain and irrigation. In a typical leaching operation, the solid mixture to be separated consists of particles, inert insoluble carrier A and solute C. The solvent, B is added to the mixture to selectively dissolve C. The overflow from the stage is free of solids and consists of only solvent B and dissolved C. The underflow consists of slurry of liquid of similar composition in the liquid overflow and solid carrier A. In an ideal leaching equilibrium stage, all the solute is dissolved by the solvent; none of the carrier is dissolved. The mass ratio of the solid to liquid in the underflow is dependant on the type of equipment used and properties of the two phases. Leaching is an environmental concern when it contributes to groundwater contamination. As water from rain, flooding or other sources seeps into the ground, it can dissolve chemicals and carry them into the underground water supply. Of particular concern are hazardous waste dumps and landfills and, in agriculture, excess fertiliser and improperly stored animal manure.

Leaching is widely used in extractive metallurgy since many metals can form soluble salts in aqueous media. Compared to pyrometallurgical operations, leaching is easier to perform and much less harmful, because no gaseous pollution occurs. The only drawback of leaching is its lower efficiency caused by the low temperatures of the operation, which dramatically affect chemical reaction rates. There is great variety of leaching processes. Usually they are classified by the types of the reagents used in the operation. The type of the reagents required depends on the type of ores (or some pretreated material) processed. A typical feed for leaching is either oxide or sulphide. In case, it is oxide, a simple acid leaching reaction can be illustrated by the zinc oxide leaching reaction,

$$ZnO + H_2SO_4 \rightarrow ZnSO_4 + H_2O \tag{8.84}$$

In this reaction solid ZnO dissolves, forming soluble zinc sulphate. In many cases other reagents are used to leach oxides. For example, in the metallurgy of aluminium, aluminium oxide is subject to leaching by alkali solutions.

$$Al_2O_3 + 3H_2O + 2NaOH \rightarrow 2NaAl(OH)_4 \tag{8.85}$$

Leaching of sulphides is a more complex process due to the refractory nature of sulphide ores. It often involves the use of pressurised vessels called *autoclaves*. A good example of the autoclave leach process can be found in the metallurgy of zinc. It is best described by the following chemical reaction:

$$2ZnS + O_2 + 2H_2SO_4 \rightarrow 2ZnSO_4 + 2H_2O + 2S \tag{8.86}$$

This reaction proceeds at temperatures above the boiling point of water, thus creating a vapour pressure inside the vessel. Oxygen is injected under pressure, making the total pressure in the autoclave more than 0.6 MPa. Some of the industrial applications of solid leaching is summarised in Table 8.2.

Hydrometallurgical treatment of copper containing materials, such as copper ores, concentrates, and the like, has been well established for many years. Currently, there exist many creative approaches to the hydrometallurgical treatment of these materials. The recovery of copper from copper sulphide concentrates using pressure leaching promises to be particularly advantageous (Phelps Dodge Corp, 2006). The mechanism by which pressure leaching releases copper from a sulphide mineral matrix, such as chalcopyrite, is generally dependent on temperature, oxygen availability, and process chemistry. In high temperature pressure leaching, typically thought of as being pressure leaching at temperatures above about 200°C, the dominant leaching reaction in dilute slurries may be written as follows:

$$4CuFeS_2 + 4H_2O + 17O_2 \rightarrow 4CuSO_4 + 2Fe_2O_3 + 4H_2SO_4 \tag{8.87}$$

According to a process described by Geobiotics (2005), a heap having dimensions of 2.5 m high and 5 m wide is constructed with hypogenic copper sulphide bearing ore. The constructed heap includes exposed sulphide mineral particles at 25 % weight of which are hypogenic copper sulphides. The concentration of the exposed sulphide mineral particles in the heap is such that the heap includes at least 10 Kg of exposed sulphide sulphur per ton of solids in the heap. Furthermore, at least 50% of the total copper in the heap is in the form of hypogenic copper sulphides. A substantial portion of the heap is then heated to a temperature of at least 50°C. The heap is inoculated

TABLE 8.2 Industrial Applications of Solid Leaching

No.	*Feed*	*Solvent*	*Product*	*Application*
1.	Sugar beets	Hot water	Sugar	Sugar industry
2.	Soyabeans/ Cottonseed	Hexane, Propane	Vegetable oil	Food industry
3.	Copper ore	Sulphuric acid/ Ammonical solution	Copper mineral	Metallurgical grade copper
4.	Gold ore	Sodium Cyanide	Gold	Gold manufacture
5.	Tree barks	Water	Tannin	Leather industry
6.	Plant roots, Leaves	Organic solvent	Drugs	Pharmaceutical industry
7.	Tea leaves/ Coffee beans	Hot water	Tea/Coffee	Beverage and food industry
8.	Bituminous coal	NMP/DMSO/TMU/ HMPA	Clean Coal	Energy industry
9.	Bauxite	Sodium hydroxide	Alumina	Aluminium industry
10.	Zinc oxide	Sulphuric acid	$ZnSO_4$	Zinc industry
11.	Pacific yew tree	Water soluble polymer such as polyaminoacid	Paclitexel anti-cancer drug	Pharmaceutical industry
12.	Chalcopyrite	Pressurised water, oxygen and sulphur slurry	Copper released	Hydrometallurgy
13.	Hypogenic copper sulphide ore	Thermophillic microorganisms, sulphuric acid with ferric iron	Copper leach solution	Bioleaching
14.	Titanium ore concentrate	Sulphuric acid	Titanium oxide	Titanium industry
15.	Laterite ore	Sulphuric acid and NaCl	Cobolt and Nickel	Metallurgy
16.	Ilmenite ore	Hot HCl	Rutile (TiO_2)	Titanium industry
17.	Molybedenum ore concentrate	HCl	Molybedenum oxide	Metallurgy
18.	Limonite ore	Sulphuric acid	Nickel	Hydrometallurgy
19.	Saprolite ore	Sulphuric acid	Cobolt	Hydrometallurgy
20.	Roasted concentrate	HCl & HNO_3	Platinum	Precious metals
21.	Molten Silicon	Ferric chloride & HCl	Silicon	Semiconductor

with a culture including a strain of thermophilic microorganisms capable of bioleaching sulphide minerals at a temperature above 50°C. A process leach solution that includes sulphuric acid and ferric iron is applied to the heap. Bioleaching is carried out so that sufficient sulphide mineral particles in the heap are *biooxidised* to oxidise at least 10 Kg of sulphide sulphur per ton of solids in the heap and to cause the dissolution of at least 50% of the copper in the heap into the process leach solution in a period of 210 days or less from the completion of the heap. A pregnant process leach solution that contains dissolved copper is collected as it drains from the heap. Copper may then be recovered from this solution.

8.7 STAGED LEACHING OPERATIONS

Consider a solid feed F containing solute C and an inert carrier solid A contacted with a leach solvent B. The extract refers to the leach solution W with some of the solute transferred from the feed. The remaining solid is the raffinate R with the solute carrier solid in the feed A and undissolved solute C. If the feed rate is F (kg/hr) the solvent rate B (kg/hr) the extract is E kg/hr and raffinate is R kg/hr, *a* overall mass balance would yield

$$F + B = E + R = M_1 \tag{8.88}$$

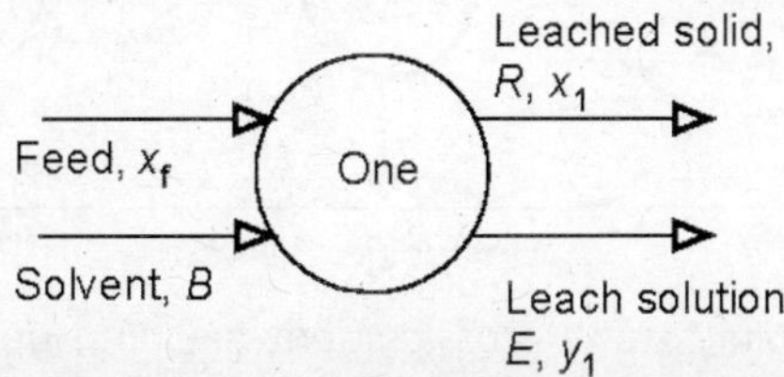

FIGURE 8.22 Single stage leaching operation.

A solute balance when fresh solvent is used, would yield;

$$x_f F = y_1 E + x_1 R = M x_{m1} \tag{8.89}$$

$$x_{m1} = x_f F/M = (y_1 E + x_1 R)(E + R) \tag{8.90}$$

The carrier solid fraction in the feed is given by N_f;

$$A/F = N_f \tag{8.91}$$

After the leaching assuming that no carrier solid went into the leach solution.

$$N_1 = A/R_1 \tag{8.92}$$

8.7.1 Multi-stage Cross-current Leaching Operations

Fresh solvent is used to leach the solute C from the feed solid mixture F.

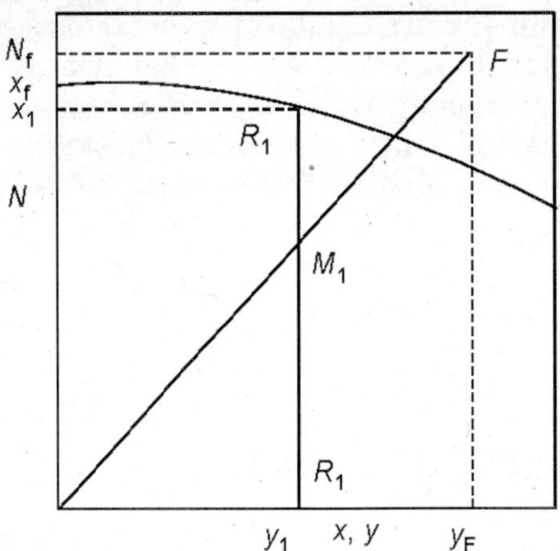

FIGURE 8.23 Operating line for a single stage leaching in an N vs x,y phase diagram.

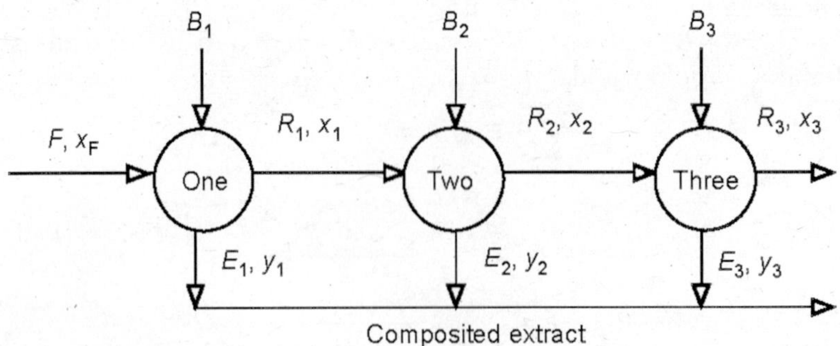

FIGURE 8.24 Multi-stage cross-current leaching operations.

The residue from the leaching step is fed into the next stage. Fresh solvent is made available at every additional stage. The extract solution is composited together as the product. The operating lines for the cross-current operation can be derived by assuming that the $R_1 = R_2 = \ldots = R$ and, $E_1 = E_2 = \ldots = E$ when equal amounts of fresh solvent B is used at every stage. Thus at the kth stage a solute balance would yield,

$$R(x_{k-1} - x_k) = E_k y_k \tag{8.93}$$

or

$$y_k = \left(\frac{R}{E}\right) x_{k-1} - x_k \left(\frac{R}{E}\right) \tag{8.94}$$

A solute balance on the first stage and the feed when fresh solvent is used can be written as:

$$R(x_f - x_1) = E y_1 \tag{8.95}$$

The operating line can be written as:

$$y_1 = -\left(\frac{R}{E}\right) x_1 + \left(\frac{R}{E}\right) x_f \tag{8.96}$$

These operating lines are shown in the distribution diagram of the solute between the extract and the raffinate phases in Figure 8.25. The slope of the operating lines is assumed to be constant and equal to *–R/E* as indicated in Eq. (8.93). As fresh solvent is used at each stage the operating line starts at the y axis. It stops upon attainment of equilibrium. This will be the case when sufficient time is allowed at each stage for the attainment of equilibrium.

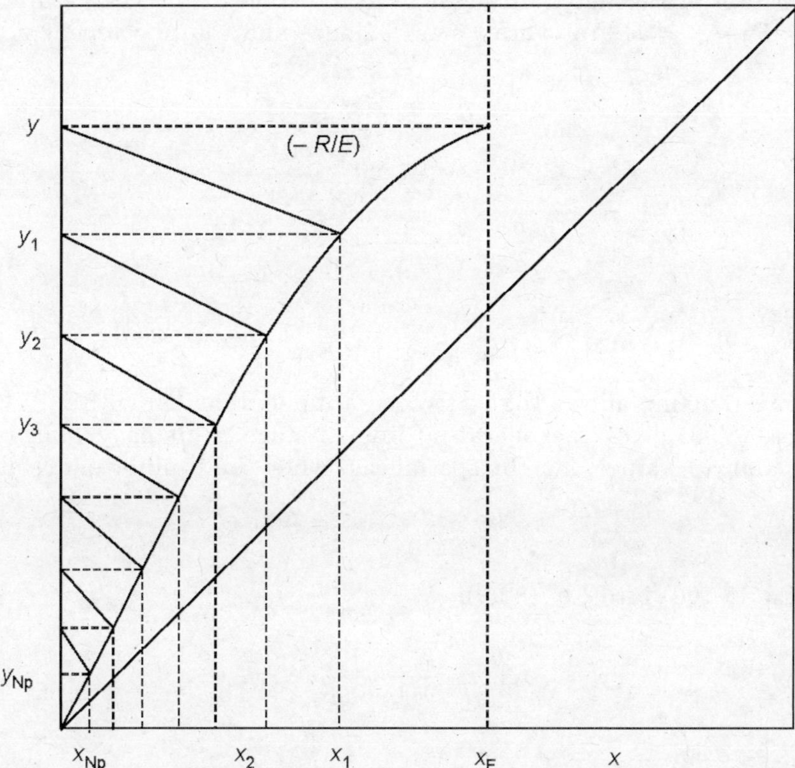

FIGURE 8.25 Operating lines for multi-stage cross-current leaching in distribution diagram of solute when fresh solvent is used at each stage.

8.7.2 Multi-stage Counter-current Continuous Leaching Operations

The counter-current multi-stage continuous leaching operations are the most efficient among different operations. This is because the fresh solvent contacts the leached solids and the feed is allowed to equilibrate with the final extract (Figure 8.26). Au Contraire, in co-current leaching operations the final extract is in equilibrium with

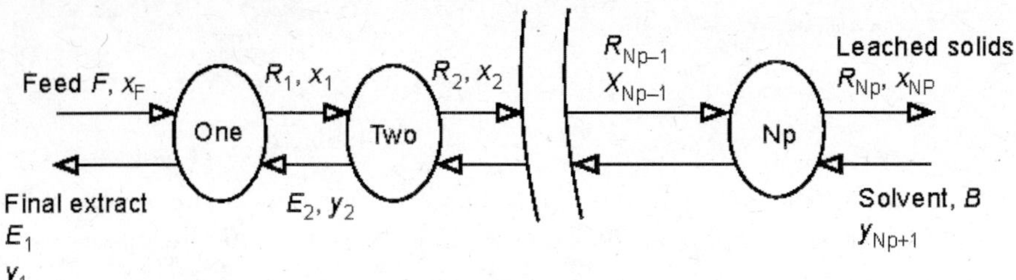

FIGURE 8.26 Counter-current continuous leaching operations.

the final leached solids. This will give a lower value. Constant extract flow rate E and raffinate flow rate R are assumed throughout all the stages. $E_1 = E_2 = \dots E_k = E$ and $R_1 = R_2 = \dots = R_k = \dots = R$. An overall mass balance and solute component balance yields,

$$F + B = E + R \tag{8.97}$$

$$x_f F = y_1 E + x_{NP} R$$

$$\frac{R}{E} = \frac{y_1}{x_f - x_{Np}} \tag{8.98}$$

At the kth stage,

$$E y_k = R(x_{k-1} - x_k) + E y_{k+1} \tag{8.99}$$

This method of contacting allows the y_k to come into equilibrium with x_k. Thus $y_k = f(x_k)$. Hence Eq. (8.99) gives the operating line at stage k giving y_{k+1} in terms of R/E and the x_k and x_{k-1} values. For the special case when the equilibrium relationship can be linear;

$$y_k = m x_k \tag{8.100}$$

Substituting Eq. (8.100) into Eq. (8.99),

$$y_{k+1} = x_k\left(m + \frac{R}{E}\right) - \left(\frac{R}{E}\right)x_{k-1} \tag{8.101}$$

At the end of stage 1,

$$(y_1 - y_2) = R/E(x_f - x_1) \tag{8.102}$$

As $y_1 = m x_1$ and $x_f = x_0$

$$y_2 = y_1(1 + R/mE) - y_0 R/mE \tag{8.103}$$

Let R/mE be the extraction factor A_f. Then,

$$y_2 = y_1(1 + A_f) - A_f y_0 \tag{8.104}$$

Thus,

$$y_3 = y_1(1 + A_f + A_f^2) - (A_f + A_f^2)\, y_0 \tag{8.105}$$

Thus,

$$y_{Np+1} = y_1\left(\frac{A_f^{Np\ 1} \quad 1}{A_f \quad 1}\right) \quad y_0 A_f\left(\frac{A_f^{Np\ 1}}{A_f \quad 1}\right) \tag{8.106}$$

Eq. (8.104) is valid when the extraction factor is not 1. For the general case,

$$F + B = E_1 + R_{Np} \tag{8.107}$$

$$x_f F = y_1 E_1 + x_{NP} R_{Np}$$

At the kth stage,

$$E_k y_k = R_{k-1} x_{k-1} - R_k x_k + E_{k+1} y_{k+1} \tag{8.108}$$

A difference point can be defined as one which gives the difference in flow at each stage such that,

$$\Delta_R = F - E_1 = R_{Np} - B \tag{8.109}$$

This is shown in the N x, y diagram and the tie lines can be used to construct the stages in the distribution diagram in the Bancroft coordinates (Figure 8.27).

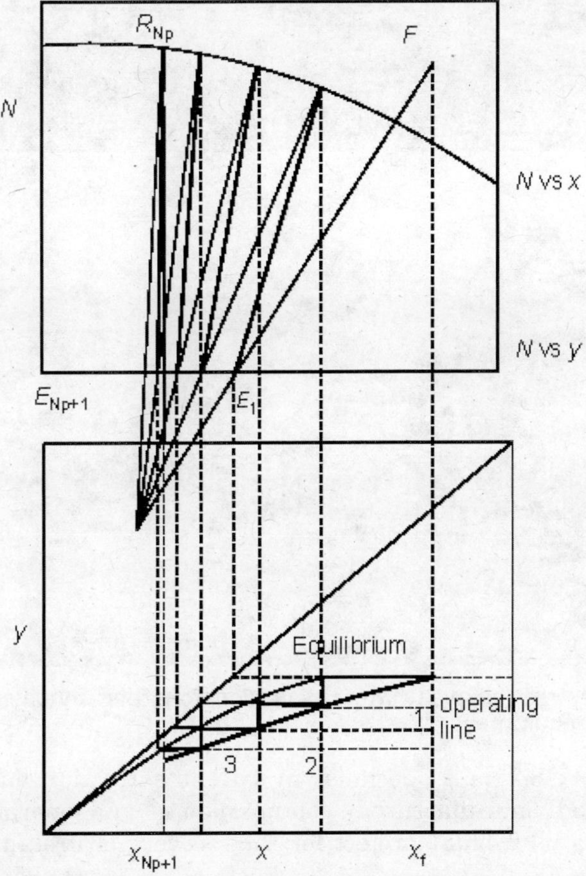

FIGURE 8.27 Multi-stage counter-current leaching operations in Bancroft coordinates.

8.8 SOLID LEACHING EQUIPMENT

A six-stage adsorption-leaching gold extraction plant located in Ghana is shown in Figure 8.28 as an example of heap leaching equipment.

FIGURE 8.28 A gold extraction plant with 1.4 million tons per annum capacity and heap leaching equipment.

SX Kinetics, CO, USA, is a specialist in solvent extraction and electrowinning pilot plants, designed, manufactured, commissioned, and operated an agitation leach circuit during a pilot plant project for the recovery of uranium and zirconium

(Figure 8.29). Its capacity was 10 lit/min of feed slurry, had 3 stages, and was operated at a temperature between 20°C–60°C. A variable speed agitator was used.

FIGURE 8.29 Agitation leach circuit for recovery of uranium and zirconium.

Leaching equipment can be broadly classified into two categories. These are agitated systems and percolated systems. Dorr agitator, thickeners are examples of agitated systems. They can be operated in the batch mode. Shank system is an example of percolation tanks. They can also be operated in closed vessels. The continuous operations are conducted in CCD, continuous counter-current decantation devices. Hydrocyclones, rotocels, basket extractor, Kennedy extractor, Bollman extractor are examples of continuous leaching equipment. Filter leaching is another method for meeting the desired objectives.

SUMMARY

Liquid extraction was defined as an operation where an external solvent is used to separate a solute from a liquid solution. This was in contrast to solid leaching where the feed mixture is in the solid form. Sixteen specific industrial examples of liquid extraction were discussed. The equilibrium line and operating line during the liquid extraction operation was shown in a ternary phase diagram. The notion of a mixture point with a composition was introduced. The different considerations for choosing a good solvent were reviewed. Some of the parameters introduced were selectivity and distribution coefficient. Single stage and multi-stage operations were described in detail.

$$\left(\frac{x_{m1} - x_1}{y_1 - x_1}\right) = \frac{E_1}{M_1} \quad \text{(singe stage operating line)}$$

$$\frac{E_1}{R_{Np}} = \left(\frac{x_M - x_{Np}}{y_1 - x_M}\right) \quad \text{(multi-stage continuous counter-current)}$$

It was pointed out that different scenarios such as the raffinate being fed into the feed of the next stage and the extract being fed into the feed of the next stage were possible. Thus ethylbenzene was extracted from a mixture of ethylbenzene and styrene. The stages stepped off are shown in Figure 8.10 forming steps of z as opposed to the usual vertical cascade. The multi-stage continuous counter-current operations were discussed. The operating lines were shown in the N vs x, y diagram as well as the Bancroft coordinates or distribution diagram. The use of reflux was shown to yield a product with high solute content as extract. Reflux ratio was introduced. The degrees of freedom available in relation to the given operating variables were discussed. The use of co-solvent during liquid extraction was discussed. The Kremser equations when the equilibrium relationship is linear were derived. Extraction factor and co-solvent factor were introduced.

$$x_{Np} = x_1\left(\frac{A_f^{Np} - 1}{A_f - 1}\right) - x_0 A_f\left(\frac{A_f^{Np-2} - 1}{A_f - 1}\right) \quad \text{(Extraction factor, Kremser equation)}$$

$$x_{Np} = x_1\left(\frac{A_f^{Np}/(1 + C_s)^{Np} - 1}{(A_f - 1)}\right) - x_0 A_f\left(\frac{A_f^{Np-2} - 1}{A_f - 1}\right) \quad \text{(Co-extraction factor)}$$

The different kinds of extraction equipment used were discussed. The Karr column, schiebel column, RDC, sieve tray and pulsed column were shown as figures. The governing equation for differential extraction was derived. 21 specific examples of solid leaching were mentioned. Bioleaching, pressurized leaching, leaching of ores, coal beneficiation using supersolvents, etc. were shown.

The single stage leaching operation and multiple stage leaching operations were discussed. The operating lines were derived and shown on the phase diagram.

$$Ey_k = R(x_{k-1} - x_k) + Ey_{k+1} \quad \text{(multi-stage operating line, counter-current)}$$

In multiple stages both cross-current and counter-current operations were discussed. The Kremser equations for counter-current operations when the equilibrium relationship is linear were derived.

$$y_{Np+1} = y_1\left(\frac{A_f^{Np+1} - 1}{A_f - 1}\right) - y_0 A_f\left(\frac{A_f^{Np} - 1}{A_f - 1}\right) \quad \text{(Kremser equation, } A_f \neq 1\text{)}$$

EXERCISES

1. When a solvent is used to separate a partially soluble solute in the feed liquid, what is the name of the operation used?
2. In calculating the degrees of freedom for a system that is selected for liquid extraction operation, when can two liquid phases be selected? Comment on the miscibility of the solvent with the liquid in the feed that is in majority.
3. In worked example 8.1, can the raffinate composition be reduced down to 10% styrene using cross-current contacting. How?
4. What is the difference between the derivations in cross-current contacting when the liquid A and solvent B are partially miscible and when A and B are immiscible. What is the physical meaning behind the changes?
5. Show that $x_M = (Fx_F + Sy_S)/(F + S)$ for continuous counter-current multi-stage liquid extraction operations whose schematic is shown in Figure 8.11.
6. In schematic in Figure 8.5 for a single stage liquid extraction operations, show that x_{m1} is given by $(x_1R_1 + E_1y_1)/(E_1 + R_1)$.
7. How are concentrated solutions treated? Say a feed solution has 65% solute C. Will the techniques discussed in the above sections be applicable to the concentrated feed?
8. Show using schematic the difference between a cross-current train of operations when the raffinate from the previous stage is given as a feed in the next stage and a cross-current multi-stage set of operations when the extract from the previous stage is given as a feed in the subsequent stage.
9. When there are two solutes present in the system, show how the distribution diagram and phase diagram need to be modified for a counter-current set of operations.
10. Show using a schematic the train of stages during a continuous counter-current liquid extraction operations when two solutes are to be extracted and when reflux is used to obtain the product as high yielding solvents?
11. Given the number of stages required for the extract and raffinate composition and when the feed conditions are specified, how is the reflux ratio, product to raffinate ratio and raffinate to solvent ratio calculated?
12. What modifications are needed to be able to step off the stages from a ternary diagram instead of the Bancroft X, Y rectangular coordinates?
13. Why is liquid extraction considered more scientific compared with other operations?
14. The feed mixture to be extracted usually is a solution, i.e. the solute C is dissolved in liquid A. There must be some heat of dissolution associated with the formation of the initial solution. Where can this figure into the calculations when the solvent B is selected and used to extract the solute?

15. What is the physical significance of the cosolvent factor?
16. What is the physical significance of the extraction factor?
17. Like dehumidification to humidification, desorption to adsorption, wetting to drying, what is the process to liquid extraction?
18. What coordinates would be a good choice to represent Figure 8.16, i.e. the use of cosolvent in the multi-stage counter-current liquid extraction operations?
19. What is the difference between solid leaching operation and desorption of the adsorbed solute using a solvent in adsorption, chromatography and ion exchange?
20. What effect will the osmotic pressure have when the extract phase and raffinate phase in liquid extraction continuous operations are brought in contact with each other? Can this explain flooding found in sieve tray and spray columns?

PROBLEMS

1. *Purification of Sucralose*

Tate and Lyle (2006) have a patent on extractive methods for purifying sucralose. Sucralose, 4,1′6′-trichloro-4,1′6′-trideoxygalactosucrose, a sweetener with a sweetness intensity several hundred times that of sucrose is derived from sucrose by replacing the hydroxyl groups in the 4, 1′ and 6′ positions with chlorine. Crystallisation of sucralose directly from the synthesis mixture yields material with high impurity levels. Purification is done by silica gel chromatography. This is not well suited for large-scale applications. These impurities can have an adverse impact on the sweetness, taste, and flavour-modifying properties of sucralose. The product is eluted from silica gel chromatography using ethyl acetate. The patent from Tate and Lyle discusses a multi-step extraction process. The solvents are immiscible with water. Solvents that can be used are n-heptane, n-hexane and n-pentane, cyclohexane etc. As shown in Figure 8.30, at first the aqueous sucralose solution with impurities (1000) may be fed into the first Karr extraction column (1500) where it may be combined with water-saturated ethyl acetate (2000) in a ratio of 0.35:1, ethyl acetate to water. Two distinct phases may be obtained from the first Karr column (1500): *aqueous phase* containing sucralose and *ethyl acetate phase* containing residual sucralose and impurities.

The ethyl acetate stream (3000) may be introduced to the second Karr column (2500) where it may combined with water (4000) in a ratio of about 0.7:1, ethyl acetate to water to recover the residual sucralose. Again two distinct phases are formed in the second Karr column (2500): the aqueous phase containing recovered sucralose and the ethyl acetate phase containing impurities and some sucralose.

The aqueous stream (5000) from the second Karr column (2500) may be combined with the aqueous phase from the first Karr column (1500). The less polar impurities may then be purged from the system with ethyl acetate solvent

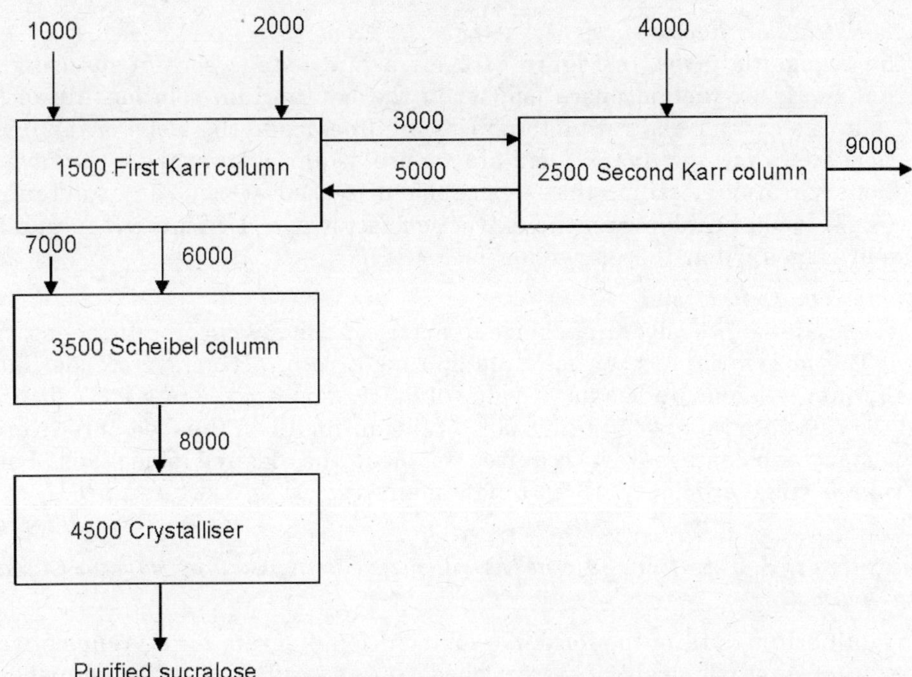

FIGURE 8.30 Purification of sucralose and use of liquid extraction.

stream (9000). The combined aqueous stream (6000) may be introduced into the Scheibel column (3500) and combined with ethyl acetate (7000) in a ratio of about 3:1, ethyl acetate:water. The sucralose from the aqueous stream (6000) will be transferred to the less polar ethyl acetate (7000). The purified sucralose in the ethyl acetate stream (8000) may be fed into the crystalliser (4500) from which purified sucralose may be recovered. Impurities are 4,6′-dichlorgalactosucrose, 1,6′-dichlorosucrose, etc. at a ratio of 0.14 and 0.04 to sucralose respectively. The recovered sucralose crystals have 99.6% purity. 1000 kg/hr of feed solution contains 30% sucralose, what is the solvent feed rate used to obtain the desired objectives. For a counter-current multi-stage operation of 7 stages, what would the extraction factor be for an m value of 0.13?

(**Ans:** A_f = 0.991; Solvent rate: 2100 kg/hr)

2. *Steroid Extraction*
   Sitosterol is isolated from a vegetable extract. 6.5 kg per hour of the extract is pumped upward through a packed column 0.61 m high and 10 cm in diameter. 3 kg/hr of methylene chloride is sprayed downward through the bed. The equilibrium relationship is linear and the m value is 0.14. 50% recovery was found. How high should the tower be made for a 90% recovery?

   (**Ans:** 2.1 m)

3. *When Extraction Factor = 1*
In the schematic given in Figure 8.11 for a multi-stage continuous counter-current liquid extraction operation, when the equilibrium relationship of the solute in the extract and raffinate phases are linear and the slope of the linear relationship is m, how many ideal stages are required to achieve the desired level of separation. Assume that fresh solvent is used without any solute in it. Eq. (8.47) is applicable when the extraction factor $A_f \neq 1$. What is the number of ideal stages when the extraction factor is 1?

4. *Actinomycin Extraction*
Butyl acetate is the solvent of choice to extract actinomycin from fermentation beer. The fermentation beer contains 260 mg/litre of actinomycin. The linear equilibrium relationship has an m value of 0.018. The extract rate is 37 litres/hr and the raffinate rate is 450 lit/hr. 90% of the antibiotic in the feed is recovered. How many ideal stages are required to meet the desired objectives. For a Murphree stage efficiency of 60%, how many real stages are required?
(**Ans:** 2.9, 21)

5. *Separation of Styrene from Styrene/Ethylbenzene using Counter-current Cascade with Reflux*
The equilibrium data in the form of a distribution diagram for styrene between the styrene/di-ethylene glycol extract phase and the raffinate styrene/ethylbenze phases are given as follows:

X	0.087	0.188	0.288	0.384	0.458	0.464	0.561	0.573	0.781
Y*	0.143	0.273	0.386	0.480	0.557	0.565	0.655	0.674	0.833

For a product with 96% styrene, and a raffinate with 4% styrene, show that the minimum solvent reflux ratio needed is 1.875 and the minimum solvent ratio needed is R/B = 3.125 to produce a product at 5000 kg/h. Operate at a reflux ratio of twice the minimum. Show that 21 ideal stages are needed to meet the desired objectives.

6. *When the Extraction Factor = 1*
In the schematic given in Figure 8.31 for a multi-stage continuous counter-current liquid extraction operation, when the equilibrium relationship of the solute in the extract and raffinate phases are linear for both the solvent and co-solvent used and the slopes of the linear relationships are m_1 and m_2 respectively for the solvent and cosolvent respectively, how many ideal stages are required to achieve the desired level of separation. Assume that fresh solvent and fresh co-solvent is used without any solute in it. Eq. (8.66) is applicable when the extraction factor $A_f \neq 1$. What is the number of ideal stages when the extraction factor is 1?

7. *Decaffeination*
An aqueous extract of roast and ground coffee of approximately 12% solids was first stripped of any volatile flavour and/or aroma compounds by passing the extract through a centitherm before decaffeination, the inlet to the centitherm

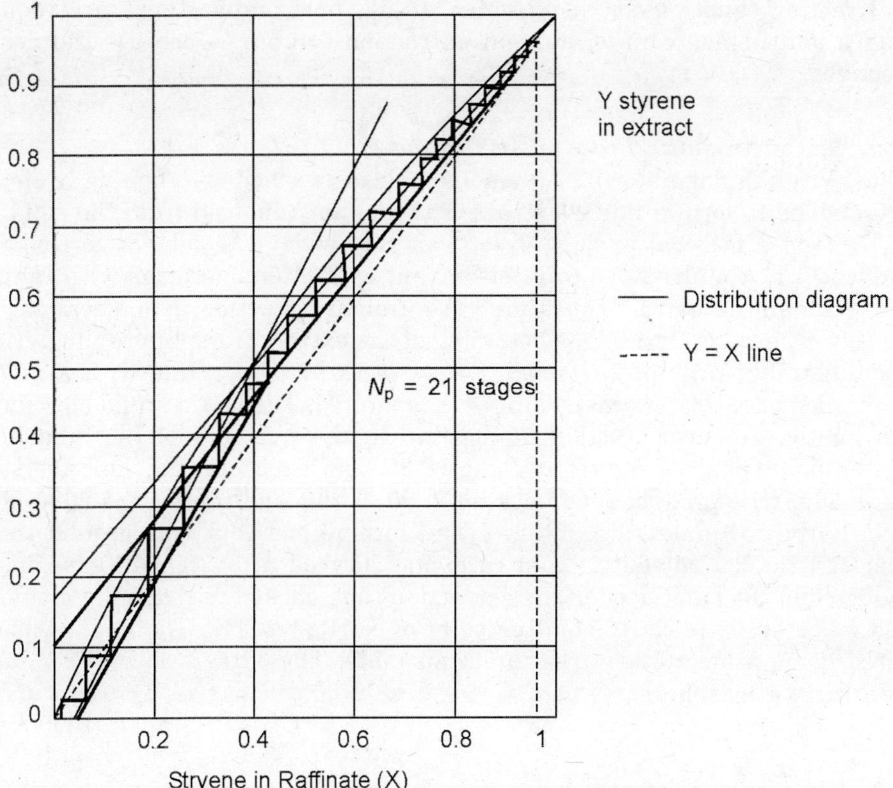

FIGURE 8.31 Distribution diagram and combined graph for counter-current continuous multi-stage liquid extraction operations with reflux.

being at about 90°F and the outlet at about 130°F, the extract being concentrated to 55% solids; the solids solution was thereafter diluted to about 20% soluble solids and contained about 1.0% caffeine. The flavour compounds were collected and processed by conventional means for later addition to the decaffeinated extract prior to spray drying. The loaded methylene chloride was utilised to decaffeinate the extract to achieve an average per cent decaffeination of 96.3%. The extract operating conditions during decaffeination were as follows—the 2" Karr column as above was employed and the column was agitated at 120 spm at a 1" stroke length; the extract for decaffeination and the solvent were both fed at 110°F to the Karr column; The extract flowed upwardly through the column at 12 cc per minute and the loaded methylene chloride flowed downwardly at 220 cc per minute; this resulted in a derivable solvent/extract ratio of 4.8/1 by volume and 5.8/1 by weight. The decaffeinated extract was further processed to remove residual solvent and constituted the useful product of the present invention; thus, the extract drawoff was stripped of methylene chloride by passage through a flash evaporation to achieve methylene chloride levels in the decaffeinated extract of less than 50 ppm and thereafter spray dried. Using the Kremser equation for an m value of 1.3, how many stages are required using a multi-stage counter-current extraction column to achieve the desired objectives?

(**Ans:** 2 stages)

8. *Extraction of Aromatic Fraction in Bituminous Coal*
A West Virginia Bakerstown bituminous coal was washed with boiling N-Methyl-2-Pyrolidone to form a liquid extract. The amount removed from the solid coal was 70% on a mineral and ash-free basis buy weight. A solvent to coal ratio used was 3:1. A multi-stage counter-current extraction operation with reflux is used to obtain an extract containing the aromatic fraction in the coal extract. The solvent used is Pyridine. About 50% of the extract transfers to the pyridine phase. A reflux ratio of 2 is used. For 1000 kg/hr of pyridine with a product composition of 0.973, how many stages are required. The liquid-liquid equilibrium data for the aromatic fraction of the coal in the pyridine extract phase and NMP raffinate phase are given in Figure 8.32. The ash content of the Bakerstown coal is 26%. The product removal rate is 500 kg/hr and the feed is 2000 kg/hr. What is the raffinate rate of flow? The bituminous coal was ground to 200 mesh and the 3:1 solvent to coal ratio was arrived at by careful experiments conducted in the batch mode. The extraction time needed to reach the maximum yield was 20 minutes at the boiling point of NMP at 202°C. The batch was well stirred using a magnetic stirrer on a hot plate. The extraction was conducted in an inert atmosphere.

(**Ans:** 7 stages; 1500 kg/hr)

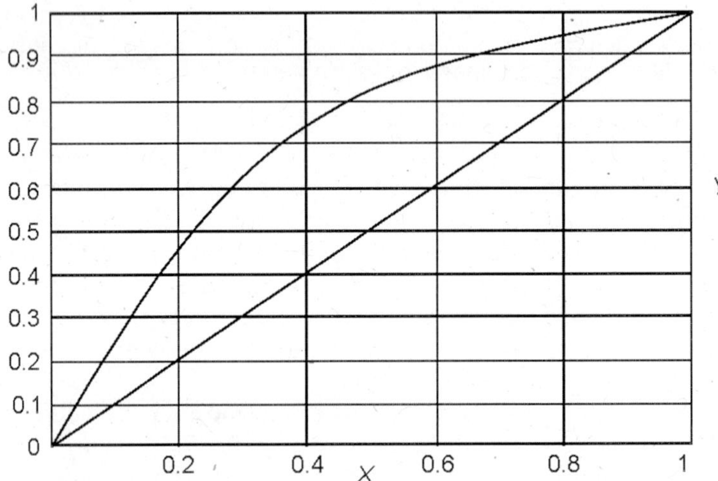

FIGURE 8.32 Distribution liagram for aromatic coal extract between the pyridine extract and NMP raffinate phases.

9. *Extraction Process for Gelatin*
Gelatin is obtained by the partial hydrolysis of collagen, the chief protein component in skins, bones, hides, and white connective tissue of the animal

body. Gelatin is produced by either acid conditioning or by alkaline or lime conditioning followed by hot water extraction of the gelatin from the collagen material. Most acid treated gelatin is made from pork skins yielding grease as a byproduct which is also marketable. The process employed includes comminuting the skins (collagen material), washing the skins to remove extraneous material, treating with a 1–5% acid solution of mineral acid in water, neutralisation of the skins by adjustment of the pH with a base followed by water washing and finally water extraction in 7–8 batches at increasing temperatures up to about 212°F. Each extraction is typically conducted for approximately 2–4 hours. Grease is removed from the gelatin extract which is then filtered, concentrated, chilled and dried by air on wire mesh belts. The dry gelatin is then ground and blended to specification. For a counter-current extraction of 8 stages, what would the extraction factor be for a 95% extraction of gelatin using fresh hot water from a feed acid solution at 0.18 weight fraction gelatin? Take an m value of 0.026 and the extract composition of gelatin as 0.003.

(**Ans:** extraction Factor: 0.655)

10. *Kremser Equation for Co-current Liquid Extraction*
As shown in worked example 8.2, find the number of stages required to perform a continuous co-current multi-stage liquid extraction operation in terms of the compositions of feed, extract and raffinate. Assume that the equilibrium relationship is linear and use a extraction factor definition to simplify the expression.

11. *Kremser Equation for Co-current Liquid Extraction with Co-solvent*
As derived in section 8.4.6, find the number of stages required to perform a continuous co-current multi-stage liquid extraction operation in terms of the compositions of feed, extract and raffinate when a co-solvent is also used in the extraction train. Assume that the equilibrium relationship is linear between the raffinate and extract phases for both the solvent and co-solvent and use an extraction factor and a co-solvent factor definition to simplify the expression.

12. *Kremser Equation for Cross-current Liquid Extraction*
As in worked example 8.2, find the number of stages required to perform a cross-current multi-stage liquid extraction operation in terms of the compositions of feed, extract and raffinate. Assume that the equilibrium relationship is linear and use an extraction factor definition to simplify the expression. Assume that the raffinate from the previous stage is fed as the feed to the next stage and the extract is removed from each stage as a cross-operation in addition to the solvent feed as a cross-operation.

13. *Kremser Equation for Cross-current Liquid Extraction with Co-solvent*
In a similar fashion to the derivation in section 8.4.6, find the number of stages required to perform a cross-current multi-stage liquid extraction operation in terms of the compositions of feed, extract and raffinate when a co-solvent is also used in the extraction train. Assume that the equilibrium relationship is linear between the raffinate and extract phases for both the solvent and co-solvent and use an extraction factor and a co-solvent factor definition to simplify the expression. Assume that the raffinate from the previous stage is fed as the

feed to the next stage and the extract is removed from each stage as a cross-operation in addition to the solvent feed as a cross-operation.

14. *Kremser Equation for Cross-current Liquid Extraction*
As in worked example 8.2, find the number of stages required to perform a cross-current multi-stage liquid extraction operation in terms of the compositions of feed, extract and raffinate. Assume that the equilibrium relationship is linear and use an extraction factor definition to simplify the expression. Assume that the extract from the previous stage is fed as the feed to the next stage and the raffinate is removed from each stage as a cross-operation in addition to the solvent feed as a cross-operation.

15. *Kremser Equation for Cross-current Liquid Extraction with Co-solvent*
As derived in section 8.4.6, find the number of stages required to perform a cross-current multi-stage liquid extraction operation in terms of the compositions of feed, extract and raffinate when a co-solvent is also used in the extraction train. Assume that the equilibrium relationship is linear between the raffinate and extract phases for both the solvent and co-solvent and use an extraction factor and a co-solvent factor definition to simplify the expression. Assume that the extract from the previous stage is fed as the feed to the next stage and the raffinate is removed from each stage as a cross-operation in addition to the solvent feed as a cross-operation.

16. *Potable Water from Sea Water*
In a similar fashion to that shown in Figure 8.14, rather than reflux of the extract, in order to desalinate sea water by extraction of NaCl, the liquid water in the feed solution is used as reflux in Figure 8.33. The sea water has 3.6% by weight fraction NaCl. Deodecyl amine is used as a solvent for sodium chloride. The partition coefficient of NaCl between the amine solvent and water is 0.69. The potable water can be collected at the stripping section after separation from the deodecyl amine. Show that the operating line for the stripping section can be written as assuming equimolal flow rates of raffinate and extract,

$$y_{k+1} = R/Ex_k - x_w W/E$$

For an exit potable water concentration of 100 ppm and a discharge rate of

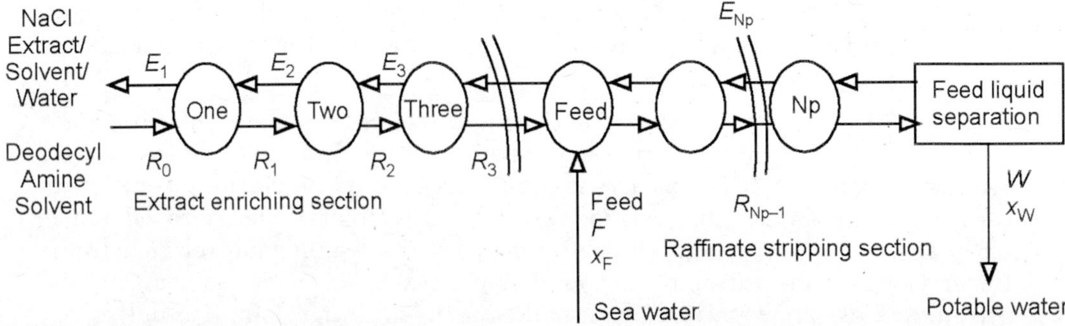

FIGURE 8.33 Reflux of feed liquid in counter-current liquid extraction operations.

potable water, W of 100 kg/hr for a feed of 1000 kg/hr of sea water and a solvent of 300 kg/hr how many ideal stages are required? The water partitions into the raffinate and extract phases with a ratio of 3. The weight fraction in the extract phase is 0.55. It can be assumed that fresh solvent was used.

(**Ans:** 7 stages)

17. *Purex Liquid Extraction Process*

Purex process is used for the recovery and reuse of uranium in the nuclear industry. The nuclear waste is washed with nitric acid and the nitrate of uranium is dissolved in water. Tributyl phosphate is used as a solvent and kerosene is used as a co-solvent in a multi-stage counter-current liquid extraction operation to remove the uranyl nitrate from the aqueous phase to the organic phase. This is later recovered using water. For an extraction factor and cosolvent factor of 0.32 and 0.5 respectively, how many ideal stages are required to extract 95% of the uranyl nitrate from a 1000 kg/hr of aqueous solution? The aqueous solution concentration of uranyl nitrate is 30%. The solvent and co-solvent rates are 500 kg/hr and 100 kg/hr respectively. What are m_1 and m_2?

(**Ans:** 6 stages; 6.25; 15.3)

18. *Co-solvent in Leaching*

Eq. (8.104) was derived for multi-stage counter-current continuous leaching operations for a linear equilibrium relationship. Derive the relationship once again when a cosolvent is used. Use a cosolvent factor if necessary. Derive the equation that relates the raffinate compositions as well.

19. *Continuous Counter-current Decantation*

400 kg/h of sodium hydroxide is manufactured by reaction of sodium carbonate and calcium hydroxide. Pure water is added to wash the calcium carbonate. A solution containing 10% sodium hydroxide results as overflow from the first stage. Should three stages be used, determine the amount of wash water required and the % of hydroxide lost in the discharged sludge. How many thickeners would be required to reduce the loss to 0.1% of that made? The equilibrium data is given below:

N (kg $CaCO_3$/ kg soln)	y(NaOH fraction in raffinate)	x^*(NaOH fraction in extract)
0.495	0.09	0.092
0.525	0.07	0.076
0.568	0.047	0.061
0.600	0.033	0.045
0.620	0.021	0.03
0.650	0.012	0.02
0.659	0.01	0.014
0.666	0.005	0.01

(**Ans:** 3596 kg/hr; 4331 kg/hr; 6 stages)

20. *Soyabean Oil*

Hexane is used as a leach solvent to treat flaked soyabeans. A 0.3 m thick layer of the flakes with 250 μm flake thickness, will be fed onto perforated conveyor belt which passes under a series of continuously operating sprays. The soyabean flakes enter containing 20% oil and are to be leached to 0.5 % oil on a solvent-free basis. The net forward flow of solvent is to be 1.0 kg hexane introduced as fresh solvent per kg flakes and the fresh solvent is free of oil. The solvent drained from the flake is free of solid. The miscella contains 10% of insoluble solid in the feed as a suspended solid which falls through the performance of the belt during loading. How many stages are required? The equilibrium data is given below:

N	$100x^*$
1.752	0
1.515	20
1.429	30

(**Ans:** 4.5)

REFERENCES

Belter, P.A., Cussler, E.L. and Hu, W.S., 1988, *Bioseparations*, John Wiley, New York.

Catani, S.J., Vernon, N.M., Neiditch, D.S., Wiley, J.E. and Micinski, E, 2006, US Patent 7,049,435, *Extractive Methods for Purifying Sucralose,* Tate and Lyle, UK.

Cusler, E.L., 1997, *Diffusion: Mass Transfer in Fluid Systems,* Cambridge University Press, UK.

Internet Website, Extraction Equipment, Koch Modular Process Systems, 2006, *Extraction Technology Group*, Paramus, New Jersey.

Kohr, W.J., Shrader, V. and Johansson, C., 2005, US Patent 6802888, *High temperature heap bioleaching process*, Geobiotics, CO, USA.

Laddha, G.S. and Degaleesan, T.E., 1976, *Transport Phenomena in Liquid Extraction*, Tata-McGraw Hill, New Delhi.

Lo, T.C., Baird, M.H.I. and Hanson, C., 1983, *Handbook of Solvent Extraction,* John Wiley, New York.

Marsden, J.O., Brewer, R.E., Robertson, J.M., Hazen, W., Thompson. P.B., and David, R., 2006, US Patent 7041152, *Method for processing elemental sulphur-bearing materials using high temperature pressure leaching*, Phelps Dodge Corp., AZ.

McCabe, W.L., Smith, J.C., and Harriott, P., 1993, Unit Operations of Chemical Engineering, Fifth Edition, McGraw Hill, New York.

Meinhold, J.F., Musto, J.A., Kramer, K.C. and Gottesman, M., (deceased), 1987, *Method for the decaffeination of roasted coffee extracts*, US patent 4,659,577, General Food Corp., White Plains, New York.

Rao, G. and Sittig, M., 1997, *Dryden's Outlines of Chemical Technology – For the 21st Century*, East-West Press, New Delhi.

Sharma, K.R., March 2002, *Recycle of Uranium in Fission Products*, AIChE Spring National Meeting, New Orleans.

________, 2005, *Damped Wave Transport and Relaxation,* Elsevier, Amsterdam.

Treybal, R.E., 1963, *Liquid Extraction,* McGraw Hill, New York.

________, 1980, *Mass Transfer Operations,* Third Edition, McGraw Hill, New York.

CHAPTER 9

Gas Absorption, Foam Separation and Solution Stripping

Nomenclature

a	interfacial area (m^2/kg)
A_f	absorption factor ($L_s/m'G_s$)
C_i	interfacial concentration of solute (mol/m^3)
C	total concentration (mol/m^3)
$\underline{C}$	effective concentration in ascending stream (mol/m^3)
C_F	concentration of colligend in feed (mol/m^3)
C_Q	concentration of colligend in foamate (mol/m^3)
C_s	concentration of surfactant (mol/m^3)
C_W	concentration of colligend in bottoms (mol/m^3)
d_b	bubble diameter (m)
F	packing factor
G	gas flow rate ($mol/m^2/hr$)
G_s	carrier gas of solute A flow rate (m/s)
k'''	first order reaction rate constant (1/s)
k''	first order reaction rate constant (k'''C) ($mol/m^3.hr$)
K_G	overall gas phase mass transfer coefficient ($mol/m^2/s$)
L	liquid flow rate ($mol/m^2/hr$)
L_s	absorbent solvent flow rate ($mol/m^2/hr$)
p_A	partial pressure of solute A (N/m^2)
p_A^*	equilibrium vapour pressure of pure solute A at a temperature (N/m^2)
P	system total pressure (N/m^2)
Q	foamate volumetric flow rate (m^3/hr)
m	Henry's law constant
m'	slope of the equilibrium line between Y & X when linear
m''	$dy/dx - L/G = m''$
M_g	average molecular weight of the gas (kg/mol)
M_l	average molecular weight of the liquid (kg/mol)

N_p	number of stages
R	universal gas constant (J/mol/K)
S	surface to volume ratio of bubble (1/m)
T	temperature (K)
U	volumetric flow rate of interstitial liquid upflow (m^3/hr)
v_g	superficial gas velocity (m/s)
v_l	superficial liquid velocity (m/s)
x	liquid mole fraction of solute A
x_0	liquid mole fraction of A at the entrance of tower
x_{NP}	liquid mole fraction of A at the exit of tower
X	solute ratio $x/(1-x)$
X_{Np}	liquid mole ratio A
y	vapour phase mole fraction of solute A
y_1	vapour phase mole fraction of A at the exit of tower
$y_{N_{p+1}}$	vapour phase mole fraction of A at the entrance of tower
$Y_{N_{p+1}}$	vapour phase mole ratio of solute A at the entrance of the tower
Y	mole ratio of solute A $y/(1-y)$
Z	height of tower (m)

Greek

ρ_g	density of gas (kg/m^3)
ρ_l	density of liquid (kg/m^3)
μ	viscosity of liquid (kg/m/s)
ψ	ratio of density of water to density of liquid
σ	surface tension (N/m)
Γ	surface excess (mol/m^2)
Γ_I	surface excess from i^{th} species (mol/m^2)
γ	activity of solute
γ_i	activity of ith solute

9.1 INDUSTRIAL APPLICATIONS OF GAS ABSORPTION

One of the first recognised unit operations in chemical engineering profession is gas absorption. The process of removal of impurities from a mixture of gases or to separate in bulk a desired species present in a feed mixture of gases by contact with a liquid is called *gas absorption*. The liquid is called *absorbent* liquid. It can be distinguished from *adsorption* when the desired species adsorbs onto the solid adsorbent and from *liquid extraction* where the feed mixture is all liquid and in *solid leaching* where the feed mixture is all solid. Depending on the concentration of the species of interest that transfers across the gas-liquid interface, the liquid can be referred to as *solvent* when the concentration of the species is high. When the process of transfer is entirely physical such as dissolution, the solvent is called a *physical solvent*. When a chemical reaction takes place the solvent is called a *chemical solvent*. *Stripping* refers to the operations where the species from a liquid solution is removed upon contact with gaseous stream. This is the reverse operation of gas absorption. The principles and methods applicable to gas absorption apply to solution stripping as well. The gas may be bubbled through the liquid, or it may pass over streams of the liquid, arranged to provide a large surface through which the mass transfer can occur. The liquid film in the latter case can flow down the sides of columns or over packing, or it can cascade from one tray to another with the liquid falling and the gas rising in the counter flow. The gas, or its components, either dissolves in the liquid (absorption) or extracts a volatile component from the solution (stripping).

In the history of mankind, after the onset of the industrial revolution, a nation's industrial might was gauged by its ability to manufacture sulphuric acid. From 1749 onwards, the lead chamber method was used. This required air, water, SO_2, sulphur dioxide and nitrate and a lead container. This is not a well understood procedure. The nitrate was expensive. During the production some nitrate was lost to the atmosphere as NO_x emission and added to the cost of sulphuric acid. John Glover designed a tower that was named after him as the *Glover tower*, where the burning gases stripped the sulphuric acid off nitrous oxide. The chamber process is obsolete. These days, absorption towers are used to manufacture sulphuric acid when the SO_3 obtained after the catalytic conversion of SO_2 is contacted with water to form the sulphuric acid of high concentration, oleum.

Thomas Kilgore Sherwood was born in Columbus, Ohio on July 25, 1903 and became one of America's great chemical engineers. His energy, research contributions, applied engineering achievements and influence on chemical engineering education were prodigious. Sherwood came to the Massachusetts Institute of Technology in 1923 to do his graduate work in the chemical engineering department and completed his doctoral thesis on *The Mechanism of the Drying of Solids* under Warren K. Lewis in 1929. From 1930 to 1969 he was professor at MIT and contributed decisively to the standards of excellence of this famous institution. Sherwood's primary research area was mass transfer and its interaction with flow, chemical reaction and industrial process operations in which those phenomena played an important part. His rapid rise to the position of world authority in the field of mass transfer was accelerated by the publication of his book *Absorption and Extraction* in 1937 as the first significant text in this area. Although completely rewritten, with Pigford and Wilke in 1974 under the title *Mass Transfer*, the book has maintained enormous influence and worldwide use of the 'Sherwood Number' is a memorial to that effort. In addition to three honorary doctorates, Sherwood received many honors and awards, including the U.S. Medal for Merit in 1948 and the Lewis Award in 1972. He died on January 14, 1976.

It can be noted that one of the first books was on gas absorption. In the first few years during the origin of chemical engineering at Massachussets Institute of Technology in 1888, a set of lectures were given in unit operations. These were largely empirical studies. During World War II, the technology departments of leading chemical companies such as Du Pont, Dow chemical had files on solubility and heat exchange data. These were later assembled into the modern subjects of mass transfer, heat transfer, reaction engineering, momentum transfer, thermodynamics, transport phenomena, etc. Over the years a lot more understanding at a fundamental scientific level has gone into the development of theories and models that can be used in lieu of the empirical correlations that were instructed in the initial years of the profession. Typical chemical plant skylines these days consist of tall thin distillation column, second tallest are the fat gas absorption columns and shortest ones are reactors.

Some industrial applications of gas absorption are given in Table 9.1. Carbon dioxide (CO_2) and hydrogen disulphide (H_2S) are separated by the process of gas absorption. They occur in concentrations of 5–50%. Sulphur dioxide (SO_2), mercaptans, carbonyl sulphide (COS) form constituents in gas stream that can be removed. Water as an impurity can be removed from gas mixture either by adsorption or dehumidification or even absorption. SO_3 sulphur trioxide, prussic acid, HCN, NO_x, nitrogen oxides, etc. participate in gas absorption.

TABLE 9.1 Industrial Applications of Gas Absorption

No.	*Gases to be absorbed*	*Target*	*Solvent used*	*Application*
1.	SO_3		Water	Sulphuric acid plant
2.	H_2S	< 4 ppm	Ethanolamine	LPG treatment
3.	CO_2	< 500 ppm	Hot K_2CO_3	Synthesis gas
4.	Acetylene, Acetylene dimer		DMF	Sachsse process for acetylene
5.	NO_2		Water	Nitric acid plant
6.	NH_3		NaCl solution	Solvay process for high purity Na_2CO_3
7.	Br		Water	Bromine from sea water
8.	HCl		Water	Direct chlorination of methane
9.	Ethylene		Fuel oil	Steam cracking process for ethylene and acetylene
10.	CO_2, H_2S, NH_3	< 16 ppm CO_2		Ammonia manufacture
11.	H_2S, CO_2	< 1 ppm		Ethylene manufacture
12.	SO_2	90% removal		Flue gas desulphurisation
13.	CO_2	< 1000 ppm		Hydrogen manufacture
14.	CO_2, H_2S, COS	10 ppm H_2S		Refinery desulphurisation
15.	Benzene		Oil (hydrocarbon)	Coke oven gas

Benzene and toluene are removed from the absorption oil using steam stripping. Most of the leached solutions from Table 8.2 can be steam stripped to obtain the solute that has been leached from the solid feed mixture. This can be said of the examples given for liquid extraction in Table 8.1. Stripping is a tool used widely in the industry to separate the solute from the solution.

The concentration of the solution is an important parameter in the design of absorption. Absorption equipment is selected to provide large interfacial area and is modified to suit a given application. Gas absorption is used to abate air pollution. This is a means to treat the culprit prior to discharge. This is different from the methods used to identify the culprit in the raw materials and remove it prior to reaction or capture it during the reaction.

As discussed in Chapter 3 during gas absorption, the mass transfer occurs across the gas-liquid interface. The solute at the bulk gas concentration will diffuse to the interface. The interfacial gas composition and liquid composition are in equilibrium. The liquid phase transfer occurs between the interface to the bulk liquid concentration. Sometimes the species may react with the liquid. On other times the species may react in the gas phase. In such cases suitable terms to account for the reaction rate have to be included to obtain the operating lines.

9.2 EQUILIBRIUM SOLUBILITY, RAOULT'S LAW AND SOLVENT PROPERTIES

The rate of mass transfer of the solute from gas phase to the liquid phase during gas absorption will depend on the extent to which the solute can dissolve in the liquid. The fugacity of the species in the gas phase can be equated with the fugacity of the species in the solution phase at equilibrium. A gas law such as ideal gas or real gas law can be used to obtain the fugacity in the vapour phase. In the liquid phase the mole fraction and activity coefficient as shown earlier, Poynting correction factor can be used to obtain the equilibrium line.

The equilibrium gas solubility of a species in a given liquid varies with temperature and pressure of the system and the nature of the chemical solute species and liquid. In Figure 9.1 is given the equilibrium solubility curve for some gases as a function of the mole fraction of the solute in the gas phase. Higher the solute concentration in the gas phase, higher is the solubility of the gas solute.

If for a given concentration of solute in liquid, the equilibrium mole fraction of solute in vapour and so the equilibrium partial pressure of the solute in the vapour phase be high then the solute is said to be relatively insoluble in the liquid. On the other hand for the same solute concentration in the liquid should the mole fraction of solute in the gas phase be high or the partial pressure of the vapour be high then the species is said to be soluble in the liquid.

When temperature of the system is changed, van't Hoff's law is obeyed. Should the system temperature be decreased, the change would be in the direction where heat will be liberated. Usually the solubility of vapours in liquid will increase with the decrease in temperature. Experimental measurements are needed to plot the solubility curves. Unless the mechanism of dissolution is understood to a level that reliable models can be used in lieu of the experimentation.

Ideal solutions are first considered as a model for the liquid phase in the absorption operations. It is good for several systems but may not be for all the systems. Later special provisions can be made for dealing with liquid phase systems that deviate

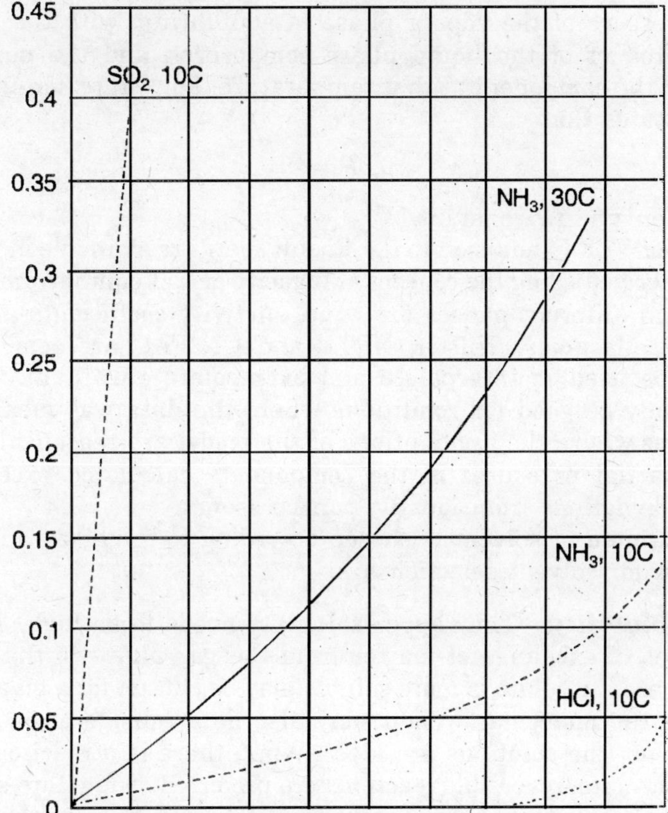

FIGURE 9.1 Equilibrium solubility of gases in water at atmospheric pressure.

from the ideal solutions. For ideal solutions, the vapour phase partial pressure can be calculated from the knowledge of the composition of the liquid phase alone *a priori* to any experimentation. This can be useful for systems where data is not readily available. Four characteristics that can be mentioned for ideal solutions are:

1. The average intermolecular forces of attraction and repulsion in the solution are unchanged on mixing of the constituents.
2. Volume of solution varies linearly with composition.
3. The mixing is adiabatic.
4. The total vapour pressure of the system varies linearly with the composition expressed as mole fraction.

Real systems only approach the ideal solution in the limits. Search can be made from among isomers, or compounds from a series with similar structure differing only with a methyl group. The Raoults law states that

$$p_A = p_A^* x \tag{9.1}$$

The partial pressure of the vapour phase in equilibrium with the ideal solution is given by the product of the liquid phase composition and the pure component vapour pressure of the component at that temperature. For dilute non-ideal solutions, the Henry's law states that

$$y^* = p_A/P = mx \tag{9.2}$$

where m is the Henry's law constant.

Other models for VLE in addition to the Raoult's law are available in the literature. These models are derived using the concept of fugacity and at equilibrium the fugacities of the component in different phases are equal. Activity coefficients are defined for liquids. Van Laar, Margules, Wilson, NRTL and UNIFAC are some of the other models that can be used to interpolate and extrapolate VLE data. The resulting information will only be good for conditions where the data was measured and for those circumstances where the assumptions of the model are reasonably good. When the sum of the partial pressures of the components calculated exceeds the total system pressure, deviations from ideality can be seen.

The absorbent liquid or solvent is chosen according to the given application. The salient parameters for solvent selection are:

1. ***Vapour Solubility:*** The vapour solubility needs to be high. This will offer higher rates of exchange of matter from the gas phase to the liquid phase. Solvent where the solute is more soluble may be chosen for a given application. Sometimes the molecular weight may also be a consideration. This may be the case when the solutions are ideal. When there is a reaction between the solvent and solute, reversible reactions are preferred. The solute-solvent system should be amenable for stripping operations later, only then would the separation be complete.
2. ***Volatality:*** The solvent must be less volatile. This will minimise the solvent losses thus lowering the overall cost of the product.
3. ***Co-solvent:*** A co-solvent can be used to increase the overall yield of the absorption operations.
4. ***Corrosion:*** The materials of construction required for the equipment should not be corrosive.
5. ***Cost:*** The solvent should be low in cost. The losses must not make or break the project. The solvent must be available commercially.
6. ***Viscosity:*** Low viscosity is preferred. High viscous systems may need specialised equipment.
7. ***Ignition Point:*** The solvent should be non-flammable.
8. ***Chemical Stability:*** The solvent must be chemically stable.
9. ***Freezing Point:*** The freezing point of the solvent must be low.
10. ***Weatherability:*** Solvent should be able to be stored for large periods of time in hot weather.

9.3 SINGLE-COMPONENT SINGLE-STAGE COUNTER-CURRENT ABSORPTION

The gas from which the vapour is absorbed is contacted with the absorbent liquid in a counter-current fashion in a single stage as shown in Figure 9.2. The gas flow rate is G mol/m^2/hr of which G_s is the carrier gas that is insoluble in the absorbent liquid and the rest is the solute A that needs to be absorbed. A mole ratio of solute is defined as:

$$Y = \frac{y}{1 - y} \tag{9.3}$$

where y is the mole fraction of solute A in the gas mixture.

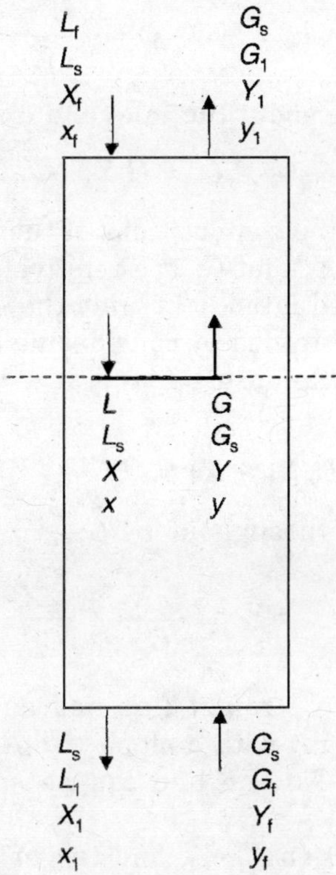

FIGURE 9.2 Schematic of single component single stage counter-current absorption.

This gives the ratio of the solute present in the gas stream with respect to the non-solute portion of the mixture. In a one-solute system, the non-solute portion of the gas will be the carrier gas G_s. Mole ratios in vapour phase can be replaced with ratios of partial pressures. Thus,

$$Y = \frac{p_A}{P - p_A} \tag{9.4}$$

For a one-component solute in the feed mixture,

$$G_s = G(1 - y) = \frac{G}{1+Y} \tag{9.5}$$

In a similar fashion, the liquid absorbent stream consists of x mole fraction of the solute, L_s mole/m^2/hr flow rate of the liquid absorbent and L mole/m^2/hr of the total solution. The mole ratio of the solute in the solution is defined as:

$$X = \frac{x}{1 - x} \tag{9.6}$$

For a one-component solute in the solution mixture,

$$L_s = L(1 - x) = \frac{L}{1+X} \tag{9.7}$$

A solute component balance about the inlet and exit streams in Figure 9.2 yields,

$$Y_f G_s + X_f L_s = G_s Y_1 + L_s X_1 \tag{9.8}$$

It is convenient to express the compositions of the solute in terms of solute ratios as the solvent and carrier gas L_s and G_s are relatively unchanged during the entire operation. These can be grouped later to provide the slope of the operating line, etc. A similar relationship can be obtained by considering the solute balance between the compositions at the bottom of the absorber and at a cross-section of the absorber as shown in Figure 9.2.

$$L_s(X_1 - X) = G_s(Y_f - Y) \tag{9.9}$$

Thus by dividing Eq. (9.9) throughout by G_s,

or

$$Y = Y_f - \frac{L_s X_1}{G_s} + \frac{L_s X}{G_s} \tag{9.10}$$

Equation (9.10) is that of a straight line and as can be seen in Figure 9.3, it passes through the points (X_f, Y_1) with a slope of L_s/G_s. The intercept on the Y axis can be seen to be $Y_f + X_f\, L_s/G_s$. The line can be seen to pass through the point (X_1, Y_f).

For a given feed concentration X_f, Y_f, and target composition Y_1, it can be seen that a minimum solvent required can be calculated. This calculation does depend on the curvature of the equilibrium line near the operating region of the absorber. If the curvature near the operating region is concave upwards as is the case in Figure 9.3, the minimum solvent is obtained by the slope of the line that passes through the points (X^*, Y_f) and (X_f, Y_1). This line passes through the equilibrium composition in the liquid with the feed. Should the curvature be convex upwards near the operating line of the absorber, the line that is tangential to the curve is selected for obtaining the slope at the operating line with minimum solvent needed.

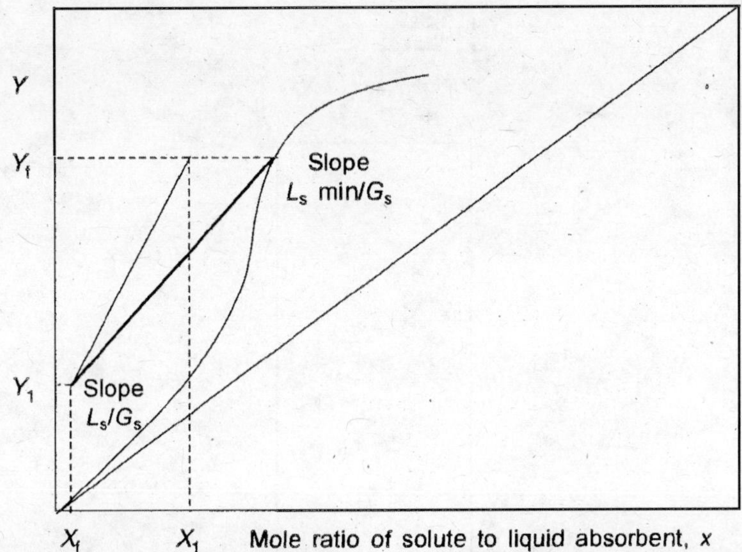

FIGURE 9.3 Combined graph of equilibrium line and operating line for counter-current absorption of one component.

9.4 SINGLE-COMPONENT SINGLE-STAGE CO-CURRENT ABSORPTION

Sometimes co-current flow is preferred especially when the solute forms the entire vapour phase.

The solute component balance for single component co-current absorption can be written from Figure 9.4 as:

$$G_s Y_f + L_s X_f = G_s Y_1 + L_s X_1 \tag{9.11}$$

The solute component balance for a single solute component can be written in terms of the feed compositions and for a set of compositions in the cross-section of the tower as follows:

$$G_s Y_f + L_s X_f = G_s Y + L_s X \tag{9.12}$$

The operating line can be obtained from Eq. (9.12) by dividing throughout by G_s.

$$Y = Y_f + \left(\frac{L_s}{G_s}\right) X_f - \left(\frac{L_s}{G_s}\right) X \tag{9.13}$$

Equation (9.13) can be seen to be that of a straight line with slope of $-L_s/G_s$ and passing through the points, (X_f, Y_f) and (X_1, Y_1). The maximum solute removal is obtained in this method when the line touches the equilibrium line.

FIGURE 9.4 Single stage co-current absorption.

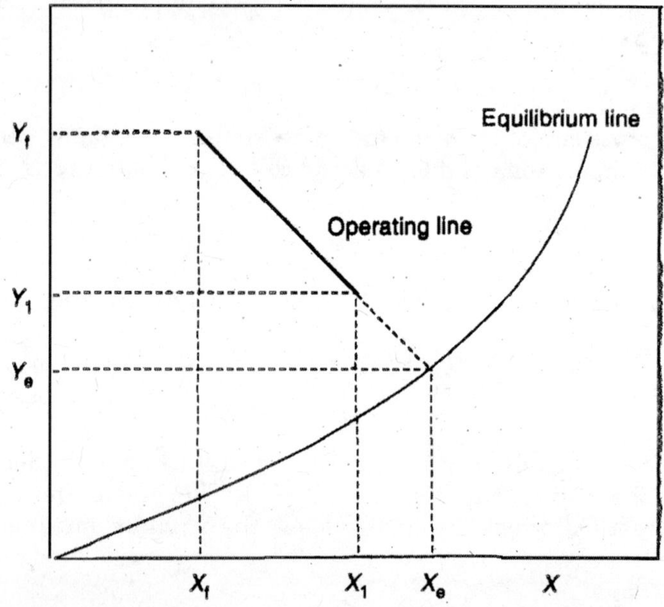

FIGURE 9.5 Combined graph of equilibrium line and operating line for single component co-current absorption.

9.5 MULTI-STAGE COUNTER-CURRENT ABSORPTION

A schematic of a multi-stage continuous counter-current absorption is shown in Figure 9.6.

A solute component balance between the feed and the kth can be written as:

$$L_s X_k + Y_1 G_s = L_s X_0 + G_s Y_{k+1} \tag{9.14}$$

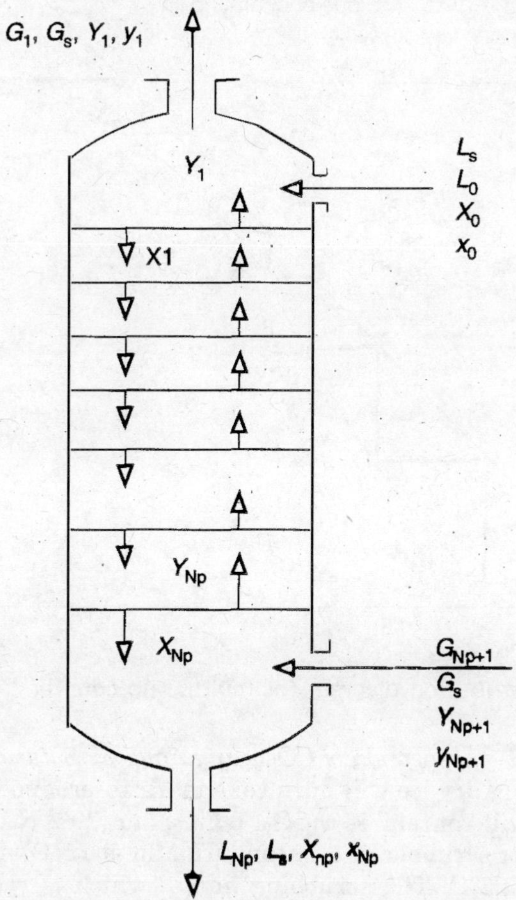

FIGURE 9.6 Multi-stage continuous counter-current absorption.

The equation of the operating line which can be seen to be a straight line is obtained by dividing Eq. (9.14) throughout by G_s.

$$\left(\frac{L_s}{G_s}\right) X_k + \left(Y_1 - \left(\frac{L_s}{G_s}\right) X_0\right) = Y_{k+1} \tag{9.15}$$

The slope of the operating line is L_s/G_s. This is shown in Figure 9.7. Also shown is the number of stages required to obtain the desired level of separation and the way to step them off the combined graph. The minimum solvent required can be calculated in a similar fashion to that shown for a single stage counter-current absorption in section 9.3. In Figure 9.7 it can be seen that fresh solvent was used at an initial zero solute concentration.

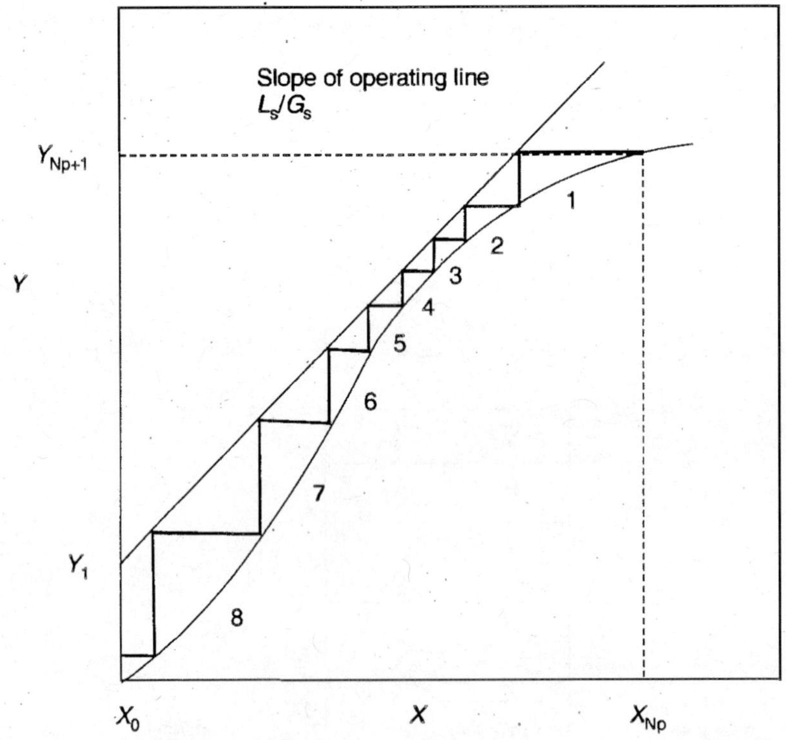

FIGURE 9.7 Distribution diagram for multi-stage counter-current absorption.

Worked Example 9.1 *Absorption of CO_2 using Aqueous Solution of Monoethanolamine*
A plant manufacturing dry ice will burn coke in air to produce a flue gas which when cleaned and cooled will contain 15% CO_2, 6% O_2 and 79% N_2. The gas will be blown into a sieve tray tower scrubber at 1.2 standard atm at 25°C with a 30% ethanolamine solution entering at 25°C. The scrubbing liquid, which is recycled from a stripper, will contain 0.058 mol CO_2/mol solution. The gas leaving the scrubber is to contain 2% CO_2. Assume isothermal operation, and

(a) Determine the minimum liquid/gas ratio.
(b) Operate the absorber at 1.2 times the minimum and calculate the number of theoretical stages to achieve the desired objectives. The equilibrium data at 25° is as follows:

The equilibrium data is given in the X, Y coordinates at 25°C in Table 9.2:

TABLE 9.2 Equilibrium Data of Solute Ratios for CO_2 and Aqueous Ethanolamine Solution

X	Y
0.061571	0.006178
0.06383	0.014235
0.066098	0.032843
0.068376	0.065421
0.070664	0.121357
0.072961	0.204756
0.075269	0.341176

The mole fraction of CO_2 in the feed mixture,

$$= 0.15/44/(0.15/44 + 0.06/32 + 0.79/28)$$

$$= 0.1018 \tag{9.16}$$

Mole ratio of CO_2 $Y_{Np+1} = 0.1133$

The mole fraction of CO_2 at exit, $Y_1 = 0.013$

The mole ratio of solute in absorbent liquid at the exit = 0.0628 (9.17)

The minimum solvent needed can be read from the slope of the line joining the equilibrium composition (X^*_{Np+1}, Y_{Np+1}) and the exit point (X_0, Y_1) and is found to be 9. When the absorber is operated at 1.2 times the minimum, it is found to be 10.8. Thus the minimum solvent rate to gas flow rate ratio used is 10.8. The number of stages stepped off from the graph is 3. (Figure 9.8)

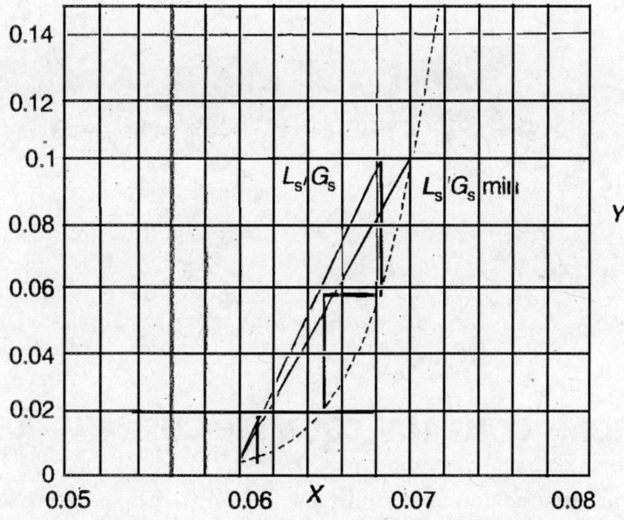

FIGURE 9.8 Distribution diagram for worked example 9.1—counter-current multi-stage absorption of CO_2 in aqueous monoethanolamine.

9.5.1 Linear Equilibrium Relationship

Rather than using the combined graph, an analytical solution can be derived for the number of stages required to perform the separation when the equilibrium relationship between the solute ratios is linear. Here,

$$Y_k = m'X_k \tag{9.18}$$

A solute mass balance on stage 1 can yield,

$$L_sX_0 + G_sY_2 = G_sY_1 + X_1L_s \tag{9.19}$$

or

$$Y_2 = Y_1(1 + A_f) - A_fY_0 \tag{9.20}$$

where, A_f is the absorption factor, $A_f = L_s/m'G_s$.

A solute balance on stage two yields,

$$Y_3 = Y_2(1 + A_f) - A_fY_0 = Y_1(1 + A_f + A_f^2) - A_fY_0(A_f + A_f^2) \tag{9.21}$$

Thus for $Y_{N_{p+1}}$,

$$Y_{N_{p+1}} = Y_1(1 + A_f + A_f^2 + \ldots + A_f^{Np}) - Y_0A_f(1 + A_f + \ldots + A_f^{Np-1}) \tag{9.22}$$

Re-arranging and taking the natural logarithms on both sides,

$$N_{Np+1}\ln(A_f) = \ln\left(\frac{Y_{Np+1} - Y_1 - A_f(Y_0 - Y_{Np+1})}{Y_0 - Y_1}\right) \tag{9.23}$$

Eq. (9.23) is valid when the absorption factor is not equal to 1.

For the special case when the absorption factor is 1, Eq. (9.20) can be rewritten as;

$$Y_2 = 2Y_1 - Y_0 \tag{9.24}$$

$$Y_3 = 2Y_2 - Y_1 = 2(2Y_1 - Y_0) - Y_1 = 3Y_1 - 2Y_0 \tag{9.25}$$

Thus for Y_{Np+1},

$$Y_{Np+1} = (N_p+1)Y_1 - N_pY_0 \tag{9.26}$$

or

$$N_p = \frac{Y_{Np+1} - Y_1}{Y_1 - mX_0} \tag{9.27}$$

9.6 CONTINUOUS CONTACT COUNTER-CURRENT ABSORPTION

In the industry, gas absorption is usually carried out in a packed tower (Figure 9.9). The packed tower is a cylinder much like a pipe set on one end and is filled with packing material. Intimate contact between the liquid and solid phases are achieved by pouring the liquid into the top of the tower and letting it trickle down through the

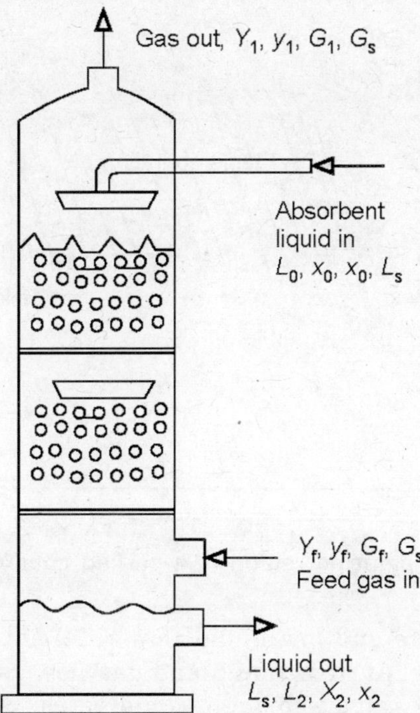

FIGURE 9.9 Packed tower for gas absorption by counter-current continuous contacting.

packing and the gas is pumped into the tower from its bottom and allowed to flow counter-currently upward. The flow rate of the gas cannot be high enough to cause fluidisation of the packing material.

9.6.1 Tower Fluid Mechanics

The information from fluid mechanics is used to determine the cross-sectional area of the packed tower. The mass transfer analysis leads to the calculation of the tower height needed to attain the desired level of separation.

The tower packing material must be inert to the chemical compounds and substances used during the absorption operation. The packing may be either structured or random. Structured packing results in an improvement of about 30% in efficiency. Over 80 years of engineering art has gone into the development of the packing material. Raschig rings and berl saddles are regarded as the first generation packings. The second generation materials are intalox saddles and pall rings and the Nuttler rings form the third generation of materials.

The objective of the packing materials is to offer improved flow as well as increased interfacial area of contact between the gas and liquid during absorption operations. Usually there is a trade-off between the two. The feasible operating regions of the packed tower and possible problems are shown in Figure 9.10.

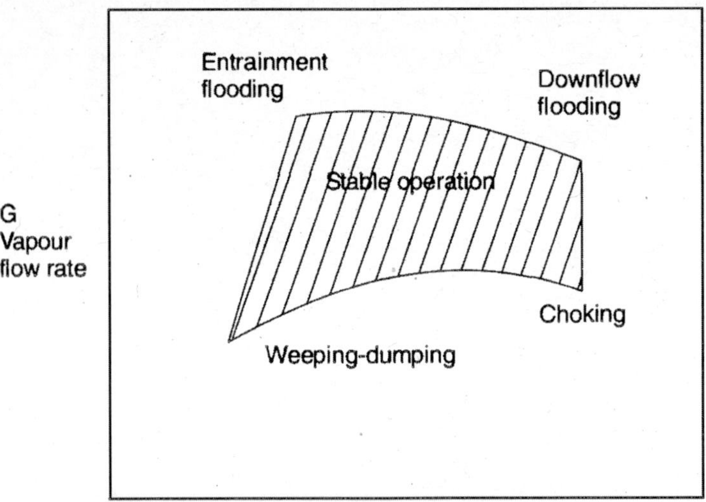

FIGURE 9.10 Stable operating region in a packed column (Smith, 1963).

Thus at high gas flow rate and low liquid flow rate, entrainment problems will be found in the packed tower. At high liquid and gas flow rates, flooding conditions have been found. At low gas rates and high liquid rates, choking has been found. Weeping is another state that needs to be avoided in addition to channeling, slugging, bubbling, jetting, etc.

The operating liquid and gas flow rates fall in a window of operating range as shown in Figure 9.10. The flow rates must be sufficient to load the tower and not flood the tower. The tower's cross-sectional area is an important parameter in achieving this objective. The gas and liquid fluxes change with change in the tower's cross-sectional area.

In Figure 9.11 is given a empirical correlation that is used in the industry for estimating tower's cross-sectional areas. The quantities are a mixture of dimensionless and dimensional items. The flow parameter is defined as:

$$\text{Flow parameter} = \frac{L}{G}\left(\frac{\rho_g}{\rho_l}\right)^{1/2}\left(\frac{M_g}{M_l}\right)$$

$$= \left(\frac{\rho_l v_l^2}{\rho_g v_g^2}\right)^{1/2}\left(\frac{M_g}{M_l}\right) \tag{9.28}$$

where v_l and v_g are the superficial velocities of the liquid and gas respectively. The flow parameter is the abscissa in Figure 9.10. The ordinate is the capacity factor. It can be regarded as a ratio of the kinetic energy in the gas and the potential energy in the liquid.

$$\text{Capacity factor} = M_g^2 G^2 F \psi \mu^{0.2} / \rho_g \rho_l g_c \tag{9.29}$$

The capacity factor is dimensional. The gas flow rate is given in terms of mol/ft/sec. The densities are given as pounds per cubic foot. The viscosity is given as centipoise. ψ is the ratio of the density of water and the density of liquid. The gravitational constant is g_c and is 32. The packing factor F is inversely proportional to its size. Values of F are given in Table 9.3. This is a mixture of metric and English and is common in the petrochemical industry.

In order to read the chart in Figure 9.11, the tower's cross-sectional area, the gas flow and liquid flow rates in mass units are needed. A packing can be chosen, followed by a pressure drop. Towers are designed to operate at pressure drops of 0.2–0.6 inches of water per foot. Lower pressure drop is selected to minimise foaming. An aliter is to choose a flooding condition and choose half its value. This method is good when the density differences are high. For systems with closer densities, other methods need to be employed.

TABLE 9.3 Packing Factors F and Area per Volume for Random Packings Nominal Packing Size (inches)

	1/2	5/8	3/4	1	11/2	2	3
Raschig Rings (Ceramic)	580(111)	380(100)	255(80)	179(58)	93(38)	65(28)	37(19)
Raschig Rings 1/32 (metal)	300(128)	170	155(84)	115(63)			
Raschig Rings (1/16 metal)	410(118)	300	220(72)	144(57)	83(41)	57(31)	32(21)
Berl Saddles Ceramic	240(142)		170(82)	110(76)	65(44)	45(32)	
Pall Rings Metal		81(104)		56(63)	40(39)	27(31)	18
Pall Rings Plastic		95(104)		55(63)	40(39)	26(31)	17(26)
Intalox Saddles Ceramic	200(190)		145(102)	92(78)	52(60)	40(36)	22
Hy-PakRings Metal				45(69)	29(42)	26(33)	16(31)

( ) area/volume (Stringle, 1987)

9.6.2 Height of Transfer Unit, Number of Transfer Units

The tower height can be calculated as follows. Consider a thin slice of the tower across its diameter in Figure 9.9. Let the height of the slice be Δz. Let the mole fractions in the liquid solution and gas mixture be x and y respectively on a solution

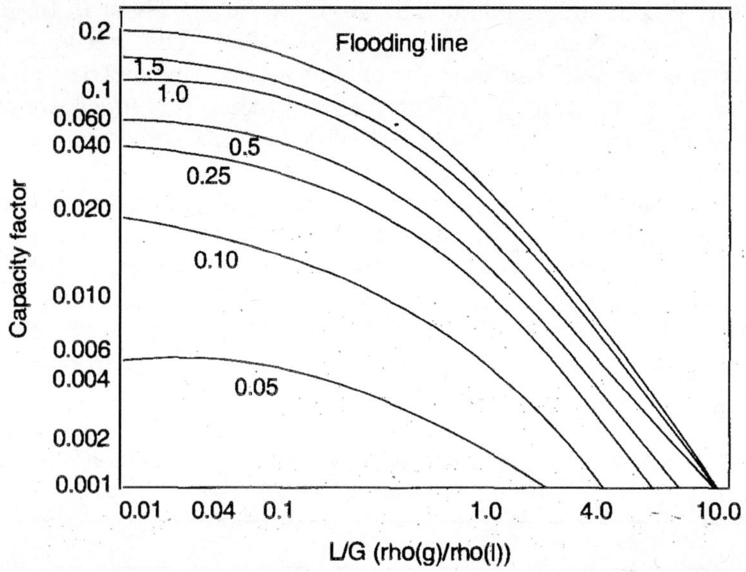

FIGURE 9.11 Correlation for estimating tower cross-sectional area in absorption packed tower design.

basis. A solute component balance at steady state assuming no reaction in the differential volume sliced up gives,

(solute in) – (solute out) – (solute reacted) = (solute accumulated)

$$GA\ (\Delta y) = LA\ (\Delta x) \tag{9.30}$$

Dividing Eq. (9.30) by the differential volume, $A\Delta z$ and obtaining the limits as Δz goes to zero.

$$\frac{Gdy}{dz} = \frac{Ldx}{dz} \tag{9.31}$$

It can be recognised that

$$\frac{L}{G} = \frac{dy}{dx} \tag{9.32}$$

Integrating Eq. (9.32),

$$y = (L/G)x + c' \tag{9.33}$$

The integration constant in Eq. (9.33) can be solved for by recognising that at $z = 0$, $x = x_0$, $y = y_1$ as shown in Figure 9.9,

$$y_1 = (L/G)x_0 + c' \tag{9.34}$$

Subtracting Eq. (9.34) from Eq. (9.33),

$$y = (L/G)x + y_1 - (L/G)x_0 \tag{9.35}$$

Eq. (9.35) is the operating line and as discussed earlier, it is the first step in the design process. The second step is to obtain the equilibrium line. For dilute systems, Henry's law may be applicable and the equilibrium line can also be given by another straight line passing through the origin,

$$y^* = mx \tag{9.36}$$

The third step is to calculate an efficiency from the mass transfer rate effects. The transient effects are neglected as a first approximation. At steady state when there is no chemical reaction, the solute balance on the differential volume slice taken *on the gas phase only* yields,

(solute flow in) – (solute flow out) – (solute lost by absorption) = (solute accumulation)

$$0 = GA(\Delta y) - K_G a\rho(A\Delta z)(C - C_i^*) \tag{9.37}$$

where K_G is the overall mass transfer coefficient in the gas phase and a is the interfacial area. Dividing Eq. (9.37) by $A\Delta z$ and obtaining the limits as the differential element becomes zero,

$$0 = -Gdy/dz - K_G \rho a C(y - y^*)$$

$$0 = -Gdy/dz - K_y \rho a(y - y^*) \tag{9.38}$$

Integration of Eq. (9.38) will result in an expression for the height of the tower.

$$H = \int_0^H dz = -\frac{G}{K_y a\rho} \int_{y_1}^{y_f} \frac{dy}{(y - y^*)} \tag{9.39}$$

When the equilibrium relationship is linear, Eq. (9.39) can be integrated to obtain an analytical solution,

$$H = \int_0^H dz = -\frac{G}{K_y a\rho} \int_{y_1}^{y_f} \frac{dy}{(y - mx)} \tag{9.40}$$

The x can be eliminated between Eq. (9.40) and the operating line given in Eq. (9.35) to yield,

$$\left(\frac{mG}{L}\right) y - \left(\frac{mG}{L}\right) y_1 + mx_0 = mx$$

or

$$mx = \frac{y}{A_f} - \frac{y_1}{A_f} + mx_0 \tag{9.41}$$

$$H = \int_0^H dz = -\frac{G}{K_y a\rho} \int_{y_1}^{y_f} \frac{dy}{y(1 - 1/A_f) + (y_1/A_f) - mx_0} \tag{9.42}$$

$$H = \left(\frac{1}{1 - 1/A_f}\right)\left(\frac{G}{K_y \rho a}\right) \ln\left(\frac{y_0 - mx_0}{y_f - mx_f}\right) \tag{9.43}$$

Thus,

$$\text{H} = \text{HTU . NTU}$$

where, HTU = $G/K_y \rho a$

$$\text{NTU} = \int_{y_1}^{y_f} \frac{dy}{(y - y^*)} \tag{9.44}$$

where HTU is the height of 1 transfer unit and NTU is the number of transfer units.

9.6.3 Gas Absorption with Simultaneous Reaction

During absorption simultaneous reaction may occur in the liquid phase or gas phase. This can be taken into account as follows. A solute component balance at steady state assuming a first order reaction rate, in the differential volume sliced up gives,

(solute in) – (solute out) – (solute reacted) = (solute accumulated)

$$GA(\Delta y) - A\Delta z\, k'''C_A = LA(\Delta x) \tag{9.45}$$

Assuming that the reaction is predominantly in the liquid phase C_A may be written as xC. Dividing Eq. (9.30) by the differential volume, $A\Delta z$ and obtaining the limits as Δz goes to zero.

$$\frac{Gdy}{dz} = \frac{Ldx}{dz} + k''x \tag{9.46}$$

where $k'' = k'''C$ and has the units of mol/m^3/hr.

or
$$\frac{Gdy}{dx} \cdot \frac{dx}{dz} - \frac{Ldx}{dz} = k''x \tag{9.47}$$

or
$$\frac{dx}{dz}\left(\frac{dy}{dx} - \frac{L}{G}\right) = \left(\frac{k''}{G}\right)x \tag{9.48}$$

$$\frac{G}{k''}\frac{dx}{x} = \frac{dz}{\left(\dfrac{dy}{dx} - \dfrac{L}{G}\right)}$$

Integrating both sides,

$$\frac{G}{k''}\ln\left(\frac{x}{x_0}\right) = \int \frac{dz}{\left(\dfrac{dy}{dx} - \dfrac{L}{G}\right)} \tag{9.49}$$

The denominator in RHS gives the deviation of the slope of the operating line from the straight line for the case of no reaction. For relatively slow reactions and realising that $dy/dx = L/D$ when there was no reaction it can be assumed that through the height of the tower the variation of y with x is linear in the operating line. Thus let $dy/dx - L/G = m''$. Hence the RHS can be integrated as $\int dz/m'' = Z/m''$ and,

$$\ln\left(\frac{x}{x_0}\right) = \frac{k''}{G}\frac{Z}{m''}$$

or

$$x = x_0 \exp\left(\frac{k''Z}{Gm''}\right) \tag{9.50}$$

Eq. (9.35) is the variation of the liquid mole fraction along the height of the tower. It can be seen to be a exponential decay.

9.6.4 Concentrated Vapour Mixture

The analysis in the above sections were performed for dilute systems. For concentrated systems, the flow rates L and G cannot be assumed to be constant through the height of the packed tower as done in Eq. (9.30). For concentrated systems without any reaction at steady state, the solute balance on the differential slice cab be written as:

$$\Delta(Gy) = \Delta(Lx) \tag{9.51}$$

The flow rate of gas and liquid G and L can be written in terms of the carrier gas rate G_s and liquid absorbent rate L_s as:

$$G = \frac{G_s}{(1-y)} \text{ and } L = L_s(1-x) \tag{9.52}$$

Combining Eqs. (9.52) and (9.51);

$$\frac{L_s}{G_s} = \frac{d(y/1-y)}{d(x/1-x)} \tag{9.53}$$

$$= \left(\frac{1-x}{1-y}\right)^2 \frac{dy}{dx}$$

Integrating,

$$\frac{1}{1-y} = \frac{L_s}{G_s}\left(\frac{1}{1-x}\right) + c' \tag{9.54}$$

The integration constant can be solved for by realising that at $x = x_0$, $y = y_1$.

$$\frac{1}{1-y_1} = \frac{L_s}{G_s}\left(\frac{1}{1-x_0}\right) + c' \tag{9.55}$$

Subtracting Eq. (9.55) from Eq. (9.54);

$$\frac{(y - y_1)}{(1 - y_1)(1 - y)} = \left(\frac{L_s}{G_s}\right) \frac{(x - x_0)}{(1 - x_0)(1 - x)} \tag{9.56}$$

When fresh solvent is used, $x_0 = 0$ and Eq. (9.56) becomes,

$$\frac{(y - y_1)}{(1 - y_1)(1 - y)} = \left(\frac{L_s}{G_s}\right) \frac{x}{(1 - x)} \tag{9.57}$$

$$\frac{x}{(1 - x)} = X; \frac{y}{(1 - y)} = Y$$

Then,

$$\frac{L_s}{G_s} X + Y_1 = Y \tag{9.58}$$

Thus the operating line given by Eq. (9.57) and rewritten in terms of the mole ratios in Eq. (9.58) is linear in terms of solute mole ratios. The slope of the straight line is (L_s/G_s). The ordinate intercept is at Y_1 the final solute ratio in the exit gas.

9.7 FOAM SEPARATION

Foam separation is a special case of a broader class of operations called *adsubble methods* (Lemlich, 1966). Material gets attached to the surfaces of gas bubbles passing through the solution or suspension. Bubbles rise to form *foam or froth* which carries the identified component overhead. The identified component may be the desired material or undesirable impurity. The material is thus stripped of the liquid solution or mixture. This operation is preferred when large volumes of liquid compared with the identified component is present. The material removed is called the *colligend*. A surfactant is added called the *collector* when the colligend is not surface-active. The attachment of colligend and collector may be by chelation or by other complex formation. An aliter would be to choose the collector of the opposite charge from the colligend and get attached to each other.

The classification of available bubble separation methods are shown in Figure 9.13. *Foam fractionation* involves the removal of dissolved material from the solution. The overflowing foam after collapse is called the *foamate*. A schematic of such a process operated in the continuous mode is shown in Figure 9.12.

Flotation refers to particulate solid removal. The *ore flotation* is used in the mining industry. Macroscopic particles are removed using *macroflotation*, miroscopic units are removed by *microflotation*. *Molecular flotation* refers to the removal of surface inactive molecules by the use of collector. In *ion flotation* the collector forms a scum. In *adsorptive colloid flotation* the removal of dissolved material is in a piggy back fashion by attachment of the material to colloidal particles which is flocculated.

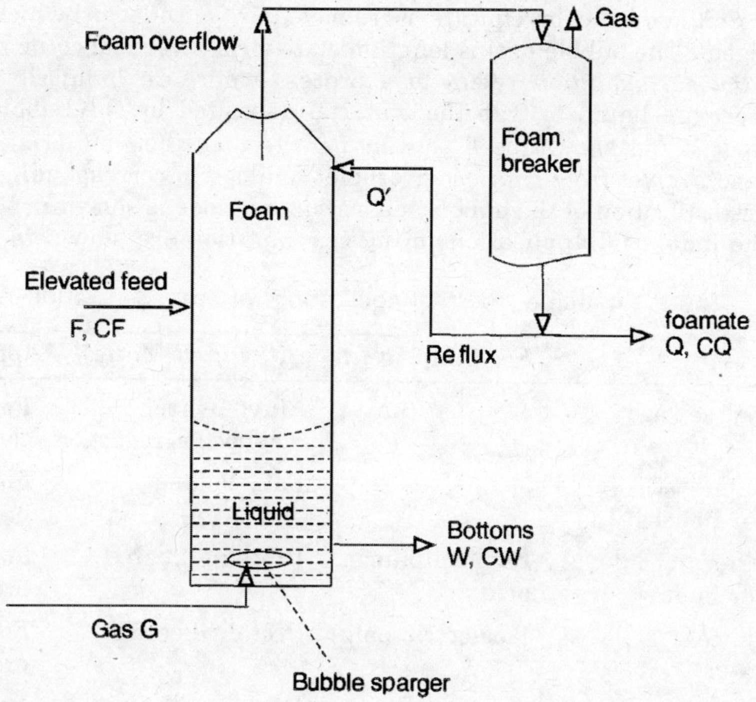

FIGURE 9.12 Schematic of continuous foam separation operations.

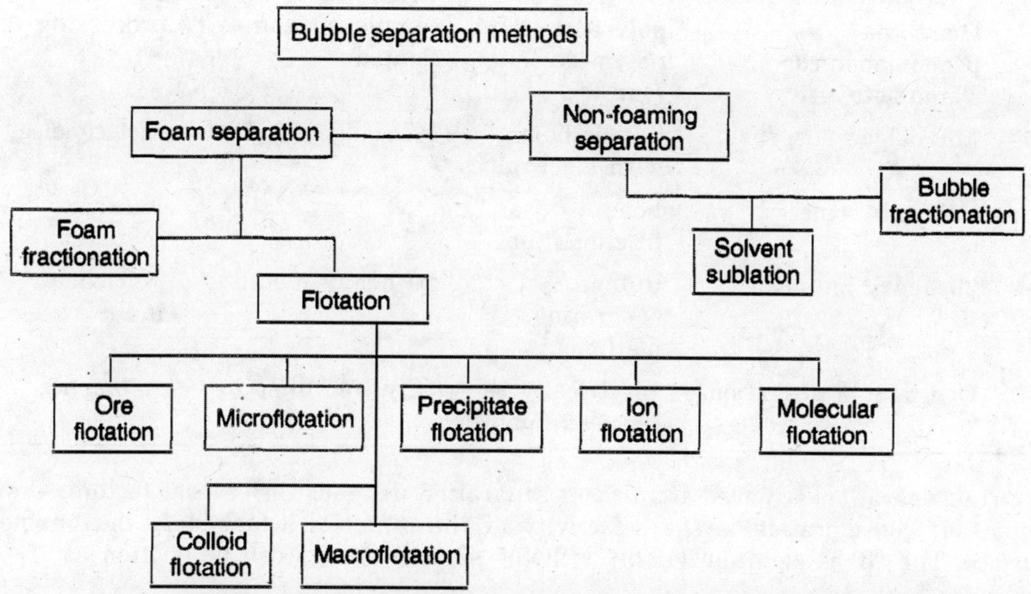

FIGURE 9.13 Classification of bubble separation methods.

In *precipitate flotation*, the precipitate is removed by a collector which is not a precipitating agent. The bubble pool is lengthened to form a vertical column in *bubble fractionation*. *Solvent sublation* refers to a process where an immiscible liquid is placed atop the main liquid to trap the material deposited by the bubbles as they leave. The upper liquid should be a solvent for the identified material. *Emulsion fractionation* and *droplet fractionation* or droplet analogs to corresponding adsubble methods. The classification of the bubble separation methods is shown in Figure 9.13.

Some of the industrial applications of foam separation are shown in Table 9.4.

TABLE 9.4 Some Industrial Applications of Foam Separation

No.	*Feed*	*Foaming Agent*	*Identified Matter*	*Application*
1.	Municipal sewage		Alkyl Benzene Sulphonate reduction	Pollution control
2.	Wastewaters of slaughter houses		Fats floated	Food processing
3.	Wastewaters of slaughter houses	Lignosulphuric acid	Proteins floated	Food processing
4.	Potato juice wastewater	Isoelectric point	Protein floated	Food processing
5.	Refinery Wastewater	Additives	Oil content lowered	Petroleum refinery
6.	Yam starch Dioscorea pseudojaponica Yamamoto	Recovery of polysaccharide polymers that foam	Glycoprotein complexed and floated	Food processing
7.	Microalgae suspension	CO_2 bubbles with electroflocs	Microalgae floated	Biotechnology
8.	^{14}N with some ^{15}N	Foam fractionation	^{15}N	Nuclear industry
9.	29 heavy metals	Bubble separation methods	29 heavy metals	Nuclear industry
10.	Overhaul of airline base	Electrically released bubbles	Oily iron dust	Aerospace

The separation achieved from foam separation depends on several factors. One important consideration is the selectivity of the material attached to the bubble surface. The Gibbs equation (1928) with the surface energy can be written as:

$$d\sigma = -RT\ \Gamma d\ \ln(\gamma) \tag{9.59}$$

where σ is the surface tension, γ is the activity of the identified matter or dissolved species the solute and Γ is the surface excess with units of surface concentration, i.e. mol/m^2. The minus sign indicates that the material that concentrates at the surface lowers the surface tension. If more than one solute is present then the contribution from n solutes can be written by summing the contributions of activities from n solutes as:

$$d\sigma = -\,RT\ \Gamma_i \Sigma d\ \ln(\gamma_i) \tag{9.60}$$

where i denotes the ith species. When applied to a non-ionic surfactant in pure water of concentrations below the critical micelle concentration, Eq. (9.59) simplifies to,

$$\Gamma_s = -\frac{1}{RT}\frac{d\sigma}{d\ln C_s} \tag{9.61}$$

where C_s is the concentration of the surfactant. The major surfactant in the foam can be assumed to be in the form of an attached monolayer with a constant Γ_s of the order of magnitude of 3 E-10 mol/cm^2 for a molecular weight of several hundred. Trace materials have been found to follow the linear isotherm,

$$\Gamma_i = K_i C_i \tag{9.62}$$

Langmuir type isotherm as found in adsorption systems have also been found. K_i for the colligend can be adversely affected by the presence or absence of the collector.

For a pool feed into the liquid in schematic in Figure 9.12, without the reflux,

$$C_Q = C_W + GS\Gamma_W/Q \tag{9.63}$$

$$C_W = C_F - GS\Gamma_W/F \tag{9.64}$$

where C_F, C_W and C_Q are the concentrations of the identified material in the feed stream, bottoms reside stream and foamate stream respectively. G, F and Q are the volumetric flow rates in m^3/hr of the gas, feed and foamate flow rates respectively. Γ_W is the surface excess in equilibrium with C_W. S is the surface to volume ratio of a bubble. For a spherical bubble, $S = 6/d_b$ where, d_b is the bubble diameter. The surface excess is obtained from experimentation or the equilibrium line is obtained from thermodynamics. Correlations exist for bubble sizes in the literature. The operating line is given by the Eqs. (9.63 and 9.64).

The foam-column theory was developed by treating the operation shown in Figure 9.12 as analogous to distillation with entrainment. The two streams in the foam separation operation identified are the descending stream of interfacial liquid and an ascending stream of bubble surface and interstitial fluid. The ascending stream is analogous to the vapour phase in distillation. An effective concentration of colligend/collector in the ascending stream is defined as:

$$\underline{C} = C + GS\Gamma/U \tag{9.65}$$

where U is the interstitial volumetric rate of interstitial liquid upflow, C is the concentration of liquid at that level and Γ is the surface excess in equilibrium with

C. An equilibrium curve can be constructed with $\underline{C}$ vs C in Figure 9.14. The operating lines are found from the material balance of the solute. The slope of the line is found to be L/U where L is the downflow rate.

$$\frac{L}{U} = \frac{\Delta \underline{C}}{\Delta C} \tag{9.66}$$

The combined graph showing the operating line, equilibrium line and the cascade stepping of stages are shown in Figure 9.14. Thus three stages were needed to achieve the desired objectives. The NTU, number of transfer units, can be estimated as:

$$\mathrm{NTU} = \int_{\underline{C}_W^0}^{C_Q} \frac{dC}{\underline{C}^0 - \underline{C}} \tag{9.67}$$

$\underline{C}^0$, $\underline{C}_W^0$ are equated to C and C_W by the effective equilibrium curve and $\underline{C}$ is related to C by the operating line.

For a system that obeys the linear isotherm (Eq. (9.62)), the integration in Eq. (9.67) can be performed analytically. In this case,

$$\mathrm{NTU} = \frac{F}{(GSK - W)} \ln \left[\frac{FW + F(GSK - W)\, C_F/C_W}{GSK(GSK + F - W)}\right] \tag{9.68}$$

If the height of the foam fractionation unit is increased sufficiently, a concentration pinch will develop. The separation attained will be that of a single stage as given by Eq. (9.64). In order to assure intimate contact between the counter flowing interstitial streams, the fraction of the liquid in the foam should be maintained below 10%. Uniform bubble sizes are desirable. Foam overflow on a gas free basis can be given as:

$$\frac{Q}{G} = 22\left(\frac{v_G^3 \mu \mu_s^2}{g^3 \rho^3 d_b^8}\right)^{1/4} \tag{9.69}$$

The drainage theory is not applicable for columns with tortuous cross-sections. Foam coalescence occurs due to the growth of the larger foam bubbles at the expense of the smaller bubble due to interbubble diffusion. The coalescence rate can be assumed to obey first order rate kinetics. Foam breaking is desirable to collapse the overflowing foam. Foam can be broken with a rotating performated basket. If foamate is aqueous, the operation can be improved by discharging into Teflon instead of glass. Foam that is not stable can be broken by running foamate on it. Dephelmagation is the partial collapse of foam to give reflux.

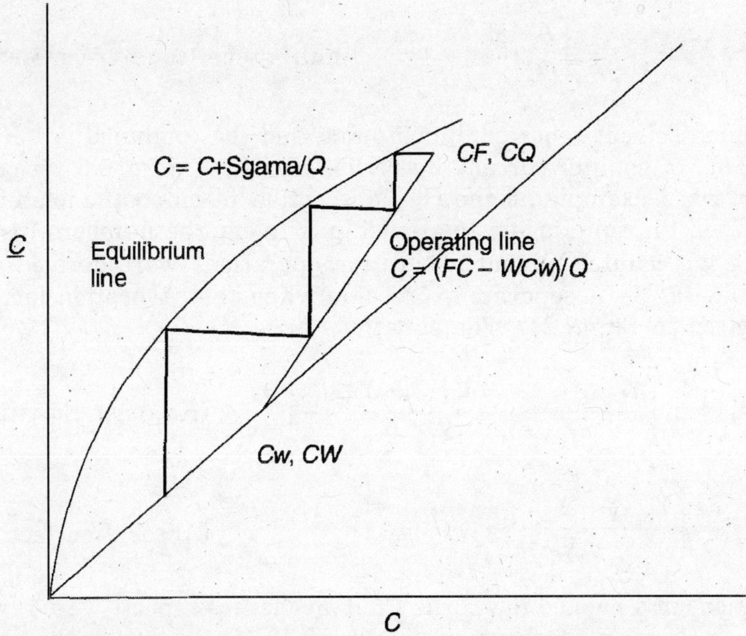

FIGURE 9.14 Combined graph of equilibrium line and operating line for foam separation stripping operations.

SUMMARY

Separation from gaseous form a desired species by contact with a solvent is called absorption. The importance of Glover tower from the lead chamber process for sulphuric acid production from a history of technology point of view was pointed out. 15 specific examples from the industry with the details of the feed, solvent used, solute removed and application used in were given in Table 8.1.

The solubility data for the solute in the absorbent liquid solvent was shown to be used in the construction of the equilibrium line. The Raoult's law was used for the VLE model. The desired properties of the solvent were identified as good vapour solubility, solvent less volatile, low cost, nonflammable, low in viscosity, less corrosive, chemically stable, weatharable and low freezing point.

The component mass balance and overall mass balance was used to obtain the operating line for a single stage and multistage countercurrent absorption operation for dilute systems. It can be seen to be:

$$Y = Y_f - \left(\frac{L_s}{G_s}\right)X_1 + X\left(\frac{L_s}{G_s}\right) \quad \text{(counter-current, 1 component)}$$

$$Y = Y_f + \left(\frac{L_s}{G_s}\right)X_f - X\left(\frac{L_s}{G_s}\right) \quad \text{(co-current, single stage)}$$

$$\left(\frac{L_s}{G_s}\right)X_k + \left(Y_1 - \left(\frac{L_s}{G_s}\right)X_0\right) = Y_{k+1} \quad \text{(multi-stage, counter-current)}$$

The minimum solvent required, pinch point, and the combined graph was shown in Figure 9.3 for a counter-current operation and in Figure 9.5 for a co-current operation. A worked example on the CO_2 absorption in monoethanolamine solution was shown. When the equilibrium relationship is linear the number of trays needed for completing the counter-current absorption operations were derived analytically and given in Eq. (9.23). A separate expression when the absorption factor is 1 was derived and given in Eq. (9.27). The equations are:

$$N_{p+1} \ln (A_f) = \ln \left(\frac{Y_{Np+1} - Y_1 - A_f(Y_0 - Y_{Np+1})}{Y_0 - Y_1}\right) \quad \text{(Kremser Equation)}$$

$$N_p = \left(\frac{Y_{Np+1} - Y_1}{Y_1 - mX_0}\right) \quad \text{(Absorption factor = 1)}$$

For operations in a packed tower the fluid mechanics aspects were reviewed. The feasible operating region is shown in Figure 9.10. Operating within the window shown keeps the tower from flooding, weeping, choking, etc. The correlations between the capacity factor and flow parameter for several different packings were given in Figure 9.11. The characteristics of the packings are provided in Table 9.4. The height of a transfer unit and the number of transfer units needed to obtain a desired degree of separation was derived and is:

$$H = \left(\frac{1}{1 - 1/A_f}\right)\left(\frac{G}{K_y \rho a}\right) \ln \left(\frac{y_0 - mx_0}{y_f - mx_f}\right)$$

The modifications needed for analysis when the solute undergoes simultaneous reaction was shown in Section 9.6.3. For concentrated systems, the change in gas and liquid flow rates with change in the concentration in the gas and liquid phases was shown to be taken into account by using the coordinates of X, Y. This is the solute ratio of the solute with the solute-free solvent liquid and solute-free carrier gas respectively. The analysis used for gas absorption are equally applicable to solution stripping.

The schematic of a continuous foam separation process was shown in Figure 9.12. By foaming the solute from a liquid solution can be removed by gases much like the solution stripping operations. Ten specific examples from the industry of foam separation was given as Table 9.4. The classification of bubble separations was given in Figure 9.13. The interfacial tension and Gibbs surface energy was used to develop the equilibrium line for the excess surface concentration of solute vs the solute concentration. The details of the foam column theory was discussed. The combined graph of the equilibrium line, operating line and the cascade for the number of ideal stages are shown in Figure 9.14.

EXERCISES

1. What is the difference between stripping and liquid extraction operations?
2. Can more than one solvent be used during gas absorption?
3. What is the difference between a scrubber and a gas absorption tower?
4. What is the difference between a wash tower and an absorption tower?
5. How is the operation of a cooling tower different from that of an absorption tower?
6. When we discuss quench tower, what is different about its operations compared with gas absorption tower?
7. What happens to the heat released during gas absorption?
8. Can you achieve 100% removal of the solute from the feed stream by any contact method? Why?
9. How will you convert Figure 9.11 into a dimensionless chart?
10. Can the packed tower be made horizontal? What modifications to your calculations will have to be made?
11. Can you have a packed tower for the initial portion of the solute removal and followed by the multi-stage tray column to achieve the target removal of solute?
12. Should the accumulation of solute be taken into account in Eq. (9.30), how will it affect the results of the calculations for the height of the packed tower needed?
13. Can solute accumulation effects be taken into account in Eq. (9.37)? If so, is absorption an unsteady process?
14. In Eq. (9.27), can the exit vapour and the entering absorbent liquid be in equilibrium? What are the implications?
15. Can the mole fraction of solute in the gas phase be less that in the liquid phase during gas absorption?
16. What is the effect of osmosis during continuous contact operations? Will the liquid absorbent flow from a region of low solute concentration, the liquid phase to a region of high solute concentration the gas phase. Can this osmotic flow cause flooding?
17. Why is multi-stage cross-flow not considered seriously for gas absorption operations?
18. When steam is used as a stripping agent, how are equilibrium line and other relevant equations derived?
19. In the derivation of Eq. (9.27) the number of plates needed to achieve absorption when the equilibrium relationship is linear, what is the effect of the concentration of the vapour, i.e. for dilute systems vs. concentrated systems? What is the difference in the calculating procedure? How come?

20. In Figure 9.12, problem 7, what is the minimum solvent needed? How will you obtain this?

21. In pinch analysis for stripping, how can there be both a minimum gas rate and minimum liquid rate depending on the operating parameters? Explain.

22. Under what circumstances can the operating line be equal to 1 in absorption or solution stripping operations?

23. Under what circumstances can the operating line have a slope of zero during gas absorption or solution stripping?

24. Under what circumstances can the operating line have a slope of infinity during gas absorption and solution stripping?

25. Can the treatment developed for concentrated vapour be used for dilute vapour? How? What happens to the equations developed for dilute systems?

26. Three absorption towers were built. The first one was co-current absorption. The second absorption tower was counter-current absorption and the third absorption tower was co-current absorption. In the combined graph, show qualitatively the operating lines in reference to a typical equilibrium line for a solution. By this approach can you reach the maximum removal of the solute form vapour, i.e. the lowest achievable concentration of solute in the gas phase or is it by direct counter-current absorption? Discuss.

27. How can the concept of recycle be used to improve the efficiency of operations of gas absorption, solution stripping?

28. Can foam separation be used to remove NaCl from sea water? Would you use ion flocculation?

29. Can you have Freundlich isotherm be applied to any of the foaming systems?

30. Can you have Langmuir isotherm applied to the surface excess and concentration of the solute relationships?

31. Can you have BET isotherm applied to the surface excess and concentration of the solute relationships?

32. When more than one of the operations discussed in previous chapters be applicable to the given problem, how will you select one of them?

33. Is surfactant action used in washing machines be treated with the equations developed in section 9.7 on foam separation? What is needed?

34. What is the role of foam in precipitate flotation? Why cannot the precipitate be separated by decantation which usually is less costly an operation?

35. Can foam separation separate two solutes from a multi-component solution?

PROBLEMS

1. *Multiple Stage Steam Stripping of Taints from Cream*
Steam-stripping process can be used to remove a taint from cream. In a single stage operation, the conditions were that stage-contact desorption was to be used to remove a taint that was present at a concentration of 10 ppm in the cream, by contact with a counter flow current of steam. Now consider, the case of a rather more difficult taint to remove in which the equilibrium concentration of the taint in the steam is only 7.5 times as great as that in the cream. If the relative flow rates of cream and steam are given in the ratio 1: 0.75, how many contact stages would be required to reduce the taint concentration in the cream to 0.3 ppm assuming (a) 100% stage efficiency and (b) 70% stage efficiency? The initial concentration of the taint is 10 ppm. **(Ans:** 2 stages; 3 stages)

2. *Tower's Cross-sectional Area*
A natural gas stream flows at 23 lbs/sec and contains 2% CO_2. This is contacted with an aqueous diethylamine solution flowing at 40 lbs/sec. Choose 1.5" Raschig rings and a pressure drop of 0.25 inches of water per foot to avoid foaming problems. The densities of gas and liquid at the operating temperature are 2.8 and 63 lb/ft^3 respectively. The liquid viscosity is 2 cp. What is the absorption tower's cross-sectional area? **(Ans:** 26 sq. ft.)

3. *NTU and HTU for Concentrated Vapour*
The number of transfer units and height of a transfer unit was calculated in section 9.6.2 for dilute vapour gas mixtures. Show that for the concentrated system:

$$H = \frac{G_s}{K_y \rho a} \int_{y_1}^{y_f} \frac{dy}{(1-y)^2 (y-y^*)} = \text{HTU.NTU}$$

4. *Height of the Packed Tower*
An fresh organic amine is used as a absorbent liquid in a packed tower with pall rings to absorb carbon dioxide. The feed gas with a CO_2 concentration 1.26% is set to leave from the tower at a concentration of 0.04%. If the exit liquid is in equilibrium with the entering gas, it would contain 0.8% CO_2. The gas flow rate is 2.3 gmol/s, the liquid flow rate is 4.8 gmol/s, the tower's diameter is 40 cm, and the overall mass transfer coefficient times the interfacial area is 5 E-5 gmol/cm^3/sec. Calculate the height of the packed tower needed to obtain the desired level of separation. **(Ans:** 3.2 m)

5. *Stripping in a Packed Tower*
Oxygen dissolved in water is stripped using excess nitrogen in a packed tower. The height of the tower is 2 m and its diameter is 0.6 m. The tower is filled with Hy-Pak rings. The transfer coefficient in the liquid is 0.0022 cm/sec. The water flow rate is 300 cm^3/sec. How much oxygen can be removed by this operation? What is the nitrogen flow rate? **(Ans:** 98% O_2 removal)

6. *Concentrated Vapour Removal in a Packed Tower*
A gas mixture contains 37% NH_3 and flows at 1.2 m^3/sec. This is absorbed with water at 1 atm pressure at 0°C The exit ammonia concentrations in the liquid and gas phases are 23% and 1% respectively. Using 2" Berl saddles, what diameter and height of a packed tower is needed? What is the flow rate of water? It operated at 50% flooding. The overall gas mass transfer coefficient was found from experimental data to be 0.032 m/s.
(**Ans:** H = 5.3 m., d = 0.84 m., L_s = 65.2 mol/sec)

7. *Absorption of Acetone Vapour in MEK*
The equilibrium data for acetone vapour and methyl ethyl ketone as absorbent liquid on the basis of solute ratios in the liquid and gas phases are given in Figure 9.15. The feed gas contains 9.1% acetone by mole fraction and the rest is air. The absorbent liquid, MEK entering the tower on the top plate, has an acetone mole fraction of 2.5%. When the exit gas acetone concentration is 1% by mole fraction, show that the number of ideal stages required for the separation is 11 at 31°C and atmospheric pressure. For a gas flow rate of 1000 mol/hr, how much solvent is needed? (**Ans:** 145.5 mol/hr)

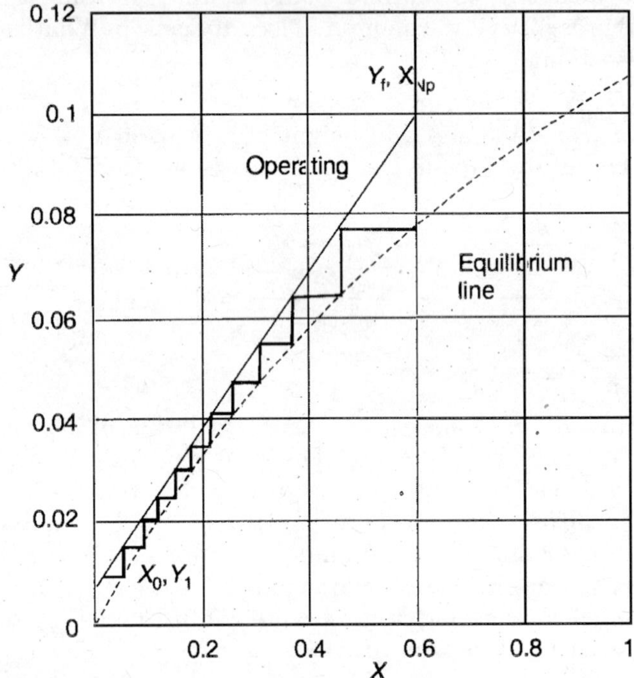

FIGURE 9.15 Absorption of acetone in methyl ethyl ketone by counter-current multi-stage operations.

8. *Absorption of HCl Vapour using Water*
The HCl water equilibrium solubility data at equilibrium is given below in Figure 9.16. A feed with 90% HCl vapour in a counter-current multi-stage

absorption tray tower is reduced to 10% HCl vapour. The feed water has 26% by weight HCl. The spent water exits at a HCl concentration of 40% by weight HCl. For a flow rate of 1100 mol/hr of gas, how much water is needed for a 3-stage operation? **(Ans:** 8800 mol/hr)

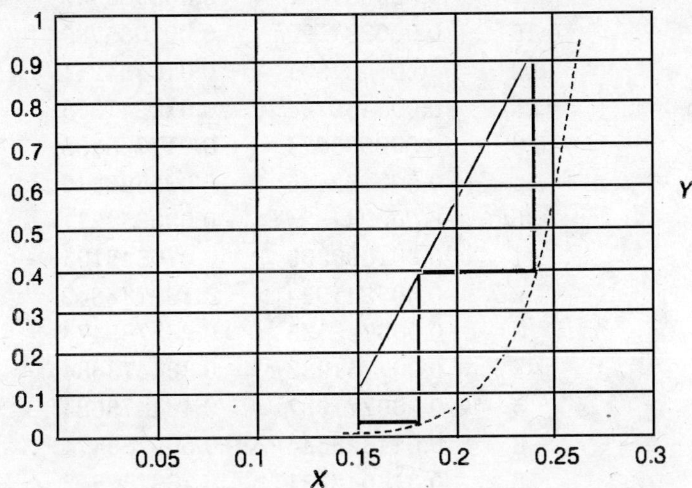

FIGURE 9.16 Combined graph of equilibrium line and operating line during the absorption of HCl vapour.

9. *Concentrated SO_2 Vapour Absorption*
The concentrated vapour absorption principles may be applied to design a counter-current multi-stage absorption tower for a feed gas mixture with 97.5% mole fraction of sulphur dioxide. The liquid absorbent used is fresh water without any sulphur dioxide present in it. The target is 82.5% removal of sulphur dioxide from the gas mixture. An inert atmosphere of nitrogen is maintained at atmospheric pressure at 20°C. The equilibrium data for the mole fraction of sulphur dioxide in the vapour vs the weight fraction of sulphur dioxide in the liquid is given below in Figure 9.17. Show that 2 stages are needed to achieve the desired separation. Show the calculations on a solute ratio coordinates, i.e. Y vs X plot. Confirm that the operating line in this graph given by Eq. (9.58) on a semi-log graph is curvilinear. This is although Eq. (9.58) is that for a straight line. What is the L_s/G_s ratio needed? How much solvent is needed? What is the amount of water needed to absorb 10,000 mol/hr of feed mixture?

(Ans: 2 stages; 5.4 lakh mol/hr of water)

$gSO_2/100gH_2O$	x	y
0.01	1.87515E-05	9.21053E-05
0.05	9.37881E-05	0.001407895
0.1	0.000187652	0.003986842
0.15	0.000281593	0.007065789
0.2	0.00037561	0.010434211
0.25	0.000469704	0.013947368
0.3	0.000563874	0.017763158
0.4	0.000752445	0.025526316
0.5	0.000941324	0.033684211
1	0.001890359	0.076842105
2	0.003811944	0.169736842
3	0.005765535	0.265789474
4	0.007751938	0.364473684
5	0.009771987	0.464473684
6	0.011826544	0.565789474
8	0.016042781	0.769736842
10	0.020408163	0.975

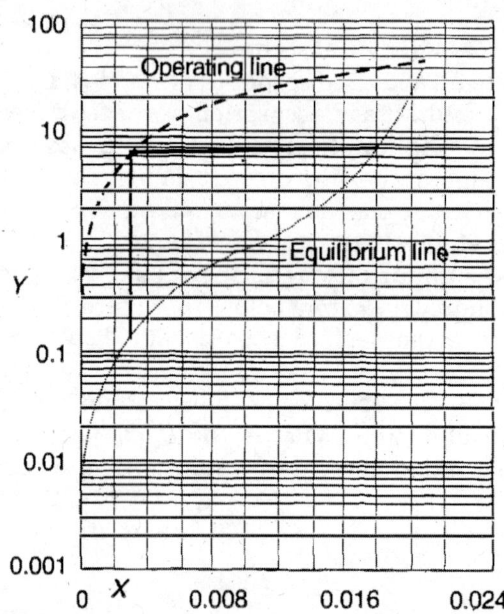

FIGURE 9.17 Combined graph of equilibrium line and operating line in a semi-log ordinate axis of solute ratios during absorption of concentrated sulphur dioxide vapour using water absorbent liquid at 20°C.

10. *Concentrated SO_2 Vapour Absorption*

The concentrated vapour absorption principles may be applied to design a counter-current multi-stage absorption tower for a feed gas mixture with 97.5% mole fraction of sulphur dioxide. The liquid absorbent used is fresh water without any sulphur dioxide present in it. The target is removal of 61.0% sulphur dioxide from the gas mixture. An inert atmosphere of nitrogen is maintained at atmospheric pressure at 90°C. The equilibrium data for the mole fraction of sulphur dioxide in the vapour vs the weight fraction of sulphur dioxide in the liquid is given below. Show that little over 1 stage is needed to achieve the desired separation. Show the calculations on a solute ratio coordinates, i.e. Y vs X plot. Confirm that the operating line in this graph given by Eq. (9.58) on a semi-log graph is curvilinear. This is although Eq. (9.58) is that for a straight line. What is the L_s/G_s ratio needed? How much solvent is needed? What is the amount of water needed to absorb 10,000 mol/hr of feed mixture?

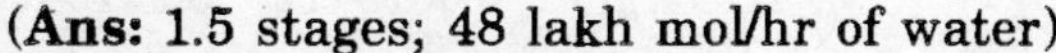

(Ans: 1.5 stages; 48 lakh mol/hr of water)

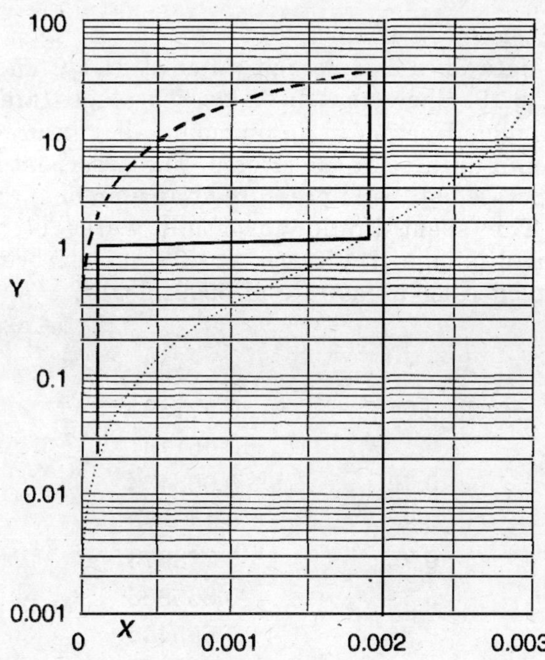

FIGURE 9.18 Combined graph of equilibrium line and operating line in a semi-log ordinate axis of solute ratios during absorption of concentrated sulphur dioxide vapour using water absorbent liquid at 90°C.

$gSO_2/100gH_2O$	x	y	X	Y	*operating line*
0.01	1.87515E-05	0.001592105	1.87519E-05	0.001594644	0.960036
0.05	9.37881E-05	0.016973684	9.37969E-05	0.017266765	2.4009
0.1	0.000187652	0.041710526	0.000187688	0.043526019	4.203604
0.15	0.000281593	0.068684211	0.000281673	0.073749647	6.008112
0.2	0.00037561	0.096973684	0.000375752	0.10738744	7.814429
0.25	0.000469704	0.126052632	0.000469925	0.144233665	9.622556
0.3	0.000563874	0.155263158	0.000564193	0.183800623	11.4325
0.4	0.000752445	0.215789474	0.000753012	0.275167785	15.05783
0.5	0.000941324	0.277631579	0.000942211	0.384335155	18.69045
1	0.001890359	0.597368421	0.001893939	1.483660131	36.96364
1.5	0.0028472	0.921052632	0.00285533	11.66666667	55.42234
1.6	0.003039514	0.986842105	0.00304878	75	59.13659

11. *Absorption of Ammonia in Water*

The equilibrium data for ammonia and water at 21.1°C and 2 atm pressure is given in Figure 9.19. Show that little over 5 stages are needed to absorb ammonia at 90% mole fraction in an ammonia-air mixture down to 10% mole fraction of ammonia in the exit gas stream. The absorbent liquid water enters the tray tower operated at 2 atm pressure at an ammonia concentration of 10% mole fraction. The spent liquid absorbent water is removed from the countercurrent multi-stage tray tower at 40% mole fraction ammonia. How much water is needed for a gas rate of 5000 mol/hr?

(**Ans:** 13,333.33 mol/hr)

x	y
0.05	0.027891
0.1	0.051701
0.15	0.088435
0.2	0.145578
0.25	0.233673
0.3	0.365986
0.35	0.555442
0.4	0.810884

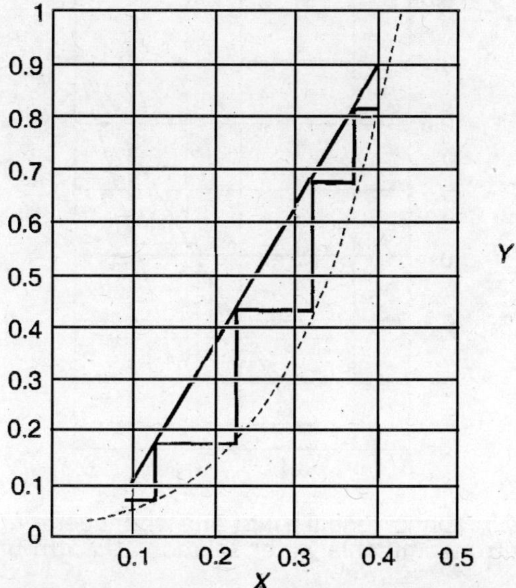

FIGURE 9.19 Combined graph of equilibrium line and operating line during absorption of ammonia vapour using water absorbent liquid at 21.1°C.

12. *Stripping of Ammonia from Water*

A counter-current tray tower operated at 2 atm pressure and 48.9°C is used to strip a 30% by mole fraction ammonia water solution of ammonia. The stripping agent used is steam. What is the maximum gas flow rate allowed for a liquid flow rate of 4833.33 mol/hr? When the stripping operations is operated at 1.8 times the $(L/G)_{\min}$, what is the gas flow rate used? The exit gas mole fraction ammonia is 70% and the inlet gas ammonia concentration is 10 mole %. Show that little more than 2 stages are needed to complete the stripping operation. The equilibrium data at 48.9°C and 2 atm pressure for ammonia water system is given in Figure 9.20.

(**Ans:** 1611 mol/hr; 895 mol/hr)

x	y
0.05	0.090816
0.1	0.168367
0.15	0.279252
0.2	0.445238
0.25	0.690476
0.3	1.038776

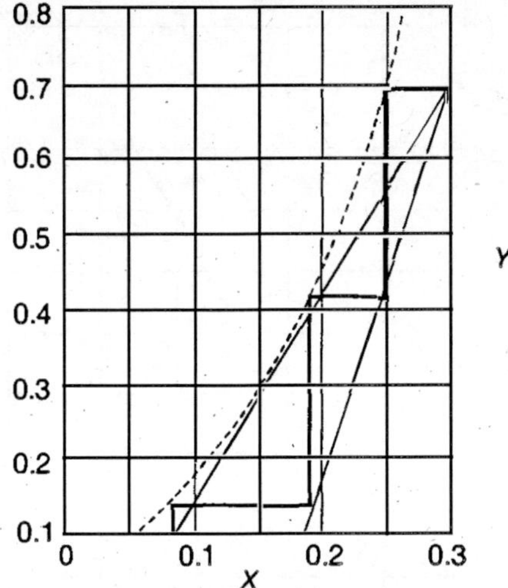

FIGURE 9.20 Combined graph of equilibrium line and operating line during stripping of ammonia from ammonia water solution at 2 atm pressure and 48.9°C.

13. Foam *Separation of Yam Starch*

The yam (Dioscorea pseudojaponica Yamamoto) tuber is abundant in both starch and mucilage. It is difficult to separate yam starch from its tuber because of the presence of viscous polysaccharide polymers (glycoprotein). The surface-active polysaccharide-containing complexes are capable of forming foams. The goal of this study was to develop a continuous foam separation process for separating and recovering starch and mucilage from yam tubers in the absence of undesirable chemical additives or treatments. This method is simple, low in

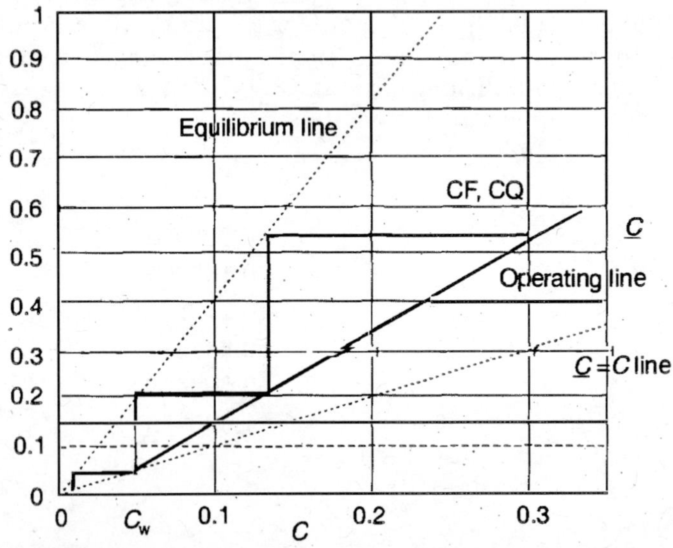

FIGURE 9.21 Combined graph for the separation of polysaccharide from yam starch.

operational and capital costs, and especially suitable for polysaccharide-protein-containing complexes recovering from a starch-containing system. How many stages are needed for a continuous foam separation operation of stripping of the polysaccharide? The feed concentration is 0.3 mol/m^3 of polysaccharide and the residue concentration is 0.05 mol/m^3. The bubble surface concentration excess follows a linear isotherm with a K value of 2 mm (Figure 9.21). Spherical bubbles with an average bubble diameter of 1 cm was used to create the foam. The upward superficial velocity is 200 m^3/hr and the gas flow rate is 500 m^3/hr. The feed rate is 600 m^3/hr. The foamate rate is 200 m^3/hr. What is the bottoms discharge rate when the product removal rate is 500 m^3/hr? What is the concentration of the polysaccharide in the foamate?

(**Ans:** 400 m^3/hr; 3 stages; 0.54 mol/m^3)

14. *Foam Separation of Microalgae*

Microalgae are unicellular organisms, which produce oxygen by photosynthesis. Over 100,000 species of microalgae are known and discovering new uses for them is a major component in the development of industries based on biotechnology. Microalgae are particularly useful because of their high growth rate and tolerance to varying environmental conditions. Microalgae have uses in the production of vitamins, pharmaceuticals, natural dyes, as a source of fatty acids, proteins and other biochemicals in health food products.

Factors derived from microalgae have also been claimed to prevent neuro-degenerative diseases such as Alzheimer's and macular degeneration, which leads to blindness. They are effective in the biological control of agricultural pests; as soil conditioners and biofertilisers in agriculture; for the production of oxygen and removal of nitrogen, phosphorus and toxic substances in sewage treatment; and in biodegradation of plastics. Microalgae also have use as a renewable biomass source for the production of a diesel fuel substitute (biodiesel) and for electricity generation. Burning of fossil fuels in power plants is a primary contributor to excess carbon dioxide in the atmosphere, which has been linked to global climatic change. Release of carbon dioxide into the atmosphere can be significantly reduced by operation of microalgae fuel farms in tandem with fossil fuel plants to scrub CO_2 from flue gases.

If the microalgae are used to produce fuel, a mass culture facility reduces the CO_2 emission from the power plant by approximately 50%. Due to the wide range of uses of microalgae and microalgae-based products, an effective method of harvesting microalgae is essential. The effective separation of microalgae from water is a crucial step in this process. An efficient and cost-effective method of obtaining dry, concentrated biomass from an aqueous solution of microalgae, without causing the cells to be ruptured. A three-stage process, comprising flocculation, flotation and dehydration, is used to separate the biomass. Enterprises engaged in growing microalgae of all types and therefore for all applications, including food and pharmaceutical products, can use this process. It can be adapted towards specific species if necessary. The system is cheaper and faster than the currently available methods and retains many of the properties of the microalgae which are otherwise lost in conventional technologies.

The system is simple to use and inexpensive to maintain. The separator has no internal moving parts. No special operator training is required in order to operate and maintain the system.

Microalgae suspension from a reservoir is passed to a mixer unit where flocculation occurs. The flocculated suspension is then directed to a flotation column of adjustable height into which CO_2 (or air) is fed through a disperser, producing bubbles of uniform size. The bubbles carry electrostatically adsorbed flocs to the surface of the liquid, forming a foam layer, which is skimmed off at the top through an overflow outlet. Purified water is discharged through the bottom. Microalgae are filtered through cloth, dried and packed. Solid biomass is passed through a filtration unit and further dried in a drying chamber. A feature of the process is the telescopic design of the column, which allows the height to be adjusted so that the position of the overflow outlet corresponds to the position of the foam layer, resulting in efficient removal of foam.

The surface excess obeys the linear isotherm of a K value of 0.011 m (Figure 9.22). The reflux ratio used was 3.5. Given a foamate rate of 600 m^3, and an initial feed concentration of 0.8 mol/m^3, residue composition of 0.1 mol/m^3, how many stages are needed for the operation. Bubble sizes are 1 cm. What is the interstitial flow rate? The feed rate is 1000 m^3/hr and the residue rate is 500 m^3/hr. A gas flow rate of 500 m^3/hr is used. What is the reflux return to the column? What is the foamate concentration?

(**Ans:** 1 stage; U = 400 m^3/hr; 215 m^3/hr; 1.5 mol/m^3)

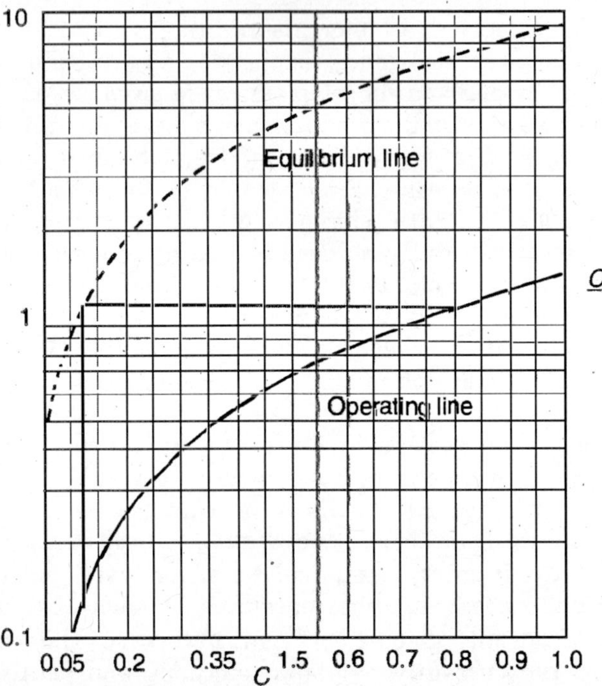

FIGURE 9.22 Combined graph of equilibrium line and operating line for foam separation of microalgae from a suspension.

REFERENCES

Astarita, G., Savage, D.W. and Bisio, A., 1983, *Gas Treating with Chemical Solvents*, Wiley, New York.

Carleson, 1989, *Surfactant Based Separation Processes,* Marcell Dekker, New York.

Cussler, E.L, 1997, *Diffusion: Mass Transfer in Fluid Systems*, Cambridge University Press, UK.

Dognon and Dumontet, 1941, *Comptes Rendus*, **135**, 884.

Dorman and Lemlich, 1965, *Nature*, **207**, 145.

Gaudin, 1957, *Flotation*, McGraw Hill, New York.

Gibbs, 1928, *Collected Works*, Longmans Green, New York.

Kaiser, V., 1994, *Chem. Eng. Prog.*, **90**, 6, 55.

Karger, Grieves, Lemlich, Rubin and Sebba, 1967, *Separation Science*, **2**, 401.

Kobe and McKetta, 1963, *Advances in Petroleum Chemistry and Refining*, **7**, Interscience, New York.

Kohl, A.L. and Riesenfield, F.C., 1985, *Gas Purification,* Gulf Publishing, Houston.

Lemlich, 1966, *Chem. Eng.*, **73**, 7.

Lemlich, 1972, *Adsorptive Bubble Separation Techniques*, Academic, New York.

Leva, M., 1954, *Chem. Eng. Prog. Symp. Ser.*, **50**, 51.

Mahne and Pinfold, 1968, *J. Appl. Chem.*, **18**, 52.

Perry, R.H. and Green, D.W., 1997, *Perry's Chemical Engineers Handbook*, Seventh Edition, McGraw Hill, New York.

Rousseau, 1987, *Handbook of Separation Processes,* Wiley, New York.

Sebba, 1959, *Nature*, **184**, 1062.

————, 1962, *Ion Flotation*, Elsevier, Amsterdam.

Sharma, K.R., October, 2000, *Overall Efficiency in Multicomponent One Component Countercurrent Absorption Tray Tower using Calculus of Finite Differences, 52nd Southeast Regional Meeting of the ACS, SERMACS*, New Orleans, LA, USA.

————, June, 2001, *Gas Absorption with Rapid Chemical Reaction, 33rd ACS Central Regional Meeting / Great Lakes Meeting*, Grand Rapids, MI, USA.

————, May, 2002, *Distribution Diagram Represented with Error Function, Middle Atlantic Regional Meeting of ACS, MARM 02*, Fairfax.

————, September, 2003, *Can the Mole Fraction in Gas Phase be Less than Mole Fraction in Liquid Phase during Absorption?, 226th ACS National Meeting*, New York.

Sherwood, T.K., 1937, *Absorption and Extraction,* McGraw Hill, New York.

________, T.K., Pigford, Wilke, 1974, *Mass Transfer,* McGraw Hill, New York.

________, T.K., Shipley, G.H., and Halloway, F.A.L., 1938, *Ind. Eng. Chem.*, **30**, 765.

Smith, 1963, *Design of Equilibrium Stage Processes*, McGraw Hill, New York.

Strigle, R.F., 1987, *Random Packings and Packed Towers*, Gulf Publishing, Houston.

Wadsworth and Davis, 1964, *Unit Processes in Hydrometallurgy*, Gordon and Breach, New York.

CHAPTER 10

Solution Crystallisation and Metal Alloying

Nomenclature

A_N	Avagadro number (6.023 E23 molecules/gmol)
B^0	nucleation rate (number/cm^3/s)
c'	pre-exponential frequency factor
C	concentration of solute in solution (mol/m^3)
C_s	concentration of solute in saturated solution (mol/m^3)
ΔC	molar supersaturation (mol/m^3)
L	crystal size (m)
PD	particle diameter (m)
R	universal gas constant (J/mol/K)
$100s$	percentage supersaturation
v_m	molar volume of crystal (m^3/mol)
y	mole fraction of solute in solution
y_s	mole fraction of solute in saturated solution

Greek

ρ_M	molar density of solution (mol/m^3)
ρ_s	molar density of saturated solution (mol/m^3)
σ	interfacial tension between solid and liquid (N/m)
v	number of ions per molecule of solute
χ	concentration ratio (C/C_s)
ψ	slope of the temperature vs solubility curve (K)

10.1 INTRODUCTION TO SOLUTION CRYSTALLISATION

A *crystal* is the most highly ordered, organised type of non-living matter. The constituents, be it particles, atoms, molecules or ions, are arranged in a regular, three-dimensional array called *space lattices*. The interatomic distances in a crystal of any material is a constant and a signature of the material. The arrangement of the constituents is the same in all the three directions. When crystals are allowed to form without hindrance from other crystals or outside bodies, they appear as polyhedrons having sharp corners and flat sides or faces. *Crystallisation* refers to a process where solid particles that are highly ordered fall out from a supersaturated solution. The formation of the well ordered crystalline phase occurs in a homogeneous phase. It is one of the methods of obtaining a solute from a solution in the solid form at a controlled level of purity and yield. The *supersaturation* is eliminated by the formation and growth of the crystals. The driving force for crystallisation is the concentration difference between the supersaturated solution and saturated solution that ensues after the precipitation of the crystal particles. These can be removed by mechanical means. Although solid particles can sublime from the gas phase such as in snow fall, of particular interest is the liquid-solid systems.

Crystallisation is becoming an important operation in the chemical and biotechnological industry. The chemicals that can be marketed and distributed in the form of crystals are increasing. Its highly purified state attracts the attention of many a customer. These can be obtained in a single step from impure solutions. Proteins form crystalline structure. In the downstream processing from fermentation, crystallisation is an important unit operation. It is used in places where distillation, absorption methods may not be sufficient to achieve the desired objectives. Metal crystallisation is also an important industrial application of crystallisation. This is a solidification process compared with common industrial crystallisation processes that are performed from solutions.

Proteins are present in many commercially available preparation forms. In particular, proteins are present as the active ingredient in some pharmaceutical medicaments. For the preparation of these preparation forms, it is beneficial to use the proteins in crystalline form. Besides being easier to handle, proteins as dry crystals are more stable than, for example, proteins in dissolved form. As an example, the preparation of a pharmaceutical preparation containing the hormone *insulin* as an active ingredient, uses the *protein insulin in crystalline form*. Crystallised insulin is stable, for example, at a temperature of –20°C over a number of years. Crystalline forms of proteins having a molecular weight of up to several hundred thousand daltons and peptides having a lower molecular weight are known. The amino acid sequence of the proteins can either be identical to a naturally occurring sequence or it can be changed relative to the natural form. In addition to containing amino acid chains, the proteins can also contain sugar radicals or other ligands as side chains. The proteins may be isolated from natural sources, or the protein may be prepared by genetic engineering or synthetically, or the proteins may be obtained by a combination of these processes. In aqueous solution, proteins have a three-dimensional structure of greater or lesser complexity which is based on a specific spatial folding of the amino acid chains. The intact structure of a protein is essential for its biological action. During crystallisation of proteins from aqueous solutions, this structure is largely retained. The crystal structure of a protein is predominantly determined by its amino acid sequence. Another factor determining crystal structure is, inclusion of low molecular weight substances such as metal ion salts and water molecules. In particular, the presence of a certain amount of intracrystalline water molecules (water of crystallisation) is necessary for the stability of the crystal structure of proteins.

The driving force in the crystallisation is the difference between the concentration of the dissolved substance in the supersaturated solution and that in the just saturated solution, i.e. *the disturbance of the solution equilibrium*. The pre-concentrated solution must therefore first be saturated and then supersaturated beyond the saturation point. The supersaturation can be achieved in practice by one of the following four ways:

(a) Cooling Crystallisation
(b) Vacuum Crystallisation
(c) Evaporative Crystallisation
(d) Reactive Crystallisation

In the case of high temperature dependence of the solubility, the supersaturation is achieved by simply cooling the saturated solution by surface cooling such as in *cooling crystallisation*. If there is only a slight temperature dependence of the solubility, the solution is supersaturated by evaporating the solvent such as in *evaporative crystallisation*. If the solubility depends to a marked extent on the temperature or if the solution has to be treated in a gentle manner thermally, solution cooling and solvent evaporation are combined such as in *vacuum crystallisation*. In *reactive crystallisation* the solvent is consumed in the reaction such as polymerisation of styrene leaving the dissolved polybutadiene in a supersaturated state.

As a rule, the elimination of supersaturation takes place in two simultaneous steps. In the first step, crystal nuclei are formed; in the second step, those crystal nuclei which are larger than a critical minimum size grow into very coarse product crystals by taking up solid from the supersaturated solution. The nucleation rate increases with increasing supersaturation. As a rule, spontaneous formation of many small nuclei is observable after a specific supersaturation has been exceeded. This effect is evident in the formation of fine crystal showers. The crystallisation process is, as a rule, to prepare a crystalline, saleable product of uniform quality, the product acquiring this quality in particular through the crystal size distribution.

There are 5 main types of crystals and seven classes of crystallographic systems. These 7 systems are: Cubic, Hexagonal, Trigonal, Tetragonal, Orthorhombic, Monoclinic and Triclinic.

One substance can exist in two forms. For example, calcium carbonate $CaCO_3$ can be in hexagonal calcite form or in orthorhombic aragonite form. *Crystallography* refers to the study of the spatial arrangement and structure formation. *Magma* refers to the two-phase mixture, in solution crystallisation, of the mother liquor and crystals of all sizes which occupy the crystalliser and is withdrawn as a product. An *invariant* crystal denotes a growing crystal that maintains geometric similarity during growth.

10.2 SUPERSATURATION, PHASE DIAGRAM AND CONSIDERATIONS

The solubility curves of solutes in solution can be used as equilibrium data for bulk crystals. The solubility may change with the size of the solute and care should be taken to ascertain the experimental conditions under which the solubility curve was developed. The solubility as a function of temperature is also valuable information. The solubility change with an increase in temperature cannot be generalised. Substances such as potassium nitrate dissolve in increasing measure in water when the temperature of the system is increased whereas other substances such as the monohydrate of manganese sulphate decreases in solubility with increase in temperature. The solubility of sodium chloride does not change much with temperature in the range of 0–100°C for water. Many a inorganic substance crystallise with water of crystallisation. In some systems a few different hydrates are formed. The phase equilibriums in such systems are not simple.

For example, the phase diagram of the system Zinc Sulphate—water is shown in Figure 10.1. The concentration in mass fraction of anhydrous zinc sulphate is plotted against the equilibrium temperature. The region to the right of *pabcdq* is one homogeneous solution. The solidification of the different phases can be seen to the left of *eagfhj*. The solid formed at *a* is the eutectic. It consists of an intimate mixture of ice and $ZnSO_4.7H_2O$. The area *pae* represents a mixture of ice and saturated solution. Ice is formed in the region to the left of *pa*.

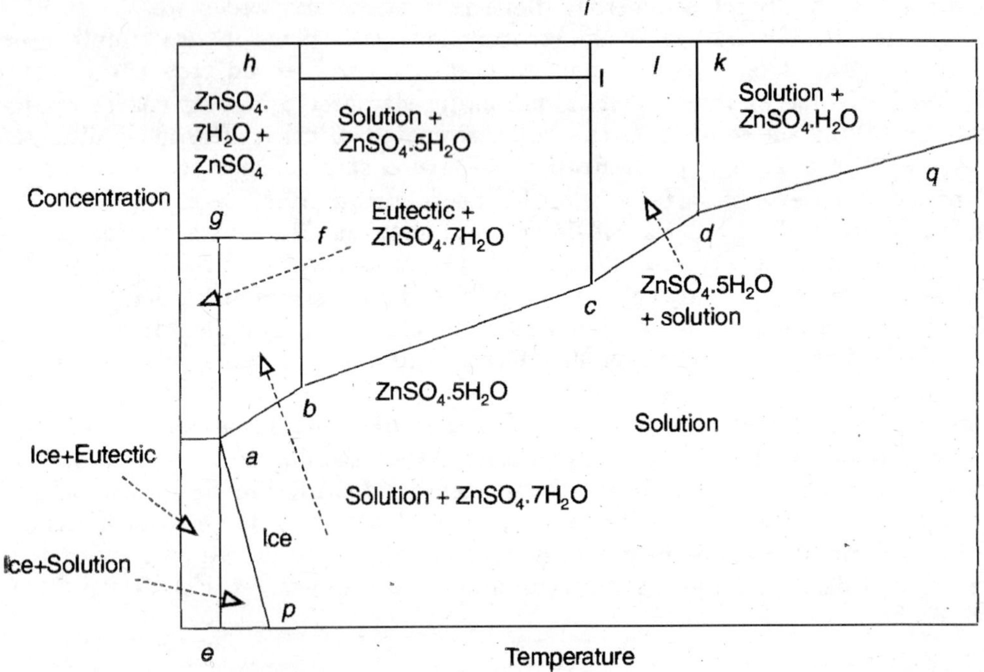

FIGURE 10.1 Phase dagram of zinc sulphate.$7H_2O$.

Some of the salient considerations in the design of crystallisation operations are:

(a) Good yield
(b) High purity
(c) Size range of crystals
(d) CCD, Crystal size distribution
(e) Control of crystal size during the process
(f) Energy requirements
(g) Rates of nucleation and growth

(a) *Yield from Crystallisation Operation:* As usually the case is, mother liquor and crystals are in contact with each other for sufficient period of time and as a result equilibrium condition can be assumed. The yield of the process can be calculated from the concentration of the original solution, and the solubility at the final system temperature. When calculating, water of crystallisation can be taken into account. For evaporative crystallisers, water lost can be factored into the analysis.

Worked Example 10.1 *Yield of $ZnSO_4$ Crystals*
A solution consisting 27% $ZnSO_4$ and 73% water is cooled to 15°C. During cooling, 7% of the total water in the system evaporates. How many kg of crystals are obtained per kg of original mixture?

Crystals are $ZnSO_4.7H_2O$
From Figure 10.1, at 15°C, 23% anhydrous $ZnSO_4$ and 77% water is present.

Molecular weight of $ZnSO_4$ = 161.4
Molecular weight of $ZnSO_4.7H_2O$ = 287.4

For 1000 kg of solution, total water = 730 kg
Water evaporated = 51.1 kg

$ZnSO_4.7H_2O$ present = 1000 × .27 × (287.4)/161.4 = 480.8 kg

Free water = 1000 – 51.1 – 480.8 = 468.1 kg

In 100 kg of mother liquor = .23 × 100(287.4/161.4) = 12.9 kg $ZnSO_4.7H_2O$

Free water = 100 – 12.9 = 87.1 kg
= 12.9/87.1 × 468.1 = 69.3

Final Crop = 480.8 – 69.3 = 411.4 kg.

(b) *Purity of the Product:* As usually is the case, the rate processes are low and the products are usually 99.5% to 99.8% pure. The remaining 2000 ppm to 5000 ppm comprises the mother liquor and occlusions within the crystal. The occlusions within the crystal can be viewed under a microscope as dislocations in the crystal. The mother liquor adheres to the crystal as a film and forms a residual component that is not completely removed after the separation of the crystals from the liquid.

(c) *Crystal Size Distribution:* The crystal size distribution is independent of the mass and energy balance equations. Whether the crystal is a single huge one or a collection of small ones, the balance equations are the same. The crystal size distribution is an important parameter in the performance of a crystalliser. The cumulative % is plotted against the particle diameter. A new parameter called the coefficient of variation is defined as follows:

$$CV = \frac{100\,(PD_{16\%} - PD_{84\%})}{2\,PD_{50\%}} \tag{10.1}$$

The control of crystal size in the crystalliser can be achieved by removing some crystals and recycling some. This can change the equilibrium characteristics of the system.

(d) *Energy Requirements:* The heat evolved when solid forms from a solution can be calculated from latent heat information. Crystallisation is usually exothermic and the heat of crystallisation varies with temperature and concentration. Enthalpy vs concentration charts can be constructed as shown in Figure 10.2. Sometimes solid phases are also included in plots similar to Figure 10.2. The information in Figure 10.2 was

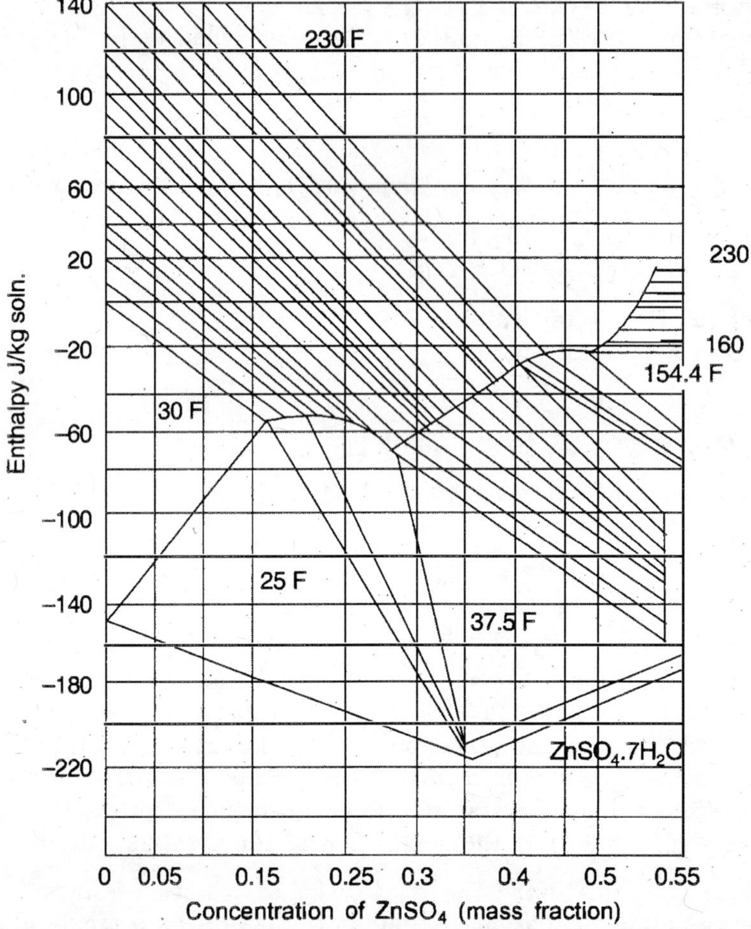

FIGURE 10.2 Enthalpy concentration diagram for $ZnSO_4.7H_2O$.

extrapolated from the information available from $MgSO_4$ and must be used with care. This figure is consistent with the earlier graph shown in Figure 10.1.

(e) *Degree of Supersaturation:* As discussed earlier, supersaturation can be generated in one of the four methods. In the case of high temperature dependence of the solubility, the supersaturation is achieved by simply cooling the saturated solution by surface cooling such as in cooling crystallisation. If there is only a slight temperature dependence of the solubility, the solution is supersaturated by evaporating the solvent such as in evaporative crystallisation. If the solubility depends to a marked extent on the temperature or if the solution has to be treated in a gentle manner thermally, solution cooling and solvent evaporation are combined such as in vacuum crystallisation. In reactive crystallisation, the solvent is consumed in the reaction such as polymerisation of styrene leaving the dissolved polybutadiene in a supersaturated state.

Supersaturation is the concentration difference between that of the supersaturated solution in which the crystal is growing and that of the solution in equilibrium with the crystal. The two phases are nearly at the same temperature. The supersaturation can be given by:

$$\Delta C = C - C_s \tag{10.2}$$

$$\Delta y = y - y_s \tag{10.3}$$

The two supersaturations defined in terms of the concentration difference and difference in mole fractions between the solution and saturated solution can be related to each other by using the molar densities as follows:

$$\Delta C = \rho_M y - \rho_s y_s \tag{10.4}$$

In crystallisers usually, ρ_M and ρ_s can be considered equal and Eq. (10.4) becomes

$$\Delta C = \rho_M \Delta y \tag{10.5}$$

The concentration ratio χ and fractional supersaturation are defined by

$$\chi = \frac{C}{C_s} = 1 + \frac{\Delta C}{C_s} = \frac{y}{y_s} = 1 + \frac{\Delta y}{y_s} = 1 + s \tag{10.6}$$

When the solubility of a solute in solution increases with increase in temperature in significant amount, the supersaturation can be expressed as an equivalent temperature difference rather than concentration difference. The change of solubility with temperature is assumed to be linear and,

$$\Delta T_c = T_s - T_c = \psi(y - y_s) = \frac{\psi}{\rho_m \Delta C} \tag{10.7}$$

Figure 10.3 shows the crystal growth and quality nucleation type in $ZnSO_4.7H_2O$.

Temperature °C Saturation, Ts	Growth	Nucleation	
		Absence of crystal-solid contact	Presence of crystal-solid contact
T ↓	Good growth	No nucleation	Contact nucleation
	Veiled growth		
	Dendritic spikewise brooming growth	Splintering	Attrition of colliding crystals
		Heterogeneous nucleation	

FIGURE 10.3 Crystal growth and quality and nucleation type in $ZnSO_4.7H_2O$.

10.3 RATE OF NUCLEATION

The kinetic parameter that can be attributed as a means to control the CSD, crystal size distribution, in the product is the *nucleation rate*. Its units are number of new particles formed per unit time per unit volume of magma or solids-free mother liquor. Four kinds of nucleation can be identified. These are:

(a) Homogeneous Nucelation
(b) Heterogeneous nucleation
(c) Crystal Contact with Wall
(d) Crystal Contact with Pump Impeller

Nucleation is the birth of small particles of a new phase within a supersaturated homogeneous existing phase. The nucleation phenomenon remains the same regardless of whether it is solution crystallisation, melt crystallisation, condensation of fog drops in a supercooled vapour or generation of bubbles in a superheated liquid. Nucleation is a consequence of local fluctuations that is rapid on a molecular scale within a homogeneous phase that is in a state of metastable equilibrium. This phenomenon is called the *homogeneous nucleation*. The formation of particles is not on account of foreign particle present or by collisions with the wall of the container. When foreign particles catalyze the nucleation rate, it is heterogeneous nucleation.

In industrial crystallisers, homogeneous nucleation seldom occurs. The sequence of stages in the evolution of a crystal are:

$$\begin{array}{ccc} \text{Cluster} \rightarrow \rightarrow & \text{Embryo} \\ & \Downarrow \\ \text{Crystal} \Leftarrow & \text{Nucleus} \end{array}$$

Thermodynamically, the difference between a small particle and a large one at the same temperature is that the small particle possesses a significant amount of surface energy per unit mass and the large ones do not. *Ostwald ripening* is when the large crystal grows at the expense of small crystals. The solubility dependence on particle size is given by the Kelvin equation,

$$\ln(\chi) = \frac{4v_m\sigma}{\nu RTL} \tag{10.8}$$

The rate of nucleation from the kinetic theory is given by,

$$B^0 = c' \exp\left(\frac{-16\pi\sigma^3 v_m^2 A_N}{3\nu^2(RT)^3 (\ln\chi)^2}\right) \tag{10.9}$$

where B^0 is the nucleation rate in number/cm^3/s, A_N is the Avagadro number, c' the frequency factor and R the molar gas constant. The factor c' is a statistical measure of the rate of formation of embryos that reach the critical size. It is proportional to the concentration of the individual particles and to the rate of collision of these particles with an embryo of the critical size required to form a stable nucleus. From analogy with nucleation of water drops from supersaturated water vapour, it is of the order of 10^{25} nuclei/cm^3/s. The experimental determination of solid-liquid interfacial tension is difficult. The values of interfacial tension σ in Eq. (10.9) is 80–100 ergs/cm^2.

In heterogeneous nucleation, the catalytic effect of the foreign particles on nucleation rate is the reduction of the energy required for nucleation. Here,

$$B^0 = 10^{25} \exp\left(\frac{-16\pi\sigma^3 v_m^2 A_N}{3\nu^2(RT)^3 s^2}\right) \tag{10.10}$$

where 100s is the percentage of supersaturation.

It has been known for a long time that contact nucleation is influenced by the intensity of agitation.

10.4 CRYSTALLISATION EQUIPMENT

Over time with different application needs and development in technology different types of crystallisation equipment have been developed in the industry in the past several years. Some of the main types of equipment that are popular are as follows:

(a) Mixed Suspension, Mixed Product Removal Crystalliser, MSMPR
(b) Forced Circulation Evaporator Crystalliser
(c) Draft Tube Baffle Evaporator Crystalliser, DTB
(d) Draft Tube Crystalliser, DT
(e) Surface Cooled Crystalliser
(f) Direct Contact Refrigeration Crystalliser
(g) Reaction Type Crystalliser

(h) Mixed Suspension, Classified Product Removal Crystalliser
(i) Classified Suspension Crystalliser
Growth or Oslo Crystalliser
Scraped Surface Crystalliser
(j) Double Pipe Scrapped Surface Crystalliser
Armstrong Crystalliser
Batch Crystallisation
(k) Recompression Evaporation Crystallisation

A continuous forced circulation crystalliser is shown in Figure 10.4. It is much like a simple forced circulation evaporator, but it includes specific features to allow correct crystallisation, namely:

- an 'active volume' designed case by case, to get both required residence time for crystal growth and mother liquor desupersaturation;
- a given agitation (recirculation rate) rated to control the extent of supersaturation arising from the evaporation, and to keep the temperature difference in the heat exchanger within reasonable limits;
- a special design of the liquid-vapour separation area to minimise the carry over losses and avoid the formation of an excessive amount of fines, which is highly detrimental to crystal growth.

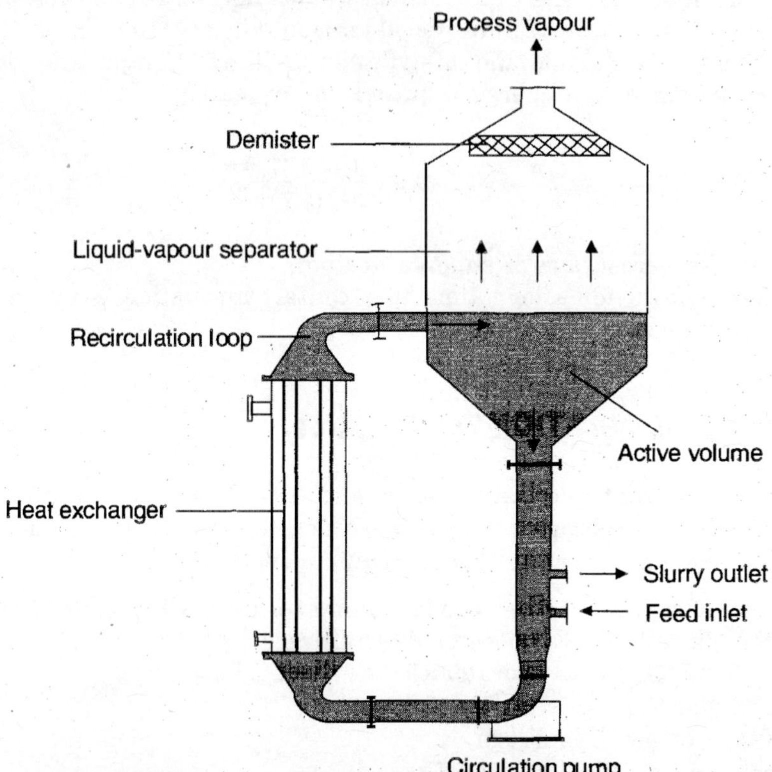

FIGURE 10.4 Forced circulation crystalliser.

Depending upon specific process requirements, additional devices can be provided. They are:

- internal baffles, used mainly for excess mother liquor overflow and/or withdrawal of fines when crystal growth is slow or disturbed by impurities build-up (see Figure 10.4),
- elutriation leg to improve product purity and deliver a narrow crystal size distribution,
- an internal scrubbing section to reduce to very low value the carry over losses, or even to provide stripping or absorption devices when a volatile compound must be recovered.

Forced circulation crystallisers are of the MSMPR type (Mixed Suspension Mixed Product Removal) and operate either on controlled or 'natural' slurry density depending upon process requirements and/or unit material balance. These units can be either single or multiple effect and the vapour recompression concept (either thermal or mechanical) is often applied. Usually, they operate from low vacuum to atmosphere pressure. As a rule, these units are used for high evaporation rates and when crystal size is not of utmost importance or if crystal grows at a fair rate. Almost any material of construction can be considered for the fabrication of these crystallisers. It is worth bearing in mind that the heating element is omitted for vacuum cooling crystallisers. Typical products are:

- NaCl (food or technical grade)
- KNO_3
- Na_2SO_4, K_2SO_4
- NH_4Cl
- $Na_2CO_3.H_2O$
- Citric acid

When the problem of scaling impedes the process of concentration, a design similar to the one described above is proposed. This applies for $CaSO_4$ saturated solutions like fertiliser grade phosphoric acid, demineralisation effluents and vinasses.

The Draft Tube Baffle Crystalliser is an elaborated Mixed Suspension Mixed Product Removal (MSMPR) design, which has proven to be well suited for vacuum cooling and for processes exhibiting a moderate evaporation rate. The concept is such that if no (or little) heat make-up is required, it results in a rather compact arrangement; therefore the initial investment is minimised. As a rule, these units operate with a rather low supersaturation, which is sometimes a limitation to crystal growth, so that very large crystals can be produced only by providing extensive and costly dissolving of fines. The Draft Tube Baffle unit (Figure 10.5) includes a baffled area (settling zone), peripheral to the active volume, from where excess of mother liquor and/or fines are removed for further processing. The necessary agitation of the suspension mixed with the incoming feed solution is provided by a bottom entry agitation at moderate energy consumption. Draft Tube Baffle crystallisers are often equipped with an elutriation leg below the body to classify the crystals. When destruction of fines is not needed or wanted, baffles are omitted and the internal circulation rate

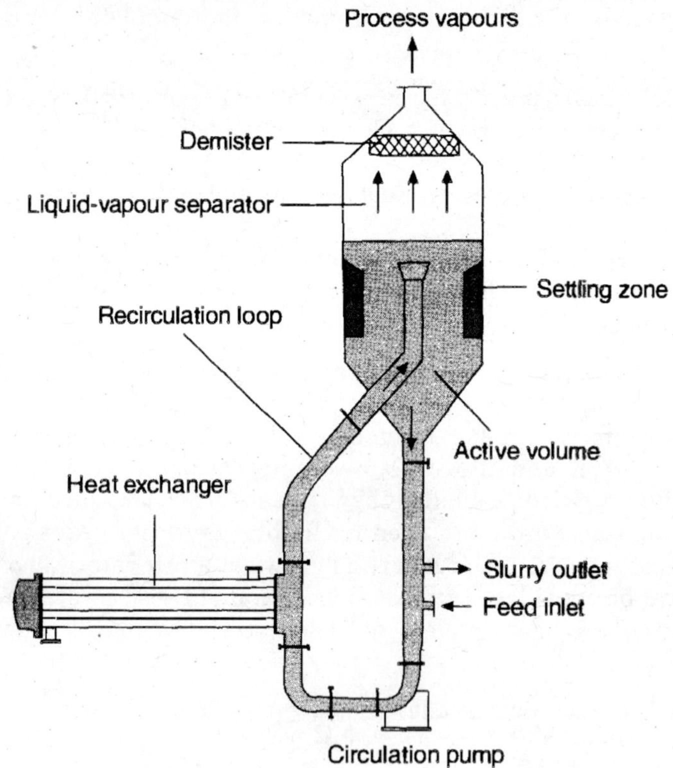

FIGURE 10.5 Draft tube Baffle crystalliser.

is set to have the minimum nucleating influence on the suspension (Draft Tube design, draft-tube crystalliser). When large evaporation rates are required, an external heating loop must be provided, making the arrangement less competitive from an initial cost standpoint. The Draft Tube Baffle Crystalliser, which can be considered when crystallisation can be achieved with natural suspension, has proven to be well suited to many applications such as:

- boric acid
- $Na_2SO_4.10H_20$ (Glauber salt)
- melamine
- citric acid
- $NaClO_3$

The Kestner induced circulation crystalliser design has been recently developed to provide additional agitation of the active volume of forced circulation crystallisers with the use of only one pump. This Kestner unit, as shown in Figure 10.6, operates similarly to a Draft Tube Baffle crystalliser but without the internal agitation device. The main applications are for evaporative crystallisation cases. The unit also operates according to the Mixed Suspension Mixed Product Removal principle and all options described for the other designs are of course available for this concept. The equipment

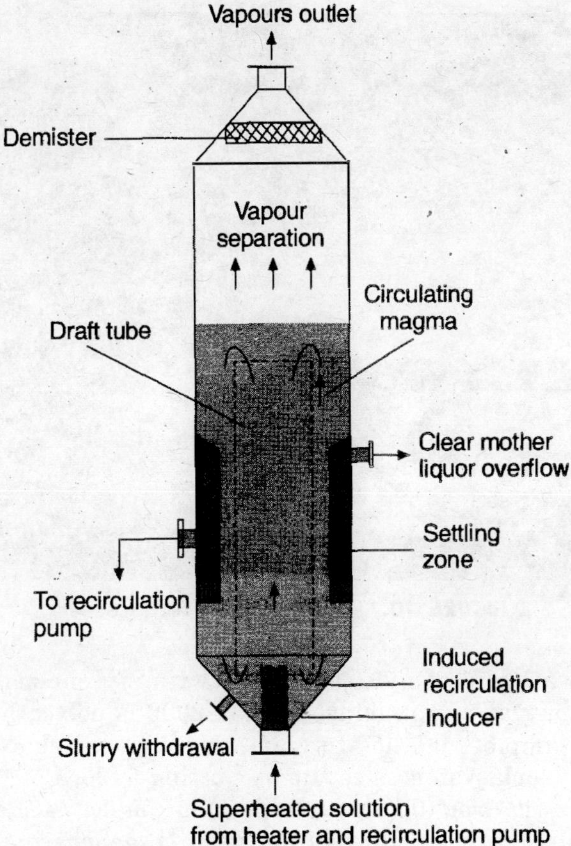

FIGURE 10.6 Induced circulation crystalliser.

is able to produce a narrow crystal size distribution. Like other designs, it can be fabricated in almost any material of construction. Performances and product quality are equivalent to those of a Draft Tube Baffle unit designed to the same specification, but appear to be limited to non-aqueous solutions as the induced flow would be quite limited when the mother liquor exhibits high viscosity. It has been successfully applied to the following productions:

- $NaCl$
- NH_4ClO_4
- NH_4Cl

10.5 METAL ALLOYING

Diffusion is used to prepare alloys of different compositions in the metal industry (Figure 10.7). Recently polymer alloys have also been reported in the industry and shown to have attractive performance properties. For example, the Triax 1000 alloy

FIGURE 10.7 Steel making furnace.

from Monsanto used a blend of nylon and ABS with styrene-acrylonitrile-glyceidel methacrylate terpolymer as compatibiliser. Triax 2000 is polycarbonate/SAN blends.

Difffusion is the simplest method to create a wide spectrum of compositions at one time. A diffusion block system is set up by putting a block of 1.2 wt% carbon and another block of 99.75 wt% iron (0.25 wt% Carbon) in contact as shown in Figure 10.8. At a given temperature, carbon diffuses into the iron, generating a carbon gradient across the system as shown in Figure 10.9. The location of different compositions can be determined through mathematical modeling of the experiment. Once a general position for the desired composition is determined, the cross-section of the block is removed and tested through X-ray fluorescence to determine the relative per cents of the two elements present. This method of determining the composition is preferable since it is non-destructive. Other method is the wet chemical analysis in which chemistry is used to find the composition and other types of spectroscopy such as the mass spectrometer. Once the composition of the sample is determined, it can be examined at different temperatures to find the phases present. Since this method produces a limited amount of sample, it lends itself very well for the determination of these phases by *X*-ray diffraction.

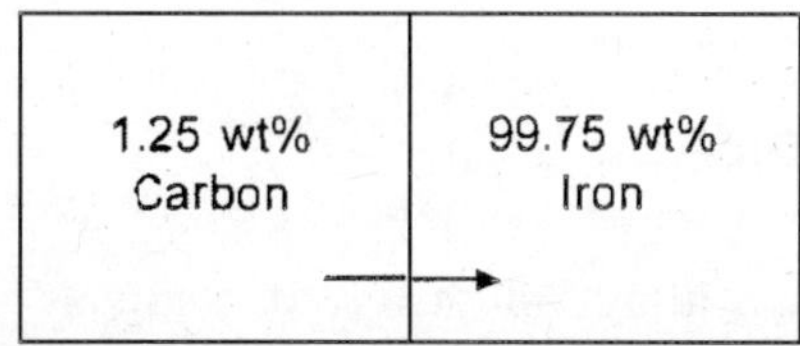

FIGURE 10.8 Diffusion block set-up before diffusion takes place.

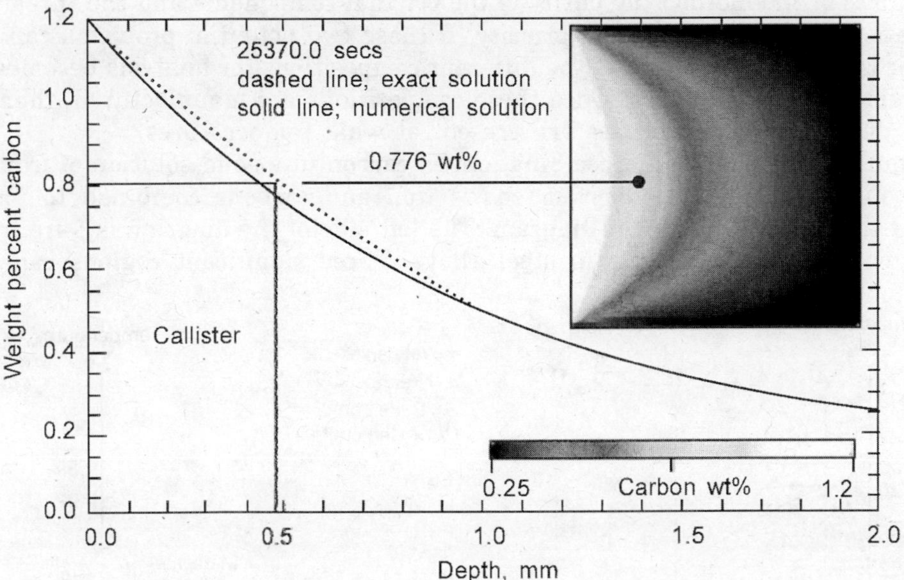

FIGURE 10.9 Composition gradient for iron/carbon diffusion.

Before any determination of steel phases or microstructure can be attempted, there must be a way to obtain samples of different iron/carbon compositions. Industry generally uses two methods: diffusion and individual composition production. If the compositions are not correct then any data gathered will be faulty. Constructing a full phase diagram by these methods can take one year for a two-component system, three years for three components, and up to ten years for four components, so workers must show great care in getting exact composition for a small margin for error.

In either method starting material purity is very important. For example steel production requires graphite carbon and iron. There must be no impurities in either of the constituents. Even the smallest impurity can cause the sample to be much different than that of the phase-equilibrium relations that were intended to be studied. Care must also be taken to assure that none of the instruments used to mix the different compositions contaminates the samples, so any material that comes in contact with the sample should not react with the sample. After the preliminary handling of the ingredients, it is time to create the different compositions. More detail on a certain composition requires a large amount of that composition. The specific per cents of carbon and iron can be measured and then melted to form that composition. For example if a 50 wt% carbon/50 wt% iron composition is needed even parts by weight would be mixed in a manner minimising impurities, such as oxygen and nitrogen, possible by carrying out the melting in a vacuum furnace and making sure the crucible or other melting vessel is free of contaminants. This method can be used when large volumes of a certain composition are needed to do experimentation such as dilatometry or differential thermal analysis.

Regardless of the method the purity of the original components and the threat of contamination must be a very high priority. If these two potential problems can be kept under control, the creation of the different compositions for analysis becomes a very straight forward operation. Once these compositions are manufactured, then it is time to determine which phases are present at what temperatures.

The equilibrium diagram for combinations of carbon in a solid solution of iron is shown in Figure 10.10. The diagram shows iron and carbons combined to form Fe-Fe_3C at the 6.67%C end of the diagram. The left side of the diagram is pure iron combined with carbon, resulting in steel alloys. Three significant regions can be

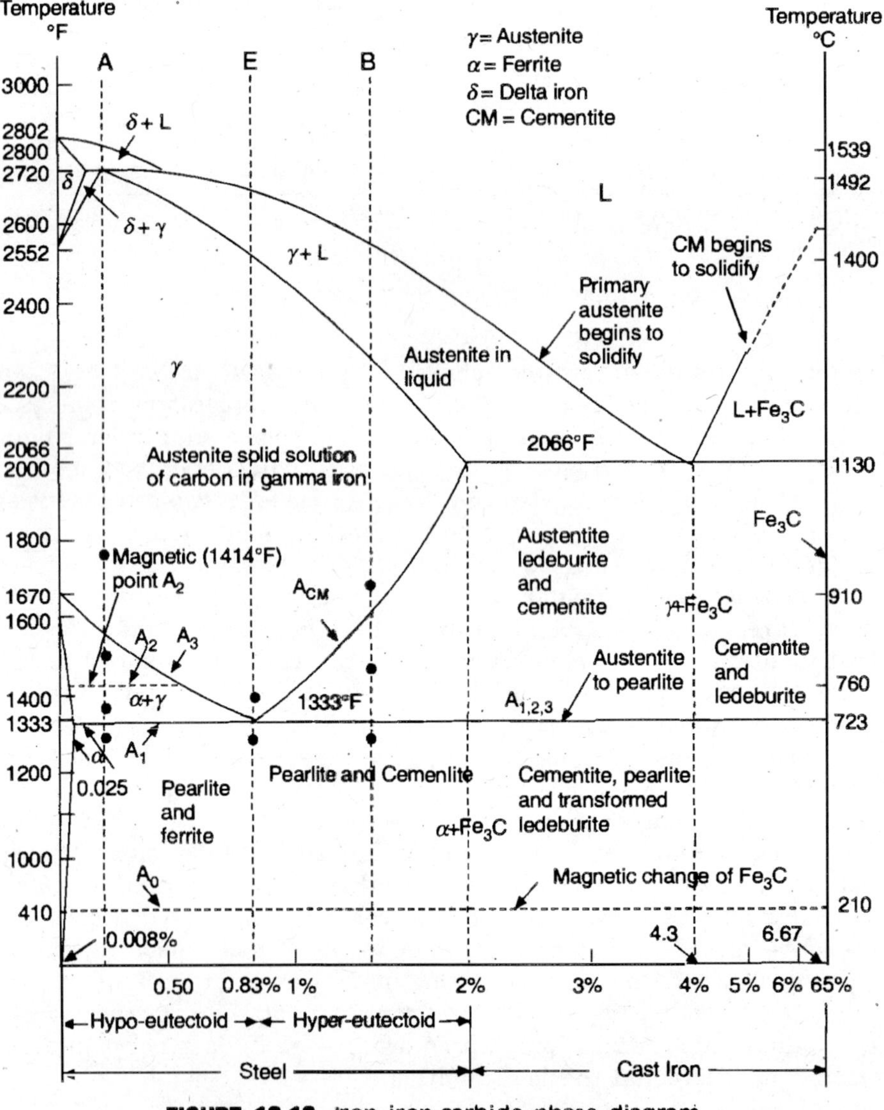

FIGURE 10.10 Iron iron-carbide phase diagram.

made relative to the steel portion of the diagram. They are the eutectoid E, the hypoeutectoid A, and the hypereutectoid B. The right side of the pure iron line is carbon in combination with various forms of iron called alpha iron (ferrite), gamma iron (austenite), and delta iron. The black dots mark clickable sections of the diagram. Allotropic changes take place when there is a change in crystal lattice structure. From 2802°–2552°F the delta iron has a body-centered cubic lattice structure. At 2552°F, the lattice changes from a body-centered cubic to a face-centered cubic lattice type. At 1400°F, the curve shows a plateau but this does not signify an allotropic change. It is called the *Curie temperature*, where the metal changes its magnetic properties. Two very important phase changes take place at 0.83%C and at 4.3%C. At 0.83%C, the transformation is eutectoid, called pearlite.

$$\text{gamma (austenite)} \longrightarrow \text{alpha} + Fe_3C \text{ (cementite)}$$

At 4.3%C and 2066°F, the transformation is eutectic, called ledeburite.

$$\text{L(liquid)} \longrightarrow \text{gamma (austenite)} + Fe_3C \text{ (cementite)}$$

10.5.1 Decarburising Steel

Steel is incredibly versatile because it can be heat-treated in order to produce a vast range of microstructures and associated mechanical properties. The heat-treatment usually involves the steel being heated into a temperature in the austenite phase field. This temperature is quite high in the range 800–1200°C, depending on the details of the chemical composition. Commercial heat treatments are generally carried out in electrical (resistance) furnaces or natural gas fired furnaces. The size of the furnace may range from a moderately sized building to handle many thousands tonnes of steel to a furnace the size of a microwave oven. It is inevitable that the furnace atmosphere contains oxygen. More accurately, the chemical potential of carbon in the atmosphere may be lower than that in the steel being heat treated. Carbon will therefore be removed from the steel by the process commonly known as *Decarburisation*. The chemical composition of steel is approximately Fe-0.8C wt%. It has been heated in an electric furnace, without any particular protection, at 1200°C for 2 hours and then cooled slowly to ambient temperature. The sample of steel is then ground flat on one surface using SiC grinding paper lubricated with water, followed by polishing with fine diamond paste. Once an acceptable polish is obtained, the sample is etched in 2% nital (a mixture of nitric acid and methanol) for 20 seconds before washing with methanol and drying using warm air. The resulting cross-section reveals the extent of decarburisation at the surface, with remarkable changes in microstructure with distance away from the surface.

Decarburised surface (low magnification)

The microstructure at the free surface of the steel. Note that the amount of ferrite (light etching phase) increases as the surface is approached. The ferrite nucleates at the austenite grain boundaries and hence appears as layers. The dark etching regions

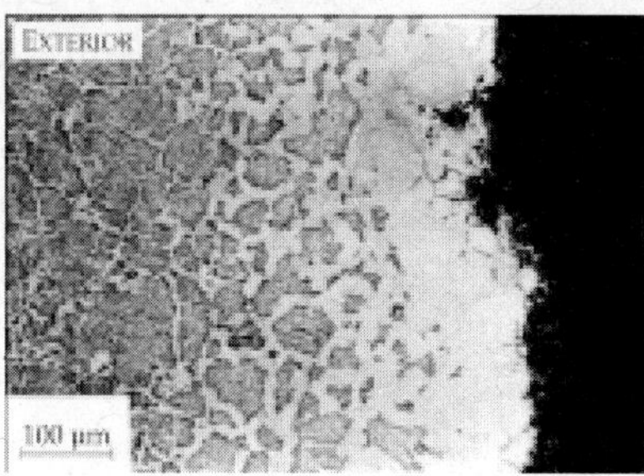

in the figure are mixtures of Widmanstatten ferrite and pearlite which form after the grain boundary layers of ferrite.

Away from surface (low magnification)

The microstructure away from the surface of the steel. There is a much smaller quantity of ferrite.

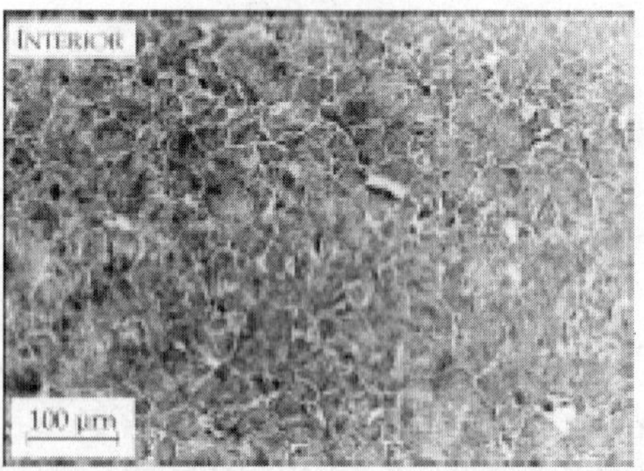

Decarburised surface (high magnification)

The microstructure at the free surface of the steel. At higher magnification, it becomes clear that the top surface is almost completely denuded of carbon. The oxygen has therefore started to oxidise the iron. The oxidation is penetrating the prior austenite grain boundaries because these are high-energy sites. A much higher magnification image is also available to illustrate this phenomenon.

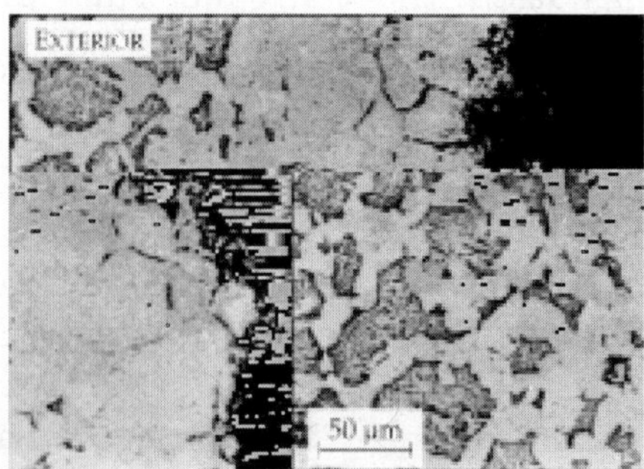

Between surface & interior (high magnification)

An image taken in a region between the surface and the unaffected interior of the sample. The microstructure on the left is representative of a low-carbon steel whereas

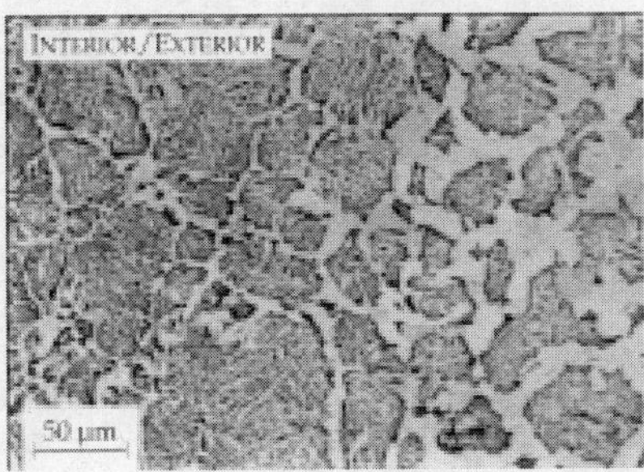

that on the right is of a higher carbon steel. Thus, the allotriomorphic ferrite content decreases, to be replaced by Widmanstatten ferrite as the interior regions are approached.

Away from surface (High magnification)

The microstructure away from the surface of the steel. The Widmanstatten ferrite is more apparent at this higher magnification. Notice that the steel cannot be of exactly the eutectoid composition since there is some ferrite even in the regions which are

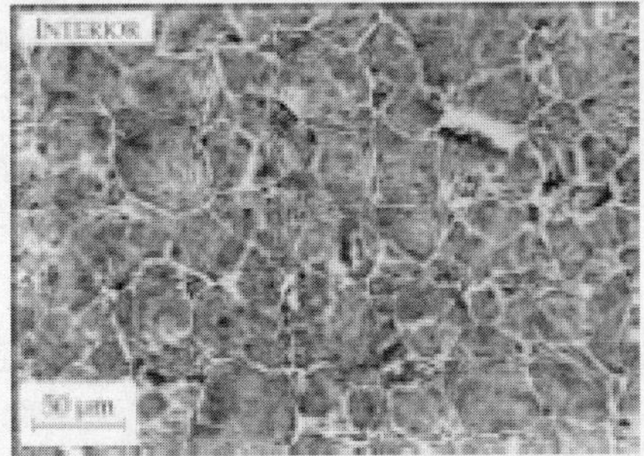

not decarburised. It is better described as hypoeutectoid. A much higher magnification image is also available to illustrate this phenomenon.

SUMMARY

Crystallisation process was defined according to its formation from a supersaturated solution from which it falls out. Nucleation was recognized as an important precursor to crystallisation. Solution crystallisation and melt crystallisation was distinguished from each other.Four different methods of attaining supersaturation state were identified. These are by evaporation of solvent, reaction of solvent, temperature change, and vacuum vaporization of solvent. Industrial applications in crystallisation were given. Five types of crystal and 7 classes of crystallographic systems were identified. The solubility information can be tapped into to develop the phase diagrams. The phase diagram of heptahydrate of $ZnSO_4$ was shown in Figure 10.1. The salient considerations in the design of crystallizer are good yield, high purity, size range of crystals, CCD, Crystal Size Distribution, control of crystal size during the process, energy requirements and rates of nucleation and growth. The enthalpy concentration diagram for the heptahydrate of $ZnSO_4$ was given in Figure 10.2. Degree of supersaturation can be estimated from the temperature difference or mole fraction difference.The quality of the crystal dependence on nucleation type was shown in Figure 10.3. Homogeneous and heterogeneous nucleation rates were recognized and the equations to calculate them were provided.

$$B^0 = c' \exp\left(\frac{-16\pi\sigma^3 v_m^2 A_N}{3v^2 (RT)^3 \ln(\chi)^2}\right) \qquad \text{(homogeneous)}$$

$$B^0 = 10^{25} \exp\left(\frac{-16\pi\sigma^3 v_m^2 A_N}{3v^2 (RT)^3 s^2}\right) \qquad \text{(heterogeneous)}$$

The different types of crystallisation equipment were discussed. Specific examples from the industry were given. Their operating range, benefits, etc. were discussed. The role of diffusion in metal alloying was highlighted. The Iron–Iron Carbide phase diagram was shown in Figure 10.10. Be it the slab casting of iron in manufacture of stainless steel or decarburization of steel to control the content of carbon, the role of mass transfer and diffusion were clearly identified.

EXERCISES

1. In the formation of snow, where is the impure solution?
2. What is the difference between melting point and crystalline temperature of a substance?
3. Can two crystalline substances co-exist in a mixture?
4. Does the crystal formation occur before or after the precipitation?
5. What is the difference between melt crystallisation and solution crystallisation?
6. Why is crystallisation not affected from a solid form?
7. Polybutadiene gets grafted onto during continuous polymerisation process for the manufacture of ABS thermoplastic. The SAN chains get attached to the PBd that is initially in a dissolved state in a solution of styrene and other monomers and solvent. The solvent also polymerises. As a result the state of supersaturation is achieved on account of the reduction in solvent concentration as well as the increase in molecular weight of the PBd phase. It falls out of the solution as small spheres of diameter of 300 nm. Are these crystalline? Nucleation did precede the growth of the particles.
8. Is crystallisation a reversible phenomenon? Elaborate.
9. Four different methods of achieving supersaturation state were discussed. Is there a fifth method when the solute reacts and the product has a different solubility characteristic with the solvent?

10. Is there a surface tension between the crystalline phase and amorphous phase that surrounds it?
11. What is the difference between the predictions of Eq. (10.8) and Eq. (10.10) for predicting the homogeneous nucleation rate and heterogeneous nucleation rate?
12. How can the dependence of crystallisation rate on agitation rate be interpreted?
13. Which is better for the quality of crystal made—a continuous operation or a batch operation?
14. In terms of fugacity relationships, can the crystalline phase be treated as a separate phase compared with the amorphous solid phase?
15. Why is the word *magma* used in crystallisation?

PROBLEMS

1. *Production of Vacuum Salt Based on Sea Water as Raw Material*
Nowadays, there is the possibility to combine solar evaporation with modern salt crystallisation for improving salt quality. Figure 10.11 shows the flow sheet of a certain salt crystallisation unit, which is operated with concentrated liquor from a solar pond.

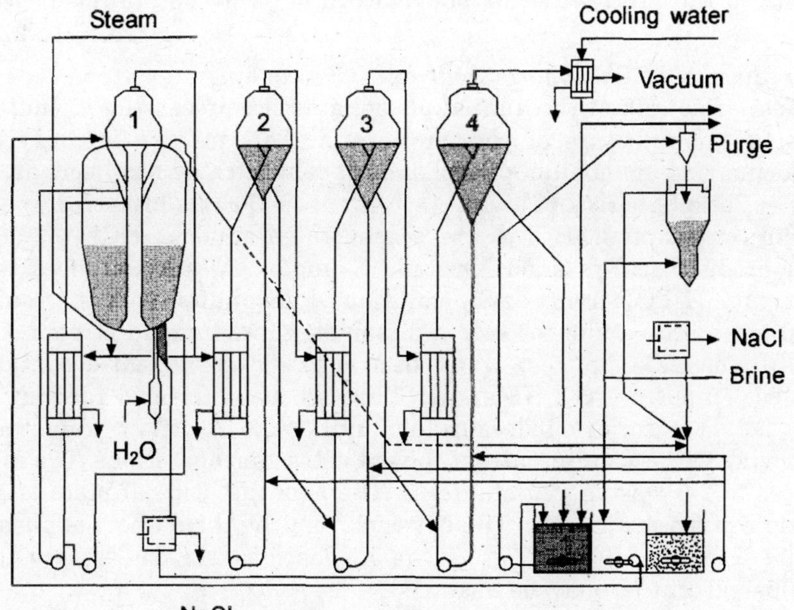

FIGURE 10.11 Multiple-effect evaporation crystallisation for NaCl.

Beside normal FC-types there is installed one fluidised-bed crystalliser for producing a coarse quality as well. The plant is operated as multiple-effect evaporation unit in sequence. The Oslo-type crystalliser is set as the first effect ensuring best conditions for crystal growth at the highest temperature. The plant is fed with concentrated liquor from the solar pond into tank B, from where the Oslo-type crystalliser gets fed. This tank is operated undersaturated ensuring crystal-freeness in the feed to the Oslo-type crystalliser, otherwise the production of coarse crystals gets endangered. Solution not used for the Oslo-type overflows into tank C, from where the FC-type crystallisers get fed. The slurry withdrawal from the FC-type crystallisers is taken from stage to stage and from the last effect to tank A.

From there the separation station is served which consists of a hydrocyclone, a washing-thickener and a centrifuge. In the washing thickener, the highly impure mother liquor is exchanged against the fresh feed liquor before the suspension is given to the centrifuge finally. On the centrifuge the salt is washed with water, then dried and packed. The separation of the grain-size product is effected by a special pusher-type centrifuge, ensuring a soft treatment of this sensitive product.

A plant of this type was erected for the treatment of 60 t/h concentrated liquor from a solar pond on the coast of the Mediterranean Sea. It is producing 2.5 t/h grain-size salt of ≤ 2 mm mean crystal size and 6.5 t/h of normal vacuum salt. Per hour, it consumes 11 t of steam and evaporates 34 t of water. By harvesting the concentrated liquor in special tanks over the summer period, the plant is able to be operated the entire year. What is the maximum nucleation rate in the process? A 5% supersaturation may be assumed.

(**Ans:** 1e-22 nuclei/cm^3/s)

2. *Production of Phosphoric Acid by HCl Leaching*
Most, if not all, of the annual phosphate rock processing in India is attributed to the manufacture of phosphoric acid. This in turn is converted largely to calcium and ammonium phosphates for concentrated fertilisers at the production site. Phosphate rock of high P_2O_5 content (> 30%) is preferred to avoid excessive acid consumption. HCl gas or concentrated aqueous HCl (> 30%) as waste or by-product acid is sued. Organic C and/or C_5 alcohol solvents are used to extract H_3PO_4 from $CaCl_2$ solution. Phosphate rock is ground to pass a 20 mesh screen and fed into a dissolver where an acid stream of concentrated HCl plus make up fresh water from the counter-current decantation system is used. Fumes of CO_2, HF and HCl are scrubbed for acid recovery. The mixture is fed to a series of decantation units with overflow from the first settler moving to the counter-current solvent extraction operations. The solids underflow goes to 2-3 washing crystallisers. The aqueous acid raffinate is separated in a triple effect evaporator. The bottoms from the third effect is phosphoric acid at 80% concentration. What is the nucleation rate in the crystalliser? A 3% supersaturation may be assumed.

3. *Process for Producing Bisphenol A*
A feed stream 1 of 5840 kg/hr molten crude bisphenol-A containing 95.4% p, p′-bisphenol-A and other impurities of a temperature of 180°C was combined with a feed stream 2 of 8700 kg/hr of water at a temperature of 100°C under

pressure sufficient to prevent partial vaporisation of the water and fed to a bisphenol-A-crystalliser 3. The crystalliser 3 cooled the organic/water mixture to 95°C by evaporative cooling. Purified bisphenol-A crystals in a liquid phase were formed. The bisphenol-A crystals were separated from the liquid phase and washed with hot water in a centrifuge 4, and were then dried and passed to a storage device 5 for bisphenol-A. The liquid phase from the centrifuge 4 was separated in a liquid/liquid separator 6 to obtain a water-rich phase 7 and a bisphenol-rich organic phase 8.2030 kg/hr of the bisphenol-rich organic phase 8 were combined in a mix tank with 61370 kg/hour of recycle liquor 14. The isomerised recycle liquor 14 contained 80 per cent phenol and 11.6 per cent p,p′-bisphenol-A, the balance being impurities. The remaining 1300 kg/hr of the bisphenol-rich organic phase 8 was removed from the process. 63400 kg/hr of the produced mixture in the mix tank was fed to a continuous adduct crystalliser 11 operating at 50°C.

The adduct crystals from the crystalliser were separated from the mother liquor and washed with phenol in an adduct centrifuge 12. The mother liquor and the phenol leaving the adduct centrifuge 12 were combined to a recycle liquor 10 and distilled in a first distillation device 16 to remove water and optionally other light impurities and then cooled to 75°C 41400 kg/hr, of distilled recycle liquor 10 was fed to an isomerisation reactor 13 which contained an isomerisation catalyst and which was operated at 65°C.

As the isomerisation catalyst a sulphonated macroporous cross-linked polystyrene resin comprising 4 weight per cent of divinylbenzene groups was used. The remaining distilled recycle liquor was combined with the isomerisation reactor effluent and the bisphenol-A rich organic phase 8 from the liquid/liquid separator 6. The adduct crystals 15 recovered from the adduct centrifuge 12 were melted, stripped of a phenol stream 9 which was used to wash the adduct crystals in the adduct centrifuge 12. A bisphenol-A recycle stream 19 of 2070 kg/hr of molten recovered bisphenol containing 97.3 p,p′-bisphenol-A was combined with the feed stream 1 of 5840 kg/hr of molten crude bisphenol-A and the feed stream 2 of 8700 kg/hr of hot water and returned to the bisphenol-A crystalliser 3. 4500 kg/hr of purified bisphenol-A crystals were produced by this process. The product was analysed and found to contain 99.6 per cent p,p′-bisphenol-A and to have an alcohol color of 2.3 APHA. What is the maximum nucleation rate during the process? (**Ans:** 1e-21 nuclei/cm^3/s)

4. *Vitamin C*

Vitamin C preparations are currently available commercially in crystalline ascorbic acid form or pressed tablets (Figure 10.12). L-ascorbic acid is a powerful water soluble antioxidant that is vital for the growth of tissue types in humans. Ascorbic acid helps regulate blood pressure, reduce blood cholesterol levels and aids in the removal of blood cholesterol level from the arterial walls. It participates in the metabolic cycle of folic acid, regulates the uptake of iron, and is needed for the conversion of L-phenylalanine, and L-tyrosine into noradrenaline. The conversion of tryptophan into serotonin requires adequate supplies of vitamin C. This is the neuroharmone responsible for sleep, pain control and well being.

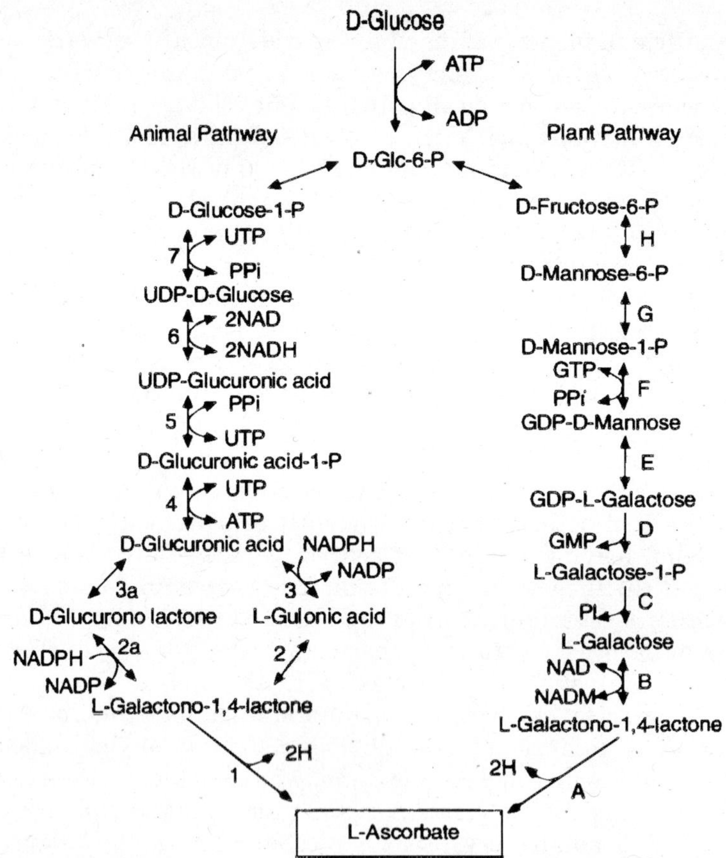

FIGURE 10.12 Biosynthetic pathway to produce ascorbic acid.

Vitamin C can be manufactured by developing yeast and then islotaing the product in a continuous crystalliser. Calculate the nucleation rate in the crytalliser assuming a supersaturation of 7%. **(Ans:** 1e-18 nuclei/cm^3/s)

5. *Sweetener*

[N-(3,3-dimethylbutyl)-L-α-aspartyl]-L-phenylalanine methyl ester, has a sweetening potency, on the weight basis, of at least 50 times that of aspartame and about 10,000 times that of sucrose (table sugar) so that it can constitute a high intensity sweetener. Since sweetening agents are mainly employed in foods for human consumption, they must be prepared using a method which can provide a highly purified product substantially free from impurities or decomposition products. Furthermore, in the case of a sweetening substance which tends to be decomposed relatively easily, like N-(3,3-dimethylbutyl)-APM, some countermeasures are required against the decomposition thereof after forwarded as a product.

These crystals are monohydrate crystals as a result of the X-ray crystal structure analysis, and that they show the specific peaks of diffracted X-rays

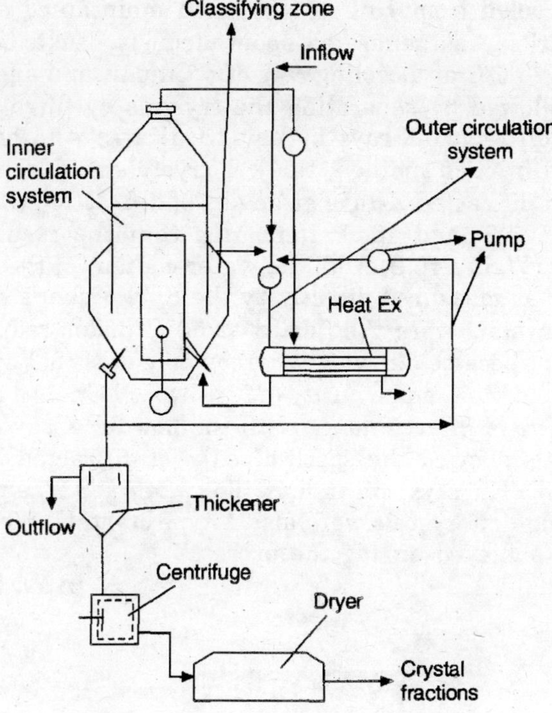

FIGURE 10.13 Novel BASF crystalliser.

at angles of diffraction of at least 6.0°, 24.8°, 8.2° and 16.5° when measured by a powder X-ray diffractometer using CuKa radiation.

In a reactor being equipped with an agitating blade and permitting markedly smooth transfer of gaseous hydrogen into the liquid layer, the following substances were charged with stirring: 550 mL of ion-exchanged water, 1100 mL of methanol, 61 g of aspartame, 20 g of 10% palladium-carbon (having a water content of 50 wt %) and 19 g of 3,3-dimethylbutyl aldehyde.

After the completion of charging, hydrogen gas was introduced at a flow rate of 200 mL/min while the mixture was continued to be stirred at room temperature. The progress of the reaction was monitored by sampling the reaction mixture and analysing the resulting products in the samplers by high-performance liquid chromatography (HPLC). After six hours' reaction, the reactor was filled with a nitrogen gas stream, followed by filtration through a microporous filter (0.50 μm) to remove the catalyst. The filtrate (1,460 g), when analysed, revealed that the N-(3,3-dimethylbutyl)-APM had been produced in an amount of 64 g (yield: 85%). The filtrate was then concentrated to 497 g, whereby crystals were precipitated. The resulting slurry was heated at 70°C to dissolve the crystals. Then, the resulting solution was analysed by gas chromatography, and revealed that the methanol content in the solution was 8.15 wt%. Subsequently, this uniform solution of N-(3,3-dimethylbutyl)-APM was gradually cooled from 70°C to 40°C and maintained at 40°C for 1 hour, whereby nucleation was generated spontaneously. Next, the resulting slurry was then cooled to 5°C at a cooling rate of 5°C/hour, and aged overnight at this temperature, followed by separating the crystals by filtration. As a result of measuring the diffracted X-rays by using CuKa rays in accordance with the powder X-ray diffraction method, the wet crystals thus separated showed the specific peaks of diffracted X-rays at angles of diffraction (2θ, CuKa rays) of at least 6.0°, 24.8°, 8.2° and 16.5°, indicating that the resulting crystals were A-type crystals. (Figure 10.9) And, the water content of the crystals was found to be 51.14 wt% as measured directly by the Karl Fisher's method. The above-described wet crystals were then dried at 50°C under reduced pressure until the water content became 5.8 wt%, whereby 58 g of N-(3,3-dimethylbutyl)-APM was obtained (yield: 72%, and a purity of 97% by HPLC). As a result of measuring, the diffracted X-rays in accordance with the powder X-ray diffraction method, the dried crystals showed the specific peaks of diffracted X-rays at angles of diffraction (2θ, CuKa rays) of at least 6.0°, 24.8°, 8.2° and 16.5°, indicating that the resulting dry crystals were also A-type crystals. What is the maximum nucleation rate achieved during the process?

(**Ans:** 1E-24 nuclei/cm^3/s)

REFERENCES

Ballentine, K., 1996, Internet website, *http://www.eng.vt.edu/eng/materials/classes/MSE2094_NoteBook.*

Clontz, J.C.S. and McCAbe, W.L., 1971, *AIChE Symp. Ser.*, **67**, 100, 6.

Deusser, R, Kraemer, P, Thurow, H, 2002, *Process for drying protein crystals,* US Patent 6408536, Aventis Pharma Deutschland GmbH., Germany.

GEA Niro Inc., Internet website, Columbia, MD, USA.

Kawahara, S, Kishishita, A, Nagashima, K, and Takemoto, T, 2005, *Crystallisation processes for the formation of stable crystals of aspartame derivative*, United States Patent 6914151, Ajinomoto Co., Inc., Tokyo, JP.

McCabe, W.L., Smith, J.C. and Harriott, P., 1993, *Unit Operations of Chemical Engineering,* Fifth Edition, McGraw Hill, New York.

Perry, R.H. and Green, D.W., 1997, *Perry's Chemical Engineers Handbook,* Seventh Edition, McGraw Hill, New York.

Porro, D. and Sauer, M., 2003, *Ascorbic Acid Production from Yeast,* US Patent 6630330, Biopolo, Milan.

Randolph and Larson, 1988, *Theory of Particulate Processes,* Academic, New York.

Sharma, K.R., November 2001, *Mesoscale Simulation of Spinodal Nucleation, 93rd AIChE_Annual Meeting*, Reno, NV, USA.

Sharma, K.R., April 2002, *Nucleation, Particulation in Fluidised Bed Degasifier, 223rd ACS National Meeting*, Orlando, FL.

________, October 1998, *Kinetics of Spinodal Decomposition and Nucleation of Bituminous Coal Extract in Polycarbonate Matrix, 50th ACS Southeast Regional Meeting*, Research Triangle Park, NC, USA.

________, 2005, *Damped Wave Transport and Relaxation*, Elsevier, Amsterdam.

Young, T.C., Feord, D.M. and Frey, J.W., 2006, Process for Producing Bisphenol A, US Patent 7,067,704, Dow Global Technologies, MI, USA.

CHAPTER 11

Emerging Applications in Mass Transfer

Nomenclature

a	particle radius (m)
C_A	concentration of solute A (mole/m^3)
C_T	concentration of the solute in the tissue space (mole/m^3)
D_{AB}	binary diffusivity (m^2/s)
D_A	binary diffusivity of the drug (m^2/s)
D_e	effective diffusivity (m^2/s)
D_n	diffusivity of the neurotransmitter (m^2/s)
D_{ip}	diffusion coefficient of the polymer within the polymeric material (m^2/s)
$\partial E/\partial x$	electric force (V/cm)
F	Faraday's constant (Coulomb/gm)
k_+	encounter rate constant (sec.lit/mol)
k_-	encounter rate constant (sec.lit/mol)
k_b	rate constant of backward reaction (1/s)
k_f	rate constant of forward reaction (1/s)
I_0	modified Bessel function of zeroth order and first kind
I_1	modified Bessel function of the first order and first kind
I_2	modified Bessel function of the second order and first kind
I_3	modified Bessel function of the third order and first kind
J_0	Bessel function of the zeroth order and first kind
J_1	Bessel function of the first order and first kind
J_2	Bessel function of the second order and first kind
J_3	Bessel function of the third order and first kind
K_0	modified Bessel function of the zeroth order and second kind
K_a	equilibrium association constant ($K_a = 1/K_d$)
K_b	binding constant (C_B/C_A)
K_d	equilibrium dissociation constant ($=1/K_a$) $= k_b/k_f$
K_e	equilibrium rate constant ($K_e = C_{Be}/C_{Ae}$)
K_m	Michaelis constant

M_t	total mass of drug released from the external surface at time t, (moles)
r	distance (m)
r_c	radius of the blood capillary (m)
r_T	radius of the tissue cylinder (m)
R	radius of cylinder (m)
t	time (sec)
t_m	thickness of the capillary (m)
u_A	ionic mobility under the action of a unit force per mole (m/s)
V_m	velocity of mass (m/s) sqrt(D/τ_{mr})
x	distance (m)
X	dimensionless distance $x/\sqrt{(D\tau_r)}$
X_c	critical radius beyond which there is no mass transfer
X_{pen}	penetration distance, $X_{pen} = (4\pi^2 + \tau^2)^{1/2}$
Y	dimensionless distance $y/\sqrt{(\alpha\tau_r)}$
z_A	valency of species A

Greek

β	first zero of $J_{1/2}(\beta) = 0$
ε	relative permittivity
ε_0	permittivity of free space 8.854 E-14 C/V/cm
ρ	density (kg/m^3)
τ	dimensionless time (t/τ_r)
ψ	Fick number $(D_{gs}t/h^2)$
τ_{mr}	relaxation time, mass (s)
τ_{mom}	relaxation time, momentum (s)
Ω	effectiveness factor for drug delivery = $\langle C_A \rangle / C_o$
λ_A	ionic conductance
ϕ	potential x on interface, (mv)
ϕ_0	potential at surface, (mv)
κ	characteristic (cm^{-1})
•	Zeta potential, mv
η	viscosity, Kg/m/s

Dimensionless Groups

$1/X_L$	reciprocal of dimensionless length $(D_{ip}\tau_{mr})^{1/2}/L$
k^*	dimensionless rate constant $(k_b\tau_{mr})$
k_b^*	dimensionless rate constant $(k_b^* = k_b\tau_{mr})$
k_f^*	dimensionless rate constant $(k_f^* = k_f\tau_{mr})$
K^*	dimensionless equilibrium rate constant $K^* = (k_b^* + k_f^* + 1)$
M^*	dimensionless release amount $M_t/AL(c_{ext} + c_0)$
Pe_m	Peclect number (mass), $V_z/\sqrt{D_{gs}/\tau_{mr}}$
Rel_m	relaxation number, mass, $(D_{ip}\tau_{mr}/L^2)$

11.1 DRUG DELIVERY

11.1.1 Controlled Release

The better understanding of the process of drug ingestion and assimilation needs the application of the laws of diffusion described in chapters 1 and 2. The study of effectiveness of drug therapy is an interdisciplinary activity. For example, consider the localised delivery of drug to a tissue by controlled release from an implanted, drug loaded polymer. In the period following implantation, drug molecules are slowly released from the polymer into the extracellular space of the tissue. The availability of drug to the tissue, i.e. the rate of release into the extracellular space into the implant is controlled by varying the chemical or physical properties of the polymer. The relationship between polymer chemistry, implant structure and the rate of drug release needs to be better understood.

Prescribed rates of drug delivery offer improved efficacy, safety and convenience. One example is the drug dosed by periodic pills and that by controlled release technology. When the drug is administered, the concentration of the drug rises sharply in the blood (Figure 11.1). This rise can carry the drug concentration past the effective and above the toxic levels. The concentration then drops below the effective level. In contrast, when the drug delivered is by controlled release its concentration rises above the level required to be effective and stays there, without sudden excursions to toxic or ineffective levels. This steady controlled release is achieved by diffusion.

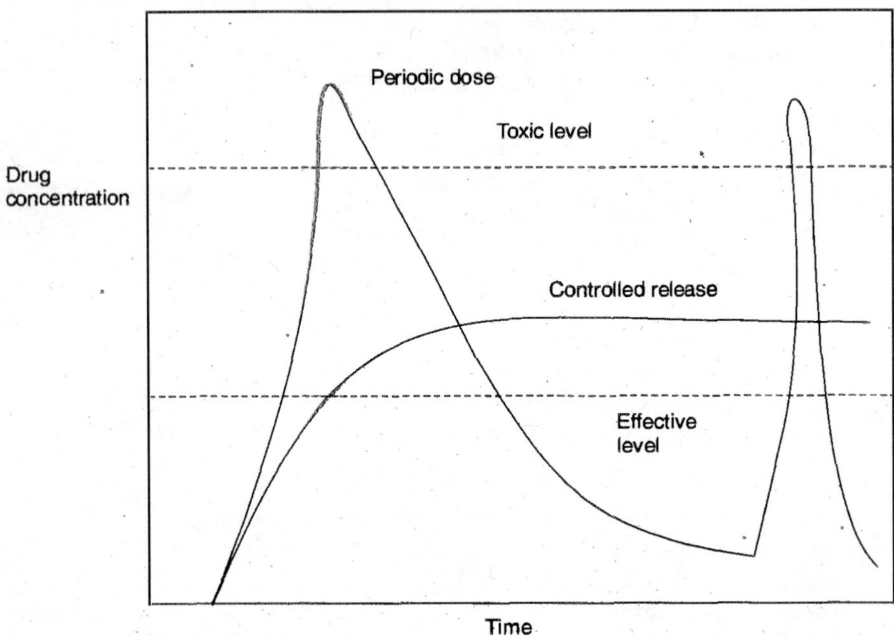

FIGURE 11.1 Controlled release of drugs vs periodic dosage.

The objective is to obtain the release rate of a solute like a drug to be constant with time. A *zero order release* is said to occur when the solute concentration varies linearly with time. However, once the drug is released into the extracellular space, rates of drug migration through the tissue, uptake into the circulatory system, and elimination from the body depend on the characteristics of the drug and physiology of the delivery site. To fathom the barriers to drug delivery, various length scales are considered:

1.0–2.0 m—length scale of an organism
1–100 mm—length scale of an organ
1–100 μm—length scale of a tissue segment
1–100 nm—within extracellular matrix

Implantable drug delivery system design requires good understanding of the molecular mechanisms for drug transport and elimination, particularly at the site of delivery. Multiple forms of drug administration are available such as pills, injections, lotions and suppositories. Oral dosage is preferred as they are painless, uncomplicated, and self-administered. Many drugs are degraded within the gastrointestinal tract or are not absorbed in sufficient quantities to be effective. The route and method of drug administration will influence the kinetic of biodistribution and elimination and, therefore the effectiveness of the therapy. Delivery of identical amounts of drug to tissue sites differing in local anatomy and physiology can result in measurable differences in the pattern of delivery. For instance, when identical preparations of human immunoglobin G (IgG) were administered orally, subcutaneously, intramuscularly or intraveneously to human patients different patterns of IgG concentration in the plasma were observed over time.

11.1.2 Delivery of Drugs and Finite Speed Diffusion

Response to a Pulse in Mass Flux. The concentration of A as a function of time and distance from the site of initial injection can be predicted by solving the finite speed damped wave diffusion and relaxation equation in one dimension in cartesian coordinate system. Writing Eq. [C.21 in Sharma 2005], neglecting bulk flow and the y and z components, subject to the following initial and boundary conditions,

$$\tau_{\text{mr}} \frac{\partial^2 C_\text{A}}{\partial t^2} + \frac{\partial C_\text{A}}{\partial t} = D_{\text{AB}} \frac{\partial^2 C_\text{A}}{\partial x^2} \tag{11.1}$$

$$-\frac{\partial C(x,t)}{\partial x} = \delta(x)\frac{M}{A} \tag{11.2}$$

where $J''' = M/A$ is a pulse of mass flux injected at time zero divided by the diffusivity and has units of (mol/m^4). M/A is the amount in moles injected divided by the cross-sectional area through which the mass diffusion occurs.

$$C(x,\ t) = 0,\ x = -\infty \tag{11.3}$$

$$C(x,\ t) = 0,\ x = \infty \tag{11.4}$$

Obtaining the Laplace transform of Eq. (11.1), assuming that at $t = 0$, the concentration everywhere in the medium is zero,

$$\tau_{mr}\, s^2 C_A + sC_A = D_{AB}\frac{d^2C_A}{dx^2} \tag{11.5}$$

$$C_A(s^2\tau_{mr} + s) = D_{AB}\frac{d^2C_A}{dx^2} \tag{11.6}$$

or

$$C_A = c_1 \exp\left(\frac{(s^2\tau_{mr} + s)^{1/2}x}{D_{AB}^{1/2}}\right) + c_2 \exp\left(-x\left(\frac{(s^2\tau_{mr} + s)}{D_{AB}^{1/2}}\right)^{1/2}\right) \tag{11.7}$$

The concentration can first be solved for in the domain $0 \le x \le \infty$, and then by symmetry it can be reflected for the domain $-\infty \le x \le 0$. At $x = \infty$, $C_A = 0$ and hence, c_1 need be set to zero.

or

$$C_A = c_2 \exp\left(\frac{-x(s^2 + s/\tau_{mr})^{1/2}}{v_m}\right) \tag{11.8}$$

From the boundary condition given in Eq. (11.2),

$$J''' = \frac{c_2\sqrt{s^2 + s/\tau_{mr}}}{v_m} \tag{11.9}$$

or

$$c_2 = \frac{J''' v_m}{\sqrt{s^2 + s/\tau_{mr}}} \tag{11.10}$$

Thus,

$$C_A = \frac{J''' v_m}{\sqrt{s^2 + s/\tau_{mr}}} \exp\left(\frac{-x(s^2 + s/\tau_{mr})^{1/2}}{v_m}\right) \tag{11.11}$$

where,

$$v_m = \sqrt{\frac{D}{\tau_{mr}}} \tag{11.12}$$

Inverting the Laplace transform as available from the tables in Mickley, Sherwood and Reid (1975),

$$\frac{C_A}{J''' v_m \tau_{mr}} = \exp\left(-\frac{t}{2\tau_{mr}}\right) I_0\left[\frac{\sqrt{t^2 - x^2/v_m^2}}{2\tau_{mr}}\right] \tag{11.13}$$

Cylindrical Coordinates. Consider the movement of solute A from the surface of the cylinder of radius R into a homogeneous tissue. For example, the cylinder might represent the external surface of a capillary that contains high concentration of a drug. The concentration within the tissue, in the region $r > R$, can be determined by solving for the governing equation in polar coordinates from Appendix C from Eq. (C.22):

$$\tau_{mr}\frac{\partial^2 C_A}{\partial t^2} + \frac{\partial C_A}{\partial t} = D_{AB}\left(\frac{1}{r}\frac{\partial}{\partial r}\right)\left(\frac{r\partial C_A}{\partial r}\right) \tag{11.14}$$

If the concentration of A in the tissue outside the cylinder is initially zero, and the concentration at the outer surface of the cylinder maintained at C_0, the initial and boundary conditions can be written as:

$$C(r,\ t) = 0;\ \text{for } R \le r;\ t = 0 \tag{11.15}$$

$$C(r,\ t) = C_0;\ \text{for } r = R;\ t > 0 \tag{11.16}$$

$$C(r,\ t) = 0;\ \text{for } r = \infty;\ t > 0 \tag{11.17}$$

Let

$$u = \frac{C}{C_0};\quad X = \frac{r}{\sqrt{D\tau_{mr}}};\quad \tau = t/\tau_{mr} \tag{11.18}$$

$$\frac{\partial^2 u}{\partial \tau^2} + \frac{\partial u}{\partial \tau} = \left(\frac{1}{X}\frac{\partial}{\partial X}\right)\left(\frac{X\partial u}{\partial X}\right) \tag{11.19}$$

Let $u = w\exp(-\tau/2)$. As shown in the preceding sections, the damping term is removed from the governing equation.

$$\frac{\partial^2 w}{\partial \tau^2} - \frac{w}{4} = \left(\frac{1}{X}\frac{\partial}{\partial X}\right)\left(\frac{X\partial w}{\partial X}\right) \tag{11.20}$$

Let $\eta = \tau^2 - X^2$ for $\tau > X$
Then Eq. (11.20) is transformed into,

$$\eta^2\frac{\partial^2 w}{\partial \eta^2} + \frac{3\eta}{2}\frac{\partial w}{\partial \eta} - \frac{\eta w}{16} = 0 \tag{11.21}$$

Comparing Eq. (11.21) with the Bessel equation as given in (Eq. (A.30 of Sharma, 2005),

$$a = 3/2;\quad b = 0;\quad c = 0;\quad d = -1/16;\quad s = 1/2 \tag{11.22}$$

The order p of the solution is then $p = 2\sqrt{1/16} = 1/2$ (11.23)

or

$$W = \frac{c_1 I_{-1/2}\left(\sqrt{\dfrac{\tau^2 - X^2}{2}}\right)}{(\tau^2 - X^2)^{1/4}} + \frac{c_1 I_{-1/2}\left(\sqrt{\dfrac{\tau^2 - X^2}{2}}\right)}{(\tau^2 - X^2)^{1/4}} \tag{11.24}$$

c_2 can be seen to be zero as W is finite and not infinitely large at $\eta = 0$. An approximate solution can be obtained by eliminating c_1 between Eq. (11.16) and the above equation. However, it can be noted that this is a mild function of time. As the general solution of PDE consists of n arbitrary functions when the order of the PDE is n compared with n arbitrary constants for ODE. From the boundary condition at $X = X_R$,

$$1 = \exp\left(-\frac{\tau}{2}\right) c_1 \frac{I_{1/2}\left(\dfrac{\tau^2 - X_R^2}{2}\right)}{(\tau^2 - X_R^2)^{1/4}} \tag{11.25}$$

or

$$u = \left(\frac{\tau^2 - X_R^2}{\tau^2 - X^2}\right)^{1/4} \left(\frac{I_{1/2}\left(\frac{1}{2}\sqrt{\tau^2 - X^2}\right)}{I_{1/2}\left(\frac{1}{2}\sqrt{\tau^2 - X_R^2}\right)}\right) \tag{11.26}$$

In terms of elementary functions, Eq. (11.26) can be written as:

$$u = \left(\frac{\tau^2 - X_R^2}{\tau^2 - X^2}\right)^{1/2} \left[\frac{\sin h\left(\frac{1}{2}\sqrt{\tau^2 - X^2}\right)}{\sin h\left(\frac{1}{2}\sqrt{\tau^2 - X_R^2}\right)}\right] \tag{11.27}$$

In the limit of X_R going to zero, the expression becomes

$$u = \left(\frac{\tau}{(\tau^2 - X^2)^{1/2}}\right) \left[\frac{\sin h\left(1/2\sqrt{\tau^2 - X^2}\right)}{\sin h\,(\tau/2)}\right] \tag{11.28}$$

This is for $\tau > X$. For $X > \tau$,

$$u = \left(\frac{X_R^2 - \tau^2}{X^2 - \tau^2}\right)^{1/4} \left[\frac{J_{1/2}\left(\frac{1}{2}\sqrt{X^2 - \tau^2}\right)}{I_{1/2}\left(\frac{1}{2}\sqrt{\tau^2 - X^2}\right)}\right] \tag{11.29}$$

Eq. (11.29) can be written in terms of trigonometric functions as:

$$u = \left(\frac{X_R^2 - \tau^2}{X^2 - \tau^2}\right)^{1/2} \left[\frac{\sin\left(\frac{1}{2}\sqrt{X^2 - \tau^2}\right)}{\sin h\left(\frac{1}{2}\sqrt{\tau^2 - X^2}\right)}\right] \tag{11.30}$$

In the limit of X_R going to zero, the expression becomes

$$u = \left(\frac{\tau}{X^2 - \tau^2}\right)^{1/2} \left[\frac{\sin\left(\frac{1}{2}\sqrt{X^2 - \tau^2}\right)}{\sin h\left(\frac{\tau}{2}\right)}\right] \tag{11.31}$$

The time lag at a point in the cylindrical medium to realize the disturbance can be calculated as:

$$\tau_{lag}^2 = X_p^2 - 4\pi^2 \tag{11.32}$$

or

$$\tau_{lag} = \sqrt{X_p^2 - 4\pi^2} \tag{11.33}$$

For a given instant in time, the penetration distance can be calculated from Eq. (11.32):

$$X_{pen} = (4\pi^2 + \tau^2)^{1/2} \tag{11.34}$$

For times greater than the time lag and less than X_p, the dimensionless concentration is given by Eq. (11.26). For dimensionless times greater than X_p, the dimensionless concentration is given by Eq. (11.29). For distances *closer to the surface compared with 2π, the time lag will be zero.*

For a solute with $D_A = 10^{-11}$ m²/s and a cylinder with radius $R = 0.001$ m, and a mass relaxation time of 30 seconds, the solute will penetrate in the first 6 hrs no further than a distance of 12.5 mm, in the first 24 hrs no further than a distance of 50 mm and in the first 100 hrs no further than a distance of 207.8 mm.

Spherical Coordinates. Biological systems sometimes exhibit spherical symmetry. For instance, certain cells are assumed to be spherical as are vesicles within cells and synthetic vesicles. The transport of solutes is often sought in the circular coordinate system. The governing equation from (Appendix C, Eq. (C.23), in Sharma, 2005) neglecting convective effects, can be written as:

$$\tau_{mr}\frac{\partial^2 C_A}{\partial t^2} + \frac{\partial C_A}{\partial t} = D_{AB}\left[\frac{1}{r^2}\frac{\partial}{\partial r}\left(r^2\frac{\partial C_A}{\partial r}\right)\right] \tag{11.35}$$

The gradients in the θ and ϕ directions are neglected. Eq. (11.35) is rendered dimensionless by the following substitutions,

Let

$$u = \frac{C}{C_0};\quad X = \frac{r}{\sqrt{D\tau_{mr}}};\quad \tau = \frac{t}{\tau_{mr}} \tag{11.36}$$

$$\frac{\partial^2 u}{\partial \tau^2} + \frac{\partial u}{\partial \tau} = \left[\frac{1}{X^2}\frac{\partial}{\partial X}\left(X^2\frac{\partial u}{\partial X}\right)\right] \tag{11.37}$$

Consider the flux of solute A towards a spherical cell in suspension. If the solute is consumed at the surface of the cell, and the rate of consumption is rapid, solute concentrations in the vicinity of the cell. The time and space conditions are as follows:

$$u = 1, \text{ for } X \geq X_R, \tau = 0 \tag{11.38}$$

$$u = 0, X = X_R \tag{11.39}$$

$$u = 1, X = \infty \tag{11.40}$$

At steady state,

$$u = -C_1/X + C_2 \tag{11.41}$$

From Eq. (11.40), $C_2 = 1$. From Eq. (11.39),

$$C_1 = X_R \tag{11.42}$$

Thus,

$$u^{ss} = 1 - R/r \tag{11.43}$$

In order to solve for the transient portion of the solution, the initial value problem is changed to a boundary value problem by the following substitution:

Let
$$V = 1 - u \tag{11.44}$$

The governing Eq. (11.37) can then be written as:

$$\frac{\partial^2 V}{\partial \tau^2} + \frac{\partial V}{\partial \tau} = \left[\frac{1}{X^2}\frac{\partial}{\partial X}\left(X^2\frac{\partial V}{\partial X}\right)\right] \tag{11.45}$$

The damping term is removed from the governing equation. This is done realising that the transient temperature decays with time in an exponential fashion. The other reason for this maneuver is to study the wave equation without the damping term. Let $u = w \exp(-n\tau)$.

$$\exp(-n\tau)\, w(-n + n^2) + \exp(-n\tau)\frac{\partial w}{\partial \tau}(1 - 2n) + \exp(-n\tau)\frac{\partial^2 w}{\partial \tau^2} = \text{RHS} \tag{11.46}$$

By choosing $n = 1/2$, the damping component of the equation is removed. Thus for $n = 1/2$,

$$-\frac{w}{4} + \frac{\partial^2 w}{\partial \tau^2} = \frac{\partial^2 w}{\partial X^2} + \frac{2}{X}\frac{\partial w}{\partial X} \tag{11.47}$$

Eq. (11.47) can be solved using the method of relativistic transformation of coordinates. Consider the transformation variable η as:

$$\eta = \tau^2 - X^2 \tag{11.48}$$

for $\tau > X$

$$\frac{\partial w}{\partial \tau} = \left(\frac{\partial w}{\partial \eta}\right) 2\tau \tag{11.49}$$

$$\frac{\partial^2 w}{\partial \tau^2} = \left(\frac{\partial^2 w}{\partial \eta^2}\right) 4\tau^2 + 2\left(\frac{\partial w}{\partial \eta}\right) \tag{11.50}$$

Similarly,

$$\frac{\partial w}{\partial X} = \left(\frac{\partial w}{\partial \eta}\right) 2X \tag{11.51}$$

$$\frac{\partial^2 w}{\partial X^2} = \left(\frac{\partial^2 w}{\partial \eta^2}\right) 4X^2 + 2\left(\frac{\partial w}{\partial \eta}\right) \tag{11.52}$$

Plugging Eqs. (11.49–11.52) into Eq. (11.47)

$$\left(\frac{\partial^2 w}{\partial \eta^2}\right) 4(\tau^2 - X^2) + 8\left(\frac{\partial w}{\partial \eta}\right) - \frac{w}{4} = 0 \tag{11.53}$$

or

$$\frac{4\eta^2 \partial^2 w}{\partial \eta^2} + \frac{8\eta \partial w}{\partial \eta} - \frac{\eta w}{4} = 0 \tag{11.54}$$

or

$$\frac{\eta^2 \partial^2 w}{\partial \eta^2} + \frac{2\eta \partial w}{\partial \eta} - \frac{\eta w}{16} = 0 \tag{11.55}$$

Comparing Eq. (11.55) with the generalised Bessel equation as given in (Eq. (A.30) in Sharma, 2005) the solution to the special differential equation in Eq. (11.55) is,

$$a = 2;\ b = 0;\ c = 0;\ s = 1/2;\ d = -1/16 \tag{11.56}$$

The order of the Bessel solution would be,

$$p = 2\sqrt{(1/4)} = 1;\quad \sqrt{(|d|/s)} = 1/2 \tag{11.57}$$

Hence the solution to Eq. (11.47) can be written as:

$$w = \frac{c_1 I_1\left(\frac{\sqrt{\tau^2 - X^2}}{2}\right)}{(\tau^2 - X^2)^{1/2}} + \frac{c_2 K_2\left(\frac{\sqrt{\tau^2 - X^2}}{2}\right)}{(\tau^2 - X^2)^{1/2}} \tag{11.58}$$

c_2 can be seen to be zero as W is finite and not infinitely large at $\eta = 0$. The solution is in terms of a composite modified Bessel function of the first order and first kind. Therefore the heat flux can be written as:

$$V = c_1 \exp\left(-\frac{\tau}{2}\right) \frac{I_1\left(\frac{\sqrt{\tau^2 - X^2}}{2}\right)}{\sqrt{\tau^2 - X^2}} \tag{11.59}$$

From the boundary condition at the solid surface,

$$1 = c_1 \exp\left(-\frac{\tau}{2}\right) \frac{I_1\left(\dfrac{(\tau^2 - X_R^2)^{1/2}}{2}\right)}{(\tau^2 - X_R^2)^{1/2}} \tag{11.60}$$

Dividing Eq. (11.59) by Eq. (11.60), the solution for u can be given in a more usable form as:

$$V = \left(\frac{\tau^2 - X_R^2}{\tau^2 - X^2}\right)^{1/2} \left(\frac{I_1(1/2\sqrt{\tau^2 - X^2})}{I_1(1/2\sqrt{\tau^2 - X_R^2})}\right) \tag{11.61}$$

This is valid for $\tau > X$. For $X > \tau$;

$$V = \left(\frac{\tau^2 - X_R^2}{X^2 - \tau^2}\right)^{1/2} \left(\frac{J_1(1/2\sqrt{X^2 - \tau^2})}{I_1(1/2\sqrt{\tau^2 - X_R^2})}\right) \tag{11.62}$$

For $X = \tau$, the solution at the wave front results. This can be obtained by solving Eq. (11.55) at $\eta = 0$. In the limit of X_R going to zero,

$$V = \left(\frac{\tau}{(X^2 - \tau^2)^{1/2}}\right) \left(\frac{I_1(1/2\sqrt{\tau^2 - X^2})}{I_1(\tau/2)}\right) \tag{11.63}$$

This is valid for $\tau > X$. For $X > \tau$,

$$V = \left(\frac{\tau}{(X^2 - \tau^2)^{1/2}}\right) \left(J_1\left(\frac{1/2\sqrt{(X^2 - \tau^2)}}{I_1(\tau/2)}\right)\right) \tag{11.64}$$

Three regimes can be identified. The first regime is that of the thermal lag and consists of no change from the initial concentration. The second regime is when

$$\tau_{lag}^2 = X^2 - (7.6634)^2 \tag{11.65}$$

or

$$\tau_{lag} = \sqrt{X_p^2 - (7.6634)^2} \tag{11.66}$$

The first zero of $J_1(x)$ occurs at $x = 3.8317$. The 7.6634 is twice the first root of the Bessel function of the first order and first kind. For times greater than the time lag and less than X_p, the dimensionless concentration is given by Eq. (11.68). For dimensionless times greater than X_p, the dimensionless concentration is given by Eq. (11.67). For distances closer to the surface compared with $7.6634\sqrt{D\tau_r}$ the thermal lag time will be zero. The ballistic term manifests as a thermal lag at a given point in the medium. Thus, in the limit of X_R going to zero,

$$u = 1 - \left(\frac{\tau}{(\tau^2 - X^2)^{1/2}}\right)\left(\frac{I_1(1/2\sqrt{\tau^2 - X^2})}{I_1(\tau/2)}\right) \tag{11.67}$$

The above Eq. (11.67) is valid for $\tau > X$.

For $X > \tau$,

$$u = 1 - \left(\frac{\tau}{(X^2 - \tau^2)^{1/2}}\right)\left(\frac{J_1(1/2\sqrt{X^2 - \tau^2})}{I_1(\tau/2)}\right) \tag{11.68}$$

When $X = \tau$,

$$V = c' \exp\left(-\frac{\tau}{2}\right) = c' \exp\left(-\frac{X}{2}\right) \tag{11.69}$$

$$C' = \exp\left(\frac{X_R}{2}\right) \tag{11.70}$$

or

$$u = 1 - \exp\left(\frac{X_R - X}{2}\right) \tag{11.71}$$

In the limit of X_R going to zero,

$$u = 1 - \exp\left(-\frac{X}{2}\right) \tag{11.72}$$

The dimensionless concentration profile is shown in Figure 11.2. The three regimes are piecewise continuous and the function is monotonic. For short times for distances greater than the said time, the dimensionless concentration is described by the Bessel composite function of the first kind and first order. At the occurrence of the first zero of the Bessel function, the concentration will be undisturbed.

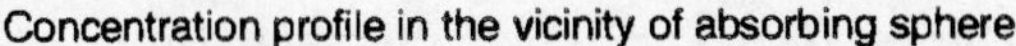

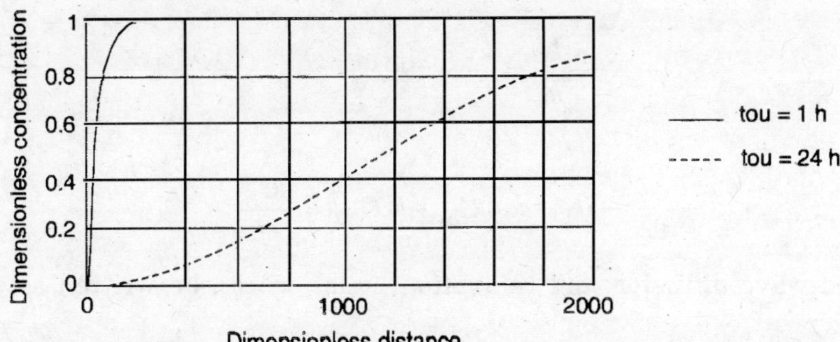

FIGURE 11.2 Dimensionless concentration profile near the absorbing sphere.

Solute Binding and Elimination. The concentration of solute/drug within a region of tissue will depend not only on damped wave diffusion and relaxation but also on physiological processes. For instance, a solute may be generated or consumed within a cell or tissue region by a chemical reaction, usually one mediated by enzymes. Alternatively, the solute could be eliminated from the diffusion process by immobilisation to some fixed element, binding to the cytoskeleton or an organelle or by partitioning from the extracellular space into a capillary where it enters the circulatory system. Similarly, enzymatic conversion can be used to eliminate the solute diffusing through the extracellular space. The elimination may be effected by immobilisation by some fixed element and by internalisation into cells or capillaries or generated by secretion from a cell. When a diffusing solute is generated or consumed homogeneously within some region of interest in a tissue, the differential mass balance in molar concentrations can be written (Eq. C.21 in Sharma, 2005) by neglecting convective effects and assuming constant diffusivity as:

$$\frac{\tau_{mr}\partial^2 C_A}{\partial t^2} + \frac{\partial C_A}{\partial t} = D_{AB}\left(\frac{\partial^2 C_A}{\partial x^2} + \frac{\partial^2 C_A}{\partial y^2} + \frac{\partial^2 C_A}{\partial z^2}\right) \tag{11.73}$$

where R_A is the molar rate of generation or consumption of the drug or solute per volume within the differential volume of the element.

Consider a solute that diffuses within the extracellular space of the tissue, but also interacts with the tissue by reversibly binding to some fixed component of the tissue. For example, the diffusing solute might bind to a protein on the cell surface or in the extracellular matrix. When bound, the solute may be considered to be completely immobilised and when released, it is free to diffuse again within the extracellular space. If the conversion of the diffusible solute A, to the bound form B is rapid and characterised by an equilibrium constant $K_e = C_{Be}/C_{Ae}$.

$$A \underset{k_b}{\overset{k_f}{\rightleftarrows}} B \tag{11.74}$$

$$R_A = -(k_f + k_b)C_A + k_b C_{A0} \tag{11.75}$$

At equilibrium, the rate becomes zero and it can be seen that,

$$\frac{k_f}{k_b} = \frac{C_{Be}}{C_{Ae}} = K_e \tag{11.76}$$

Thus,

$$-\frac{\partial J}{\partial x} - (k_f + k_b)C_A + k_b C_{A0} = \frac{\partial C_A}{\partial t} \tag{11.77}$$

The damped wave diffusion and relaxation equation can be written as:

$$J = -D_A \frac{\partial C_A}{\partial x} - \tau_{mr}\frac{\partial J}{\partial \tau} \tag{11.78}$$

Differentiating the mass balance equation with respect to time and the damped wave and relaxation equation with respect to space and eliminating the second cross-derivative of mass flux with respect to time and space,

$$D_A \frac{\partial^2 C_A}{\partial x^2} = \tau_{mr} \frac{\partial^2 C_A}{\partial t^2} + (\tau_{mr}(k_f + k_b) + 1) \frac{\partial C_A}{\partial t} + (k_f + k_b)C_A - k_b C_{A0} \quad (11.79)$$

Let $$X = x(\sqrt{D_A \tau_{mr}}); \ \tau = t/\tau_{mr}; \ k_b^* = k_b \tau_{mr}; \ k_f^* = k_f \tau_{mr} \quad (11.80)$$

$$\frac{\partial^2 C_A}{\partial X^2} = \frac{\partial^2 C_A}{\partial \tau^2} + (k_b^* + k_f^* + 1) \frac{\partial C_A}{\partial \tau} + (k_b^* + k_f^*)\, C_A - k_b^* C_{A0} \quad (11.81)$$

Let $$u = (k_b^* + k_f^*)\, C_A - k_b^* C_{A0} \quad (11.82)$$

Then,

$$\frac{1}{(k_b^* + k_f^*)} \frac{\partial^2 u}{\partial X^2} = \frac{1}{k_b^* + k_f^*} \frac{\partial^2 u}{\partial \tau^2} + \frac{(k_b^* + k_f^* + 1)}{(k_b^* + k_f^*)} \frac{\partial u}{\partial \tau} + u \quad (11.83)$$

or

$$\frac{\partial^2 u}{\partial X^2} = \frac{\partial^2 u}{\partial \tau^2} + (k_b^* + k_f^* + 1) \frac{\partial u}{\partial \tau} + u(k_b^* + k_f^*) \quad (11.84)$$

Solutions to the above equation will be similar to those of the problem without any reaction except that the diffusion coefficient is reduced by a factor equal to the binding constant plus unity. These same equations can be used to evaluate penetration into tissues when more complicated equilibrium expressions are appropriate by substituting the non-linear equilibrium expression into the governing equation and solving the rest of the equation. Numerical values for K_e can be obtained from the information on the equilibrium association or dissociation constants.

Let $$K^* = \tau_{mr}(k_f + k_b + 1) = (k_b^* + k_f^* + 1) \quad (11.85)$$

Eq. (11.84) becomes

$$\frac{\partial^2 u}{\partial X^2} = \frac{\partial^2 u}{\partial \tau^2} + \frac{K^* \partial u}{\partial \tau} + u(K^* - 1) \quad (11.86)$$

Let $$u = W \exp(-n\tau) \quad (11.87)$$

The damping term is first removed by a $u = w \exp(-n\tau)$ substitution.

$$\frac{\partial u}{\partial \tau} = -nw + w_\tau \quad (11.88)$$

$$\frac{\partial^2 u}{\partial \tau^2} = +n^2 w - 2nw_\tau + w_{\tau\tau} \quad (11.89)$$

$$w_{XX} = (n^2 - nK^* + K^* - 1)\, w + w_\tau (K^* - 2n) + w_{\tau\tau} \quad (11.90)$$

Choosing $n = K^*/2$, Eq. (11.86) becomes

$$w_{XX} = -\, w/4(K^* - 2)^2 + w_{\tau\tau} \tag{11.91}$$

Let the transformation be

$$\eta = \tau^2 - X^2 \qquad \text{for } \tau > X \tag{11.92}$$

$$\frac{4\eta\partial^2 w}{\partial\eta^2} + \frac{4\partial w}{\partial\eta} - \frac{w|K^* - 2|^2}{4} = 0 \tag{11.93}$$

or

$$\frac{\eta^2\partial^2 w}{\partial\eta^2} + \frac{\eta\partial w}{\partial\eta} - \frac{w\eta(|K^* - 2|^2)}{16} = 0 \tag{11.94}$$

Comparing Eq. (11.94) with the generalised Bessel equation (A.30 in Sharma 2005),

$$a = 1;\ b = 0;\ c = 0;\ s = 1/2;\ d = -1/16(K^* - 2)^2 \tag{11.95}$$

The order $p = 0$. sqrt(d)/s $= i\,|K^* - 2|^2$ and is imaginary. Hence the solution is,

$$w = c_1 I_0\ 1/2\ |K^* - 2|^2\ \sqrt{(\tau^2 - X^2)} + c_2 K_0\ 1/2\ |K^* - 2|^2\ \sqrt{(\tau^2 - X^2)} \tag{11.96}$$

c_2 can be seen to be zero from the condition that at $\eta = 0$, W is finite.

From the boundary condition at the wavefront, at $\eta = 0$, $w = 1$ it can be seen that $c_1 = u_s$. The solution at the wavefront can be obtained by directly solving for Eq. (11.93) at $\eta = 0$. The integration constant is solved for when $X = 0$, $u = u_s$. Thus,

$$\frac{u}{u_s} = \exp\left(-\frac{\tau K^*}{2}\right) I_0\ \frac{1}{2}\ |K^* - 2|\ \sqrt{\tau^2 - X^2} \tag{11.97}$$

This is valid for $K^* \neq 2$.

For $K^* = 2$, Eq. (11.91) reverts to the wave equation. It can be seen that the solution changes to a Bessel composite function of the first kind and first order when $X > \tau$,

$$\frac{u}{u_s} = \exp\left(-\frac{\tau K^*}{2}\right) J_0\ \frac{1}{2}|K^* - 2|\ \sqrt{(X^2 - \tau^2)} \tag{11.98}$$

When the argument of the Bessel composite function of the first kind and zeroth order becomes 2.4048, the concentration u becomes zero. Thus, for a given point in the medium, the time lag experienced prior to any appreciable change in the concentration can be estimated as:

$$\tau_{\text{lag}} = \left(X_p^2 - \frac{23.13}{|K^* - 2|^2}\right)^{1/2} \tag{11.99}$$

For a given time instant, the penetration distance beyond which the concentration is undisturbed can be estimated as:

$$X_{pen} = \left(\tau^2 + \frac{23.13}{|K^* - 2|^2}\right)^{1/2} \tag{11.100}$$

Worked Example 11.1 *Effective Diffusivity*

Rewrite Eq. (11.79) in terms of an effective diffusivity and interpret the governing equation.

Eq. (11.77) can be written as:

$$-\frac{\partial J}{\partial x} + R_A = \frac{\partial C_A}{\partial t} \tag{11.101}$$

$$R_A = -(k_f + k_b)\,C_A + k_b C_{A0} = \frac{dC_A}{dt} = \frac{1}{\tau_{mr}}\frac{dC_A}{d\tau} \tag{11.102}$$

$$R_B = k_f C_A - k_b C_B = \frac{1}{\tau_{mr}}\frac{dC_B}{d\tau} \tag{11.103}$$

Let

$$K_b = \frac{C_B}{C_A} \tag{11.104}$$

$$R_B = k_f C_A - k_b C_B = \frac{K_b}{\tau_{mr}}\frac{dC_A}{d\tau}$$

$$R_A = -R_B \tag{11.105}$$

Thus,

$$-\tau_{mr}\frac{\partial J}{\partial x} = (1 + K_b)\frac{\partial C_A}{\partial \tau} \tag{11.106}$$

Let

$$X = \frac{x}{\sqrt{D\tau_{mr}}} \tag{11.107}$$

$$-\frac{\partial J}{\partial X} = (1 + K_b)\,v_m\frac{\partial C_A}{\partial \tau} \tag{11.108}$$

The damped wave diffusion and relaxation equation can be written as:

$$J = -\frac{v_m \partial C_A}{\partial X} - \frac{\partial J}{\partial \tau} \tag{11.109}$$

Differentiating Eq. (11.108) wrt τ and Eq. [11.109] wrt X,

$$-\frac{\partial^2 J}{\partial X \partial \tau} = (1 + K_b)v_m \frac{\partial^2 C_A}{\partial \tau^2} \tag{11.110}$$

The damped wave diffusion and relaxation equation can be written as:

$$\frac{\partial J}{\partial X} = -v_m \frac{\partial^2 C_A}{\partial X^2} - \frac{\partial^2 J}{\partial \tau \partial X} \tag{11.111}$$

Eliminating the second cross-derivative wrt J between Eqs. [11.108] and [11.109],

$$\frac{\partial^2 C_A}{\partial X^2} = (1 + K_b)\left(\frac{\partial^2 C_A}{\partial \tau^2} + \frac{\partial C_A}{\partial \tau}\right) \tag{11.112}$$

Defining an effective diffusivity,

$$D_e = \frac{D}{(1 + K_b)} \tag{11.113}$$

and a new ordinate, $Y = X\sqrt{D/D_e}$

$$\frac{\partial^2 C_A}{\partial Y^2} = \frac{\partial^2 C_A}{\partial \tau^2} + \frac{\partial C_A}{\partial \tau} \tag{11.114}$$

The binding constant K_b is thus lumped with the diffusivity to give an effective diffusivity. Solutions to Eq. (11.114) were presented in (Sharma, 2005) for some boundary conditions.

The binding constant K_b can be obtained from information on the equilibrium association K_a or dissociation K_d constants for receptor-ligand pairs. These constants may be defined as:

$$L + R \underset{k_r}{\overset{k_f}{\Leftrightarrow}} L - R \tag{11.115}$$

$$K_d = \frac{1}{K_a} = \frac{[L][R]}{[L - R]} = \frac{k_r}{k_f} \tag{11.116}$$

where K is the ligand, R is the receptor and L–R is the ligand-receptor complex. Consider the case of a drug molecule, which can be assigned the binding ligand $[L] = C_A$ interacting with a binding site R which is present on the surface of the cells or stationary molecules in the extracellular space to form bound complexes $[L–R] = C_B$. Assuming that the receptor concentration is constant in the tissue, the binding constant K_b is related to the dissociation constant:

$$K_b = \frac{C_B}{C_A} = \frac{[L-R]}{[L]} = \frac{[R]}{K_d} \tag{11.117}$$

Some values of the dissociation constant is presented in Saltzman (2001).

Upon substitution of an appropriate kinetic expression for the rate of generation or consumption of solute within the tissue space, Eq. (11.101) can be solved to determine the spatio-temporal patterns exhibited by the concentration. For example, consider the one-dimensional damped wave diffusion and relaxation of solute from the interface where the concentration is maintained constant. If the diffusing solute is also eliminated from the tissue such that the volumetric rate of elimination is first order with a characteristic rate constant k, Eq. (11.118) can be reduced to,

$$D_A \frac{\partial^2 C_A}{\partial x^2} = \tau_{mr} \frac{\partial^2 C_A}{\partial t^2} + (k_f^* + 1) \frac{\partial C_A}{\partial t} + k_f C_A \tag{11.118}$$

This equation can be solved subject to the initial and boundary conditions:

$$C_A(x, t) = 0, \quad \text{for } x \geq 0 \tag{11.119}$$

$$C_A(x, t) = C_0; \quad \text{for } x = 0; \quad t > 0 \tag{11.120}$$

$$C_A(x, t) = 0; \quad \text{for } x = \infty, \ t > 0 \tag{11.121}$$

Eq. (11.118) can be solved as shown in (Sharma, 2005) and the solution written as:

Let,
$$u = \frac{C_A}{C_o};\ X = \frac{x}{\sqrt{D\tau_{mr}}};\ \tau = \frac{t}{\tau_{mr}} \tag{11.122}$$

Eq. (11.118) becomes,

$$\frac{\partial^2 u}{\partial X^2} = \frac{\partial^2 u}{\partial \tau^2} + (k_f^* + 1)\frac{\partial u}{\partial \tau} + k_f^* u \tag{11.123}$$

The Eq. (11.123) can be solved by the method of relativistic transformation of coordinates. The damping term is first removed by a $u = w \exp(-n\tau)$ substitution. Choosing $n = (1 + k^*)/2$ Eq. (11.123) becomes:

$$\frac{\partial^2 w}{\partial X^2} = -\frac{w(1 - k_f^*)^2}{4} + w_{\tau\tau} \tag{11.124}$$

Let the transformation be:

$$\eta = \tau^2 - X^2 \quad \text{for } \tau > X$$

Eq. (11.124) becomes;

$$\frac{4\tau^2 \partial^2 w}{\partial \eta^2} + \frac{2\partial w}{\partial \eta} - \frac{w(1 - k^*)^2}{4} = \frac{4X^2 \partial^2 w}{\partial \eta^2} - \frac{2\partial w}{\partial \eta} \tag{11.125}$$

or
$$\frac{\eta^2 \partial^2 w}{\partial \eta^2} + \frac{\eta \partial w}{\partial \eta} - \frac{w\eta(1 - k^*)^2}{16} = 0 \tag{11.126}$$

Comparing Eq. (11.126) with the generalised Bessel equation (A.30 in Sharma, 2005),

$$a = 1;\ b = 0;\ c = 0;\ s = 1/2;\ d = \frac{(1 - k_f^*)^2}{16} \tag{11.127}$$

The order $p = 0$. $\sqrt{(d)}/s = 1/2\ i(1 - k^*)$ and is imaginary. Hence the solution is:

$$w = c_1 I_0\ 1/2\ |1 - k_f^*|\ \sqrt{(\tau^2 - X^2)} + c_2 K_0\ 1/2\ |1 - k_f^*|\ \sqrt{(\tau^2 - X^2)} \tag{11.128}$$

c_2 can be seen to be zero from the condition that at $\eta = 0$, W is finite.

$$w = \exp\left(-\tau\left(\frac{1 + k_f^*}{2}\right)\right) I_0 \left[1/2\ |1 - k_f^*|\ \sqrt{(\tau^2 - X^2)}\right] \tag{11.129}$$

From the boundary condition at $X = 0$,

$$1 = \exp\left(-\tau\left(\frac{1 + k_f^*}{2}\right)\right) c_1 I_0 \left[\frac{|1 - k_f^*|\tau}{2}\right] \tag{11.130}$$

c_1 can be eliminated between Eqs. (11.129, 11.130) or calculated to yield,

$$u = \frac{I_0\left[1/2\ |1 - k_f^*| \sqrt{(\tau^2 - X^2)}\right]}{I_0 1/2\ [|1 - k_f^*|\tau]} \tag{11.131}$$

This is valid for $\tau > X$, $k_f^* \neq 1$

For $X > \tau$,

$$u = \frac{J_0\left[1/2\ |1 - k_f^*| \sqrt{(X^2 - \tau^2)}\right]}{I_0 1/2\ [|1 - k_f^*|\tau]} \tag{11.132}$$

At the wavefront, $\tau = X$, $u = \exp(-\tau(1 + k_f^*)/2) = \exp(-X(1 + k_f^*)/2)$

The mass inertia can be calculated from the first zero of the Bessel function at 2.4048. Thus,

$$\tau_{\text{inertia}} = \left(X_p^2 - \frac{23.1323}{|1 - k_f^*|^2}\right)^{1/2} \tag{11.133}$$

Three regimes can be identified in the concentration at an interior point in the semi-infinite medium (Sharma, 2005). During the first regime of mass inertia, there is no transfer of mass upto a certain threshold time at the interior point $X_p = 10$. The second regime is represented by a modified Bessel function of the first kind and zeroth order. The rise in dimensionless concentration proceeds from the dimensionless time 2.733 upto the wave front at $X_p = 10.0$. The third regime is given by Eq. (11.131) and represents the decay in time of the dimensionless concentration. It is given by the modified Bessel composite function of the first kind and zeroth order. In Figure 3.5 (in Sharma, 2005) is shown the three regimes of the concentration when $k_f^* = 2.0$. It can be seen from Figure 3.5 that the mass inertia time has increased to 8.767. The rise is nearly a jump in concentration at the interior point $X_p = 10.0$. When $k^* = 0.25$ as shown in Figure 3.6 the inertia time is 7.673. In Figure 3.6 the three regimes for the case when $k_f^* = 0.0$ is plotted. In Table 3.1 the mass inertia time for various values of k_f^* for the interior point $X_p = 10.0$ is shown. k_f^* needs to be sufficiently far from 1 to keep the inertia time positive. The corresponding steady-state solution of Eq. (11.118) can be obtained;

$$D_A \frac{\partial^2 C_A}{\partial x^2} = k_f C_A \tag{11.134}$$

$$C_A = c_1 \exp\left(-x\left(\frac{k_f}{D_A}\right)\right)^{1/2} + c_2 \exp\left(+x\left(\frac{k_f}{D_A}\right)\right)^{1/2} \tag{11.135}$$

At $x = \infty$, $C_A = 0$ and hence, $c_2 = 0$, and from the boundary condition given in Eq. (11.120), c_1 can be seen to be C_0.

$$\frac{C_A}{C_0} = \exp\left(-x\left(\frac{k_f}{D_A}\right)\right)^{1/2} \tag{11.136}$$

Maternal Effect Genes. Solutions to the damped wave diffusion and relaxation equation in 1 dimension can be used to interpret a variety of biological phenomena. Such examples are from developmental biology, drug design and neuroscience. Maternal effect genes are segregated to defined regions in the developing embryo. One of these genes, which encodes a protein called bicoid, is concentrated at the anterior end of drosophilla embryos (Driever, 1988). Bicoid produced at the anterior end diffuses toward the posterior pole. As the protein diffuses, it can be metabolised. Simultaneous diffusion and elimination produce a stable protein gradient. The cells of the embryo respond to the local concentration of bicoid by expressing certain genes. In this way, the bicoid gradient provides positional information to cells throughout the embryo. These events are critical for the formation of structures throughout the organism. They are achieved by a mechanism for gradient formation, diffusion with homogeneous elimination which can be explained by the steady state model given above.

Drug Penetration in the Tissue. The damped wave diffusion and relaxation equation can be used to develop a simple, quantitative method for predicting the extent of drug penetration into a tissue following the introduction of a local source. Using Eq. (11.123) the drug penetration is quantitated. When a gradient of drug concentration is present, the region of tissue nearest the implant will be exposed to high, possibly toxic, drug levels. The region farthest from the implant will be untreated. An effectiveness factor for drug delivery can be defined as the ratio of the average concentration at the implant/tissue interface C_o,

$$\Omega = \frac{\langle C_A \rangle}{C_0} = \frac{\int_0^L C_A \, dx}{\int_0^L dx} \tag{11.137}$$

where L is the distance into the tissue that requires treatment with the drug. For values of Ω near unity, the overall concentration profile is nearly flat, and solute delivery to this region of the tissue is effective. Substituting Eq. (11.136) into Eq. (11.137) yields,

$$\Omega = 1/\phi \, (1 - \exp(-\phi)) \tag{11.138}$$

where the dimensionless parameter ϕ is defined after Thiele (1939) as:

$$\phi^2 = L^2 k_f / D \tag{11.139}$$

The effectiveness factor can also be defined for the transient case.

$$\frac{\partial^2 u}{\partial X^2} = \frac{\partial^2 u}{\partial \tau^2} + (k_f^* + 1)\frac{\partial u}{\partial \tau} + k_f^* u \tag{11.140}$$

Obtaining the Laplace transform of Eq. (11.140),

$$\frac{d^2 u}{dX^2} = (s^2 + s(k_f^* + 1) + k_f^*)u \tag{11.141}$$

$$u = c_1 \exp(-X(s^2 + s(k_f^* + 1) + k_f^*)) + c_2 \exp(X(s^2 + s(k_f^* + 1) + k_f^*)) \tag{11.142}$$

From the boundary condition at $X = \infty$, c_2 can be seen to be zero. From the boundary condition at $X = 0$,

$$u = \left(\frac{1}{s}\right) \exp(-X(s^2 + s(1 + k_f^*) + k_f^*)^{1/2} \tag{11.143}$$

Eq. (11.137) in the Laplace domain can be written as:

$$\Omega = \frac{\langle C_A \rangle}{C_0} = \frac{C_0 \int_0^L u \, dx}{\int_0^L dx} \tag{11.144}$$

Eq. (11.143) substituted in Eq. (11.144) yields;

$$\Omega = \frac{C_0}{L}\frac{1}{s}(s^2 + s(k_f^* + 1) + k_f^*)^{1/2} \, (1 - \exp(-L(s^2 + s(k_f^* + 1) + k_f^*)^{1/2}) \tag{11.145}$$

$$= \frac{C_0}{L} \int_0^{\tau} \exp\left(\frac{-p(k_f^* + 1)}{}\right) I_0 \frac{p}{2} \,|(k_f^* - 1|\, dp$$

$$= \frac{C_0}{L} \int_0^{\tau} \exp\left(\frac{-p(k_f^* + 1)}{2}\right) I_0 \left(\frac{|(k_f^* - 1|(\tau^2 - L^2)^{1/2}}{2}\right) dp \qquad (11.146)$$

It can be seen that as ϕ increases, the rate of drug diffusion decreases with respect to the rate of elimination. This results in steep concentration gradients or ineffective solute delivery to the tissue. The physico-chemical properties of the drug which determine ϕ on the extent of drug penetration in tissue (Saltzman and Radomsky, 1991).

Neurotransmission Across the Synaptic Cleft. These models can be used to predict the movement of molecules in more complex geometrical arrangements such as the synaptic cleft. Neurotransmitter molecules are released from vesciles in the presynaptic neuron into the synaptic cleft. The region of space between the presynaptic neuron sending a signal and the postsynaptic neuron receiving a signal can possess different characteristic geometries. The rectangular distance across the cleft, the spherical vesicle that releases the neurotransmitter and the cylindrical patch of receptors that sense the signal on the postsynaptic neuron. The concentration of neurotransmitter within the synaptic cleft n can be solved for from the damped wave diffusion and relaxation equation. Written on a molar basis in three dimensions:

$$-\nabla J + f(x, y, z, t) = \frac{\partial n}{\partial t} \qquad (11.147)$$

The damped wave diffusion and relaxation equation in three dimensions can be written as:

$$J = -\nabla C - \tau_{mr} \frac{\partial J}{\partial t} \qquad (11.148)$$

Differentiating Eq. (11.147) wrt to t and operating Eq. (11.148) wrt ∇ and eliminating $\nabla \partial J/\partial t$,

$$D_n \left(\frac{\partial^2}{\partial x^2} + \frac{\partial^2}{\partial y^2} + \frac{\partial^2}{\partial z^2}\right) n + f(x, y, z, t) = (f' + 1) \frac{\partial n}{\partial t} + \tau_{mr} \frac{\partial^2 n}{\partial t^2} \qquad (11.149)$$

A continuous source function that is consistent with experimental data is,

$$f(x, y, z, t) = q.\ \exp(-1/b(x^2 + y^2) - z^2/c)\ .\ t^{\alpha} \exp(-\beta t) \qquad (11.150)$$

where q is related to the number of neurotransmitter molecules initially in the vesicle, b and c are Gaussian variances in the lateral and z directions, limited by the size of the vesicle opening and cleft width d_{syn}, and α and β are parameters.

Neurotransmitter molecules diffuse in the cleft and do not cross the membrane boundaries:

$$\frac{\partial n}{\partial z} = 0, \text{ at } \quad z = 0 \text{ and } z = d_{\text{syn}} \tag{11.151}$$

These equations were solved for the special case when $\tau_{\text{mr}} = 0$ by Klienle et al (1996).

Enzymatic Reactions. Biological systems are chemically and structurally complex with slow rates of diffusion. As the diffusing molecules become larger, it is obstructed by physical structures. Growth and metabolism, the key processes of life, occur through an orderly, regulated, coupled array of biochemical reactions. The reaction between any two biochemical substrates A and B requires contact or collision between the two reactants. The rate of the reaction is often determined by the frequency of collisions. When the frequency of collisions is an important and significant consideration, the finite speed diffusion effects may become important. Most cellular functions depend on enzymatic reactions. Intracellular enzymes participate in several hundred reactions per second. A typical intracellular compartment with dimensions of the order of magnitude of 1 μm is mixed by the process of damped wave diffusion and relaxation. Every two molecules collide at a frequency of 1 second in an intracellular compartment with a diffusion time of 10 ms and typical diffusivity 10^{-10} m^2/s.

The enzyme-substrate is often analyzed by a mechanism as shown in Figure 11.3. The overall reaction of solutes A and B occurs in reversible steps. The first step involves collision in which the two solutes come in proximity with each other to allow

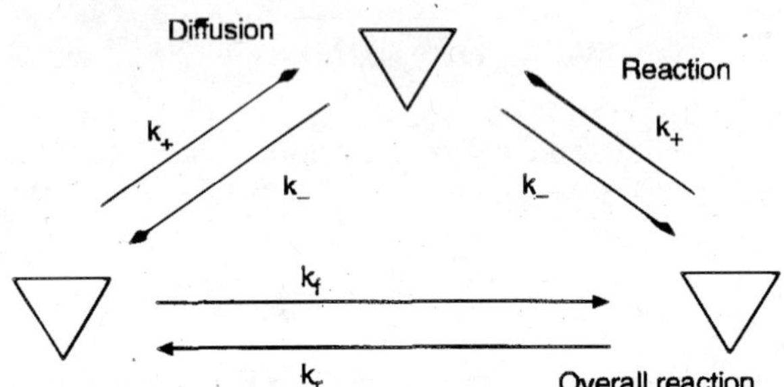

FIGURE 11.3 Mechanisms for diffusion limited reactions.

for the reaction. The second step involves the intrinsic rate of reaction. Mass balance equations for the reaction kinetics are written for the species A, B, A–B and A–B. The intermediate species A–B reaches steady state concentration rapidly compared to the other species.

$$\frac{dC_{\text{AB}}}{dt} = k_f C_A C_B - k_r C_{\text{AB}} \tag{11.152}$$

where

$$k_f = k_{+1} k_+ /(k_- + k_1)$$

$$k_r = k_- k_{-1} /(k_1 + k_-) \tag{11.153}$$

The overall apparent rates of the forward and reverse reaction (k_f and k_r) can be written in terms of rates of encounter (k_+ and k_-) and the intrinsic forward and reverse reaction rates (k_+ and k_-). The encounter step involves the simultaneous diffusion of two separated solute species that collide during random motion. The kinetics of this event (k_+ and k_-) can be determined by analysis of the diffusion equation. The concept of binding constant was discussed in the previous section to describe the diffusion of solutes to the surface of the cell. If the diffusion is toward the solute molecule which is fixed in space and is surrounded by solute B that can diffuse with diffusion coefficient D equal to D_A and D_B, the rate of collision is equal to the steady-state flux of B molecules to the surface of the A molecule:

$$k_+ \, C_{B,\infty} = -D(4\pi a^2)\frac{\partial C_B}{\partial t} \tag{11.154}$$

where $C_{B,\infty}$ is the bulk concentration of solute B and a the radius of the encounter complex A–B that is formed upon collision of B with A. Substituting the steady state concentration profile C_B,

$$k_+ = 4\pi a D \tag{11.155}$$

This reaction rate is for a single molecule of A. The reverse rate constant can be estimated in a similar fashion. The reverse reaction starts with a single B molecule bound in the encounter complex with volume 4 $\pi a^3/3$, which must diffuse subsequently into an unbounded fluid in which the concentration is negligible. The rate of dissociation of the complex is equated to the rate of diffusion of B molecules away from the complex surface:

$$k_-\left[\frac{1}{4}\,\frac{\pi a^3}{3}\right] = -\,D(\pi a^2)\,\frac{\partial C_B}{\partial t} \tag{11.156}$$

Equation (5.322) after substitution of C_B yields;

$$k_- = 3D/a^2 \tag{11.157}$$

The encounter rate constants k_+ and k_- can usually be estimated from physical properties of the two reactants such as the diffusion coefficient and encounter radius. The relative importance of diffusion vs reaction in determining the overall rate of reaction can be estimated by using Eq. (5.319).

The relative rates of diffusion and reaction are important in the regulation of biochemical pathways. Transcription factors are proteins that bind to specialised regions of DNA and thereby facilitate transcription by RNA polymerase. The binding of one class of transcription factors the basic leucine zipper, bZIP proteins has been found to be diffusion limited and the wave diffusion may be an important consideration. The bZIP factors have a C terminal leucine zipper domain, a basic region that binds to DNA, and a domain that is important for transcriptional regulation. Stable association of the bZIP dimer with DNA occur via a monomer of dimer pathway. Binding of both monomer and dimer was found to be diffusion limited. Analysis of the kinetics of both pathways suggests that the monomer pathway may have an overall kinetic advantage. The bZIP factors must dimerize for transcriptional activity and therefore it is assumed that dimerization occurs before DNA binding.

The monomer and dimer binding, the reactions 3 and 2, were found to be rapid so that the dimerization reaction is not necessary for the function of bZIP factors. The monomer pathway as shown in Figure 11.4, may provide kinetic advantages in transcriptional regulation. Monomeric bZIP can bind to DNA more rapidly than dimeric bZIP. Monomers can slide along the side of the DNA backbone. Silding represents an opportunity for bZIP to perform a one-dimensional search of the DNA strand. Faster searching allows the transcription factor to find its target rapidly. Binding to the target is stabilised by dimerization of bZIP at the target site. The one-dimensional diffusion in the function of the DNA binding proteins has been studied with the restriction endonuclease EcoRV, which cleaves the DNA polymer after recognising a specific restriction site on the DNA strand. EcoRV recognises and cleaves GAT↓ATC sites on DNA. The overall rate of cleavage was determined for PCR amplified DNA sequences of different length, but each containing one restriction site. The role of linear diffusion in the overall kinetics of cleavage can be modeled [Jeltsch and Pingoud, 1998]:

$$\frac{k_{on}}{k_{+1}} = \sum_{1}^{L} \exp\frac{(-(n - s)^2)}{P} \tag{11.158}$$

where k_{on} is the overall rate of association with the target site at position s. k_{+1} is the rate of association of EcoRV with a site n, and P is the probability that the enzyme will diffuse one basepair along the DNA rather than dissociate from the DNA polymer. P is equal to the ratio of the rate constant for diffusion (k_{diff}) to the rate constant for dissociation (k_{off}).

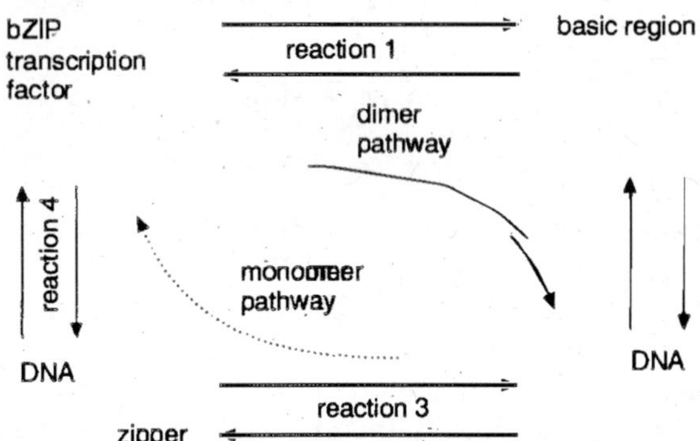

FIGURE 11.4 Mechanism for bZIP transcription factor.

In most forms of drug delivery, spatial localisation and duration of drug concentration are constrained by organ physiology and metabolism. Drugs administered orally will distribute to tissues based on the principles of wave diffusion, relaxation, permeation and flow. Controlled delivery systems offer an alternative approach to regulating both the duration and spatial localisation of therapeutic agents. In controlled delivery, the active agent is combined with other components to produce delivery system. Controlled drug delivery systems frequently involve combinations of active agents with inert polymeric materials. Controlled delivery systems are distinguished from sustained release drug formulations. Sustained release is often achieved by mixing an active agent with binders that alter the agent's rate of dissolution in the intestinal tract or adsorption from a local injection site. Controlled delivery systems include a component that can be engineered to regulate an essential characteristic, e.g. duration of release, rate of release or targeting and have a duration of action longer than a day.

Transdermal Delivery Systems. Many classes of polymeric materials are now available for use as biomaterials. Silicone, polyethylene, polyurethanes, PMMA and EVAc account for the majority of polymeric materials currently used in clinical applications. Reservoir drug delivery devices in which a liquid reservoir of drug is enclosed in a silicone elastomer tube were demonstrated to provide controlled release of small molecules several decades ago [Folkman and Long, 1964]. Lipophilic drugs to the eye or skin such as Ocusert, Alza corp., Estraderm and Trasderm Nitro, Ciba Giegy, offer a number of advantages. In most common reservoir and transdermal systems, release from the reservoir into the external solution occurs in three steps:

1. dissolution of the drug in the polymer
2. diffusion of drug across the polymer membrane
3. dissolution of the drug into the external phase

Assuming that the rate of diffusion across the membrane is much slower than the rate of dissolution portioning at either interface so that portioning can be assumed to be at equilibrium, the kinetics of release can be modeled by performing a mass balance on the drug within the polymer membrane.

For a slice control volume in the membrane with thickness Δx, a mass balance on the diffusing drug molecule yields,

$$\frac{\tau_{\rm mr}\partial^2 c_{\rm p}}{\partial t^2} + \frac{\partial c_{\rm p}}{\partial t} = D_{\rm ip}\frac{\partial^2 c_{\rm p}}{\partial x^2} \tag{11.159}$$

where $D_{\rm ip}$ is the diffusion coefficient for the drug within the polymer material and $c_{\rm p}$ is the concentration of the drug (mg/mL) within the polymer. Eq. (11.159) implies that there is no bulk flow or generation/consumption of drug within the polymer (Fig. 11.5).

For a drug diffusing within a polymeric material, the density does not change with time, and the drug is always present at low concentrations within the material. The concentration within the membrane can be solved for subject to the following conditions;

$$t = 0,\ 0 < x < L,\ c_{\rm p} = c_{\rm p,0} \tag{11.160}$$

$$t > 0,\ x = 0,\ c_{\rm p} = c_{\rm p,1} \tag{11.161}$$

$$t > 0,\ x = L,\ c_{\rm p} = c_{\rm p,2} \tag{11.162}$$

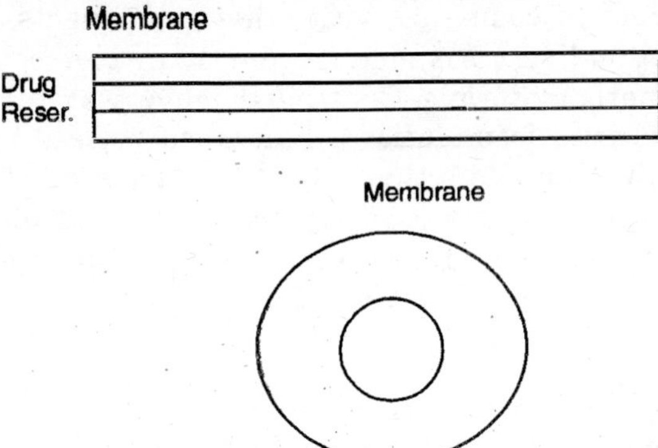

FIGURE 11.5 Reservoir delivery systems based on rate limiting polymer membranes.

The initial condition is given by Eq. (11.160) and the boundary conditions are given by Eqs. (11.161, 11.162). The concentration may be assumed to be the sum of the transient and steady state part;

Let
$$c_p = c_p^{ss} + c_p^t \tag{11.163}$$

$$0 = D_{ip} \frac{\partial^2 c_p^{ss}}{\partial x^2} \tag{11.164}$$

$$c_p^{ss} = c_1 x + c_2$$

From Eq. (11.161), c_2 can be seen to be $c_{p,1}$. From Eq. (11.162),

$$c_1 = (c_{p,2} - c_{p,1})/L \tag{11.165}$$

Thus,

$$\frac{(c_p^{ss} - c_{p,1})}{(c_{p,2} - c_{p,1})} = \frac{x}{L} \tag{11.166}$$

The rest of the problem is in solving for the transient portion of the solution,

$$\tau_{mr} \frac{\partial^2 c_p^t}{\partial t^2} + \frac{\partial c_p^t}{\partial t} = D_{ip} \frac{\partial^2 c_p^t}{\partial x^2} \tag{11.167}$$

The boundary conditions will now be homogeneous;

$$c_p^t = 0, \; x = 0 \tag{11.168}$$

$$c_p^t = 0, \; x = L \tag{11.169}$$

Let
$$\tau = t/\tau_{mr}; \; X = x/(D_{ip}\tau_{mr})^{1/2} \tag{11.170}$$

$$\frac{\partial^2 c_p^t}{\partial \tau^2} + \frac{\partial c_p^t}{\partial \tau} = \frac{\partial^2 c_p^t}{\partial X^2} \tag{11.171}$$

Let $c_p = w \exp(-\tau/2)$ and Eq. (11.171) less the damping term can be written in terms of w as shown in the earlier sections as:

$$\frac{\partial^2 w}{\partial \tau^2} - \frac{w}{4} = \frac{\partial^2 w}{\partial X^2} \tag{11.172}$$

Eq. (11.172) can be solved by the separation of variables as shown in Sharma (2005).

Let

$$w = g(X)\, V(\tau) \tag{11.173}$$

$$\frac{V''}{V} - \frac{1}{4} = \frac{g''}{g} = -\lambda_n^2 \tag{11.174}$$

$$g = c_1 \sin(\lambda_n X) + c_2 \cos(\lambda_n X) \tag{11.175}$$

At $X = 0$, $g = 0$, so c_2 can be seen to be zero.
At $X = L/(D_{ip}\tau_{mr})^{1/2}$, $g = 0$

$$\frac{\lambda_n L}{(D_{ip}\tau_{mr})^{1/2}} = n\pi, \qquad n = 1, 2, 3, \ldots \tag{11.176}$$

$$V = c_3 \exp\left(+\tau\sqrt{1/4 - \lambda_n^2}\right) + c_4 \exp\left(-\tau\sqrt{1/4 - \lambda_n^2}\right) \tag{11.177}$$

When

$$\tau = \infty,\ c_p^t = w \exp(-\tau/2) = 0 \tag{11.178}$$

Thus it can be seen that c_3 can be set to zero in Eq. (11.177).

Thus,

$$c_p = \sum_1^{\infty} c_n \sin(\lambda_n X) \exp\left(-\tau\sqrt{1/4 - \lambda_n^2}\right) \tag{11.179}$$

From the initial condition,

$$c_{p,0} = \sum_1^{\infty} c_n \sin(\lambda_n X) \tag{11.180}$$

multiplying both sides by $\sin(\lambda_m X)$ and integrating between $X = 0$, X_L, it can be seen that all the terms other than the one when $\lambda_n = \lambda_m$ will become zero. Thus,

$$2c_{p,0}\,\frac{1 - (-1)^n}{n\pi} = c_n \tag{11.181}$$

Further it can be seen that when $L < \pi(D_{ip}\tau_{mr})^{1/2}$ subcritical damped oscillations in concentration can be seen. The solution for the concentration can be seen as:

$$c_p = c_{p,1} + (c_{p,2} - c_{p,1})\frac{x}{L} + 2c_{p,0}\sum_1^{\infty}\frac{1 - (-1)^n}{n\pi}\sin(\lambda_n X)\exp\left(-\frac{\tau}{2}\right)\exp\left(-\tau\sqrt{(1/4 - \lambda_n^2)}\right) \tag{11.182}$$

For small membranes, $L < \pi(D_{ip}\tau_{mr})^{1/2}$, subcritical damped oscillations in concentration can be found.

$$c_p = c_{p,1} + (c_{p,2} - c_{p,1})\frac{x}{L} + 2c_{p,0}\sum_1^\infty \frac{1-(-1)^n}{n\pi}\exp\left(-\frac{\tau}{2}\right)\sin(\lambda_n X)\cos\left(\tau\sqrt{(\lambda_n^2 - 1/4)}\right) \tag{11.183}$$

$$\text{or}\quad c_p = c_{p,1} + (c_{p,2} - c_{p,1})\frac{x}{L} + c_{p,0}\sum_1^\infty \frac{1-(-1)^n}{n\pi}\exp\left(-\frac{\tau}{2}\right)\sin(\lambda_n X + \tau(\lambda_n^2 - 1/4)^{1/2})$$
$$+ \sin(\lambda_n X - \tau(\lambda_n^2 - 1/4)^{1/2}) \tag{11.184}$$

The gradient of concentration at the external surface of the membrane for small systems are,

$$-\left.\frac{\partial c_p}{\partial X}\right|_{XL} = \frac{c_{p,2} - c_{p,1}}{L} + \left(c_{p,0}\left(\frac{D_{ip}\tau_{mr}}{L}\right)^{1/2}\right)\sum_1^\infty(1-(-1)^n)\exp\left(-\frac{\tau}{2}\right)$$
$$\cos\left(n\pi + \tau\left(D_{ip}\tau_{mr}\left(\frac{n\pi}{L}\right)^2 - \frac{1}{4}\right)^{1/2}\right) + \cos\left(n\pi + \tau\left(D_{ip}\tau_{mr}\left(\frac{n\pi}{L}\right)^2 - \frac{1}{4}\right)^{1/2}\right) \tag{11.185}$$

The mass flux is then given by

$$J = \exp\left(-\frac{t}{\tau_{mr}}\right) + \frac{c_{p,2} - c_{p,1}}{L} + \left(D_{ip}c_{p,0}\left(\frac{D_{ip}\tau_{mr}}{L}\right)^{1/2}\right)\sum_1^\infty(1+(-1)^n)\exp\left(-\frac{\tau}{2}\right)$$
$$\sin\left(\tau\left(D_{ip}\tau_{mr}\left(\frac{n\pi}{L}\right)^2 - \frac{1}{4}\right)^{1/2}\right) + \sin\left(\tau\left(D_{ip}\tau_{mr}\left(\frac{n\pi}{L}\right)^2 - \frac{1}{4}\right)^{1/2}\right) \tag{11.186}$$

The total amount of material leaving the membrane M_t can be found by integrating the mass flux through the external surface with respect to time:

$$M_t = A\int_0^t J_{x|x=L}\,dt = \left(\frac{At}{L}\right)D_{ip}(c_{p,2} - c_{p,1}) - \tau_{mr}\exp\left(-\frac{t}{\tau_{mr}}\right)$$
$$+ \left(D_{ip}c_{p,0}\left(\frac{D_{ip}\tau_{mr}}{L}\right)^{1/2}\right)\sum_1^\infty(-1+(-1)^{n+1})\left(D_{ip}\tau_{mr}\left(\frac{n\pi}{L}\right)^2 - \frac{1}{4}\right)^{1/2}\exp\left(-\frac{\tau}{2}\right).$$
$$\cos\left(\tau\left(1 + D_{ip}\tau_{mr}\left(\frac{n\pi}{L}\right)^2 - \frac{1}{4}\right)^{1/2}\right) - 2 + \cos\left(\tau\left(D_{ip}\tau_{mr}\left(\frac{n\pi}{L}\right)^2 - \frac{1}{4}\right)^{1/2}\right) \tag{11.187}$$

For small systems, such as when $L < \pi(D_{ip}\tau_{mr})^{1/2}$ as shown in Eq. (11.187) the total amount of material leaving the membrane is subcritical damped oscillatory. A is the cross-sectional area of the external surface of the device which is equal to the cross-sectional area for the transdermal system and twice the area of the transdermal system and twice the cross-sectional area for the planar reservoir system.

Diffusion through Cylindrical Membranes. For implantable reservoir devices, a cylindrical geometry is more practical than a planar arrangement. Consider a cylindrical reservoir surrounded by a polymeric membrane. The cylinder has length L, cross-sectional radius b, and a wall thickness $b–a$. The rate of drug release from this cylinder can be modified by changing the geometry of the device, i.e. by changing b, b/a or L or by changing the drug/polymer combination. The governing equation,

$$\tau_{mr}\frac{\partial^2 c}{\partial t^2} + \frac{\partial c}{\partial t} = D_{ip}\frac{1}{r}\left(\frac{\partial}{\partial r}\frac{r\partial c}{\partial r}\right) \tag{11.188}$$

If the inside of the cylinder is maintained at constant concentration of drug, $c = c_1$ at $r = a$, and the outside of the cylinder is free of drug, $c = 0$ at $r =$ b and the cylinder wall is initially saturated with drug, $c = c_1$ at $a < r < b$, then Eq. (11.188) can be solved analytically to obtain c as a function of position in the cylinder wall.

Let
$$\tau = \frac{t}{\tau_{mr}};\quad u = \frac{c}{c_1};\quad X = \frac{r}{\sqrt{(D_{ip}\tau_{mr})}} \tag{11.189}$$

$$\frac{\partial^2 u}{\partial \tau^2} + \frac{\partial u}{\partial \tau} = \frac{1}{X}\frac{\partial}{\partial X}\left(\frac{X\partial u}{\partial X}\right) \tag{11.190}$$

The dimensionless concentration may be assumed to consist of the sum of the steady state and transient parts;

Let
$$u = u^{t} + u^{ss} \tag{11.191}$$

$$u^{ss} = c' \ln X + c'' \tag{11.192}$$

From the boundary conditions,

$$X = X_a,\ u = 1 \tag{11.193}$$

$$X = X_b,\ u = 0 \tag{11.194}$$

$$1 = c' \ln X_a + c'' \tag{11.195}$$

$$0 = c' \ln X_b + c'' \tag{11.196}$$

$$-1 = c' \ln(b/a) \tag{11.197}$$

$$c' = 1/\ln(a/b) \tag{11.198}$$

$$c'' = \ln X_b/\ln(b/a) \tag{11.199}$$

$$u^{ss} = \frac{\ln(X/X_b)}{\ln(a/b)} = \frac{\ln(r/b)}{\ln(a/b)} \tag{11.200}$$

The steady state concentration profile can be seen to be independent of the diffusivity. The total mass of the drug released from the external surface of the cylinder released at time t, M_t is found by first calculating the mass flux at the external surface $r = b$, and then integrating with respect to time.

$$M_t = 2\pi bL \int_0^t J_{r=b} dt = \frac{2\pi bLt\, v_m c_1}{\ln(b/a)} \tag{11.201}$$

where v_m is the velocity of mass diffusion. Eq. (11.201) assumes that the mass flux is at steady state during the entire time of consideration. This may be an idealisation. For short times the mass flux can be expected to be transient. The transient portion of the concentration profile can be obtained as follows:

$$\frac{\partial^2 u^t}{\partial \tau^2} + \frac{\partial u^t}{\partial \tau} = \frac{1}{X}\frac{\partial}{\partial X}\left(\frac{X \partial u^t}{\partial X}\right) \tag{11.202}$$

The boundary conditions are now homogeneous;

$$X = X_a,\ u^t = 0 \tag{11.203}$$

$$X = X_b,\ u^t = 0 \tag{11.204}$$

$$\tau = 0,\ u^t = 1 \tag{11.205}$$

The damping term in Eq. [11.202] can be removed by a $u = w \exp(-\tau/2)$ to yield:

$$\frac{\partial^2 w}{\partial \tau^2} - \frac{w}{4} = \frac{1}{X}\frac{\partial}{\partial X}\left(\frac{X \partial w}{\partial X}\right) \tag{11.206}$$

Equation (11.206) can be solved for by the method of separation of variables.

Let
$$w = V(\tau)g(X) \tag{11.207}$$

$$\frac{V''}{V} - \frac{1}{4} = \frac{g'/X + g''}{g} = -\lambda_n^2 \tag{11.208}$$

$$X^2 g'' + Xg' + X^2 g \lambda_n^2 = 0 \tag{11.209}$$

$$g = c_1 J_0(\lambda_n X) + c_2 Y_0(\lambda_n X) \tag{11.210}$$

Eq. (11.210) will become infinity at $X = 0$. The dimensionless concentration is always finite and hence c_2 can be set to zero. From the boundary condition,

$$X = X_a,\ u^t = 0 \tag{11.211}$$

$$\lambda_n X_a = 2.4048 + (n-1)\pi \qquad n = 1, 2, 3, \ldots \tag{11.212}$$

$$V = c_3 \exp\left(+\tau\sqrt{1/4 - \lambda_n^2}\right) + c_4 \exp\left(-\tau\sqrt{(1/4 - \lambda_n^2)}\right) \tag{11.213}$$

As time becomes infinitely large, the transient portion of the solution becomes zero and hence c_3 needs to be set to zero.

Thus,

$$u^t = \Sigma_1^\infty c_n J_0(\lambda_n X) \exp\left(-\frac{\tau}{2}\right) \exp\left(-\tau\sqrt{(1/4 - \lambda_n^2)}\right) \tag{11.214}$$

when λ_n subcritical damped oscillations can be seen;

$$a < 4.8096(D_{ip}\tau_{mr})^{1/2} \tag{11.215}$$

Then using De Movrie's theorem and taking the real parts;

$$u^t = \Sigma_1^\infty c_n J_0(\lambda_n X) \exp\left(-\frac{\tau}{2}\right) \cos\left(\tau\sqrt{(\lambda_n^2 - 1/4)}\right) \tag{11.216}$$

The c_n can be solved for from the initial condition and using the principle of orthogonality and multiplying the infinite series with $J_0(\lambda_m X)$ and integrating between the limits of X_a and X_b.

$$c_n = \int_{X_a}^{X_b} \frac{[1 - \ln(X/X_b)]}{\ln(a/b)} \frac{J_0(\lambda_n X)\, dX}{\int_{X_a}^{X_b} J_0^2(\lambda_n X)\, dX} \tag{11.217}$$

It can be seen that the λ_n was solved for from the boundary condition at $X = X_a$. When $\lambda_n X$ is increased from X_a for large b prior to reaching X_b, the zero of the Bessel function may arise and the concentration will become zero from that point onwards. This is the critical region beyond which there is no transfer of concentration. In such cases, the total amount of drug released from the external surface will be less than predicted from calculations. As can be seen from the above analysis; the integrand when solving for the total drug released passes from zero time through a transient stage prior to reaching steady state. For common situations, the time t is often less than within a constant multiple of the time taken to reach steady state. It may contain a zero contribution during the transient stage. Thus for large membrane systems there can be expected a time lag prior to the origin of the external release of the drug. Thus $(b-a)/v_m$ is an estimate of the time taken for the external surface to see the drug from the reservoir prior to its release (Figure 11.5). This time lag can be calculated. At the region where the concentration ceases to move, $\partial c/\partial r = 0$,

$$\frac{c_1}{(D_{ip}\tau_{mr})^{1/2}} \frac{\partial u^t}{\partial X} = \Sigma_1^\infty c_n \lambda_n J_1(\lambda_n X) \exp\left(-\frac{\tau}{2}\right) \cos\left(\tau\sqrt{\lambda_n^2 - \frac{1}{4}}\right) = 0 \tag{11.218}$$

when $(2.4048 + (n-1)\pi)\, X_c = 3.8317 + (n-1)\pi$ (11.219)

or

$$X_c = \frac{(3.8317 + (n - 1)\pi)}{(2.4048 + (n - 1)\pi)} \tag{11.220}$$

For $n = 1$, when $X_c > 1.5934$, the first term in the infinite series has to be set to zero. For $n = 2$, when the zero of the Bessel occurs at $X_c > 1.258$ the second term in the infinite series has to be set to zero in order to keep the predictions for concentration from becoming negative. Thus for $X_c > 1.5934$, or

$$a < \frac{r_c}{(D_{ip}\tau_{mr})^{1/2}} < b \tag{11.221}$$

where $r_c > 1.5934(D_{ip}\tau_{mr})^{1/2}$ and less than b the concentration will be zero.

Matrix Delivery System. Proteins, for instance, do not diffuse readily through any of the hydrophobic biocompatible polymers that are usually used for implantable reservoir systems or they diffuse slowly. Membrane materials cannot be found for some therapeutic agents to provide adequate permeability to permit release from a reservoir device. Matrix system for delivery consists of dissolution and dispersion of drug molecules out of a solid polymer phase. Biodegradable materials disappear after implantation. The design of the material and device can control the rate of polymer degradation and dissolution and the rate of drug delivery. In matrix drug delivery system, molecules of drug are dissolved in a biocompatible polymer, producing a homogeneous device with drug molecules uniformly dispersed throughout the material. The drug molecules are released by diffusing through the polymer to the surface of the device from which they are released into the external environment.

The release of a drug that is initially dissolved within a polymer matrix can be predicted by solving for the damped wave diffusion and relaxation equation within the polymer slab;

$$\frac{\partial^2 u}{\partial \tau^2} + \frac{\partial u}{\partial \tau} = \frac{\partial^2 u}{\partial X^2} \tag{11.222}$$

where

$$u = \frac{c - c_{ext}}{c_0 - c_{ext}};\ \tau = \frac{t}{\tau_{mr}};\ X = \frac{x}{(D_{ip}\tau_{mr})^{1/2}} \tag{11.223}$$

and c_{ext} is the concentration of the drug in the external reservoir. The damping term can be removed from Eq. (11.222) by $u = w\exp(-\tau/2)$,

$$\frac{\partial^2 w}{\partial \tau^2} - \frac{w}{4} = \frac{\partial^2 w}{\partial X^2} \tag{11.224}$$

By the method of separation of variables, the solution for w can be obtained for the following time and space conditions as shown in Chapter 2 for transient temperature in a finite slab.

$$u = 1,\ \tau = 0 \tag{11.225}$$

$$u = 0,\ X = 0,\ L/(D_{ip}\tau_{mr})^{1/2} \tag{11.226}$$

$$u = 0,\ \tau = \infty \tag{11.227}$$

$$\partial u/\partial X = 0,\ X = 0 \tag{11.228}$$

A bifurcated solution results. For small width of the slab, $L < 2\pi\sqrt{D_{ip}\tau_{mr}}$, the transient concentration is subcritical damped oscillatory. An exact well bounded solution that is bifurcated depending on the width of the slab is provided. For $L \geq 2\pi\sqrt{D_{ip}\tau_{mr}}$,

$$u = \Sigma_1^\infty c_n \exp\left(-\frac{\tau}{2}\right)\exp\left(-\sqrt{(1/4 - \lambda_n^2)}\ \tau\right)\cos(\lambda_n X) \tag{11.229}$$

where
$$c_n = \frac{4(-1)^{n+1}}{(2n-1)\pi} \quad \text{and} \quad \lambda_n = (2n-1)\pi\frac{\sqrt{D_{ip}\tau_{mr}}}{L} \tag{11.230}$$

For $L < 2\pi\sqrt{D_{ip}\tau_{mr}}$,

$$u = \Sigma_1^\infty c_n \exp\left(-\frac{\tau}{2}\right)\cos\left(\sqrt{(\lambda_n^2 - 1/4)}\ \tau\right)\cos(\lambda_n X) \tag{11.231}$$

The total amount of drug released from the matrix can be determined by integration;

$$M_t = c_0 AL - \int_{-L/2}^{L/2} c(x, t)\, A\, dx \tag{11.232}$$

$$M_t/AL(c_{ext} + c_0) = M^*$$

$$M^* = \frac{1 - 8C^*}{(\pi^2 \mathrm{Rel_m})}\ \Sigma_1^\infty \exp\frac{(-\tau/2)}{(2n-1)^2}\cos\left(\tau\sqrt{\lambda_n^2 - \frac{1}{4}}\right) \tag{11.233}$$

Eq. (11.233) is valid for small membranes, when $L < 2\pi\sqrt{D_{ip}\tau_{mr}}$. For large membrane systems, when $L \geq 2\pi\sqrt{D_{ip}\tau_{mr}}$,

$$M^* = \frac{1 - 8C^*}{(\mathrm{Re}\, l_m \pi^2)}\ \Sigma_1^\infty \exp\left(\frac{-\tau/2}{(2n-1)^2}\right)\exp\left(-\tau\sqrt{1/4 - \lambda_n^2}\right) \tag{11.234}$$

where $C^* = (c_0 - c_{ext})/(c_{ext} + c_0)$

Relaxation number (mass), $\quad \mathrm{Rel_m} = D_{ip}\tau_{mr}/L^2 \qquad (11.235)$

Eq. (11.234) is shown in Figure 11.6. The time taken to steady state or the value at infinite time can be read off from the graph. In this case at $\tau = 0.35$, M^* reaches 1.

This approach to model the release of dispersed or dissolved drugs from a matrix can be easily extended to cylindrical or spherical shapes by solving the governing equation in the cylindrical and spherical coordinates. The macroscopic geometry of a matrix can influence the rate and pattern of protein release.

Krogh Tissue Cylinder. A microscopic view of capillaries in tissue indicates a repetitive arrangement of capillaries surrounded by a cylindrical layer of tissue. An

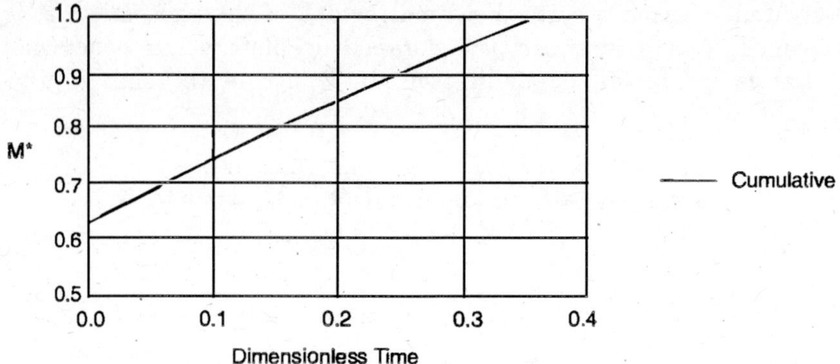

FIGURE 11.6 Dimensionless total drug released from matrix system.

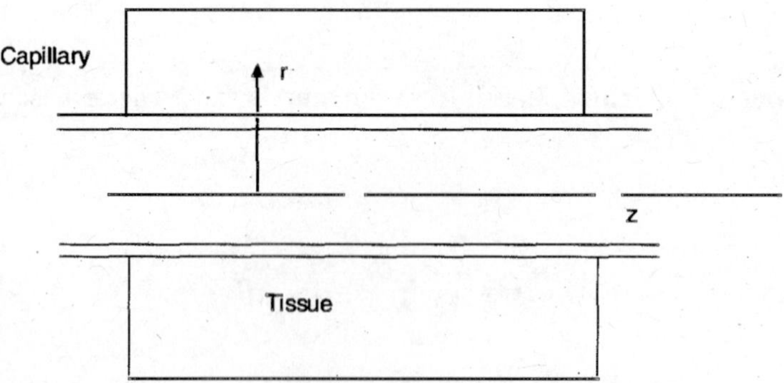

FIGURE 11.7 Geometry of Krogh tissue cylinder.

idealised sketch of the capillary bed and the corresponding layer of tissue idealised into a cylinder is shown in Figure 11.7. Let the radius of the tissue layer be r_T. The residence time of the blood in the capillary is in the order of 1 second. The wave diffusion and relaxation time is comparable in magnitude to the residence time in the blood. Krogh [1919] showed this cylindrical capillary tissue model to study the supply of oxygen to muscle. The tissue space surrounding the capillary is considered a continuous phase albeit it consists of discrete cells. An effective diffusivity D_T can be used to represent the diffusion process in the tissue. The driving force for the diffusion is driven by the consumption of the solute by the cells within the tissue space.

The Michaelis-Menten equation can be used to describe the metabolic consumption of the solute in the tissue space. The equation may be written as:

$$R = \frac{V_m C_T}{K_m + C_T} \tag{11.236}$$

where C_T is the concentration of the solute in the tissue space. For consumption of the solute, R will have a positive value and for solute production, it will have a negative value. V_m represents the maximum reaction rate. The maximum reaction rate occurs when $\langle C \rangle$ is greater than K_m. The reaction rate is then in zero order in solute concentration. The blood flows through the capillary with an average velocity of V. A steady state shell balance on the solute in the blood from z to $z + \Delta z$ can be written as:

$$-V\frac{dC}{dz} = \frac{2}{r_c} K_0 (C - C_T|_{rc+tm}) \tag{11.237}$$

where K_o is represented by an overall mass transfer coefficient. The overall mass transfer coefficient represents combined resistance of fluid flowing through the capillary k_m and the permeability of the solute in the capillary wall P_m. A steady state shell balance at a given value of z from r to $r + \Delta r$ may also be written for the solute concentration in the tissue space,

$$\frac{D_T}{r}\frac{d}{dr}\left(r\frac{dC_T}{dt}\right) - R = 0 \tag{11.238}$$

The boundary conditions for Eqs. (11.237, 11.238), are

$$z = 0,\ C = C_o \tag{11.239}$$

$$r = r_c + t_m,\ C_T = C_T|_{rc+tm} \tag{11.240}$$

$$r = r_T,\ dC_T/dr = 0 \tag{11.241}$$

The axial diffusion is neglected in the tissue space in comparison with the radial diffusion. From the zero order rate of reaction $R = R_o$ is a constant. Solving for Eq. (11.238) and the boundary conditions given in Eqs. (11.240, 11.241),

$$C_T - C_T|_{rc+tm} = (r^2 - (r_c + t_m)^2)\frac{R_o}{4D_T} - \frac{r_T^2 R_o}{2D_T}\ln\left(\frac{r}{r_c + t_m}\right) \tag{11.242}$$

The variation of concentration as a function of z can be calculated by equating the change in solute concentration within the blood to the consumption of solute in the tissue space,

$$C = C_o - \frac{R_o}{Vr_c^2}(r_T^2 - (r_c + t_m)^2)z \tag{11.243}$$

Eq. (11.242) is combined with Eq. (11.243);

$$C_T|_{rc+tm} - C_o = -\frac{R_o}{Vr_c^2}(r_T^2 - (r_c + t_m)^2)\,z - \frac{R_o}{2r_cK_o}(r_T^2 - (r_c + t_m)^2) \tag{11.244}$$

Combining Eqs. (11.242–11.244),

$$C_T - C_o = (r^2 - (r_c + t_m)^2)\frac{R_o}{4D_T} - \frac{r_T^2 R_o}{2D_T}\ln\left(\frac{r}{r_c + t_m}\right) - \frac{R_o}{Vr_c^2}(r_T^2 - (r_c + t_m)^2)z - \frac{R_o}{2r_cK_o}(r_T^2 - (r_c + t_m)^2) \tag{11.245}$$

It can be deduced that under certain conditions some regions may not receive any solute. A critical radius of tissue can be identified, $r_{critical}$ and defined as the distance beyond which no solute is present in the tissue.

At $$r = r_{critical}, \quad \frac{dC_T}{dr} = 0 \quad \text{and} \quad C_T = 0 \tag{11.246}$$

This can be solved for from Eq. (11.245) after replacing r_T with $r_{critical}$. The equation is nonlinear.

Worked Example 11.2 *Krogh Tissue Cylinder in Cartesian Coordinates*
Idealise Figure 11.7 in the cartesian coordinates and obtain the solution for the concentration of the solute in the tissue space.

The governing equations for the concentration of the solute in the capillary and in the tissue can be written after taking the r in Figure 11.7 as x,

$$-V\frac{dC}{dz} = \frac{2}{r_c}K_0(C - C_{T|rc+tm}) \tag{11.247}$$

Considering the effects of diffusion in x direction only in the tissue and assuming a zeroth order reaction rate,

$$D_{AB}\frac{\partial^2 C_T}{\partial x^2} = R_o \tag{11.248}$$

Integrating, and substituting for the boundary conditions,

$$x = x_c + t_m, \; C_T = C_{T|xc+tm} \tag{11.249}$$

$$x = x_T, \; dC_T/dx = 0 \tag{11.250}$$

$$-\frac{R_o}{D_{AB}x_T} = c_1 \tag{11.251}$$

$$C_T - C_{T|xc+tm} = \left(\frac{R_o}{2D_{AB}}\right)(x^2 - (x_c + t_m)^2) - \frac{R_o x_T}{D_{AB}}(x - (x_c + t_m)) \tag{11.252}$$

The variation of concentration as a function of z can be calculated by equating the change in solute concentration within the blood to the consumption of solute in the tissue space:

$$VAC_o - VAC = R_o z A_T \tag{11.253}$$

$$C = C_o - \frac{R_o z A_T}{VA} \tag{11.254}$$

Eq. (11.254) is combined with Eq. (11.253);

$$\frac{R_o A_T}{A} = \frac{2}{x_c} K_0 (C - C_{T|rc+tm}) \tag{11.255}$$

$$C_T|_{rc+tm} = C - K_0 x_c R_o A_T / 2A \tag{11.256}$$

Therefore,

$$C_T - C_o = \frac{R_o z A_T}{VA} + K_0 x_c \frac{R_o A_T}{2A} + \left(\frac{R_o}{2D_{AB}}\right)(x^2 - (x_c + t_m)^2) - \frac{R_o x_T}{D_{AB}}(x - (x_c + t_m)) \tag{11.257}$$

At a critical distance from the capillary wall the concentration in the solute will become zero. This can be solved for from the above equation. At and beyond the critical distance,

$$\frac{dC_T}{dx} = 0 = C_T \tag{11.258}$$

Replacing x_T with $x_{critical}$,

$$0 = C_0 + \frac{R_o z A_T}{VA} + K_0 x_c \frac{R_o A_T}{2A} + \left(\frac{R_o}{2D_{AB}}\right)(x^2 - (x_c + t_m)^2) - \frac{R_o x_{critical}}{D_{AB}}(x - (x_c + t_m)) \tag{11.259}$$

$$x_{critical}^2 \left(-\frac{R_o}{2D_{AB}}\right) = C_0 + \frac{R_o z A_T}{VA} + K_0 x_c \frac{R_o A_T}{2A} - \left(\frac{R_o}{2D_{AB}}\right)(x_c + t_m)^2 - \frac{R_o x_{critical}}{D_{AB}}(x - (x_c + t_m)) \tag{11.260}$$

The quadratic equation in $x_{critical}$ is then,

$$A\, x_{critical}^2 + B x_{critical} + C = 0 \tag{11.261}$$

where,

$$A = \left(-\frac{R_o}{2D_{AB}}\right) \tag{11.262}$$

$$B = +(x_c + t_m)\,\frac{R_o}{D_{AB}} \tag{11.263}$$

$$C = C_0 + \frac{R_o z A_T}{VA} + K_0 x_c \left(\frac{R_o A_T}{2A}\right) - \left(\frac{R_o}{2D_{AB}}\right)(x_c + t_m)^2 + \frac{R_o (x_c + t_m)}{D_{AB}} \tag{11.264}$$

When the solution of the quadratic expression for the critical distance in the tissue are real, and found to be less than the thickness of the tissue then the onset of zero concentration will occur prior to the periphery of the tissue. This zone can be seen as the anorexic or oxygen depleted regions in the tissue.

11.2 ELECTROPHORESIS

The differences in mobility of different species under an electric field is used to obtain a separation between two or more species. The term *electrophoresis* refers to the movement of a solid particle through a stationary fluid under the influence of an electric field. The constituent that migrates under the field can be large molecules, colloids, fibers, clay particles and latex spheres. Electrophoresis is often applied to polymeric and biological samples (see Figure 11.8). It is applied frequently in the analysis of proteins and DNA fragment mixtures. The advancement of biotechnology was in some measure due to the electrophoresis as a tool. Variations of this method are used in obtaining the nucleic acid sequences of DNA, deoxyribonucleic acid, isolating active biological factors associated with diseases such as cystic fibrosis, sickle cell anemia, myelomas, leukemia and establishing immunological reactions between samples on the basis of individual compounds. The technique is sensitive to small differences in molecular charge and mass. It does not interfere with the species during the investigation. Figure 11.8 depicts a schematic of electrophoresis apparatus.

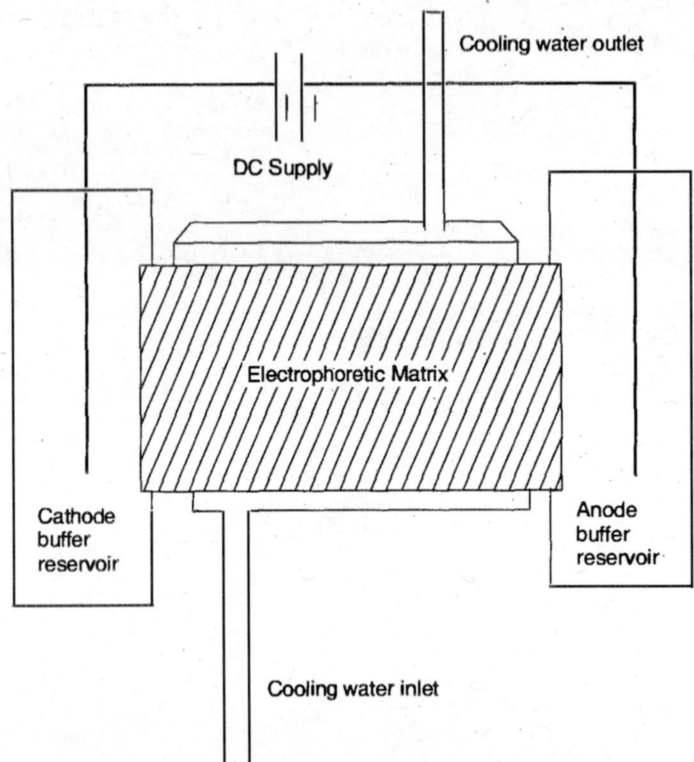

FIGURE 11.8 Schematic of Electrophoresis Apparatus.

The charge separation between the surface of the particle and the fluid surrounding it is tapped in the electrophoresis technique. The particle is caused to move due to the electric force it experiences from the electric field and resulting charge on the particle. The electric field also generates heat. There are different modes of electrophoresis such as:

1. Zone Electrophoresis
2. Isoelectric Focusing
3. Isotachophoresis

Different types of gel matrix can be employed. These are agarose, polyacrylamide, paper, capillaries, flowing buffers. The three modes can be used alone or in combination in the different matrices listed to achieve the target objectives for a given application. The molar N flux after taking into account the electric field effects can be written as:

$$N_A = -z_A u_A F C_A \frac{\partial E}{\partial x} - \frac{D_i \partial C_A}{\partial x} + C_{AV} \tag{11.265}$$

where u_A is the ionic mobility under the action of a unit force per mole, v is the stream velocity, z_AF is the charge per molecule and F is the Faraday's constant in Coulomb/gm. z_A is the valency of the species A.

$$u_A = \frac{D_A}{RT} = \frac{\lambda_A}{|z_A|F^2} \tag{11.266}$$

The ionic mobility u_A is related to the ionic diffusion coefficient. λ_A is the ionic conductance. Eq. (11.265) can be applied to a problem in equimolar counter diffusion without any convection. The rate of change of electric field is taken to be linear with respect to the distance x.

$$N_A + N_B = 0;\ C = C_A + C_B \tag{11.267}$$

$$-F\frac{\partial E}{\partial x}(C_A u_A z_A + (C - C_A)u_B z_B) = (D_{AB} - D_{BA})\frac{\partial C_A}{\partial x} \tag{11.268}$$

$$E = m'x + c \tag{11.269}$$

$$-Fm' \int dx = (D_{AB} - D_{BA}) \int \frac{\partial C_A}{(C_A(u_A z_A - u_B z_B) + C u_B z_B)} \tag{11.270}$$

$$\frac{D_{AB} - D_{BA}}{u_A z_A - u_B z_B} \ln(C_A\ (u_A z_A - u_B z_B) + C u_B z_B) = -Fm'x$$

$$\frac{(C_A(u_A z_A - u_B z_B) + C u_B z_B)}{(C_{A0}(u_A z_A - u_B z_B) + C u_B z_B)} = \exp\left[\frac{-Fm'x(u_A z_A - u_B z_B)}{D_{AB} - D_{BA}}\right] \tag{11.271}$$

The theory of electrophoretic motion has the electric potential on the surface of the object as its basis. A relation between the electric potential and the velocity of the particle can be written. An electric potential can be developed in an interface between two immiscible phases. When a beaker contains salt water solution, the walls have electric potential. Some ions bind onto the glass and the silanol groups present in glass that ionizes in water causing the wall to be charged up. There is a separation of charge that exists in a small layer of the glass and a small layer of salt water adjacent to the glass wall. These two layers are called electric double layer.

Electric potential in the diffuse layer extends into fluid phase and drops off by Poisson distribution;

$$\phi = \phi_0 \exp(-\kappa x) \tag{11.272}$$

where the characteristic κ is given by;

$$\kappa^2 = \left(\frac{F^2 \Sigma C z^2}{\varepsilon \varepsilon_0 RT}\right) \tag{11.273}$$

The velocity of the particle can be written as:

$$\mu = \frac{\varepsilon \varepsilon_0 \xi}{\eta f(\kappa a)}$$

When a >> double layer thickness $f(\kappa a)$ is about 1. It is 1, 5 for very low ionic strength solutions. Electrostatic flow is also dependent on the zeta potential at the immobilised surface and the strength of the electric field. The flow rate F is given by

$$F = \frac{\varepsilon \varepsilon_0 \xi}{\eta (\pi r^2 E)} \tag{11.274}$$

where F is the flow rate in ml/s and r is the radius of flow channel in cm.

The size of the molecule is obtained by calibration with the migration of a reference molecule in electrophoresis. Over a period of time, a number of different types of electrophoresis methods have been developed. Some of them are as follows:

1. Disc Electrophoresis
2. Zone Electrophoresis
3. Native Zone Electrophoresis
4. Reduced SDS Electrophoresis
5. Pulsed Field Gel Electrophoresis
6. Isoelectric Focusing
7. Isotachophoresis
8. Agarose Electrophoresis
9. Polyacrylamide Electrophoresis
10. Paper Electrophoresis
11. Capillary Electrophoresis
12. Force Flow Electrophoresis

11.3 BLEACHING

To bleach something is to remove or lighten its colour; a *bleach* is a chemical that can produce these effects, often via oxidisation. Pulp for the paper and pulp industry must often be bleached in order to produce an end product of suitable high-quality. In the production of chemical pulp, a cellulose-containing material is cooked with suitable cooking chemicals, and the raw pulp thus obtained is delignified and bleached with oxidising chemicals. The purpose of the conventional bleaching of chemical pulp is to complete the removal of lignin from the raw pulp obtained from the cook. The most commonly used bleaching agents today are chlorine and oxygen. There is a tendency to avoid the use of chlorine or at least limit its use to a minimum because of damage caused thereby to the environment. Oxygen is a suitable bleaching agent but its reaction selectivity is not always adequate so that other additional chemicals must be used. For these reasons, new bleaching agents have been sought. Ozone is one of them.

Ozone bleaching has been extensively studied in laboratory and on pilot scale. Ozone has proved to be a satisfactory but expensive bleaching agent and is difficult to use as the consistency of the pulp to be bleached has to be very low or very high due to the high reactivity of the ozone. For example, at low consistencies, i.e. below 5%, ozone is dissolved in water and thus satisfactory transfer of mass between the ozone and the fibres in water is achieved as the ozone containing water can freely flow between the fibres (Figure 11.9). It has also been found that ozone, being a gaseous substance, reacts well directly with a dry fibre surface which presupposes that the consistency is so high, in most cases over 30%, that there is practically no water on the surface of the fibre or between the fibres. In these circumstances the ozone containing gas can freely flow between the fibres.

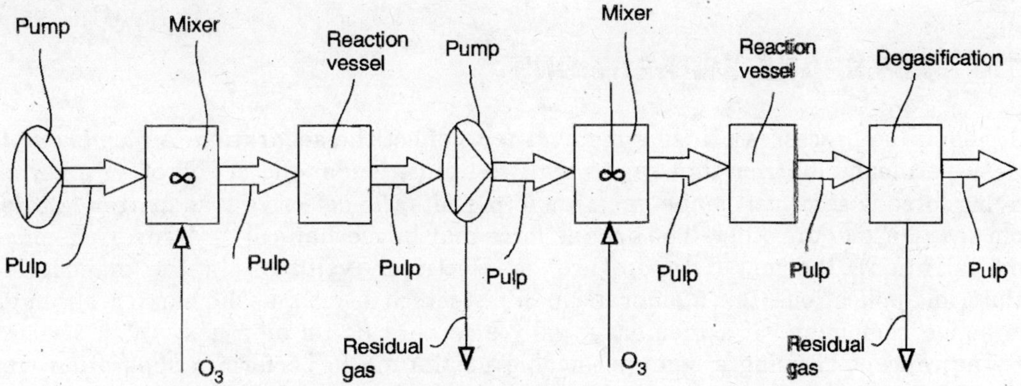

FIGURE 11.9 Ozone Bleaching of Paper Pulp in Kraft Process.

Another example of industrial bleaching is in the food industry. The white bread manufacture involves a bleaching step. The vegetable oils also use a bleaching step. The neutralised oils are mixed with adequate quantity of bleaching earth dosed through an accurate dosing system (Figure 11.10). The mixed oil is then transferred

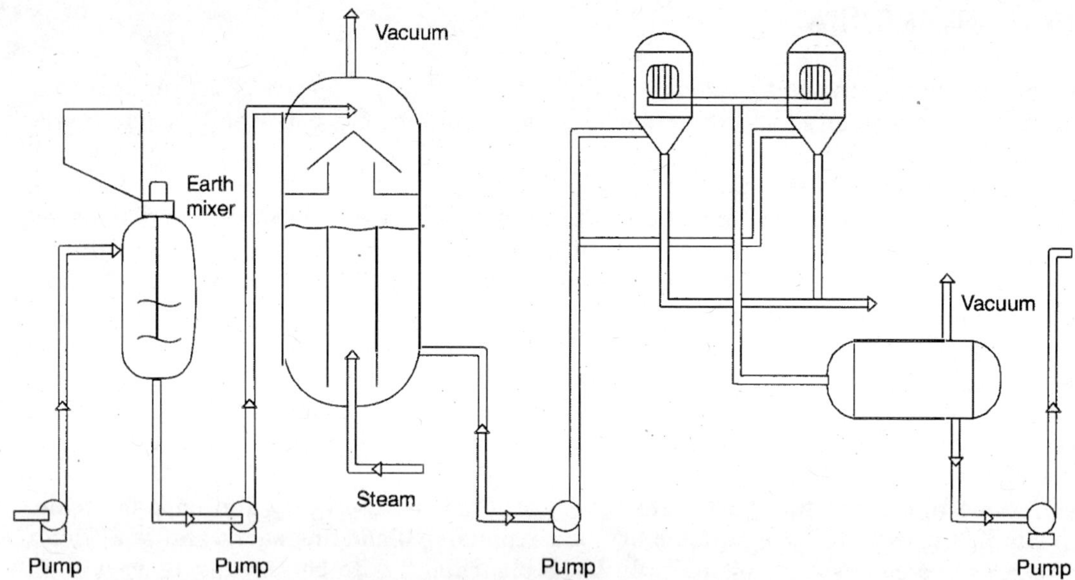

FIGURE 11.10 Example of Industrial Bleaching in Manufacture of Edible Oil.

to a bleacher where oil under vacuum is agitated with steam. This also imparts the wet bleaching effect. The oil is then filtered through pressure leaf filters. The filter leaves used are in special stainless construction with the mesh imported from reputed suppliers. The filter have special pneumatic vibrators for discharge of the cake, and special butterfly valve for easy opening and discharge. The bleached oil produced has very low phosphorous levels and no soap.

11.4 MEMBRANE SEPARATIONS

All membrane processes utilise a membrane to effect the separation. A *membrane* is a semi-permeable barrier that separates two fluids. Under the action of an external driving force, a semi-permeable membrane permits the selective transport of species from one phase to the other. The driving force may be mechanical (pressure), chemical (concentration), thermal (temperature) or electrical (voltage). In the majority of industrial applications, this membrane manifests itself as a thin solid barrier, although the active component of a membrane may also be a liquid or a gas.

The concept of using a semi-permeable membrane to perform a separation was formally discovered 200 years ago. But the use was limited to materials derived from naturally occurring substances. For instance, Abbe Nollet, who is considered to be one of the first to formally report and discuss the process of selective transport through membranes, used the bladder from a pig and the studies conducted by Thomas Graham were performed using 'vegetable parchment'. These and other natural materials demonstrated the phenomenon of selective transport but proved unsuitable for conversion into workable 'industrial-scale' devices. Thus, separation at the molecular level by means of a semi-permeable barrier remained a small scale laboratory technique and did emerge as a potential industrial scale process until the second half of the twentieth century. It was the emergence of suitable synthetic polymeric materials in the 60's that changed the situation. Processes for forming thin membranes combining high selectivity, high flux, chemical stability and operational durability were developed and this set the basis for the commercial development of membrane separation processes.

Synthetic membranes can now be cast in thin films or spun as hollow fibres, free from significant defects. The new synthetic membranes demonstrate a high level of chemical toughness, mechanical strength and durability. As a result, there now exists a range of commercially available membranes that possess the chemical and physical properties, which fit them for long-term operation on an industrial scale.

11.4.1 Industrial Membrane Separations

The food and dairy industry has created a number of significant applications for membrane separations. The dairy industry has employed membrane processes to improve the recovery of products from milk. In particular, membranes have been used to recover protein from whey and enhance the composition of whey protein products. Membrane processes are also used routinely to recover process water from process streams and reduce the biological oxygen demand of discharged streams. One of the latest developments currently undergoing trials in a number of applications is in the recovery of volatile flavour components that would otherwise be lost during processing.

Many of the latest therapeutic products are not manufactured by synthetic chemical routes, but are often produced from cell cultures. The products are generally highly complex molecules, often proteins, and the therapeutic effectiveness of the product depends on the configuration and folding of the molecule. Recovery of these products in their native state is a demanding separation task and represents a considerable challenge to conventional separation techniques. Generally, the products produced by cell culture methods are present only at very low concentrations. In the early stages of development, the products (typically functional proteins) were recovered by adsorption using specific ligands. However, the subsequent separation and recovery from the adsorbent required challenging chemical conditions and this resulted in a low yield of biologically active material. Membrane technology is being used as an alternative recovery technique because it can provide the mild operating conditions necessary for the efficient recovery of sensitive biological material.

The electronics and semiconductor industries need extra pure low conductivity water. On-site production by conventional multi-distillation no longer represents the most cost-effective method of producing high-grade ultra-pure water. Semiconductor plants have a need for ultra-clean environments and thus high-integrity filtration systems are required. The use of absolute membrane filters is now the routine method by which air is cleaned to the necessary standard.

Membranes also have a significant role to play in meeting the tighter restrictions on discharges into the environment. Consequently, membrane processes are now used routinely across a wide spectrum of the chemical industries. There have been some notable success stories, particularly in water and effluent treatment, and hydrogen recovery. Improved membrane production techniques have reduced the cost and increased the reliability and longevity of membranes to such an extent that for some application membranes provide an economically more favourable alternative to established technologies.

Despite the success of many membrane applications, not everything in the membrane garden is rosy. There remains a range of technical constraints and economic factors that militate against the universal adoption of membrane processes. Membranes can separate large molecules from aqueous media but the same membrane processes are generally not capable of fractionating one solute from another. By way of an example, the limitations of a membrane process can be seen in the application of reverse osmosis. Reverse osmosis has been very widely adopted for the large-scale desalination of brackish waters and is now viewed as the preferred alternative to multi-effect evaporation. However, when applied to sea water or strong brines only low concentration factors are possible because of the high osmotic pressures generated by the low molecular weight solutes. To obtain 'reasonable' concentration factors, very high trans-membrane pressures have to be used to overcome the osmotic pressure of the retained brine. This can compromise performance because membranes suffer compaction and at the extreme cause failure. In addition, the pumping costs at high transmembrane pressure can become prohibitive rendering the process uneconomic. Membrane processes are prone to irreversible fouling of the membrane surface, which can over a period of time lead to a significant drop in the flow of material through the membrane.

Because membrane processes are essentially two-dimensional, the economies of scale are low and the processes are built on a modular basis.

Although all membrane separations rely on the selective transport of species, the separations cover a wide range of fundamentally different processes, which operate on a number of different physical principles. Without a clear understanding of the fundamental principles governing the operation of each process, it can be potentially misleading to group all processes together under the single title of Membrane Processes. Table 11.1 sets out a classification of membrane processes.

TABLE 11.1 Membrane Processes

Process	*Pore size/mol wt*	*Driving Force*	*Mechanism*
Microfiltration	0.02–10 μm	ΔP 0–1 bar	Sieving
Ultrafiltration	0.001–0.02 μm mol wt 10^3–10^6	ΔP 0–10 bar	Sieving
Nanofiltration	< 2 nm mol wt 10^2–10^3	ΔP 10–25 bar	Diffusion
Reverse osmosis	non-porous	ΔP 10–100 bar	Diffusion
Gas separation	non-porous	ΔP 10–100 bar	Diffusion
Dialysis	10–30 A	Concentration difference	Sieving & Diffusion
Pervaporation	non-porous	Partial pressure difference	Diffusion
Electrodialysis	mol wt < 200	Electrical potential	Ion migration

The principles and methods for reverse osmosis, ultrafiltration, dialysis and electrodialysis were discussed in Chapter 6. For other systems the general approach employed in the methods shown can be used to calculate the material and energy needs for effecting a desired level of separation.

11.5 POLYMER DEVOLATILISATION

In processes such as the continuous polymerisation process to manufacture, polystyrene, PS, rubber modified polystyrene, HIPS, styrene-acrylonitrile copolymer, SAN, acrylonitrile, styrene, butadiene, ABS engineering thermoplastic, a *devolatiliser* is used to remove the unreacted monomers from the reactor effluent for recycle back to the reactors. The devolatilisation may be accomplished in:

(a) Falling Strand Devolatiliser
(b) Wiped Film Devolatiliser
(c) Single Screw Extruder
(d) Twin Screw Extruder

The devolatilisation temperatures are between 180°C to 250°C and operating pressures of 2–200 mm Hg. The polymerised syrups can be preheated in a shell and tube type heat exchanger prior to devolatilisation. The residual monomer and solvent in the polymer product are less than 400 ppm.

The principles and methods discussed in the section under drying may be applicable here. The vacuum pressure in the second stage of the devolatiliser will determine the equilibrium monomer content in the polymer. In the first stage of the devolatilisation process, the constant rate devolatilisation may be assumed. During this regime, a film of monomers that surrounds the polymer strand will be devolatilised. When the monomer content in the syrup is reduced to where spots of monomer are left in the polymeric solid, the falling rate of devolatilisation can be expected. The critical monomer content in the solid has to be obtained from experimental measurements. The diffusion of monomer in the polymeric solid to the surface of the solid involves diffusion coefficients that behave differently from the small molecules. This was discussed in Chapter 1.

Falling strand devolatilisation has been employed for many years to separate a volatile component from a liquid component. Such devolatilisation procedure has been especially useful in the manufacture of polymers, such as homopolystyrene where as in a continuous mass polymerisation process, a fluid mixture of homopolystyrene with styrene monomer results, which must be subjected to a post-polymerisation treatment to remove the unreacted styrene monomer there from. For such a removal, a falling strand devolatiliser is a convenient piece of apparatus (Figure 11.11).

A falling strand devolatiliser (Figure 11.11) comprises a preheater, a shell and tube heat exchanger adapted to feed heated fluid material from the tubes into a flash tank. The upper portion of the flash tank bears an outlet port which is typically interconnected with a source of vacuum. A fluid mixture being devolatilised is first subjected to a preheating step in the shell and tube heat exchanger. After this the

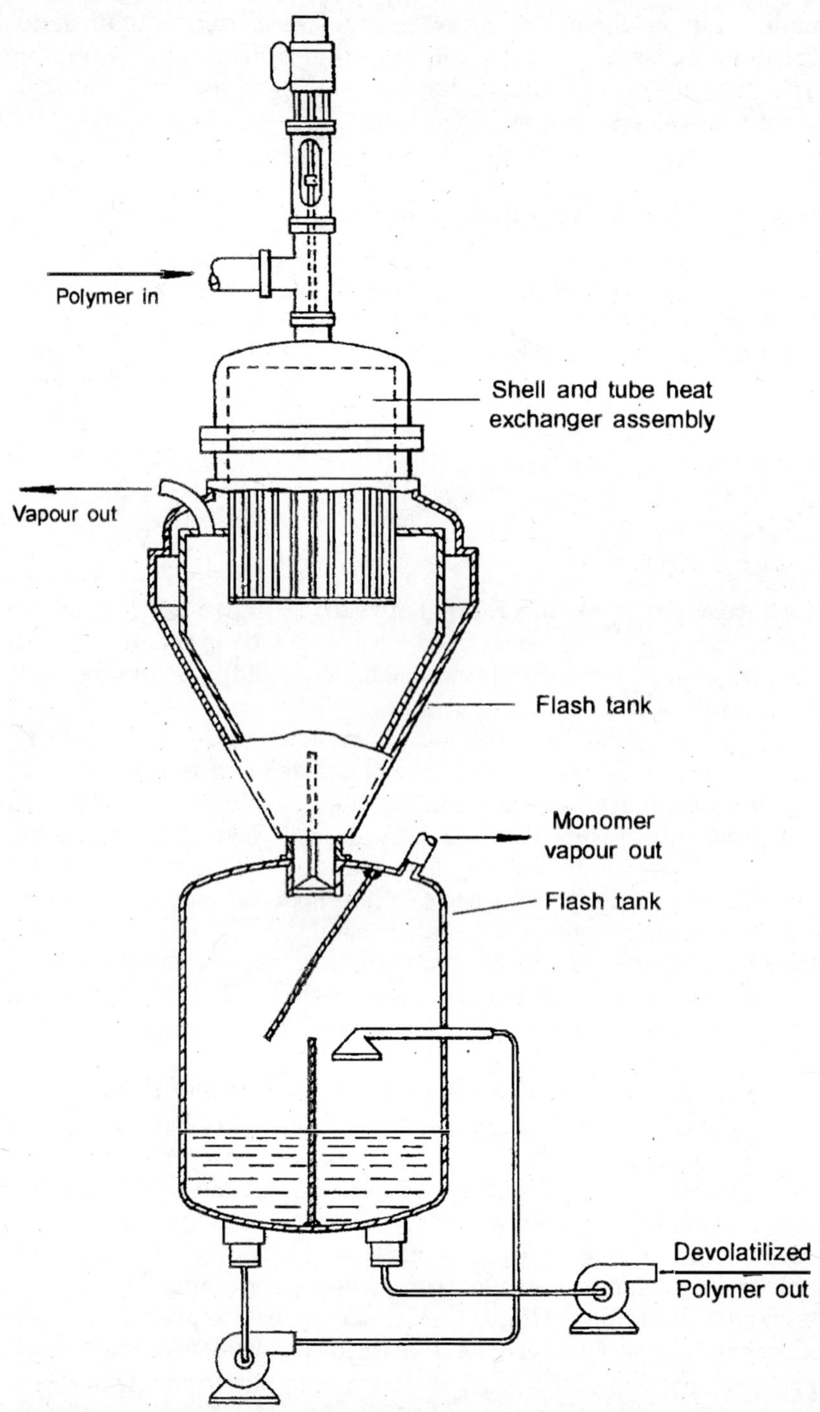

FIGURE 11.11 Two-stage falling strand devolatilisation used by Monsanto Co., St. Louis, MO, USA, in continuous mass polymerisation of HIPS, polystyrene and ABS.

heated mixture is subjected to a flashing step in which the heated mixture is discharged into the flash tank. The pressure and temperature conditions in the flash tank are regulated such that volatiles (e.g. especially monomer) are above their boiling point, while the desired component (e.g. homopolystyrene) is maintained below its boiling or decomposition point. Vaporisation of volatiles is prompted by continuously exhausting vapour of volatiles through the vacuum outlet port of the flash tank. Two successive stages of falling strand devolatilisation are sometimes used in the industry, the conditions of temperature and pressure employed in the second stage being somewhat more rigorous than those used in the first stage. Regulation of first chamber pressure can be used to control the residence time through the preheater, and the stripping effect of high boiling materials and the properties associated with the time/temperature history.

SUMMARY

Controlled drug delivery system offers a better alternative compared with the periodic one. The transient concentration profile in cylindrical and spherical coordinates was obtained using the damped wave diffusion and relaxation equation. The solute binding and elimination was studied using a reversible rate of reaction and an effective diffusivity. The mass inertial lag time associated with the propagation of concentration disturbance was calculated in cylindrical and spherical coordinates and are:

$$\tau_{lag} = \sqrt{X_p^2 - 4\pi^2} \qquad \text{(cylindrical)}$$

$$\tau_{lag} = \sqrt{X_p^2 - 7.6634^2} \qquad \text{(spherical)}$$

Protein profile in maternal effect genes, drug penetration in the tissue and neurotransmission across the synaptic cleft using the damped wave diffusion and relaxation equation was studied. Enzymatic reactions are diffusion limited and their schemes were studied.

The overall mechanism is represented by reversible reactions. Reservoir delivery systems were modelled. The cumulative mass transported across the membrane was calculated. The diffusion in cylindrical membranes was considered. Transdermal systems were mentioned. The drug released from a matrix delivery system was derived and shown as a Figure. The Krogh tissue cylinder was modelled. At steady state for the cylindrical geometry a critical radius beyond which zero mass diffusion can be calculated was derived.

The governing equation requires numerical solution. In cartesian coordinates, the quadratic equation to represent the zone of zero concentration or anorexic or oxygen depleted region was derived.

$$A\,x_{critical}^2 + Bx_{critical} + C = 0$$

where A, B and C are given in terms of the thickness of the capillary, capillary diameter, tissue diameter, binary diffusivity, entrance concentration and area of the tissue.

The technique of electrophoresis was reviewed. The laws of diffusion were modified to accommodate the electrophoretic term. The different types of electrophoresis were listed. Bleaching was distinguished form solid leaching as an operation that whitens upon oxidation. Industrial examples were given in ozone bleaching of paper pulp and whitening of modern bread and edible oil.

Industrial applications in membrane separations were identified. The membrane separations were classified based upon their solute size rejected and driving force employed. The principle and methods shown in Chapter 6 can be extended to any of these methods. Polymer devolatilisation can be treated in a similar fashion to solids drying. The devolatilisers used in the industry such as the falling strand devolatiliser and twin screw devolatiliser were discussed with schematic. The equilibrium and critical residual content in the polymeric solid will determine the maximum possible removal of volatiles.

REFERENCES

Henrickson, H., Laxen, T., Peltonen, J., 2003, *Bleaching Medium Consistency Pulp with Ozone,* US Patent 6,579,411, Andritz Oy, Helsinki, FI.

Kruse, R.L., Aliberti, V. A., Valcarce, E. M., 1983, *Mass Polymerization Process for ABS Polyblends,* US Patent 4,417,030, Monsanto Co., St. Louis, MO.

Newman, R.E., 1981, *Falling Strand Devolatilizer,* US Patent 4294,652, Monsanto, Co. St. Louis, MO.

Saltzman, W.M., 2001, *Drug delivery—Engineering Principles for Drug Therapy*, Oxford University Press, UK.

Sharma, K.R., 2005, *Damped Wave Transport and Relaxation*, Elsevier, Amsterdam.

________, April, 2006, *Finite Speed Diffusion and Delivery of Drugs and Response to Pulse Decay, 32nd Northeast Bioengineering Conference*, Easton, PA.

________, April, 2006, *Manifestation of Acceleration of Mass Flow Effects in Dissolving Pill Problem, 32nd Northeast Bioengineering Conference*, Easton, PA.

________, March, 2006, *On the Pulse Decay during Simultaneous Reaction and Damped Wave Diffusion and Relaxation, 231st ACS National Meeting*, Atlanta.

________, March, 2006, *On the Use of Final Condition in Time to Obtain Solution of Cattneo & Vernotte Damped Wave Transport and Relaxation Equation, 231st ACS National Meeting*, Atlanta.

________, March, 2006, *On the Measured Values of Relaxation Time, 231st ACS National Meeting*, Atlanta.

________, March, 2006, *Proteomics and Dimensional Separation, 231st ACS National Meeting*, Atlanta.

Sharma, K.R., March, 2006, *Errors in Gel Acrylamide Electrophoresis Due to Effect of Charge, 231st ACS National Meeting*, Atlanta.

________, March, 2006, *S Function Solution to the Dyeing Wool Problem when Reaction Modulus is 1, 231st ACS National Meeting*, Atlanta.

________, March, 2006, *On the Third Mode of Mass Transfer Called Wave Diffusion, 231st ACS National Meeting*, Atlanta.

________, March, 2006, *A Generalized Substitution to Transform Hyperbolic Damped Wave Generalized Transport Equation to a Parabolic Equation, 231st ACS National Meeting*, Atlanta.

________, March, 2006, *Accumulation at the Surface and Non-Fick Wave Diffusion Equation, 231st ACS National Meeting*, Atlanta.

________, October/November, 2005, *Regime Change in Simultaneous Wave Diffusion and Reaction, 97th AIChE Annual Meeting*, Cincinnati.

________, October/November, 2005, *Inertial Lag and Bessel Composite Function of the Third Order and First Kind Solution to the Dissolving Pill Problem, 97th AIChE Annual Meeting*, Cincinnati.

________, August/September, 2005, *Bessel Composite Function of the Third Order and First Kind: Solution to the Dissolving Pill Problem, 230th ACS National Meeting*, Washington.

________, September/October 2004, *Charge Effect and Shotgun Sequencing Errors in Gel Electrophoresis, 60th Southwest Regional Meeting of the American Chemical Society*, Fort Worth.

________, September/October, 2004, *Lorentz Transformation and Solution of Hyperbolic Damped Wave Diffusion and Relaxation Equation, 60th Southwest Regional Meeting of the American Chemical Society*, Fort Worth.

________, September/October, 2004, *Critical Region of Zero Concentration in Finite Speed Diffusion and Relaxation, 60th Southwest Regional Meeting of the American Chemical Society*, Fort Worth.

________, August, 2004, *Third Order PDE in Restriction Fragment Length Gel Electrophoresis, 228th ACS National Meeting*, Philadelphia.

________, August, 2004, *Plane of Zero Concentration in a Slab with Reacting Interface, 228th ACS National Meeting*, Philadelphia.

________, June, 2004, *Confounding Effect of Charge on Gel Electrophoresis Measurements, 59th Northwest Regional Meeting of the American Chemical Society, NORM/RMRM*, Utah State University, UT.

________, April, 2004, *Improvement of Yield of Nuclear Hydrogen using Semi-Permeable Membrane Design, 2004 AIChE Spring Meeting*, New Orleans.

________, December, 2004, *A Solution by Wave Coordinate Transformation of Hyperbolic Diffusion & Relaxation Equation with Surface Reaction, CHEMCON 2004*, Mumbai.

Sharma, K.R., November, 2003, *Simultaneous Increase of Permeability and Permselectivity in Membrane Transport, 95th AIChE Annual Meeting*, San Francisco.

________, September, 2003, *Enhanced Permselectivity by Finite Speed Mass Diffusion in Polymeric Membranes during Hydrogen Separation, 226th ACS National Meeting*, New York.

________, June, 2003, *Diffusion Issues in DNA Sequencing using Sanger's Method, 58th Northwest Regional Meeting of the ACS*, Bozeman, Montana.

________, November, 1998, *Mesoscopic Mixing Efficiency in Gas-Liquid-Solid Fluidized Bed Process for Ozone Bleaching of Pulp, 90th AIChE Annual Meeting*, Miami.

________, October/November, 1999, *Mesoscopic Mixing Efficiency of Gas in Gas-Solid Fluidized Beds in Process for Ozone Bleaching of Pulp, 91st AIChE Annual Meeting*, Dallas.

________, June, 1996, *Effect of Chain Sequence Distribution on Contamination Formation in Alphamethylstyrene Acrylonitrile Copolymers during its Manufacture in Continuous Mass Polymerization, 5th World Congress of Chemical Engineering*, San Diego.

________, September, 1997, *A Statistical Design to Define the Process Capability of Continuous Mass Polymerization of ABS at the Pilot Plant, 214th ACS National Meeting*, Las Vegas.

________, September, 1997, *Glass Transition Temperature of Copolymers and Blends: A Survey of Mathematical Models, 214th ACS National Meeting*, Las Vegas.

________, September, 1997, *Alphamethylstyrene Acrylonitrile Styrene Terpolymer using Thermal Initiation Methods, 214th ACS National Meeting*, Las Vegas.

________, June, 1998, *Polymer Polymer Interdiffusion During Coextrusion, 30th ACS Central Regional Meeting*, Cleveland.

________, June, 1998, *Improving Graft Yield in Series-Parallel Reactions in Continuous Mass Polymerization, 30th ACS Central Regional Meeting*, Cleveland.

________, June, 1998, *Dissimilar Entropy of Mixing Effects on Glass Transition Temperature of Polymer Blends, 30th ACS Central Regional Meeting*, Cleveland.

Appendix A

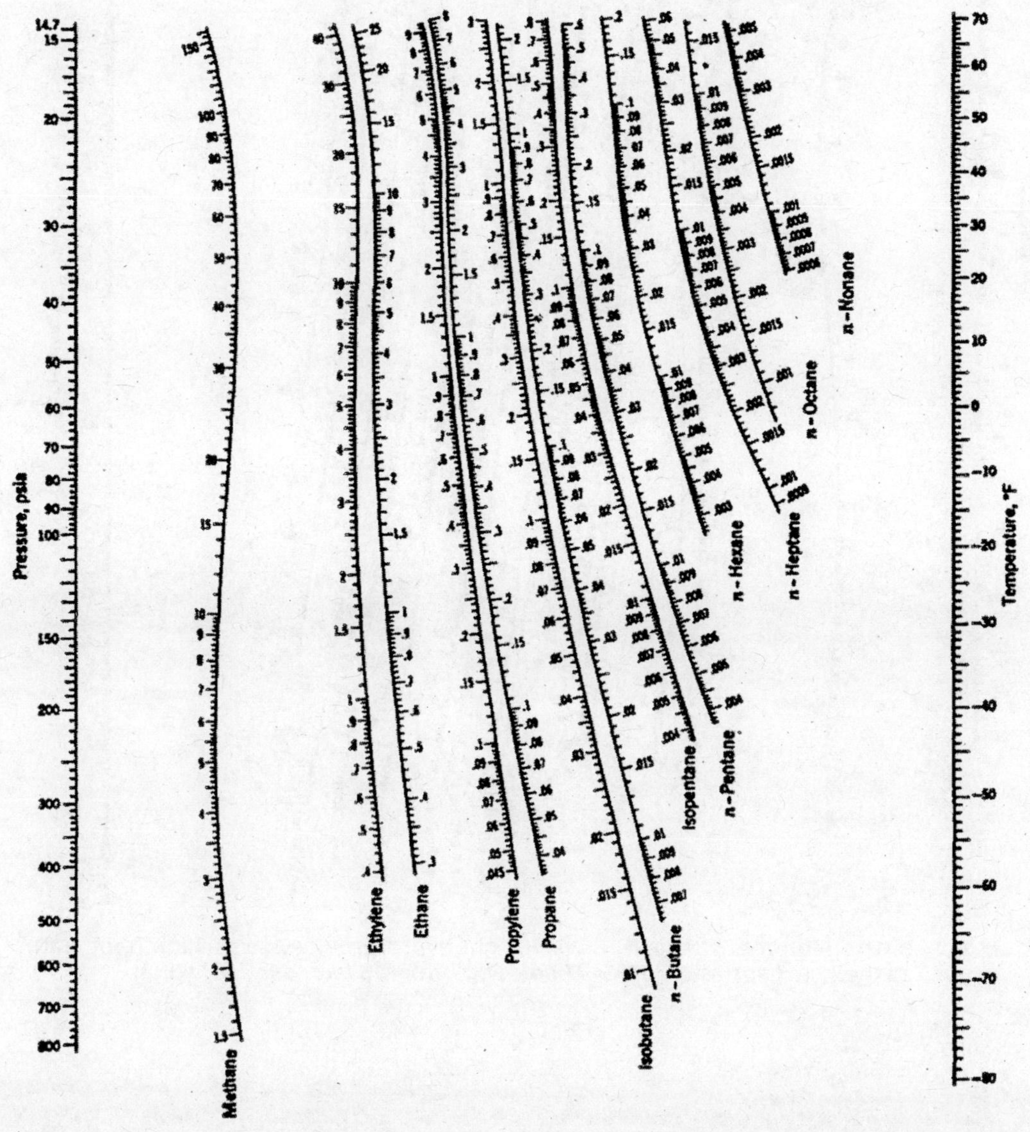

FIGURE A.1 Distribution coefficients ($K = y/x$) in light hydrocarbon systems, low temperature range (C.L. Depriester, 1953, *Chem. Eng. Progr. Symp. Ser.*, 49, (7), 1).

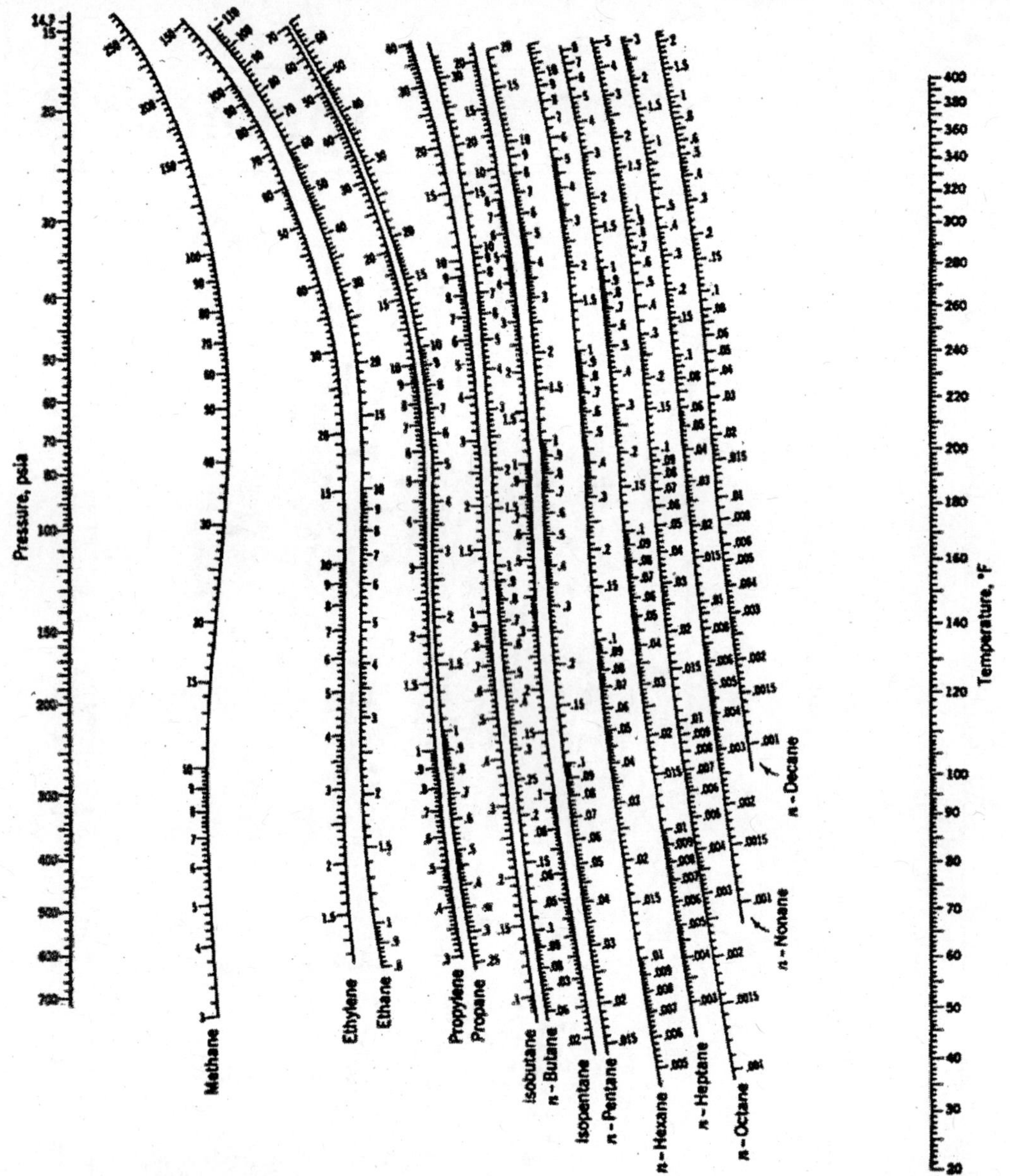

FIGURE A.2 Distribution coefficients ($K = y/x$) in light hydrocarbon systems, high temperature range (C.L. Depriester, 1953, *Chem. Eng. Progr. Symp. Ser.*, 49, (7), 1).

Appendix B

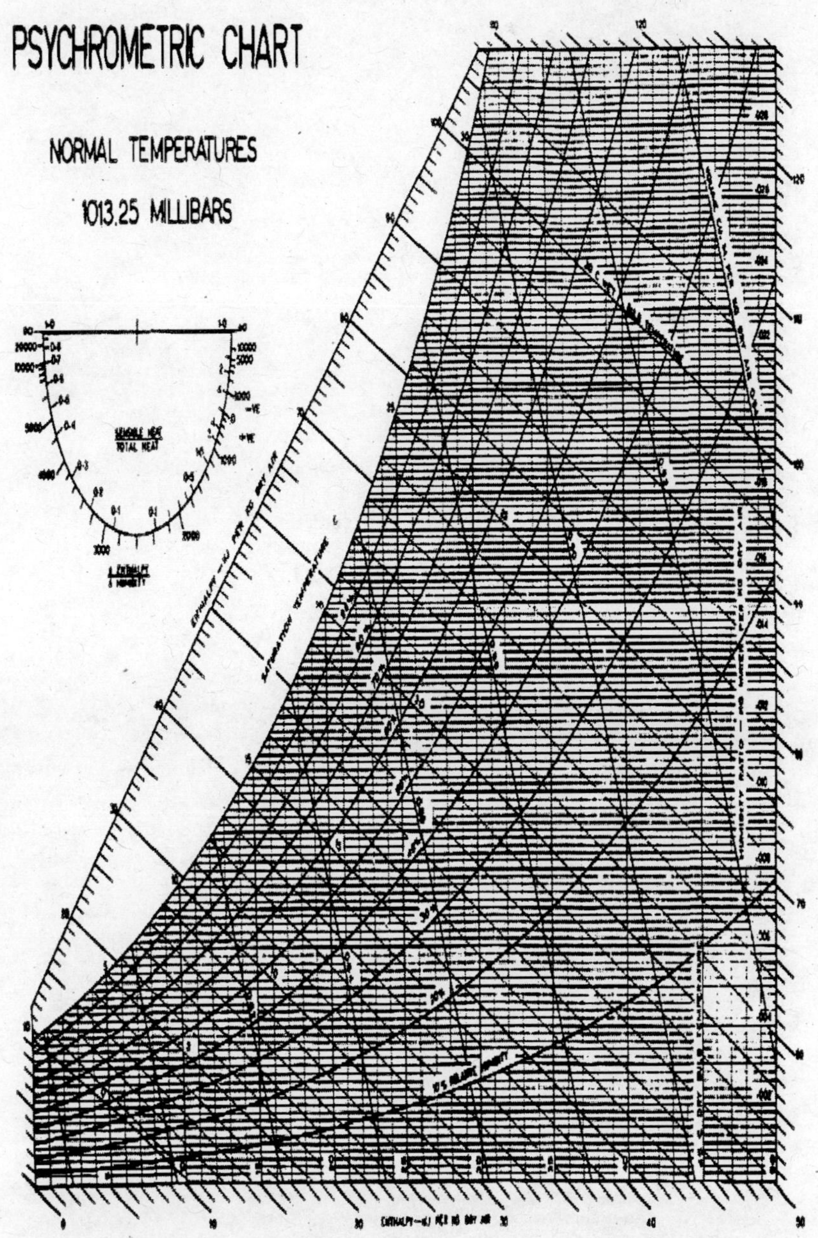

FIGURE B.1 Psychrometric chart for air water system.

Index